Ausgeschieden im Jahr 2025

Microstrip Filters for RF/Microwave Applications

WILEY SERIES IN MICROWAVE AND OPTICAL ENGINEERING

KAI CHANG, Editor
Texas A&M University

A complete list of the titles in this series appears at the end of this volume.

Microstrip Filters for RF/Microwave Applications

Second Edition

JIA-SHENG HONG

A JOHN WILEY & SONS, INC., PUBLICATION

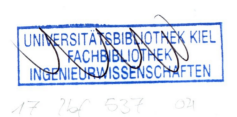

Copyright © 2011 by John Wiley & Sons, Inc. All rights reserved.

Published by John Wiley & Sons, Inc., Hoboken, New Jersey.
Published simultaneously in Canada.

No part of this publication may be reproduced, stored in a retrieval system, or transmitted in any form or by any means, electronic, mechanical, photocopying, recording, scanning, or otherwise, except as permitted under Section 107 or 108 of the 1976 United States Copyright Act, without either the prior written permission of the Publisher, or authorization through payment of the appropriate per-copy fee to the Copyright Clearance Center, Inc., 222 Rosewood Drive, Danvers, MA 01923, (978) 750-8400, fax (978) 750-4470, or on the web at www.copyright.com. Requests to the Publisher for permission should be addressed to the Permissions Department, John Wiley & Sons, Inc., 111 River Street, Hoboken, NJ 07030, (201) 748-6011, fax (201) 748-6008, or online at http://www.wiley.com/go/permissions.

Limit of Liability/Disclaimer of Warranty: While the publisher and author have used their best efforts in preparing this book, they make no representations or warranties with respect to the accuracy or completeness of the contents of this book and specifically disclaim any implied warranties of merchantability or fitness for a particular purpose. No warranty may be created or extended by sales representatives or written sales materials. The advice and strategies contained herein may not be suitable for your situation. You should consult with a professional where appropriate. Neither the publisher nor author shall be liable for any loss of profit or any other commercial damages, including but not limited to special, incidental, consequential, or other damages.

For general information on our other products and services or for technical support, please contact our Customer Care Department within the United States at (800) 762-2974, outside the United States at (317) 572-3993 or fax (317) 572-4002.

Wiley also publishes its books in a variety of electronic formats. Some content that appears in print may not be available in electronic formats. For more information about Wiley products, visit our web site at www.wiley.com.

Library of Congress Cataloging-in-Publication Data:

Hong, Jia-Sheng.
  Microstrip filters for RF/microwave applications / Jia-Sheng Hong. – 2nd ed.
    p.   cm. – (Wiley series in microwave and optical engineering ; 216)
  Includes bibliographical references and index.
  ISBN 978-0-470-40877-3 (hardback)
 1. Microwave circuits. 2. Strip transmission lines. 3. Electric filters. I. Title.
  TK7876.H66 2011
  621.381'32–dc22
                                                           2010031085

Printed in Singapore

10 9 8 7 6 5 4 3

Contents

Preface to the Second Edition		xiii
Preface to the First Edition		xv
1	**Introduction**	**1**
2	**Network Analysis**	**6**
	2.1 Network Variables	6
	2.2 Scattering Parameters	8
	2.3 Short-Circuit Admittance Parameters	10
	2.4 Open-Circuit Impedance Parameters	10
	2.5 *ABCD* Parameters	11
	2.6 Transmission-Line Networks	12
	2.7 Network Connections	13
	2.8 Network Parameter Conversions	16
	2.9 Symmetrical Network Analysis	19
	2.10 Multiport Networks	20
	2.11 Equivalent and Dual Network	23
	2.12 Multimode Networks	25
	References	27
3	**Basic Concepts and Theories of Filters**	**28**
	3.1 Transfer Functions	28
	3.1.1 General Definitions	28
	3.1.2 Poles and Zeros on the Complex Plane	29
	3.1.3 Butterworth (Maximally Flat) Response	30
	3.1.4 Chebyshev Response	31
	3.1.5 Elliptic Function Response	33
	3.1.6 Gaussian (Maximally Flat Group-Delay) Response	35
	3.1.7 All-Pass Response	36

3.2	Lowpass Prototype Filters and Elements		39
	3.2.1	Butterworth Lowpass Prototype Filters	40
	3.2.2	Chebyshev Lowpass Prototype Filters	40
	3.2.3	Elliptic-Function Lowpass Prototype Filters	43
	3.2.4	Gaussian Lowpass Prototype Filters	44
	3.2.5	All-Pass Lowpass Prototype Filters	48
3.3	Frequency and Element Transformations		49
	3.3.1	Lowpass Transformation	50
	3.3.2	Highpass Transformation	51
	3.3.3	Bandpass Transformation	52
	3.3.4	Bandstop Transformation	54
3.4	Immittance Inverters		55
	3.4.1	Definition of Immittance, Impedance, and Admittance Inverters	55
	3.4.2	Filters with Immittance Inverters	56
	3.4.3	Practical Realization of Immittance Inverters	60
3.5	Richards' Transformation and Kuroda Identities		62
	3.5.1	Richards' Transformation	62
	3.5.2	Kuroda Identities	65
	3.5.3	Coupled-Line Equivalent Circuits	65
3.6	Dissipation and Unloaded Quality Factor		69
	3.6.1	Unloaded Quality Factors of Lossy Reactive Elements	69
	3.6.2	Dissipation Effects on Lowpass and Highpass Filters	70
	3.6.3	Dissipation Effects on Bandpass and Bandstop Filters	72
	References		74

4 Transmission Lines and Components — 75

4.1	Microstrip Lines		75
	4.1.1	Microstrip Structure	75
	4.1.2	Waves In Microstrip	75
	4.1.3	Quasi-TEM Approximation	76
	4.1.4	Effective Dielectric Constant and Characteristic Impedance	76
	4.1.5	Guided Wavelength, Propagation Constant, Phase Velocity, and Electrical Length	78
	4.1.6	Synthesis of W/h	79
	4.1.7	Effect of Strip Thickness	79
	4.1.8	Dispersion in Microstrip	80
	4.1.9	Microstrip Losses	81
	4.1.10	Effect of Enclosure	82
	4.1.11	Surface Waves and Higher-Order Modes	82

	4.2	Coupled Lines	83
		4.2.1 Even- and Odd-Mode Capacitances	84
		4.2.2 Even- and Odd-Mode Characteristic Impedances and Effective Dielectric Constants	85
		4.2.3 More Accurate Design Equations	86
	4.3	Discontinuities and Components	88
		4.3.1 Microstrip Discontinuities	88
		4.3.2 Microstrip Components	91
		4.3.3 Loss Considerations for Microstrip Resonators	101
	4.4	Other Types of Microstrip Lines	103
	4.5	Coplanar Waveguide (CPW)	104
	4.6	Slotlines	107
		References	109
5	**Lowpass and Bandpass Filters**	**112**
	5.1	Lowpass Filters	112
		5.1.1 Stepped-Impedance L-C Ladder-Type Lowpass Filters	112
		5.1.2 L-C Ladder-Type of Lowpass Filters Using Open-Circuited Stubs	115
		5.1.3 Semilumped Lowpass Filters Having Finite-Frequency Attenuation Poles	119
	5.2	Bandpass Filters	123
		5.2.1 End-Coupled Half-Wavelength Resonator Filters	123
		5.2.2 Parallel-Coupled Half-Wavelength Resonator Filters	128
		5.2.3 Hairpin-Line Bandpass Filters	131
		5.2.4 Interdigital Bandpass Filters	135
		5.2.5 Combline Filters	144
		5.2.6 Pseudocombline Filters	150
		5.2.7 Stub Bandpass Filters	153
		References	160
6	**Highpass and Bandstop Filters**	**162**
	6.1	Highpass Filters	162
		6.1.1 Quasilumped Highpass Filters	162
		6.1.2 Optimum Distributed Highpass Filters	166
	6.2	Bandstop Filters	169
		6.2.1 Narrow-Band Bandstop Filters	169
		6.2.2 Bandstop Filters with Open-Circuited Stubs	177
		6.2.3 Optimum Bandstop Filters	183
		6.2.4 Bandstop Filters for RF Chokes	188
		References	191

7 Coupled-Resonator Circuits — 193

- 7.1 General Coupling Matrix for Coupled-Resonator Filters — 194
  - 7.1.1 Loop Equation Formulation — 194
  - 7.1.2 Node Equation Formulation — 198
  - 7.1.3 General Coupling Matrix — 201
- 7.2 General Theory of Couplings — 202
  - 7.2.1 Synchronously Tuned Coupled-Resonator Circuits — 203
  - 7.2.2 Asynchronously Tuned Coupled-Resonator Circuits — 209
- 7.3 General Formulation for Extracting Coupling Coefficient k — 215
- 7.4 Formulation for Extracting External Quality Factor Q_e — 216
  - 7.4.1 Singly Loaded Resonator — 216
  - 7.4.2 Doubly Loaded Resonator — 219
- 7.5 Numerical Examples — 221
  - 7.5.1 Extracting k (Synchronous Tuning) — 222
  - 7.5.2 Extracting k (Asynchronous Tuning) — 225
  - 7.5.3 Extracting Q_e — 227
- 7.6 General Coupling Matrix Including Source and Load — 228
- References — 231

8 CAD for Low-Cost and High-Volume Production — 232

- 8.1 Computer-Aided Design (CAD) Tools — 233
- 8.2 Computer-Aided Analysis (CAA) — 233
  - 8.2.1 Circuit Analysis — 233
  - 8.2.2 Electromagnetic Simulation — 238
- 8.3 Filter Synthesis by Optimization — 242
  - 8.3.1 General Description — 242
  - 8.3.2 Synthesis of a Quasielliptic-Function Filter by Optimization — 243
  - 8.3.3 Synthesis of an Asynchronously Tuned Filter by Optimization — 244
  - 8.3.4 Synthesis of a UMTS Filter by Optimization — 245
- 8.4 CAD Examples — 248
  - 8.4.1 Example One (Chebyshev Filter) — 248
  - 8.4.2 Example Two (Cross-Coupled Filter) — 252
- References — 258

9 Advanced RF/Microwave Filters — 261

- 9.1 Selective Filters with a Single Pair of Transmission Zeros — 261
  - 9.1.1 Filter Characteristics — 261
  - 9.1.2 Filter Synthesis — 263
  - 9.1.3 Filter Analysis — 266
  - 9.1.4 Microstrip Filter Realization — 267

9.2	Cascaded Quadruplet (CQ) Filters		271
	9.2.1	Microstrip CQ Filters	271
	9.2.2	Design Example	272
9.3	Trisection and Cascaded Trisection (CT) Filters		275
	9.3.1	Characteristics of CT Filters	275
	9.3.2	Trisection Filters	276
	9.3.3	Microstrip Trisection Filters	281
	9.3.4	Microstrip CT Filters	284
9.4	Advanced Filters with Transmission-Line Inserted Inverters		287
	9.4.1	Characteristics of Transmission-Line Inserted Inverters	287
	9.4.2	Filtering Characteristics with Transmission-Line Inserted Inverters	289
	9.4.3	General Transmission-Line Filter	294
9.5	Linear-Phase Filters		295
	9.5.1	Prototype of Linear-Phase Filter	296
	9.5.2	Microstrip Linear-Phase Bandpass Filters	302
9.6	Extracted Pole Filters		304
	9.6.1	Extracted Pole Synthesis Procedure	306
	9.6.2	Synthesis Example	311
	9.6.3	Microstrip-Extracted Pole Bandpass Filters	313
9.7	Canonical Filters		316
	9.7.1	General Coupling Structure	316
	9.7.2	Elliptic-Function/Selective Linear-Phase Canonical Filters	319
9.8	Multiband Filters		320
	9.8.1	Filters Using Mixed Resonators	321
	9.8.2	Filters Using Dual-Band Resonators	322
	9.8.3	Filters Using Cross-Coupled Resonators	328
	References		332
10	**Compact Filters and Filter Miniaturization**		**334**
10.1	Miniature Open-Loop and Hairpin Resonator Filters		334
10.2	Slow-Wave Resonator Filters		336
	10.2.1	Capacitively Loaded Transmission-Line Resonator	338
	10.2.2	End-Coupled Slow-Wave Resonators Filters	341
	10.2.3	Slow-Wave, Open-Loop Resonator Filters	343
10.3	Miniature Dual-Mode Resonator Filters		349
	10.3.1	Microstrip Dual-Mode Resonators	350
	10.3.2	Miniaturized Dual-Mode Resonator Filters	352
	10.3.3	Dual-Mode Triangular-Patch Resonator Filters	355
	10.3.4	Dual-Mode Open-Loop Filters	366

	10.4	Lumped-Element Filters	379
	10.5	Miniature Filters Using High Dielectric-Constant Substrates	384
	10.6	Multilayer Filters	386
		10.6.1 Aperture-Coupled Resonator Filters	386
		10.6.2 Filters with Defected Ground Structures	393
		10.6.3 Substrate-Integrated Waveguide Filters	404
		10.6.4 LTCC and LCP Filters	412
		References	421
11	**Superconducting Filters**		**433**
	11.1	High-Temperature Superconducting (HTS) Materials	433
		11.1.1 Typical HTS Materials	433
		11.1.2 Complex Conductivity of Superconductors	434
		11.1.3 Penetration Depth of Superconductors	435
		11.1.4 Surface Impedance of Superconductors	436
		11.1.5 Nonlinearity of Superconductors	440
		11.1.6 Substrates for Superconductors	440
	11.2	HTS Filters for Mobile Communications	441
		11.2.1 HTS Filter with a Single Pair of Transmission Zeros	442
		11.2.2 HTS Filter with Two Pairs of Transmission Zeros	448
		11.2.3 HTS Filter with Group-Delay Equalization	454
	11.3	HTS Filters for Satellite Communications	462
		11.3.1 C-Band HTS Filter	462
		11.3.2 X-Band HTS Filter	465
		11.3.3 Ka-Band HTS Filter	468
	11.4	HTS Filters for Radio Astronomy and Radar	469
		11.4.1 Narrowband Miniature HTS Filter at UHF Band	470
		11.4.2 Wideband HTS Filter with Strong Coupling Resonators	473
	11.5	High-Power HTS Filters	475
	11.6	Cryogenic Package	479
		References	480
12	**Ultra-Wideband (UWB) Filters**		**488**
	12.1	UWB Filters with Short-Circuited Stubs	488
		12.1.1 Design of Stub UWB Filters	488
		12.1.2 Stub UWB Filters with Improved Upper Stopband	490
	12.2	UWB-Coupled Resonator Filters	495
		12.2.1 Interdigital UWB Filters with Microstrip/CPW-Coupled Resonators	495
		12.2.2 Broadside-Coupled Slow-Wave Resonator UWB Filters	501

		12.2.3	UWB Filters Using Coupled Stepped-Impedance Resonators	505
		12.2.4	Multimode-Resonator UWB Filters	511
	12.3	Quasilumped Element UWB Filters		520
		12.3.1	Six-Pole Filter Design Example	520
		12.3.2	Eight-Pole Filter Design Example	526
	12.4	UWB Filters Using Cascaded Miniature High- And Lowpass Filters		529
		12.4.1	Miniature Wideband Highpass Filter	530
		12.4.2	Miniature Lowpass Filter	533
		12.4.3	Implementation of UWB Bandpass Filter	535
	12.5	UWB Filters with Notch Band(s)		536
		12.5.1	UWB Filters with Embedded Band Notch Stubs	537
		12.5.2	Notch Implementation Using Interdigital Coupled Lines	543
		12.5.3	UWB Filters with Notched Bands Using Vertically Coupled Resonators	550
	References			557

13 Tunable and Reconfigurable Filters 563

13.1	Tunable Combline Filters		564
13.2	Tunable Open-Loop Filters without Via-Hole Grounding		570
13.3	Reconfigurable Dual-Mode Bandpass Filters		574
	13.3.1	Reconfigurable Dual-Mode Filter with Two dc Biases	574
	13.3.2	Tunable Dual-Mode Filters Using a Single dc Bias	577
	13.3.3	Tunable Four-Pole Dual-Mode Filter	588
13.4	Wideband Filters with Reconfigurable Bandwidth		591
13.5	Reconfigurable UWB Filters		597
	13.5.1	UWB Filter with Switchable Notch	597
	13.5.2	UWB Filter with Tunable Notch	601
	13.5.3	Miniature Reconfigurable UWB Filter	602
13.6	RF MEMS Reconfigurable Filters		604
	13.6.1	MEMS and Micromachining	604
	13.6.2	Reconfigurable Filters Using RF MEMS Switches	608
13.7	Piezoelectric Transducer Tunable Filters		610
13.8	Ferroelectric Tunable Filters		610
	13.8.1	Ferroelectric Materials	611
	13.8.2	Ferroelectric Varactors	612
	13.8.3	Frequency Agile Filters Using Ferroelectrics	615
References			619

Appendix: Useful Constants and Data 625

    A.1   Physical Constants   625
    A.2   Conductivity of Metals at 25°C (298K)   625
    A.3   Electical Resistivity ρ in 10^{-8} Ωm of Metals   626
    A.4   Properties of Dielectric Substrates   626

Index 627

Preface to the Second Edition

The first edition of "Microstrip Filters for RF/Microwave Applications" was published in 2001. Over the years, this book has been well received and is used extensively in both academy and industry by microwave researchers and engineers. From its inception to publication, the book is almost 10 years old. While the fundamentals of filter circuit have not changed, further innovations in filter realizations and other applications have occurred, including changes in technology and use of new fabrication processes. There have been recent advances in RF MEMS and ferroelectric films for tunable filters; the use of liquid crystal polymer (LCP) substrates for multilayer circuits, as well as the new filters for multiband and ultra-wideband (UWB) applications.

Although the microstrip remains as a main transmission line medium for these new developments, there has been a new trend of the combined use of other planar transmission line structures, such as co-planar waveguide (CPW), slotline and defected or slotted ground structures, for novel physical implementations beyond single layer in order to achieve filter miniaturization and better performance. Over the years, practitioners have also suggested topics that should be added for completeness, or deleted in some cases, as they were not very useful in practice.

It is in response to these concerns that the 2nd edition of *Microstrip Filters for RF/Microwave Applications* has been written. The extensively revised book will offer a thoroughly up-to-date professional reference focusing on microstrip and planar filters, which find wide applications in today's wireless, microwave, communications, and radar systems. It offers a unique comprehensive treatment of filters based on the microstrip and planar structures and includes full design methodologies that are applicable to waveguide and other transmission-line filters. The new edition includes a wealth of new materials including

- CPW and slotlines
- General coupling matrix, including source and load
- Multiband filters
- Nondegenerate dual-mode filters

- Filters with defected ground structures
- Substrate-integrated waveguide filters
- Liquid crystal polymer (LCP) and LTCC multilayer filters
- HTS filters for mobile/satellite communications and radio astronomy
- Ultra-wideband (UWB) filters
- Tunable and reconfigurable filters

Microstrip Filters for RF/Microwave Applications utilize numerous design examples to demonstrate and emphasize computer-aided design with commercially available software. This intensively revised book, with cutting-edge information, remains not only a valuable design resource for partitions, but also a handy reference for students and researchers in RF and microwave engineering.

I wish to acknowledge the financial supports of the UK EPSRC, Scottish Enterprise, BAE Systems (UK), and SELEX Galileo (UK). I would like to thanks all of my former and current research associates, PhD students, and visiting scholars, including Eamon McErlean, Dr. Young-Hoon Chun, Dr. Zhang-Cheng Hao, Dr. Neil Thomson, Dr. Hussein Shaman, Dr. Sultan Alotaibi, Shuzhou Li, Wenxing Tang, and Alexander Miller, for their works, some of which are presented in the book. In addition, I would like to express my gratitude to several national and international collaborators, including Prof. Michael Lancaster and Dr. Tim Jackson (both at University of Birmingham, UK), Dr. Paul Kirby (University of Cranfield, UK), Dr. Zheng Cui (Rutherford Appleton Laboratory, UK), Prof. Yusheng He (CAS, China), Alan Burdis and Colin Bird (both at SELEX Galileo, UK), and Dr. Keren Li (NiCT, Japan). The support provided by Dr. James Rautio and other members of staff at Sonnet Software Inc., USA, is acknowledged. I also wish to thank the colleagues who I have worked with at Heriot-Watt University, including Prof. Marc Desmulliez, Prof. Alan Sangster, Dr. George Goussetis, Prof. Duncan Hand, Dr. Changhai Wang, and Dr. Paul Record.

Needless to say, I am indebted to many researchers for their published work, which have been rich sources of reference. My sincere gratitude extends to the Editor of Wiley series in microwave and optical engineering, Prof. Kai Chang; and the Executive Editor of Wiley-Interscience, George Telecki, for their encouragement in writing this new edition book. I am also indebted to my wife, Kai, and my son, Haide, without their support, writing this book would not have been possible.

JIA-SHENG HONG

2010

Preface to the First Edition

Filters play important roles in many RF/microwave applications. Emerging applications such as wireless communications continue to challenge RF/microwave filters with ever more stringent requirements — higher performance, smaller size, lighter weight, and lower cost. The recent advance of novel materials and fabrication technologies, including high-temperature superconductors (HTS), low-temperature cofired ceramics (LTCC), monolithic microwave-integrated circuit (MMIC), microelectromechanic system (MEMS), and micro-machining technology, have stimulated the rapid development of new microstrip and other filters for RF/microwave applications. In the meantime, advances in computer-aided design (CAD) tools, such as full-wave electromagnetic (EM) simulators, have revolutionized the filter design. Many novel microstrip filters with advanced filtering characteristics have been demonstrated. However, up until now there has not been a single book dedicated to this subject.

Microstrip Filters for RF/Microwave Applications offers a unique and comprehensive treatment of RF/microwave filters based on the microstrip structure, providing a link to applications of computer-aided design tools and advanced materials and technologies. Many novel and sophisticated filters using computer-aided design are discussed, from basic concepts to practical realizations. The book is self-contained — it is not only a valuable design resource, but also a handy reference for students, researchers, and engineers in microwave engineering. It can also be used for RF/microwave education.

The outstanding features of this book include discussion of many novel microstrip filter configurations with advanced filtering characteristics, new design techniques, and methods for filter miniaturization. The book emphasizes computer analysis and synthesis and full-wave electromagnetic (EM) simulation through a large number of design examples. Applications of commercially available software are demonstrated. Commercial applications are included as are design theories and methodologies, which are not only for microstrip filters, but also directly applicable to other types of filters, such as waveguide and other transmission-line filters. Therefore, this book is more than just a text on microstrip filters.

Much of work described herein has resulted from the authors' research. The authors wish to acknowledge the financial supports of the UK EPSRC and the European Commission through the Advanced Communications Technologies and Services (ACTS) program. They would also like to acknowledge their national and international collaborators, including Professor Heinz Chaloupka at Wuppertal University (Germany), Robert Greed at Marconi Research Center (U.K.), Dr. Jean-Claude Mage at Thomson-CSF/CNRS (France), and Dieter Jedamzik formerly with GEC-Marconi Materials Technology (U.K.).

The authors are indebted to many researchers for their published work, which were rich sources of reference. Their sincere gratitude extends to the Editor of Wiley series in microwave and optical engineering, Professor Kai Chang; the Executive Editor of Wiley-Interscience, George Telecki; and the reviewers for their support in writing the book. The help provided by Cassie Craig and other members of the staff at Wiley is most appreciated. The authors also wish to thank their colleagues at the University of Birmingham, including Professor Peter Hall, Dr. Fred Huang, Dr. Adrian Porch, and Dr. Peter Gardener.

In addition, Jia-Sheng Hong would like to thank Professor John Allen at the University of Oxford (U.K.), Professor Werner Wiesbeck at Kalsruhe Universiy (Germany), and Dr. Nicholas Edwards at the British Telecom (U.K.) for their many years' support and friendship. Professor Joseph Helszajn at Heriot-Watt University (U.K.), who sent his own book on filters to Jia-Sheng Hong, is also acknowledged.

Finally, Jia-Sheng Hong would like to express his deep appreciation to his wife, Kai, and his son, Haide, for their tolerance and support, which allowed him to write the book at home over many evenings, weekends, and holidays. In particular, without the help of Kai, completing this book on time would not have been possible.

JIA-SHENG HONG
M. J. LANCASTER

2001

CHAPTER ONE

Introduction

The term *microwaves* may be used to describe electromagnetic (EM) waves with frequencies ranging from 300 MHz to 300 GHz, which correspond to wavelengths (in free space) from 1 m to 1 mm. The EM waves with frequencies above 30 and up to 300 GHz are also called *millimeter waves,* because their wavelengths are in the millimeter range (1–10 mm). Above the millimeter wave spectrum is the infrared, which comprises electromagnetic waves with wavelengths between 1 μm (10^{-6} m) and 1 mm. Beyond the infrared spectrum is the visible optical spectrum, the ultraviolet spectrum, and x rays. Below the microwave frequency spectrum is the radio-frequency (RF) spectrum. The frequency boundary between RF and microwaves is somewhat arbitrary, depending on the particular technologies developed for the exploitation of that specific frequency range. Therefore, by extension, the RF/microwave applications can be referred to as communications, radar, navigation, radio astronomy, sensing, medical instrumentation, and others that explore the usage of frequency spectrums in the range, for example, 300 kHz up to 300 GHz (Fig. 1.1). For convenience, some of these frequency spectrums are further divided into many frequency bands, as indicated in Fig. 1.1.

Filters play important roles in many RF/microwave applications. They are used to separate or combine different frequencies. The electromagnetic spectrum is limited and has to be shared; filters are used to select or confine the RF/microwave signals within assigned spectral limits. Emerging applications, such as wireless communications, continue to challenge RF/microwave filters with ever more stringent requirements — higher performance, more functionalities such as tunable or reconfigurable, smaller size, lighter weight, and lower cost. Depending on the requirements and specifications, RF/microwave filters may be designed as lumped element or distributed element circuits, they may be realized in various transmission-line structures, such as waveguide, coaxial line, coplanar waveguide (CPW), slotline, and microstrip.

Microstrip Filters for RF/Microwave Applications by Jia-Sheng Hong
Copyright © 2011 John Wiley & Sons, Inc.

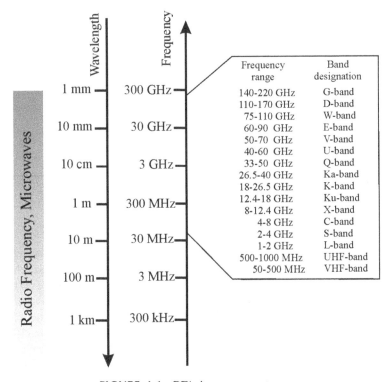

FIGURE 1.1 RF/microwave spectrums.

The recent advance of novel materials and fabrication technologies, including monolithic microwave integrated circuit (MMIC), microelectromechanic system (MEMS) or micromachining, ferroelectrics, high-temperature superconductor (HTS), low-temperature co-fired ceramics (LTCC), and liquid crystal polymers (LCP), has stimulated the rapid development of new microstrip and other filters. In the meantime, advances in computer-aided design (CAD) tools, such as full-wave electromagnetic (EM) simulators, have revolutionized the filter design. Many novel microstrip filters with advanced filtering characteristics have been demonstrated.

The main objective of this book is to offer a unique and comprehensive treatment of RF/microwave filters, based on the microstrip structure, providing a link to applications of CAD tools, advanced materials, and technologies (see Fig. 1.2). However, it is not the intention of this book to include everything that has been published on microstrip filters; such a work would be out of scale in terms of space and knowledge involved. Moreover, design theories and methods described in the book are not only for microstrip filters, but directly applicable to other types of filters, such as waveguide filters.

Although the physical realization of filters at RF/microwave frequencies may vary, the circuit network topology is common to all. Therefore, the technique content of the book begins with Chapter 2, which describes various network concepts

INTRODUCTION 3

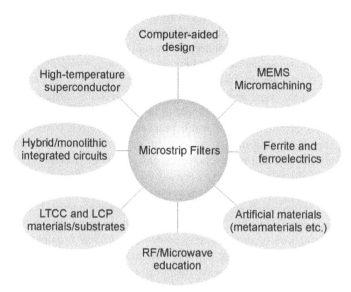

FIGURE 1.2 Microstrip filter linkage.

and equations; these are useful for the analysis of filter networks. Chapter 3 then introduces basic concepts and theories for designing general RF/microwave filters (including microstrip filters). The topics cover filter transfer functions (such as Butterworth, Chebyshev, elliptic function, all-pass, and Gaussian response), lowpass prototype filters and elements, frequency and element transformations, immittance (impedance/admittance) inverters, Richards' transformation, and Kuroda identities for distributed elements. Effects of dissipation and unloaded quality factor of filter elements on the filter performance are also discussed.

Chapter 4 summarizes basic concepts and design equations for microstrip lines, coupled microstrip lines, discontinuities, lumped and distributed components, as well as coplanar waveguide (CPW), and slotlines, which are useful for design of filters. In Chapter 5, conventional microstrip lowpass and bandpass filters, such as stepped-impedance filters, open-stub filters, semilumped element filters, end- and parallel-coupled half-wavelength resonator filters, hairpin-line filters, interdigital and combline filters, pseudocombline filters and stubline filters, are discussed with instructive design examples.

Chapter 6 discusses some typical microstrip highpass and bandstop filters. These include quasilumped element and optimum distributed highpass filters, narrow-band and wide-band bandstop filters, as well as filters for RF chokes. Design equations, tables, and examples are presented for easy reference.

The remaining chapters of the book deal with more advanced topics. Chapter 7, presents a comprehensive treatment of subjects regarding coupled resonator circuits. These are of importance for design of RF/microwave filters, in particular, the narrow-band bandpass filters that play a significant role in many applications. There is a general technique for designing coupled resonator filters, which can be applied to any type of resonator despite its physical structure. For examples, it can be applied for the design of waveguide filters, dielectric resonator filters, ceramic combline filters, microstrip filters, superconducting filters, and micromachined filters. This design method is based on coupling coefficients of intercoupled resonators and the external quality factors of the input and output resonators. Since this design technique is so useful and flexible, it would be desirable to have a deep understanding of not only its approach, but also its theory. For this purpose, the subjects cover (1) the formulation of general coupling matrix, which is of importance for representing a wide range of coupled-resonator filter topologies, and (2) the general theory of couplings for establishing the relationship between the coupling coefficient, and the physical structure of coupled resonators. This leads to a very useful formulation for extracting coupling coefficients from EM simulations or measurements. Formulations for extracting the external quality factors from frequency responses of the externally loaded input/output resonators are derived next. Numerical examples are followed to demonstrate how to use these formulations to extract coupling coefficients and external quality factors of microwave coupling structures for filter designs. In addition, a more advanced topic on general coupling matrix involving source and load is addressed.

Chapter 8 is concerned with computer-aided design (CAD). Generally speaking, any design involves using computers may be called CAD. There have been extraordinary recent advances in CAD of RF/microwave circuits, particularly in full-wave electromagnetic (EM) simulations. They have been implemented both in commercial and specific in-house software and are being applied to microwave filters simulation, modeling, design, and validation. The developments in this area are certainly stimulated by increasing computer power. Another driving force for the developments is the requirement of CAD for low-cost and high-volume production. In general, the investment for tooling, materials, and labor mainly affect the cost of filter production. Labor costs include those for design, fabrication, testing, and tuning. Here the costs for the design and tuning can be reduced greatly by using CAD, which can provide more accurate design with less design iterations, leading to first-pass or tuneless filters. This chapter discusses computer simulation and/or computer optimization. It summarizes some basic concepts and methods regarding filter design by CAD. Typical examples of the applications, including filter synthesis by optimization, are described. Many more CAD examples, particularly those based on full-wave EM simulation, can be found throughout this book.

In Chapter 9, we discuss the designs of some advanced filters, including selective filters with a single pair of transmission zeros, cascaded quadruplet (CQ) filters, trisection and cascaded trisection (CT) filters, cross-coupled filters using transmission-line inserted inverters, linear phase filters for group-delay equalization, extracted-pole filters, canonical filters, and multiband filters. These types of filters, which are different from conventional Chebyshev filters, must meet stringent requirements from RF/microwave systems, particularly from wireless communications systems.

Chapter 10 is intended to describe novel concepts, methodologies, and designs for compact filters and filter miniaturization. The new types of filters discussed include compact open-loop and hairpin resonator filters, slow-wave resonator filters, miniaturized dual-mode filters using degenerate or nondegenerate modes, lumped-element filters, filters using high dielectric constant substrates, and multilayer filters. The last topic covers aperture-coupled resonator filters, filters with defected or slotted-ground structures, substrate integrated waveguide filters, as well as low-temperature cofired ceramics (LTCC) and liquid crystal polymer (LCP) filters.

Chapter 11 introduces high-temperature superconductors (HTS) for RF/microwave filter applications. It covers some important properties of superconductors and substrates for growing HTS films, which are essential for the design of HTS microstrip filters. Typical superconducting filters with super performance for mobile and satellite communications, as well as radio astronomy and radar applications, are described in this chapter.

Chapter 12 focuses on ultra-wideband (UWB) filters, which are a key component for many promising modern applications of UWB technology. In this chapter, typical types of UWB filters are described. This includes UWB filters comprised of short-circuit stubs, UWB filters using coupled single-mode or multimode resonators, quasilumped element UWB filters, UWB filters based on cascaded highpass and lowpass filters, and UWB filters with single- or multiple-notched bands.

The final chapter of the book (Chapter 13) is concerned with electronically tunable and reconfigurable filters. In general, to develop an electronically reconfigurable filter, active switching or tuning elements, such as semiconductor p-i-n and varactor diodes, RF MEMS or other functional material-based components, including ferroelectric varactors and piezoelectric transducers need to be integrated within a passive filtering structure. Typical filters of these types are described in this chapter, which include tunable combline filters, tunable open-loop filters without using any via-hole connections, reconfigurable dual-mode filters, wideband filters with reconfigurable bandwidth, reconfigurable UWB filters, RF MEMS reconfigurable filters, piezoelectric transducer tunable filters, and ferroelectric tunable filters.

CHAPTER TWO

Network Analysis

Filter networks are essential building elements in many areas of RF/microwave engineering. Such networks are used to select/reject or separate/combine signals at different frequencies in a host of RF/microwave systems and equipments. Although the physical realization of filters at RF/microwave frequencies may vary, the circuit network topology is common to all.

At microwave frequencies, the use of voltmeters and ammeters for the direct measurement of voltages and currents do not exist. For this reason, voltage and current, as a measure of the level of electrical excitation of a network, do not play a primary role at microwave frequencies. On the other hand, it is useful to be able to describe the operation of a microwave network, such as a filter, in terms of voltages, currents, and impedances in order to make optimum use of low-frequency network concepts.

It is the purpose of this chapter to describe various network concepts and provide equations [1–10] that are useful for the analysis of filter networks.

2.1 NETWORK VARIABLES

Most RF/microwave filters and filter components can be represented by a two-port network, as shown in Figure 2.1, where V_1, V_2 and I_1, I_2 are the voltage and current variables at ports 1 and 2, respectively, Z_{01} and Z_{02} are the terminal impedances, and E_s is the source or generator voltage. Note that the voltage and current variables are complex amplitudes when we consider sinusoidal quantities. For example, a sinusoidal voltage at port 1 is given by

$$v_1(t) = |V_1|\cos(\omega t + \phi) \qquad (2.1)$$

Microstrip Filters for RF/Microwave Applications by Jia-Sheng Hong
Copyright © 2011 John Wiley & Sons, Inc.

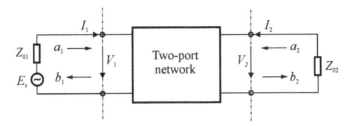

FIGURE 2.1 Two-port network showing network variables.

We can then make the following transformations

$$v_1(t) = |V_1|\cos(\omega t + \phi) = \text{Re}\left(|V_1|e^{j(\omega t + \phi)}\right) = \text{Re}\left(V_1 e^{j\omega t}\right) \tag{2.2}$$

where Re denotes "the real part of" the expression that follows it. Therefore, one can identify the complex amplitude V_1 defined by

$$V_1 = |V_1|e^{j\phi} \tag{2.3}$$

Because it is difficult to measure the voltage and current at microwave frequencies, the wave variables a_1, b_1 and a_2, b_2 are introduced, with a indicating the incident waves and b the reflected waves. The relationships between the wave variables and the voltage and current variables are defined as

$$\begin{aligned} V_n &= \sqrt{Z_{0n}}\,(a_n + b_n) \\ I_n &= \frac{1}{\sqrt{Z_{0n}}}(a_n - b_n) \end{aligned} \qquad n = 1 \text{ and } 2 \tag{2.4a}$$

or

$$\begin{aligned} a_n &= \frac{1}{2}\left(\frac{V_n}{\sqrt{Z_{0n}}} + \sqrt{Z_{0n}}\,I_n\right) \\ b_n &= \frac{1}{2}\left(\frac{V_n}{\sqrt{Z_{0n}}} - \sqrt{Z_{0n}}\,I_n\right) \end{aligned} \qquad n = 1 \text{ and } 2 \tag{2.4b}$$

The above definitions guarantees that the power at port n is

$$P_n = \frac{1}{2}\text{Re}\left(V_n \cdot I_n^*\right) = \frac{1}{2}\left(a_n a_n^* - b_n b_n^*\right) \tag{2.5}$$

where the asterisk denotes a conjugate quantity. It can be recognized that $a_n a_n^*/2$ is the incident wave power and $b_n b_n^*/2$ is the reflected wave power at port n.

2.2 SCATTERING PARAMETERS

The scattering or S parameters of a two-port network are defined in terms of the wave variables as

$$S_{11} = \left.\frac{b_1}{a_1}\right|_{a_2=0} \quad S_{12} = \left.\frac{b_1}{a_2}\right|_{a_1=0}$$
$$S_{21} = \left.\frac{b_2}{a_1}\right|_{a_2=0} \quad S_{22} = \left.\frac{b_2}{a_2}\right|_{a_1=0} \tag{2.6}$$

where $a_n = 0$ implies a perfect impedance match (no reflection from terminal impedance) at port n. These definitions may be written as

$$\begin{bmatrix} b_1 \\ b_2 \end{bmatrix} = \begin{bmatrix} S_{11} & S_{12} \\ S_{21} & S_{22} \end{bmatrix} \cdot \begin{bmatrix} a_1 \\ a_2 \end{bmatrix} \tag{2.7}$$

where the matrix containing the S parameters is referred to as the scattering matrix or S matrix, which may simply be denoted by $[S]$.

The parameters S_{11} and S_{22} are also called the reflection coefficients, whereas S_{12} and S_{21} are the transmission coefficients. These are the parameters directly measurable at microwave frequencies. The S parameters are, in general, complex, and it is convenient to express them in terms of amplitudes and phases, that is, $S_{mn} = |S_{mn}| e^{j\phi_{mn}}$ for $m, n = 1, 2$. Often their amplitudes are given in decibels (dB), which are defined as

$$20 \log |S_{mn}| \quad \text{dB} \quad m, n = 1, 2 \tag{2.8}$$

where the logarithm operation is base 10. This will be assumed through this book unless otherwise stated. For filter characterization, we may define two parameters

$$L_A = -20 \log |S_{mn}| \quad \text{dB} \quad m, n = 1, 2 (m \neq n)$$
$$L_R = 20 \log |S_{nn}| \quad \text{dB} \quad n = 1, 2 \tag{2.9}$$

where L_A denotes the insertion loss between ports n and m and L_R represents the return loss at port n. Instead of using the return loss, the voltage-standing wave ratio *VSWR* may be used. The definition of *VSWR* is

$$VSWR = \frac{1 + |S_{nn}|}{1 - |S_{nn}|} \tag{2.10}$$

Whenever a signal is transmitted through a frequency-selective network, such as a filter, some delay is introduced into the output signal in relation to the input signal. There are two other parameters that play a role in characterizing filter performance related to this delay. The first one is the phase delay, defined by

$$\tau_p = \frac{\phi_{21}}{\omega} \text{ s} \tag{2.11}$$

where ϕ_{21} is in radians and ω is in rad/s. Port 1 is the input port and port 2 is the output port. The phase delay is actually the time delay for a steady sinusoidal signal and is not necessarily the true signal delay, because a steady sinusoidal signal does not carry information; sometimes, it is also referred to as the carrier delay [1]. The more important parameter is the group delay, defined by

$$\tau_d = -\frac{d\phi_{21}}{d\omega} \text{ s} \tag{2.12}$$

This represents the true signal (baseband signal) delay and is also referred to as the envelope delay.

In network analysis or synthesis, it may be desirable to express the reflection parameter S_{11} in terms of the terminal impedance Z_{01} and the so-called input impedance $Z_{in1} = V_1/I_1$, which is the impedance looking into port 1 of the network. Such an expression can be deduced by evaluating S_{11} in Eq. (2.6) in terms of the voltage and current variables using the relationships defined in Eq. (2.4b). This gives

$$S_{11} = \left.\frac{b_1}{a_1}\right|_{a_2=0} = \frac{V_1/\sqrt{Z_{01}} - \sqrt{Z_{01}}I_1}{V_1/\sqrt{Z_{01}} + \sqrt{Z_{01}}I_1} \tag{2.13}$$

Replacing V_1 by $Z_{in1}I_1$ results in the desired expression

$$S_{11} = \frac{Z_{in1} - Z_{01}}{Z_{in1} + Z_{01}} \tag{2.14}$$

Similarly, we can have

$$S_{22} = \frac{Z_{in2} - Z_{02}}{Z_{in2} + Z_{02}} \tag{2.15}$$

where $Z_{in2} = V_2/I_2$ is the input impedance looking into port 2 of the network. Equations (2.14) and (2.15) indicate the impedance matching of the network with respect to its terminal impedances.

The S parameters have several properties that are useful for network analysis. For a reciprocal network we have $S_{12} = S_{21}$. If the network is symmetrical, an addition property, $S_{11} = S_{22}$, holds. Hence, the symmetrical network is also reciprocal. For a lossless passive network, the transmitting power and the reflected power must equal to the total incident power. The mathematical statements of this power conservation condition are

$$S_{21}S_{21}^* + S_{11}S_{11}^* = 1 \quad \text{or} \quad |S_{21}|^2 + |S_{11}|^2 = 1$$
$$S_{12}S_{12}^* + S_{22}S_{22}^* = 1 \quad \text{or} \quad |S_{12}|^2 + |S_{22}|^2 = 1 \quad (2.16)$$

2.3 SHORT-CIRCUIT ADMITTANCE PARAMETERS

The short-circuit admittance or Y parameters of a two-port network are defined as

$$Y_{11} = \left.\frac{I_1}{V_1}\right|_{V_2=0} \qquad Y_{12} = \left.\frac{I_1}{V_2}\right|_{V_1=0}$$
$$Y_{21} = \left.\frac{I_2}{V_1}\right|_{V_2=0} \qquad Y_{22} = \left.\frac{I_2}{V_2}\right|_{V_1=0} \quad (2.17)$$

in which $V_n = 0$ implies a perfect short-circuit at port n. The definitions of the Y parameters may also be written as

$$\begin{bmatrix} I_1 \\ I_2 \end{bmatrix} = \begin{bmatrix} Y_{11} & Y_{12} \\ Y_{21} & Y_{22} \end{bmatrix} \cdot \begin{bmatrix} V_1 \\ V_2 \end{bmatrix} \quad (2.18)$$

where the matrix containing the Y parameters is called the short-circuit admittance or simply Y matrix and may be denoted by $[Y]$. For reciprocal networks $Y_{12} = Y_{21}$. In addition to this, if networks are symmetrical, then $Y_{11} = Y_{22}$. For a lossless network, the Y parameters are all purely imaginary.

2.4 OPEN-CIRCUIT IMPEDANCE PARAMETERS

The open-circuit impedance or Z parameters of a two-port network are defined as

$$Z_{11} = \left.\frac{V_1}{I_1}\right|_{I_2=0} \qquad Z_{12} = \left.\frac{V_1}{I_2}\right|_{I_1=0}$$
$$Z_{21} = \left.\frac{V_2}{I_1}\right|_{I_2=0} \qquad Z_{22} = \left.\frac{V_2}{I_2}\right|_{I_1=0} \quad (2.19)$$

where $I_n = 0$ implies a perfect open-circuit at port n. These definitions can be written as

$$\begin{bmatrix} V_1 \\ V_2 \end{bmatrix} = \begin{bmatrix} Z_{11} & Z_{12} \\ Z_{21} & Z_{22} \end{bmatrix} \cdot \begin{bmatrix} I_1 \\ I_2 \end{bmatrix} \quad (2.20)$$

The matrix, which contains the Z parameters, is known as the open-circuit impedance or Z matrix denoted by $[Z]$. For reciprocal networks, we have $Z_{12} = Z_{21}$. If networks are symmetrical, then $Z_{12} = Z_{21}$ and $Z_{11} = Z_{22}$. For a lossless network, the Z parameters are all purely imaginary.

Inspecting Eqs. (2.18) and (2.20), we immediately obtain an important relation

$$[Z] = [Y]^{-1} \tag{2.21}$$

2.5 ABCD PARAMETERS

The *ABCD* parameters of a two-port network are given by

$$A = \left.\frac{V_1}{V_2}\right|_{I_2=0} \quad B = \left.\frac{V_1}{-I_2}\right|_{V_2=0}$$
$$C = \left.\frac{I_1}{V_2}\right|_{I_2=0} \quad D = \left.\frac{I_1}{-I_2}\right|_{V_2=0} \tag{2.22}$$

These parameters are actually defined in a set of linear equations in matrix notation

$$\begin{bmatrix} V_1 \\ I_1 \end{bmatrix} = \begin{bmatrix} A & B \\ C & D \end{bmatrix} \cdot \begin{bmatrix} V_2 \\ -I_2 \end{bmatrix} \tag{2.23}$$

where the matrix comprised of the *ABCD* parameters is called the *ABCD* matrix. Sometimes, it may also be referred to as the transfer or chain matrix. The *ABCD* parameters have following properties:

$$AD - BC = 1 \quad \text{for a reciprocal network} \tag{2.24}$$
$$A = D \quad \text{for a symmetrical network} \tag{2.25}$$

If the network is lossless, then A and D will be purely real and B and C will be purely imaginary.

If the network in Figure 2.1 is turned around, then the transfer matrix defined in Eq. (2.23) becomes

$$\begin{bmatrix} A_t & B_t \\ C_t & D_t \end{bmatrix} = \begin{bmatrix} D & B \\ C & A \end{bmatrix} \tag{2.26}$$

where the parameters with t subscripts are for the network after being turned around, and the parameters without subscripts are for the network before being turned around (with its original orientation). In both cases, V_1 and I_1 are at the left terminal and V_2 and I_2 are at the right terminal.

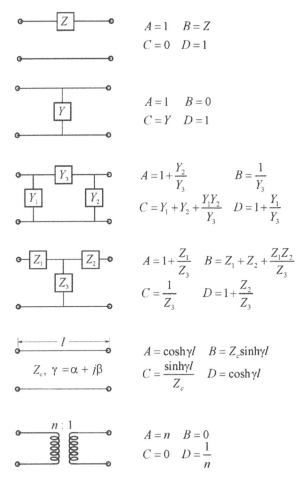

FIGURE 2.2 Some useful two-port networks and their *ABCD* parameters.

The *ABCD* parameters are very useful for analysis of a complex two-port network that may be divided into two or more cascaded subnetworks. Figure 2.2 gives the *ABCD* parameters of some useful two-port networks.

2.6 TRANSMISSION-LINE NETWORKS

Since $V_2 = -I_2 Z_{02}$, the input impedance of the two-port network in Figure 2.1 is given by

$$Z_{in1} = \frac{V_1}{I_1} = \frac{Z_{02}A + B}{Z_{02}C + D} \qquad (2.27)$$

Substituting the *ABCD* parameters for the transmission-line network given in Figure 2.2 into Eq. (2.27) leads to a very useful equation

$$Z_{in1} = Z_c \frac{Z_{02} + Z_c \tanh \gamma l}{Z_c + Z_{02} \tanh \gamma l} \tag{2.28}$$

where Z_c, γ, and l are the characteristic impedance, the complex propagation constant, and the length of the transmission line, respectively. For a lossless line, $\gamma = j\beta$ and Eq. (2.28) becomes

$$Z_{in1} = Z_c \frac{Z_{02} + jZ_c \tan \beta l}{Z_c + jZ_{02} \tan \beta l} \tag{2.29}$$

Besides the two-port transmission-line network, two types of one-port transmission networks are of equal significance in the design of microwave filters. These are formed by imposing an open- or a short-circuit at one terminal of a two-port transmission-line network. The input impedances of these one-port networks are readily found from Eqs. (2.27) or (2.28):

$$Z_{in1}|_{Z_{02}=\infty} = \frac{A}{C} = \frac{Z_c}{\tanh \gamma l} \tag{2.30}$$

$$Z_{in1}|_{Z_{02}=0} = \frac{B}{D} = Z_c \tanh \gamma l \tag{2.31}$$

Assuming a lossless transmission, these expressions become

$$Z_{in1}|_{Z_{02}=\infty} = \frac{Z_c}{j \tan \beta l} \tag{2.32}$$

$$Z_{in1}|_{Z_{02}=0} = jZ_c \tan \beta l \tag{2.33}$$

We will further discuss the transmission-line networks in the next chapter when we introduce Richards' transformation.

2.7 NETWORK CONNECTIONS

Often in the analysis of a filter network, it is convenient to treat one or more filter components or elements as individual subnetworks and then connect them to determine the network parameters of the filter. The three basic types of connection, which are usually encountered, are:

1. Parallel
2. Series
3. Cascade

14 NETWORK ANALYSIS

Suppose we wish to connect two networks N' and N'' in parallel, as shown in Figure 2.3a. An easy way to do this type of connection is to use their Y matrices. This is because

$$\begin{bmatrix} I_1 \\ I_2 \end{bmatrix} = \begin{bmatrix} I'_1 \\ I'_2 \end{bmatrix} + \begin{bmatrix} I''_1 \\ I''_2 \end{bmatrix} \quad \text{and} \quad \begin{bmatrix} V_1 \\ V_2 \end{bmatrix} = \begin{bmatrix} V'_1 \\ V'_2 \end{bmatrix} = \begin{bmatrix} V''_1 \\ V''_2 \end{bmatrix}$$

Therefore,

$$\begin{bmatrix} I_1 \\ I_2 \end{bmatrix} = \left(\begin{bmatrix} Y'_{11} & Y'_{12} \\ Y'_{21} & Y'_{22} \end{bmatrix} + \begin{bmatrix} Y''_{11} & Y''_{12} \\ Y''_{21} & Y''_{22} \end{bmatrix} \right) \cdot \begin{bmatrix} V_1 \\ V_2 \end{bmatrix} \quad (2.34a)$$

or the Y matrix of the combined network is

$$[Y] = [Y'] + [Y''] \quad (2.34b)$$

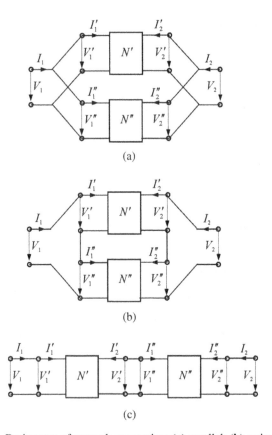

FIGURE 2.3 Basic types of network connection: (**a**) parallel; (**b**) series; (**c**) cascade.

This type of connection can be extended to more than two two-port networks connected in parallel. In that case the short-circuit admittance matrix of the composite network is given simply by the sum of the short-circuit admittance matrices of the individual networks.

Analogously, the networks of Figure 2.3b are connected in series at both their input and output terminals; consequently

$$\begin{bmatrix} V_1 \\ V_2 \end{bmatrix} = \begin{bmatrix} V_1' \\ V_2' \end{bmatrix} + \begin{bmatrix} V_1'' \\ V_2'' \end{bmatrix} \quad \text{and} \quad \begin{bmatrix} I_1 \\ I_2 \end{bmatrix} = \begin{bmatrix} I_1' \\ I_2' \end{bmatrix} = \begin{bmatrix} I_1'' \\ I_2'' \end{bmatrix}$$

This gives

$$\begin{bmatrix} V_1 \\ V_2 \end{bmatrix} = \left(\begin{bmatrix} Z_{11}' & Z_{12}' \\ Z_{21}' & Z_{22}' \end{bmatrix} + \begin{bmatrix} Z_{11}'' & Z_{12}'' \\ Z_{21}'' & Z_{22}'' \end{bmatrix} \right) \cdot \begin{bmatrix} I_1 \\ I_2 \end{bmatrix} \tag{2.35a}$$

and thus the resultant Z matrix of the composite network is given by

$$[Z] = [Z'] + [Z''] \tag{2.35b}$$

Similarly, if there are more than two two-port networks to be connected in series to form a composite network, the open-circuit impedance matrix of the composite network is equal to the sum of the individual open-circuit impedance matrices.

The cascade connection of two or more simpler networks appears to be used most frequently in analysis and design of filters. This is because most filters consist of cascaded two-port components. For simplicity, consider a network formed by the cascade connection of two subnetworks, as shown in Figure 2.3c. The following terminal voltage and current relationships at the terminals of the composite network would be obvious:

$$\begin{bmatrix} V_1 \\ I_1 \end{bmatrix} = \begin{bmatrix} V_1' \\ I_1' \end{bmatrix} \quad \text{and} \quad \begin{bmatrix} V_2 \\ I_2 \end{bmatrix} = \begin{bmatrix} V_2'' \\ I_2'' \end{bmatrix}$$

It should be noted that the outputs of the first subnetwork N' are the inputs of the followed second subnetwork N'', namely

$$\begin{bmatrix} V_2' \\ -I_2' \end{bmatrix} = \begin{bmatrix} V_1'' \\ I_1'' \end{bmatrix}$$

If the networks N' and N'' are described by the *ABCD* parameters, these terminal voltage and current relationships all together lead to

$$\begin{bmatrix} V_1 \\ I_1 \end{bmatrix} = \left(\begin{bmatrix} A' & B' \\ C' & D' \end{bmatrix} \cdot \begin{bmatrix} A'' & B'' \\ C'' & D'' \end{bmatrix} \right) \cdot \begin{bmatrix} V_2 \\ -I_2 \end{bmatrix} = \begin{bmatrix} A & B \\ C & D \end{bmatrix} \cdot \begin{bmatrix} V_2 \\ -I_2 \end{bmatrix} \tag{2.36}$$

16 NETWORK ANALYSIS

Thus, the transfer matrix of the composite network is equal to the matrix product of the transfer matrices of the cascaded subnetworks. This argument is valid for any number of two-port networks in cascade connection.

Sometimes, it may be desirable to directly cascade two two-port networks using the S parameters. Let S'_{mn} denote the S parameters of the network N', S''_{mn} denote the S parameters of the network N'', and S_{mn} denote the S parameters of the composite network for $m, n = 1, 2$. If, at the interface of the connection in Figure 2.3c,

$$\begin{aligned} b'_2 &= a''_1 \\ a'_2 &= b''_1 \end{aligned} \qquad (2.37)$$

it can be shown that the resultant S matrix of the composite network is given by

$$\begin{bmatrix} S_{11} & S_{12} \\ S_{21} & S_{22} \end{bmatrix} = \begin{bmatrix} S'_{11} + \kappa S'_{12} S'_{21} S''_{11} & \kappa S'_{12} S''_{12} \\ \kappa S'_{21} S''_{21} & S''_{22} + \kappa S''_{12} S''_{21} S'_{22} \end{bmatrix} \qquad (2.38)$$

where

$$\kappa = \frac{1}{1 - S'_{22} S''_{11}}$$

It is important to note that the relationships in Eq. (2.37) imply that the same terminal impedance is assumed at the port 2 of the network N' and the port 1 of the network N'' when S'_{mn} and S''_{mn} are evaluated individually.

2.8 NETWORK PARAMETER CONVERSIONS

From the above discussions it can be seen that for network analysis, we may use different types of network parameters. Therefore, it is often required to convert one type of parameter to another. The conversion between Z and Y is the simplest one as given by Eq. (2.21). In principle, the relationships between any two types of parameters can be deduced from the relationships of terminal variables in Eq. (2.4).

For our example, let us define the following matrix notations:

$$[V] = \begin{bmatrix} V_1 \\ V_2 \end{bmatrix} \quad [I] = \begin{bmatrix} I_1 \\ I_2 \end{bmatrix} \quad [a] = \begin{bmatrix} a_1 \\ a_2 \end{bmatrix} \quad [b] = \begin{bmatrix} b_1 \\ b_2 \end{bmatrix}$$

$$[\sqrt{Z_0}] = \begin{bmatrix} \sqrt{Z_{01}} & 0 \\ 0 & \sqrt{Z_{02}} \end{bmatrix} \quad [\sqrt{Y_0}] = \begin{bmatrix} \sqrt{Y_{01}} & 0 \\ 0 & \sqrt{Y_{02}} \end{bmatrix}$$

NETWORK PARAMETER CONVERSIONS   17

Note that the terminal admittances $Y_{0n} = 1/Z_{0n}$ for $n = 1, 2$. Thus, Eq. (2.4b) becomes

$$[a] = \frac{1}{2}\left(\left[\sqrt{Y_0}\right] \cdot [V] + \left[\sqrt{Z_0}\right] \cdot [I]\right)$$
$$[b] = \frac{1}{2}\left(\left[\sqrt{Y_0}\right] \cdot [V] - \left[\sqrt{Z_0}\right] \cdot [I]\right) \quad (2.39)$$

Suppose we wish to find the relationships between the S and the Z parameters. Substituting $[V] = [Z] \cdot [I]$ into Eq. (2.39) yields

$$[a] = \frac{1}{2}\left(\left[\sqrt{Y_0}\right] \cdot [Z] + \left[\sqrt{Z_0}\right]\right) \cdot [I]$$
$$[b] = \frac{1}{2}\left(\left[\sqrt{Y_0}\right] \cdot [Z] - \left[\sqrt{Z_0}\right]\right) \cdot [I]$$

Replacing $[b]$ by $[S] \cdot [a]$ and combining the above two equations, we can arrive at the required relationships

$$[S] = \left(\left[\sqrt{Y_0}\right] \cdot [Z] - \left[\sqrt{Z_0}\right]\right) \cdot \left(\left[\sqrt{Y_0}\right] \cdot [Z] + \left[\sqrt{Z_0}\right]\right)^{-1}$$
$$[Z] = \left(\left[\sqrt{Y_0}\right] - [S] \cdot \left[\sqrt{Y_0}\right]\right)^{-1} \cdot \left([S] \cdot \left[\sqrt{Z_0}\right] + \left[\sqrt{Z_0}\right]\right) \quad (2.40)$$

In a similar fashion, substituting $[I] = [Y] \cdot [V]$ into Eq. (2.39) we can obtain

$$[S] = \left(\left[\sqrt{Y_0}\right] - \left[\sqrt{Z_0}\right] \cdot [Y]\right) \cdot \left(\left[\sqrt{Z_0}\right] \cdot [Y] + \left[\sqrt{Y_0}\right]\right)^{-1}$$
$$[Y] = \left([S] \cdot \left[\sqrt{Z_0}\right] + \left[\sqrt{Z_0}\right]\right)^{-1} \cdot \left(\left[\sqrt{Y_0}\right] - [S] \cdot \left[\sqrt{Y_0}\right]\right) \quad (2.41)$$

Thus all the relationships between any two types of parameters can be found in this way. For convenience, these are summarized in Table 2.1–2.4 for equal terminations $Z_{01} = Z_{02} = Z_0$ and $Y_0 = 1/Z_0$.

TABLE 2.1 S Parameters in Terms of ABCD, Y, and Z Parameters

	ABCD	Y	Z
S_{11}	$\dfrac{A + B/Z_0 - CZ_0 - D}{A + B/Z_0 + CZ_0 + D}$	$\dfrac{(Y_0 - Y_{11})(Y_0 + Y_{22}) + Y_{12}Y_{21}}{(Y_0 + Y_{11})(Y_0 + Y_{22}) - Y_{12}Y_{21}}$	$\dfrac{(Z_{11} - Z_0)(Z_{22} + Z_0) - Z_{12}Z_{21}}{(Z_{11} + Z_0)(Z_{22} + Z_0) - Z_{12}Z_{21}}$
S_{12}	$\dfrac{2(AD - BC)}{A + B/Z_0 + CZ_0 + D}$	$\dfrac{-2Y_{12}Y_0}{(Y_0 + Y_{11})(Y_0 + Y_{22}) - Y_{12}Y_{21}}$	$\dfrac{2Z_{12}Z_0}{(Z_{11} + Z_0)(Z_{22} + Z_0) - Z_{12}Z_{21}}$
S_{21}	$\dfrac{2}{A + B/Z_0 + CZ_0 + D}$	$\dfrac{-2Y_{21}Y_0}{(Y_0 + Y_{11})(Y_0 + Y_{22}) - Y_{12}Y_{21}}$	$\dfrac{2Z_{21}Z_0}{(Z_{11} + Z_0)(Z_{22} + Z_0) - Z_{12}Z_{21}}$
S_{22}	$\dfrac{-A + B/Z_0 - CZ_0 + D}{A + B/Z_0 + CZ_0 + D}$	$\dfrac{(Y_0 + Y_{11})(Y_0 - Y_{22}) + Y_{12}Y_{21}}{(Y_0 + Y_{11})(Y_0 + Y_{22}) - Y_{12}Y_{21}}$	$\dfrac{(Z_{11} + Z_0)(Z_{22} - Z_0) - Z_{12}Z_{21}}{(Z_{11} + Z_0)(Z_{22} + Z_0) - Z_{12}Z_{21}}$

TABLE 2.2 ABCD Parameters in Terms of S, Y, and Z Parameters

	S	Y	Z
A	$\dfrac{(1+S_{11})(1-S_{22})+S_{12}S_{21}}{2S_{21}}$	$\dfrac{-Y_{22}}{Y_{21}}$	$\dfrac{Z_{11}}{Z_{21}}$
B	$Z_0\dfrac{(1+S_{11})(1+S_{22})-S_{12}S_{21}}{2S_{21}}$	$\dfrac{-1}{Y_{21}}$	$\dfrac{Z_{11}Z_{22}-Z_{12}Z_{21}}{Z_{21}}$
C	$\dfrac{1}{Z_0}\dfrac{(1-S_{11})(1-S_{22})-S_{12}S_{21}}{2S_{21}}$	$\dfrac{-(Y_{11}Y_{22}-Y_{12}Y_{21})}{Y_{21}}$	$\dfrac{1}{Z_{21}}$
D	$\dfrac{(1-S_{11})(1+S_{22})+S_{12}S_{21}}{2S_{21}}$	$\dfrac{-Y_{11}}{Y_{21}}$	$\dfrac{Z_{22}}{Z_{21}}$

TABLE 2.3 Y Parameters in Terms of S, ABCD, and Z Parameters

	S	ABCD	Z
Y_{11}	$Y_0\dfrac{(1-S_{11})(1+S_{22})+S_{12}S_{21}}{(1+S_{11})(1+S_{22})-S_{12}S_{21}}$	$\dfrac{D}{B}$	$\dfrac{Z_{22}}{Z_{11}Z_{22}-Z_{12}Z_{21}}$
Y_{12}	$Y_0\dfrac{-2S_{12}}{(1+S_{11})(1+S_{22})-S_{12}S_{21}}$	$\dfrac{-(AD-BC)}{B}$	$\dfrac{-Z_{12}}{Z_{11}Z_{22}-Z_{12}Z_{21}}$
Y_{21}	$Y_0\dfrac{-2S_{21}}{(1+S_{11})(1+S_{22})-S_{12}S_{21}}$	$\dfrac{-1}{B}$	$\dfrac{-Z_{21}}{Z_{11}Z_{22}-Z_{12}Z_{21}}$
Y_{22}	$Y_0\dfrac{(1+S_{11})(1-S_{22})+S_{12}S_{21}}{(1+S_{11})(1+S_{22})-S_{12}S_{21}}$	$\dfrac{A}{B}$	$\dfrac{Z_{11}}{Z_{11}Z_{22}-Z_{12}Z_{21}}$

TABLE 2.4 Z Parameters in Terms of S, ABCD, and Y Parameters

	S	ABCD	Y
Z_{11}	$Z_0\dfrac{(1+S_{11})(1-S_{22})+S_{12}S_{21}}{(1-S_{11})(1-S_{22})-S_{12}S_{21}}$	$\dfrac{A}{C}$	$\dfrac{Y_{22}}{Y_{11}Y_{22}-Y_{12}Y_{21}}$
Z_{12}	$Z_0\dfrac{2S_{12}}{(1-S_{11})(1-S_{22})-S_{12}S_{21}}$	$\dfrac{(AD-BC)}{C}$	$\dfrac{-Y_{12}}{Y_{11}Y_{22}-Y_{12}Y_{21}}$
Z_{21}	$Z_0\dfrac{2S_{21}}{(1-S_{11})(1-S_{22})-S_{12}S_{21}}$	$\dfrac{1}{C}$	$\dfrac{-Y_{21}}{Y_{11}Y_{22}-Y_{12}Y_{21}}$
Z_{22}	$Z_0\dfrac{(1-S_{11})(1+S_{22})+S_{12}S_{21}}{(1-S_{11})(1-S_{22})-S_{12}S_{21}}$	$\dfrac{D}{C}$	$\dfrac{Y_{11}}{Y_{11}Y_{22}-Y_{12}Y_{21}}$

2.9 SYMMETRICAL NETWORK ANALYSIS

If a network is symmetrical, it is convenient for network analysis to bisect the symmetrical network into two identical halves with respect to its symmetrical interface. When an even excitation is applied to the network as indicated in Figure 2.4a the symmetrical interface is open-circuited, and the two network halves become the two identical one-port even-mode networks with the other port open-circuited. In a similar fashion, under an odd excitation, as shown in Figure 2.4b, the symmetrical interface is short-circuited, and the two network halves become the two identical one-port, odd-mode networks, with the other port short-circuited. Since any excitation to a symmetrical two-port network can be obtained by a linear combination of the even and odd excitations, the network analysis will be simplified by first separately analyzing the one-port even- and odd-mode networks, and then determining the two-port network parameters from the even- and odd-mode network parameters.

For example, the one-port even- and odd-mode S parameters are

$$S_{11e} = \frac{b_e}{a_e}$$
$$S_{11o} = \frac{b_o}{a_o} \quad (2.42)$$

where the subscripts e and o refer to the even-mode and odd-mode, respectively. For the symmetrical network, the following relationships of wave variables hold

$$a_1 = a_e + a_o \quad a_2 = a_e - a_o$$
$$b_1 = b_e + b_o \quad b_2 = b_e - b_o \quad (2.43)$$

Letting $a_2 = 0$, we have from Eqs. (2.42) and (2.43) that

$$a_1 = 2a_e = 2a_o$$
$$b_1 = S_{11e}a_e + S_{11o}a_o$$
$$b_2 = S_{11e}a_e - S_{11o}a_o$$

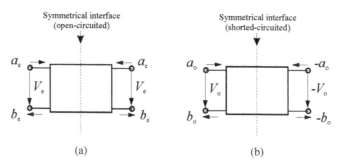

FIGURE 2.4 Symmetrical two-port networks with (**a**) even-mode excitation; (**b**) odd-mode excitation.

Substituting these results into the definitions of two-port S parameters gives

$$S_{11} = \frac{b_1}{a_1}\bigg|_{a_2=0} = \frac{1}{2}(S_{11e} + S_{11o})$$

$$S_{21} = \frac{b_2}{a_1}\bigg|_{a_2=0} = \frac{1}{2}(S_{11e} - S_{11o}) \qquad (2.44)$$

$$S_{22} = S_{11}$$

$$S_{12} = S_{21}$$

The last two equations are obvious, because of the symmetry.

Let Z_{ine} and Z_{ino} represent the input impedances of the one-port, even- and odd-mode networks. According to Eq. (2.14), the refection coefficients in Eq. (2.42) can be given by

$$S_{11e} = \frac{Z_{ine} - Z_{01}}{Z_{ine} + Z_{01}} \quad \text{and} \quad S_{11o} = \frac{Z_{ino} - Z_{01}}{Z_{ino} + Z_{01}} \qquad (2.45)$$

By substituting them into Eq. (2.44), we can arrive at some very useful formulas

$$S_{11} = S_{22} = \frac{Z_{ine} Z_{ino} - Z_{01}^2}{(Z_{ine} + Z_{01}) \cdot (Z_{ino} + Z_{01})} = \frac{Y_{01}^2 - Y_{ine} Y_{ino}}{(Y_{01} + Y_{ine}) \cdot (Y_{01} + Y_{ino})}$$

$$S_{21} = S_{12} = \frac{Z_{ine} Z_{01} - Z_{ino} Z_{01}}{(Z_{ine} + Z_{01}) \cdot (Z_{ino} + Z_{01})} = \frac{Y_{ino} Y_{01} - Y_{ine} Y_{01}}{(Y_{01} + Y_{ine}) \cdot (Y_{01} + Y_{ino})} \qquad (2.46)$$

where $Y_{ine} = 1/Z_{ine}$, $Y_{ino} = 1/Z_{ino}$, and $Y_{01} = 1/Z_{01}$. For normalized impedances/admittances such that $z = Z/Z_{01}$ and $y = Y/Y_{01}$, the formulas in Eq. (2.46) are simplified as

$$S_{11} = S_{22} = \frac{z_{ine} z_{ino} - 1}{(z_{ine} + 1) \cdot (z_{ino} + 1)} = \frac{1 - y_{ine} y_{ino}}{(1 + y_{ine}) \cdot (1 + y_{ino})}$$

$$S_{21} = S_{12} = \frac{z_{ine} - z_{ino}}{(z_{ine} + 1) \cdot (z_{ino} + 1)} = \frac{y_{ino} - y_{ine}}{(1 + y_{ine}) \cdot (1 + y_{ino})} \qquad (2.47)$$

2.10 MULTIPORT NETWORKS

Networks, that have more than two ports, may be referred to as the multiport networks. The definitions of S, Z, and Y parameters for a multiport network are similar to those for a two-port network described previously. As far as the S parameters are concerned, in general an M-port network can be described by

$$\begin{bmatrix} b_1 \\ b_2 \\ \vdots \\ b_M \end{bmatrix} = \begin{bmatrix} S_{11} & S_{12} & \cdots & S_{1M} \\ S_{21} & S_{22} & \cdots & S_{2M} \\ \vdots & \vdots & \cdots & \vdots \\ S_{M1} & S_{M2} & \cdots & S_{MM} \end{bmatrix} \cdot \begin{bmatrix} a_1 \\ a_2 \\ \vdots \\ a_M \end{bmatrix} \quad (2.48a)$$

which may be expressed as

$$[b] = [S] \cdot [a] \quad (2.48b)$$

where $[S]$ is the S-matrix of order $M \times M$ whose elements are defined by

$$S_{ij} = \left. \frac{b_i}{a_j} \right|_{a_k = 0 (k \neq j \text{ and } k=1,2,\cdots M)} \quad \text{for } i, j = 1, 2, \cdots M \quad (2.48c)$$

For a reciprocal network, $S_{ij} = S_{ji}$ and $[S]$ is a symmetrical matrix such that

$$[S]^t = [S] \quad (2.49)$$

where the superscript t denotes the transpose of matrix. For a lossless passive network, we have

$$[S]^t [S]^* = [U] \quad (2.50)$$

where the superscript * denotes the conjugate of matrix, and $[U]$ is a unity matrix.

It is worthwhile mentioning that the relationships given in Eqs. (2.21), (2.40), and (2.41) can be extended for converting network parameters of multiport networks.

The connection of two multiport networks may be performed using the following method. Assume that an M_1-port network N' and an M_2-port network N'', which are described by

$$[b'] = [S'] \cdot [a'] \quad \text{and} \quad [b''] = [S''] \cdot [a''] \quad (2.51a)$$

respectively, will connect each other at c pairs of ports. Rearrange Eq. (2.51a) such that

$$\begin{bmatrix} [b']_p \\ [b']_c \end{bmatrix} = \begin{bmatrix} [S']_{pp} & [S']_{pc} \\ [S']_{cp} & [S']_{cc} \end{bmatrix} \cdot \begin{bmatrix} [a']_p \\ [a']_c \end{bmatrix} \quad \text{and}$$

$$\begin{bmatrix} [b'']_q \\ [b'']_c \end{bmatrix} = \begin{bmatrix} [S'']_{qq} & [S'']_{qc} \\ [S'']_{cq} & [S'']_{cc} \end{bmatrix} \cdot \begin{bmatrix} [a'']_q \\ [a'']_c \end{bmatrix} \quad (2.51b)$$

where $[b']_c$ and $[a']_c$ contain the wave variables at the c connecting ports of the network N', $[b']_p$ and $[a']_p$ contain the wave variables at the p unconnected ports of the network N'. In a similar fashion $[b'']_c$ and $[a'']_c$ contain the wave variables at the c connecting ports of the network N'', $[b'']_q$ and $[a'']_q$ contain the wave variables at the q unconnected ports of the network N''; and all the S submatrices contain the corresponding S parameters. Obviously, $p + c = M_1$ and $q + c = M_2$. It is important to note that the conditions for all the connections are $[b']_c = [a'']_c$ and $[b'']_c = [a']_c$, or

$$\begin{bmatrix}[b']_c\\ [b'']_c\end{bmatrix} = \begin{bmatrix}[0] & [U]\\ [U] & [0]\end{bmatrix} \cdot \begin{bmatrix}[a']_c\\ [a'']_c\end{bmatrix} \qquad (2.52)$$

where $[0]$ and $[U]$ denote the zero matrix and unity matrix, respectively. Combine the two systems of equations in (2.51b) into one giving

$$\begin{bmatrix}[b']_p\\ [b'']_q\\ [b']_c\\ [b'']_c\end{bmatrix} = \begin{bmatrix}[S']_{pp} & [0] & [S']_{pc} & [0]\\ [0] & [S'']_{qq} & [0] & [S'']_{qc}\\ [S']_{cp} & [0] & [S']_{cc} & [0]\\ [0] & [S'']_{cq} & [0] & [S'']_{cc}\end{bmatrix} \cdot \begin{bmatrix}[a']_p\\ [a'']_q\\ [a']_c\\ [a'']_c\end{bmatrix} \qquad (2.53)$$

From Eq. (2.53) we can have

$$\begin{bmatrix}[b']_p\\ [b'']_q\end{bmatrix} = \begin{bmatrix}[S']_{pp} & [0]\\ [0] & [S'']_{qq}\end{bmatrix} \cdot \begin{bmatrix}[a']_p\\ [a'']_q\end{bmatrix} + \begin{bmatrix}[S']_{pc} & [0]\\ [0] & [S'']_{qc}\end{bmatrix} \cdot \begin{bmatrix}[a']_c\\ [a'']_c\end{bmatrix} \qquad (2.54a)$$

$$\begin{bmatrix}[b']_c\\ [b'']_c\end{bmatrix} = \begin{bmatrix}[S']_{cp} & [0]\\ [0] & [S'']_{cq}\end{bmatrix} \cdot \begin{bmatrix}[a']_p\\ [a'']_q\end{bmatrix} + \begin{bmatrix}[S']_{cc} & [0]\\ [0] & [S'']_{cc}\end{bmatrix} \cdot \begin{bmatrix}[a']_c\\ [a'']_c\end{bmatrix} \qquad (2.54b)$$

Substituting Eqs. (2.52) into (2.54b) leads to

$$\begin{bmatrix}[a']_c\\ [a'']_c\end{bmatrix} = \begin{bmatrix}-[S']_{cc} & [U]\\ [U] & -[S'']_{cc}\end{bmatrix}^{-1} \begin{bmatrix}[S']_{cp} & [0]\\ [0] & [S'']_{cq}\end{bmatrix} \cdot \begin{bmatrix}[a']_p\\ [a'']_q\end{bmatrix} \qquad (2.55)$$

It now becomes clearer from Eqs. (2.54a) and (2.55) that the composite network can be described by

$$\begin{bmatrix}[b']_p\\ [b'']_q\end{bmatrix} = [S] \cdot \begin{bmatrix}[a']_p\\ [a'']_q\end{bmatrix} \qquad (2.56a)$$

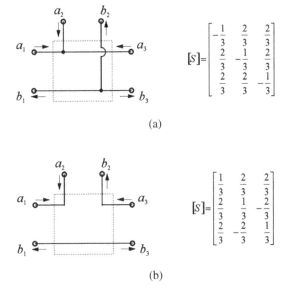

FIGURE 2.5 Auxiliary three-port networks and their S matrices: (**a**) parallel junction; (**b**) series junction.

with the resultant S matrix given by

$$[S] = \begin{bmatrix} [S']_{pp} & [0] \\ [0] & [S'']_{qq} \end{bmatrix} + \begin{bmatrix} [S']_{pc} & [0] \\ [0] & [S'']_{qc} \end{bmatrix} \cdot \begin{bmatrix} -[S']_{cc} & [U] \\ [U] & -[S'']_{cc} \end{bmatrix}^{-1}$$
$$\times \begin{bmatrix} [S']_{cp} & [0] \\ [0] & [S'']_{cq} \end{bmatrix} \quad (2.56b)$$

This procedure can be repeated if there are more than two multiport networks to be connected. The procedure is also general in such a way that it can be applied for networks with any number of ports, including two-port networks.

In order to make a parallel or series connection two auxiliary three-port networks in Figure 2.5 may be used. The one shown in Figure 2.5a is an ideal parallel junction for the parallel connection; its S matrix is given on the right. Figure 2.5b shows an ideal series-junction for the series connection along with its S matrix on the right.

2.11 EQUIVALENT AND DUAL NETWORK

Strictly speaking, two networks are said to be equivalent if the matrices of their corresponding network parameters are equal, irrespective of the fact that the networks may differ greatly in their configurations and in the number of elements possessed

24 NETWORK ANALYSIS

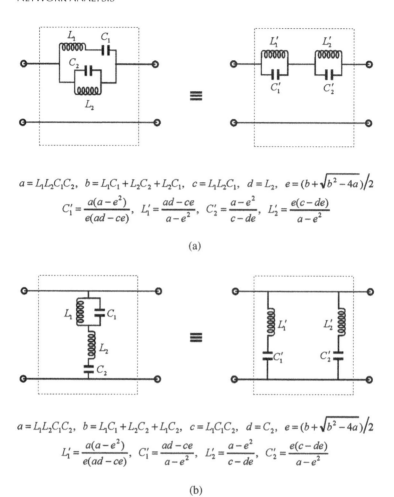

$a = L_1L_2C_1C_2$, $b = L_1C_1 + L_2C_2 + L_2C_1$, $c = L_1L_2C_1$, $d = L_2$, $e = (b + \sqrt{b^2 - 4a})/2$

$C_1' = \dfrac{a(a - e^2)}{e(ad - ce)}$, $L_1' = \dfrac{ad - ce}{a - e^2}$, $C_2' = \dfrac{a - e^2}{c - de}$, $L_2' = \dfrac{e(c - de)}{a - e^2}$

(a)

$a = L_1L_2C_1C_2$, $b = L_1C_1 + L_2C_2 + L_1C_2$, $c = L_1C_1C_2$, $d = C_2$, $e = (b + \sqrt{b^2 - 4a})/2$

$L_1' = \dfrac{a(a - e^2)}{e(ad - ce)}$, $C_1' = \dfrac{ad - ce}{a - e^2}$, $L_2' = \dfrac{a - e^2}{c - de}$, $C_2' = \dfrac{e(c - de)}{a - e^2}$

(b)

FIGURE 2.6 Equivalent networks for network transformation.

by each. In filter design, equivalent networks or circuits are often used to transform a network or circuit into another one, which can be easier realized or implemented in practice. For example, two pairs of useful equivalent networks for design of elliptic-function bandpass filters are depicted in Figure 2.6. The networks on the left actually result from the element transformation from lowpass to bandpass, which will be discussed in the next chapter, whereas the networks on the right are the corresponding equivalent networks, which are more convenient for practical implementation.

Dual networks are of great use in filter synthesis. For the definition of dual networks, let us consider two M-port networks. Assume that one network N is described by its open-circuit impedance parameters denoted by Z_{ij}, and the other N' is described by its short-circuit admittance parameters denoted by Y_{ij}'. The two networks are said to be the dual networks if

$$Z_{ii}/Z_0 = Y'_{ii}/Y'_0$$
$$Z_{ij}/Z_0 = -Y'_{ij}/Y'_0 (i \neq j)$$

where $Z_0 = 1 \, \Omega$ and $Y'_0 = 1$ mho are assumed for the normalization. As in the concept of equivalence, the internal structures of the networks are not relevant in determining of duality by use of the above definition. All that is required is dual behavior at the specified terminal pairs. In accordance with this definition, an inductance of x henries is dual to a capacitance of x farads, a resistance of x ohms is dual to a conductance of x mhos, a short-circuit is the dual of an open-circuit, a series connection is the dual of parallel connection, and so on.

It is important to note that in the strict sense of equivalence defined above, dual networks are not equivalent networks because the matrices of their corresponding network parameters are not equal. However, care must be exercised, since the term equivalence can have another sense. For example, it can be shown that $S_{21} = S'_{21}$ for two-port dual networks. This implies that two-port dual filter networks are described by the same transfer function, which will be discussed in the next chapter. In this sense, it is customary in the literature to say that two-port dual networks are also equivalent.

2.12 MULTIMODE NETWORKS

In analysis of microwave networks a single-mode operation is normally assumed. This single-mode is usually the transmission mode, like a quasi-TEM mode in microstrip or TE_{10} mode in waveguides. However, in reality, other modes can be excited in a practical microwave network like a waveguide or microstrip filter, even with a single-mode input, because there exist discontinuities in the physical structure of the network. In order to describe a practical microwave network more accurately, a multimode network representation may be used. In general, multimode networks can be described by

$$\begin{bmatrix} [b]_1 \\ [b]_2 \\ \vdots \\ [b]_P \end{bmatrix} = \begin{bmatrix} [S]_{11} & [S]_{12} & \cdots & [S]_{1P} \\ [S]_{21} & [S]_{22} & \cdots & [S]_{2P} \\ \vdots & \vdots & \cdots & \vdots \\ [S]_{P1} & [S]_{P2} & \cdots & [S]_{PP} \end{bmatrix} \cdot \begin{bmatrix} [a]_1 \\ [a]_2 \\ \vdots \\ [a]_P \end{bmatrix} \quad (2.57)$$

where P is the number of ports. The submatrices $[b]_i (i = 1, 2, \ldots P)$ are $M_i \times 1$ column matrices, each of which contains reflected wave variables of M_i modes, namely,

$$[b]_i = \begin{bmatrix} b_1 & b_2 & \cdots & b_{M_i} \end{bmatrix}_i^t$$

where the superscript t indicates the matrix transpose. Similarly, the submatrices $[a]_j$ for $j = 1, 2, \ldots P$ are $N_j \times 1$ column matrices, each of which contains incident wave variables of N_j modes, i.e., $[a]_j = \begin{bmatrix} a_1 & a_2 & \cdots & a_{N_j} \end{bmatrix}_j^t$. Thus, each of submatrices $[S]_{ij}$ is a $M_i \times N_j$ matrix, which represents the relationships between the incident modes at port j and reflected modes at port i. Equation (2.57) can be also expressed using simple notation

$$[b] = [S] \cdot [a]$$

and the scattering matrix of this type is called the generalized scattering matrix. An application of generalized scattering matrix for modeling microstrip or suspended microstrip discontinuities is described in Hong et al. [2].

Similarly, we may define the generalized Y and Z matrices to represent multimode networks. For multimode network analysis, the method described above for the multiport network connections can be extended for the connection of multimode networks.

For cascading two-port multimode networks, an alternative method that will be described below is more efficient. The method is based on a new set of network parameters that are defined by

$$\begin{bmatrix} [b]_1 \\ [a]_1 \end{bmatrix} = \begin{bmatrix} [T]_{11} & [T]_{12} \\ [T]_{21} & [T]_{22} \end{bmatrix} \cdot \begin{bmatrix} [a]_2 \\ [b]_2 \end{bmatrix} \quad (2.58)$$

In the above matrix formulation, we have all the wave variables (incident and reflected for any modes) at port 1 as the dependent variables and the all wave variables at port 2 as the independent variables. The parameters, which relate the independent and dependent wave variables, are called scattering transmission or transfer parameters, denoted by T. The matrix containing the all T parameters is referred to as scattering transmission or transfer matrix.

If we wish to connect two multimode two-port networks N' and N'' in cascade, the conditions for the connection are

$$\begin{bmatrix} [a']_2 \\ [b']_2 \end{bmatrix} = \begin{bmatrix} [b'']_1 \\ [a'']_1 \end{bmatrix} \quad (2.59)$$

These conditions state that the reflected waves at input of the second network N'' are the incident waves at the output of the first network N' and the incident waves at input of the second network N'' are the reflected waves at the output of the first network N'. More importantly, they also imply that each pair of the wave variables represents the same mode and has the same terminal impedance when the scattering transfer matrices are determined separately for the networks N' and N''. Making use of Eq. (2.59) gives

$$\begin{bmatrix} [b']_1 \\ [a']_1 \end{bmatrix} = \begin{bmatrix} [T']_{11} & [T']_{12} \\ [T']_{21} & [T']_{22} \end{bmatrix} \cdot \begin{bmatrix} [T'']_{11} & [T'']_{12} \\ [T'']_{21} & [T'']_{22} \end{bmatrix} \begin{bmatrix} [a'']_2 \\ [b'']_2 \end{bmatrix} \quad (2.60)$$

Thus, the scattering transfer matrix of the composite network is simply equal to the matrix product of the scattering transfer matrices of the cascaded networks. This procedure is very similar to that for the cascade of single-mode two-port networks described by the *ABCD* matrices and, needless to say, it can be used for single-mode networks as well. A demonstration of applying the scattering transmission matrix to microstrip discontinuity problems can be found in Hong and Shi [3].

The transformation between the generalized *S* and *T* parameters can be deduced from their definitions. This gives

$$[S]_{11} = [T]_{12}[T]_{22}^{-1} \quad [S]_{12} = [T]_{11} - [T]_{12}[T]_{22}^{-1}[T]_{21}$$
$$[S]_{21} = [T]_{22}^{-1} \quad [S]_{22} = -[T]_{22}^{-1}[T]_{21}$$
(2.61)

REFERENCES

[1] S. Haykin, *Communication Systems*, third edition, John Wiley & Sons, New York, 1994.
[2] J.-S. Hong, J.-M. Shi, and L.Sun, Exact computation of generalized scattering matrix of suspended microstrip step discontinuities, *Electron. Lett.* **25**(5), 1989, 335–336.
[3] J.-S. Hong and J.-M. Shi, Modeling microstrip step discontinuities by the transmission matrix, *Electron. Lett.* **23**(13), 1987, 678–680.
[4] C. G. Montagomery, R. H. Dicke, and E. M. Purcell, *Principles of Microwave Circuits*. McGraw-Hill, New York, 1948.
[5] E. A. Guillemin, *Introductory Circuit Theory*, John Wiley & Sons, New York, 1953.
[6] R. E. Collin, *Foundations for Microwave Engineering*, McGraw-Hill, New York, 1992.
[7] L. Weinberg, *Network analysis and synthesis*, McGraw-Hill, New York, 1962.
[8] J. L. Stewart, *Circuit Theory and Design*, John Wiley & Sons, New York, 1956.
[9] K. C. Gupta, R. Garg, and R. Ghadha, *Computer-aided Design of Microwave Circuits*, Artech House, Dedham, MA, 1981.
[10] H. J. Carlin, The scattering matrix in network theory, *IRE Trans. Circuit Theory*, **CT-3**, 1956, 88–97.

CHAPTER THREE

Basic Concepts and Theories of Filters

This chapter describes basic concepts and theories that form the foundation for design of general RF/microwave filters, including microstrip filters. The topics will cover filter transfer functions, lowpass prototype filters and elements, frequency and element transformations, immittance inverters, Richards' transformation, and Kuroda identities for distributed elements. Dissipation and unloaded quality factor of filter elements will also be discussed.

3.1 TRANSFER FUNCTIONS

3.1.1 General Definitions

The transfer function of a two-port filter network is a mathematical description of network response characteristics, namely, a mathematical expression of S_{21}. On many occasions, an amplitude-squared transfer function for a lossless passive filter network is defined as

$$|S_{21}(j\Omega)|^2 = \frac{1}{1 + \varepsilon^2 F_n^2(\Omega)} \quad (3.1)$$

where ε is a ripple constant, $F_n(\Omega)$ represents a filtering or characteristic function, and Ω is a frequency variable. For our discussion here, it is convenient to let Ω represent a radian frequency variable of a lowpass prototype filter that has a cutoff frequency at $\Omega = \Omega_c$ for $\Omega_c = 1$ (rad/s). Frequency transformations to the usual radian frequency for practical lowpass, highpass, bandpass, and bandstop filters will be discussed later.

Microstrip Filters for RF/Microwave Applications by Jia-Sheng Hong
Copyright © 2011 John Wiley & Sons, Inc.

For linear time-invariant networks, the transfer function may be defined as a rational function, that is

$$S_{21}(p) = \frac{N(p)}{D(p)} \tag{3.2}$$

where $N(p)$ and $D(p)$ are polynomials in a complex frequency variable $p = \sigma + j\Omega$. For a lossless passive network, the neper frequency $\sigma = 0$ and $p = j\Omega$. To find a realizable rational transfer function that produces response characteristics approximating the required response is the so-called approximation problem and, in many cases, the rational transfer function of Eq. (3.2) can be constructed from the amplitude-squared transfer function of Eq. (3.1) [1,2].

For a given transfer function of Eq. (3.1), the insertion loss response of the filter, following the conventional definition in Eq. (2.9), can be computed by

$$L_A(\Omega) = 10\log \frac{1}{|S_{21}(j\Omega)|^2} \text{ dB} \tag{3.3}$$

Since $|S_{11}|^2 + |S_{21}|^2 = 1$ for a lossless passive two-port network, the return loss response of the filter can be found using Eq. (2.9)

$$L_R(\Omega) = 10\log \left[1 - |S_{21}(j\Omega)|^2\right] \text{ dB} \tag{3.4}$$

If a rational transfer function is available, the phase response of the filter can be found as

$$\phi_{21}(\Omega) = \text{Arg}\, S_{21}(j\Omega) \tag{3.5}$$

The group-delay response of this network can then be calculated by

$$\tau_d(\Omega) = -\frac{d\phi_{21}(\Omega)}{d\Omega} \text{ s} \tag{3.6}$$

where $\phi_{21}(\Omega)$ is in radians and Ω is in rad/s.

3.1.2 Poles and Zeros on the Complex Plane

The (σ, Ω) plane, where a rational transfer function is defined, is called the complex plane or p-plane. The horizontal axis of this plane is called the real or σ-axis and the vertical axis is called the imaginary or $j\Omega$-axis. The values of p at which the function becomes zero are the zeros of the function and the values of p at which the function becomes infinite are the singularities (usually the poles) of the function. Therefore, the zeros of $S_{21}(p)$ are the roots of the numerator $N(p)$ and the poles of $S_{21}(p)$ are the roots of denominator $D(p)$.

These poles will be the natural frequencies of the filter whose response is described by $S_{21}(p)$. For the filter to be stable, these natural frequencies must lie in the left half of the p-plane, or on the imaginary axis. If this were not so, the oscillations would be of exponentially increasing magnitude with respect to time, a condition that is impossible in a passive network. Hence, $D(p)$ is a Hurwitz polynomial [3], that is its roots (or zeros) are in the inside of the left half-plane, or on the $j\Omega$-axis. The roots (or zeros) of $N(p)$ may occur anywhere on the entire complex plane. The zeros of $N(p)$ are called finite-frequency transmission zeros of the filter.

The poles and zeros of a rational transfer function may be depicted on the p-plane. We will see in the following that different types of transfer functions will be distinguished from their pole-zero patterns of the diagram.

3.1.3 Butterworth (Maximally Flat) Response

The amplitude-squared transfer function for Butterworth filters, which have an insertion loss $L_{Ar} = 3.01$ dB at the cutoff frequency $\Omega_c = 1$ is given by

$$|S_{21}(j\Omega)|^2 = \frac{1}{1+\Omega^{2n}} \qquad (3.7)$$

where n is the degree or the order of filter, which corresponds to the number of reactive elements required in the lowpass prototype filter. This type of response is also referred to as maximally flat, because its amplitude-squared transfer function defined in Eq. (3.7) has the maximum number of $(2n - 1)$ zero derivatives at $\Omega = 0$. Therefore, the maximally flat approximation to the ideal lowpass filter in the passband is best at $\Omega = 0$, but deteriorates as Ω approaches the cutoff frequency Ω_c. Figure 3.1 shows a typical maximally flat response.

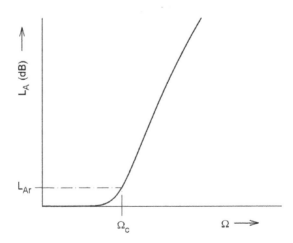

FIGURE 3.1 Butterworth (maximally flat) lowpass response.

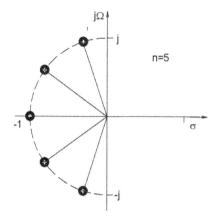

FIGURE 3.2 Pole distribution for Butterworth (maximally flat) response.

A rational transfer function constructed from Eq. (3.7) is [1,2]

$$S_{21}(p) = \frac{1}{\prod_{i=1}^{n}(p - p_i)} \qquad (3.8)$$

with

$$p_i = j \exp\left[\frac{(2i-1)\pi}{2n}\right]$$

There is no finite-frequency transmission zero [all the zeros of $S_{21}(p)$ are at infinity], and the poles p_i lie on the unit circle in the left half-plane at equal angular spacings, since $|p_i| = 1$ and $\mathrm{Arg}\, p_i = (2i-1)\pi/2n$. This is illustrated in Figure 3.2.

3.1.4 Chebyshev Response

The Chebyshev response that exhibits the equal-ripple passband and maximally flat stopband is depicted in Figure 3.3. The amplitude-squared transfer function that describes this type of response is

$$|S_{21}(j\Omega)|^2 = \frac{1}{1 + \varepsilon^2 T_n^2(\Omega)} \qquad (3.9)$$

where the ripple constant ε is related to a given passband ripple L_{Ar} in dB by

$$\varepsilon = \sqrt{10^{\frac{L_{Ar}}{10}} - 1} \qquad (3.10)$$

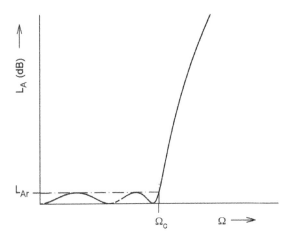

FIGURE 3.3 Chebyshev lowpass response.

$T_n(\Omega)$ is a Chebyshev function of the first kind of order n, which is defined as

$$T_n(\Omega) = \begin{cases} \cos\left(n \cos^{-1} \Omega\right) & |\Omega| \leq 1 \\ \cosh\left(n \cosh^{-1} \Omega\right) & |\Omega| \geq 1 \end{cases} \quad (3.11)$$

Hence, the filters realized from Eq. (3.9) are commonly known as Chebyshev filters.

Rhodes [2] has derived a general formula of the rational transfer function from Eq. (3.9) for the Chebyshev filter, that is

$$S_{21}(p) = \frac{\prod_{i=1}^{n} \left[\eta^2 + \sin^2(i\pi/n)\right]^{1/2}}{\prod_{i=1}^{n} (p + p_i)} \quad (3.12)$$

with

$$p_i = j \cos\left[\sin^{-1} j\eta + \frac{(2i-1)\pi}{2n}\right]$$

$$\eta = \sinh\left(\frac{1}{n} \sinh^{-1} \frac{1}{\varepsilon}\right)$$

Similar to the maximally flat case, all the transmission zeros of $S_{21}(p)$ are located at infinity. Therefore, the Butterworth and Chebyshev filters dealt with so far are sometimes referred to as all-pole filters. However, the pole locations for the Chebyshev case are different and lie on an ellipse in the left half-plane. The major axis of

TRANSFER FUNCTIONS   33

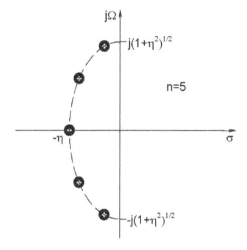

FIGURE 3.4   Pole distribution for Chebyshev response.

the ellipse is on the $j\Omega$-axis and its size is $\sqrt{1+\eta^2}$; the minor axis is on the σ-axis and is of size η. The pole distribution is shown, for $n=5$, in Figure 3.4.

3.1.5 Elliptic Function Response

The response, which is equal-ripple in both the passband and stopband, is the elliptic function response, as illustrated in Figure 3.5. The transfer function for this type of response is

$$|S_{21}(j\Omega)|^2 = \frac{1}{1+\varepsilon^2 F_n^2(\Omega)} \tag{3.13a}$$

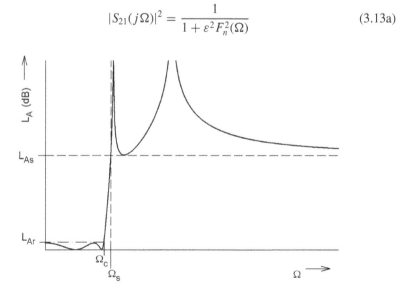

FIGURE 3.5   Elliptic-function lowpass response.

with

$$F_n(\Omega) = \begin{cases} M \dfrac{\prod_{i=1}^{n/2}(\Omega_i^2 - \Omega^2)}{\prod_{i=1}^{n/2}(\Omega_s^2/\Omega_i^2 - \Omega^2)} & \text{for } n \text{ even} \\[2ex] N \dfrac{\Omega \prod_{i=1}^{(n-1)/2}(\Omega_i^2 - \Omega^2)}{\prod_{i=1}^{(n-1)/2}(\Omega_s^2/\Omega_i^2 - \Omega^2)} & \text{for } n \;(\geq 3) \text{ odd} \end{cases} \qquad (3.13b)$$

where Ω_i $(0 < \Omega_i < 1)$ and $\Omega_s > 1$ represent some critical frequencies; M and N are constants to be defined [4,5]. $F_n(\Omega)$ will oscillate between ± 1 for $|\Omega| \le 1$ and $|F_n(\Omega = \pm 1)| = 1$.

Figure 3.6 plots the two typical oscillating curves for $n = 4$ and $n = 5$. Inspection of $F_n(\Omega)$ in Eq. (3.13b) shows that its zeros and poles are inversely proportional, the constant of proportionality being Ω_s. An important property of this is that if Ω_i can be found such that $F_n(\Omega)$ has equal ripples in the passband, it will automatically have equal ripples in the stopband. The parameter Ω_s is the frequency at which the equal-ripple stopband starts. For n even $F_n(\Omega_s) = M$ is required, which can be used to define the minimum in the stopband for a specified passband ripple constant ε.

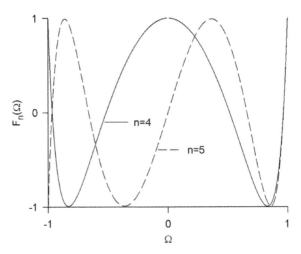

FIGURE 3.6 Plot of elliptic rational function.

The transfer function given in Eq. (3.13) can lead to expressions containing elliptic functions; for this reason, filters that display such a response are called elliptic-function filters, or simply elliptic filters. They may also occasionally be referred to as Cauer filters, after the person who first introduced the function of this type [6].

3.1.6 Gaussian (Maximally Flat Group-Delay) Response

The Gaussian response is approximated by a rational transfer function [4]

$$S_{21}(p) = \frac{a_0}{\sum_{k=0}^{n} a_k p^k} \quad (3.14)$$

where $p = \sigma + j\Omega$ is the normalized complex frequency variable, and the coefficients

$$a_k = \frac{(2n-k)!}{2^{n-k} k!(n-k)!} \quad (3.15)$$

This transfer function possesses a group delay that has maximum possible number of zero derivatives with respect to Ω at $\Omega = 0$, which is why it is said to have maximally flat group delay around $\Omega = 0$ and is, in a sense, complementary to the Butterworth response, which has a maximally flat amplitude. The above maximally flat group delay approximation was originally derived by W. E. Thomson [7]. The resulting polynomials in Eq. (3.14) with coefficients given in Eq. (3.15) are related to the Bessel functions. For these reasons, the filters of this type are also called Bessel and/or Thomson filters.

Figure 3.7 shows two typical Gaussian responses for $n = 3$ and $n = 5$, which are obtained from Eq. (3.14). In general, the Gaussian filters have a poor selectivity, as can be seen from the amplitude responses in Figure 3.7a. With increasing filter order n, the selectivity improves little and the insertion loss in decibels approaches the Gaussian form [1]

$$L_A(\Omega) = 10 \log e^{\frac{\Omega^2}{(2n-1)}} \text{ dB} \quad (3.16)$$

Use of this equation gives the 3-dB bandwidth as

$$\Omega_{3dB} \approx \sqrt{(2n-1)\ln 2} \quad (3.17)$$

which approximation is good for $n \geq 3$. Hence, unlike the Butterworth response, the 3-dB bandwidth of a Gaussian filter is a function of the filter order; the higher the filter order, the wider the 3-dB bandwidth.

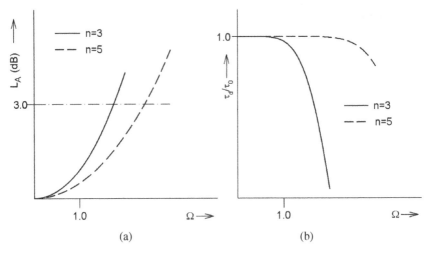

FIGURE 3.7 Gaussian (maximally flat group-delay) response: (**a**) amplitude; (**b**) group delay.

However, the Gaussian filters have a quite flat group delay in the passband, as indicated in Figure 3.7b, where the group delay is normalized by τ_0, which is the delay at the zero frequency and is inversely proportional to the bandwidth of the passband. If we let $\Omega = \Omega_c = 1$ rad/s be a reference bandwidth, then $\tau_0 = 1$. With increasing filter order n, the group delay is flat over a wider frequency range. Therefore, a high-order Gaussian filter is usually used for achieving a flat group delay over a large passband.

3.1.7 All-Pass Response

External group-delay equalizers, which are realized using all-pass networks, are widely used in communications systems. The transfer function of an all-pass network is defined by

$$S_{21}(p) = \frac{D(-p)}{D(p)} \tag{3.18}$$

where $p = \sigma + j\Omega$ is the complex frequency variable and $D(p)$ is a strict Hurwitz polynomial. At real frequencies ($p = j\Omega$), $|S_{21}(j\Omega)|^2 = S_{21}(p)S_{21}(-p) = 1$ so that the amplitude response is unity at all frequencies, which is why it is called the all-pass network. However, there will be phase shift and group delay produced by the all-pass network. We may express Eq. (3.18) at real frequencies as $S_{21}(j\Omega) = e^{j\phi_{21}(\Omega)}$, the phase shift of an all-pass network is then

$$\phi_{21}(\Omega) = -j \ln S_{21}(j\Omega) \tag{3.19}$$

and the group delay is given by

$$\tau_d(\Omega) = -\frac{d\phi_{21}(\Omega)}{d\Omega} = j\frac{d(\ln S_{21}(j\Omega))}{d\Omega}$$
$$= j\left(\frac{1}{D(-p)}\frac{dD(-p)}{dp} - \frac{1}{D(p)}\frac{dD(p)}{dp}\right)\frac{dp}{d\Omega}\bigg|_{p=j\Omega} \quad (3.20)$$

An expression for a strict Hurwitz polynomial $D(p)$ is

$$D(p) = \left(\prod_{k=1}^{n}[p-(-\sigma_k)]\right)\left(\prod_{i=1}^{m}[p-(-\sigma_i+j\Omega_i)]\cdot[p-(-\sigma_i-j\Omega_i)]\right)$$
$$(3.21)$$

where $-\sigma_k$ for $\sigma_k > 0$ are the real left-hand roots, and $-\sigma_i \pm j\Omega_i$ for $\sigma_i > 0$ and $\Omega_i > 0$ are the complex left-hand roots of $D(p)$, respectively. If all poles and zeros of an all-pass network are located along the σ-axis, such a network is said to consist of C-type sections and, therefore, referred to as a C-type all-pass network. On the other hand, if the poles and zeros of the transfer function in Eq. (3.18) are all complex with quadrantal symmetry about the origin of the complex plane, the resultant network is referred to as D-type all-pass network consisting of D-type sections only. In practice, a desired all-pass network may be constructed by a cascade connection of individual C- and D-type sections. Therefore, it is interesting to discuss their characteristics separately.

For a single-section C-type all-pass network, the transfer function is

$$S_{21}(p) = \frac{-p+\sigma_k}{p+\sigma_k} \quad (3.22a)$$

and the group delay found by Eq. (3.20) is

$$\tau_d(\Omega) = \frac{2\sigma_k}{\sigma_k^2 + \Omega^2} \quad (3.22b)$$

The pole-zero diagram and group-delay characteristics of this network are illustrated in Figure 3.8.

Similarly, for a single-section, D-type, all-pass network, the transfer function is

$$S_{21}(p) = \frac{[-p-(-\sigma_i+j\Omega_i)]\cdot[-p-(-\sigma_i-j\Omega_i)]}{[p-(-\sigma_i+j\Omega_i)]\cdot[p-(-\sigma_i-j\Omega_i)]} \quad (3.23a)$$

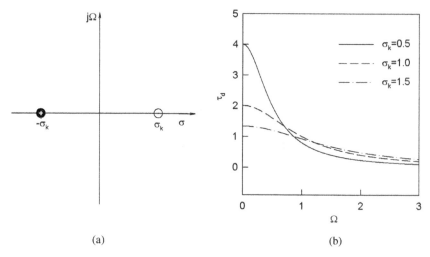

FIGURE 3.8 Characteristics of single-section C-type all-pass network: (**a**) pole-zero diagram; (**b**) group-delay response.

and the group delay is

$$\tau_d(\Omega) = \frac{4\sigma_i \left[(\sigma_i^2 + \Omega_i^2) + \Omega^2\right]}{\left[(\sigma_i^2 + \Omega_i^2) - \Omega^2\right]^2 + (2\sigma_i\Omega)^2} \quad (3.23b)$$

Figure 3.9 depicts the pole-zero diagram and group-delay characteristics of this network.

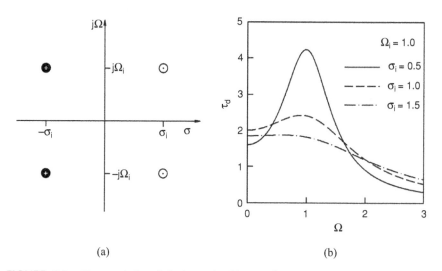

FIGURE 3.9 Characteristics of single-section D-type all-pass network: (**a**) pole-zero diagram; (**b**) group-delay response.

3.2 LOWPASS PROTOTYPE FILTERS AND ELEMENTS

Filter syntheses for realizing the transfer functions, such as those discussed in the previous section, usually result in the so-called lowpass prototype filters [8–10]. In general, a lowpass prototype filter is defined as a filter whose element values are normalized to make the source resistance or conductance equal to one, denoted by $g_0 = 1$, and the cutoff angular frequency to be unity, denoted by $\Omega_c = 1$ (rad/s). For example, Figure 3.10 demonstrates two possible forms of an n-pole lowpass prototype for realizing an all-pole filter response, including Butterworth, Chebyshev, and Gaussian responses. Either form may be used because both are dual from each other and give the same response. It should be noted that in Figure 3.10, g_i for $i = 1$ to n represents either the inductance of a series inductor or the capacitance of a shunt capacitor; therefore, n is also the number of reactive elements. If g_1 is the shunt capacitance or the series inductance, then g_0 is defined as the source resistance or the source conductance. Similarly, if g_n is the shunt capacitance or the series inductance, g_{n+1} becomes the load resistance or the load conductance. Unless otherwise specified, these g values are supposed to be the inductance in henries, capacitance in farads, resistance in ohms, and conductance in mhos.

This type of lowpass can serve as a prototype for designing many practical filters with frequency and element transformations. This will be addressed in the next section. The main objective of this section is to present equations and tables for

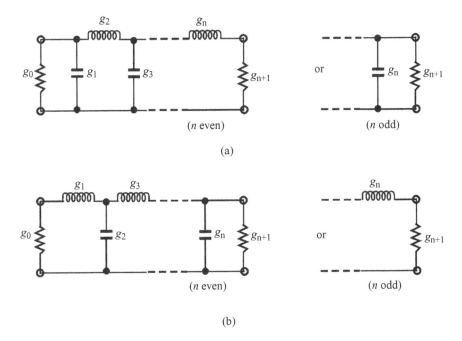

FIGURE 3.10 Lowpass prototype filters for all-pole filters with (**a**) a ladder network structure and (**b**) its dual.

40 BASIC CONCEPTS AND THEORIES OF FILTERS

obtaining element values of some commonly used lowpass prototype filters without detailing filter synthesis procedures. In addition, the determination of the degree of the prototype filter will be discussed.

3.2.1 Butterworth Lowpass Prototype Filters

For Butterworth or maximally flat lowpass prototype filters having a transfer function given in Eq. (3.7) with an insertion loss $L_{Ar} = 3.01$ dB at the cutoff $\Omega_c = 1$, the element values as referred to in Figure 3.10 may be computed by

$$g_0 = 1.0$$
$$g_i = 2\sin\left(\frac{(2i-1)\pi}{2n}\right) \text{ for } i = 1 \text{ to } n \quad (3.24)$$
$$g_{n+1} = 1.0$$

For convenience, Table 3.1 gives element values for such filters having $n = 1$ to 9. As can be seen, the two-port Butterworth filters considered here are always symmetrical in network structure, namely, $g_0 = g_{n+1}$, $g_1 = g_n$, and so on.

To determine the degree of a Butterworth lowpass prototype, a specification that is usually the minimum stopband attenuation L_{As} dB at $\Omega = \Omega_s$ for $\Omega_s > 1$ is given. Hence

$$n \geq \frac{\log\left(10^{0.1L_{As}} - 1\right)}{2\log\Omega_s} \quad (3.25)$$

For example, if $L_{As} = 40$ dB and $\Omega_s = 2$, $n \geq 6.644$, that is, a seven-pole ($n = 7$) Butterworth prototype should be chosen.

3.2.2 Chebyshev Lowpass Prototype Filters

For Chebyshev lowpass prototype filters having a transfer function given in Eq. (3.9) with a passband ripple L_{Ar} dB and the cutoff frequency $\Omega_c = 1$, the element values

TABLE 3.1 Element Values for Butterworth Lowpass Prototype Filters ($g_0 = 1.0$, $\Omega_c = 1$, $L_{Ar} = 3.01$ dB at Ω_c)

n	g_1	g_2	g_3	g_4	g_5	g_6	g_7	g_8	g_9	g_{10}
1	2.0000	1.0								
2	1.4142	1.4142	1.0							
3	1.0000	2.0000	1.0000	1.0						
4	0.7654	1.8478	1.8478	0.7654	1.0					
5	0.6180	1.6180	2.0000	1.6180	0.6180	1.0				
6	0.5176	1.4142	1.9318	1.9318	1.4142	0.5176	1.0			
7	0.4450	1.2470	1.8019	2.0000	1.8019	1.2470	0.4450	1.0		
8	0.3902	1.1111	1.6629	1.9616	1.9616	1.6629	1.1111	0.3902	1.0	
9	0.3473	1.0000	1.5321	1.8794	2.0000	1.8794	1.5321	1.0000	0.3473	1.0

for the two-port networks shown in Figure 3.10 may be computed using the following formulas:

$$g_0 = 1.0$$

$$g_1 = \frac{2}{\gamma} \sin\left(\frac{\pi}{2n}\right)$$

$$g_i = \frac{1}{g_{i-1}} \frac{4 \sin\left[\frac{(2i-1)\pi}{2n}\right] \cdot \sin\left[\frac{(2i-3)\pi}{2n}\right]}{\gamma^2 + \sin^2\left[\frac{(i-1)\pi}{n}\right]} \quad \text{for } i = 2, 3, \ldots n \quad (3.26)$$

$$g_{n+1} = \begin{cases} 1.0 & \text{for } n \text{ odd} \\ \coth^2\left(\dfrac{\beta}{4}\right) & \text{for } n \text{ even} \end{cases}$$

where

$$\beta = \ln\left[\coth\left(\frac{L_{Ar}}{17.37}\right)\right]$$

$$\gamma = \sinh\left(\frac{\beta}{2n}\right)$$

Some typical element values for such filters are tabulated in Table 3.2 for various passband ripples L_{Ar}, and for the filter degree of $n = 1$ to 9. For the required passband ripple L_{Ar} dB, the minimum stopband attenuation L_{As} dB at $\Omega = \Omega_s$ the degree of a Chebyshev lowpass prototype, which will meet this specification, can be found by

$$n \geq \frac{\cosh^{-1}\sqrt{\dfrac{10^{0.1 L_{As}} - 1}{10^{0.1 L_{Ar}} - 1}}}{\cosh^{-1}\Omega_s} \quad (3.27)$$

Using the same example as given above for the Butterworth prototype, that is, $L_{As} \geq 40$ dB at $\Omega_s = 2$, but a passband ripple $L_{Ar} = 0.1$ dB for the Chebyshev response, we have $n \geq 5.45$, that is, $n = 6$ for the Chebyshev prototype to meet this specification. This also demonstrates the superiority of the Chebyshev over the Butterworth design for this type of specification.

Sometimes the minimum return loss L_R or the maximum voltage standing-wave ratio VSWR in the passband is specified instead of the passband ripple L_{Ar}. If the return loss is defined by Eq. (3.4) and the minimum passband return loss is L_R dB ($L_R < 0$), the corresponding passband ripple is

$$L_{Ar} = -10 \log\left(1 - 10^{0.1 L_R}\right) \text{ dB} \quad (3.28)$$

TABLE 3.2 Element Values for Chebyshev Lowpass Prototype Filters ($g_0 = 1.0$, $\Omega_c = 1$)

For passband ripple $L_{Ar} = 0.01$ dB

n	g_1	g_2	g_3	g_4	g_5	g_6	g_7	g_8	g_9	g_{10}
1	0.0960	1.0								
2	0.4489	0.4078	1.1008							
3	0.6292	0.9703	0.6292	1.0						
4	0.7129	1.2004	1.3213	0.6476	1.1008					
5	0.7563	1.3049	1.5773	1.3049	0.7563	1.0				
6	0.7814	1.3600	1.6897	1.5350	1.4970	0.7098	1.1008			
7	0.7970	1.3924	1.7481	1.6331	1.7481	1.3924	0.7970	1.0		
8	0.8073	1.4131	1.7825	1.6833	1.8529	1.6193	1.5555	0.7334	1.1008	
9	0.8145	1.4271	1.8044	1.7125	1.9058	1.7125	1.8044	1.4271	0.8145	1.0

For passband ripple $L_{Ar} = 0.04321$ dB

n	g_1	g_2	g_3	g_4	g_5	g_6	g_7	g_8	g_9	g_{10}
1	0.2000	1.0								
2	0.6648	0.5445	1.2210							
3	0.8516	1.1032	0.8516	1.0						
4	0.9314	1.2920	1.5775	0.7628	1.2210					
5	0.9714	1.3721	1.8014	1.3721	0.9714	1.0				
6	0.9940	1.4131	1.8933	1.5506	1.7253	0.8141	1.2210			
7	1.0080	1.4368	1.9398	1.6220	1.9398	1.4368	1.0080	1.0		
8	1.0171	1.4518	1.9667	1.6574	2.0237	1.6107	1.7726	0.8330	1.2210	
9	1.0235	1.4619	1.9837	1.6778	2.0649	1.6778	1.9837	1.4619	1.0235	1.0

For passband ripple $L_{Ar} = 0.1$ dB

n	g_1	g_2	g_3	g_4	g_5	g_6	g_7	g_8	g_9	g_{10}
1	0.3052	1.0								
2	0.8431	0.6220	1.3554							
3	1.0316	1.1474	1.0316	1.0						
4	1.1088	1.3062	1.7704	0.8181	1.3554					
5	1.1468	1.3712	1.9750	1.3712	1.1468	1.0				
6	1.1681	1.4040	2.0562	1.5171	1.9029	0.8618	1.3554			
7	1.1812	1.4228	2.0967	1.5734	2.0967	1.4228	1.1812	1.0		
8	1.1898	1.4346	2.1199	1.6010	2.1700	1.5641	1.9445	0.8778	1.3554	
9	1.1957	1.4426	2.1346	1.6167	2.2054	1.6167	2.1346	1.4426	1.1957	1.0

For example, if $L_R = -16.426$ dB, $L_{Ar} = 0.1$ dB. Similarly, since the definition of *VSWR* is

$$VSWR = \frac{1 + |S_{11}|}{1 - |S_{11}|} \tag{3.29}$$

we can convert *VSWR* into L_{Ar} by

$$L_{Ar} = -10 \log \left[1 - \left(\frac{VSWR - 1}{VSWR + 1} \right)^2 \right] \text{dB} \quad (3.30)$$

For instance, if *VSWR* = 1.3554, then $L_{Ar} = 0.1$ dB.

3.2.3 Elliptic-Function Lowpass Prototype Filters

Figure 3.11 illustrates two commonly used network structures for elliptic-function lowpass prototype filters. In Figure 3.11a, the series branches of parallel-resonant circuits are introduced for realizing the finite-frequency transmission zeros, since they block transmission by having infinite-series impedance (open-circuit) at resonance. For this form of the elliptic- function lowpass prototype (Figure 3.11a), g_i for odd i ($i = 1, 3, \ldots$) represents the capacitance of a shunt capacitor, g_i for even i ($i = 2, 4, \ldots$) represents the inductance of an inductor, and the primed g'_i for even i ($i = 2, 4, \ldots$) are the capacitance of a capacitor in a series branch of parallel-resonant circuit. For the dual-realization form in Figure 3.11b, the shunt branches of series-resonant circuits are used for implementing the finite-frequency transmission zeros, since they short out transmission at resonance. In this case, referring to Figure 3.11b, g_i for odd i ($i = 1, 3, \ldots$) are the inductance of a series inductor, g_i for even i ($i = 2, 4, \ldots$) are the capacitance of a capacitor, and primed g'_i for even i ($i = 2, 4, \ldots$) indicate the inductance of an inductor in a shunt branch of

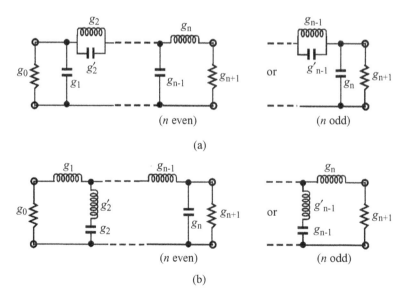

FIGURE 3.11 Lowpass prototype filters for elliptic-function filters with (**a**) series parallel-resonant branches; (**b**) dual with shunt series-resonant branches.

series-resonant circuit. Again, either form may be used, because both give the same response.

Unlike the Butterworth and Chebyshev lowpass prototype filters, there is no simple formula available for determining element values of the elliptic-function lowpass prototype filters. Table 3.3 provides some useful design data for equally terminated ($g_0 = g_{n+1} = 1$) two-port elliptic-function lowpass prototype filters shown in Figure 3.11. These element values are given for a passband ripple $L_{Ar} = 0.1$ dB, a cutoff $\Omega_c = 1$, and various Ω_s, which is the equal-ripple stopband starting frequency, referred to in Figure 3.5. Listed beside this frequency parameter is also the minimum stopband insertion loss L_{As} in dB. A smaller Ω_s implies a higher selectivity of the filter at the cost of reducing stopband rejection, as can be seen from Table 3.3. More extensive tables of elliptic- function filters are available in literature, such as found in Saal and Ulbrich [9] and Saal [11].

The degree for an elliptic-function lowpass prototype to meet a given specification may be found from the transfer function or the design tables such as Table 3.3. For instance, considering the same example as used above for the Butterworth and Chebyshev prototype, that is, $L_{As} \geq 40$ dB at $\Omega_s = 2$ and the passband ripple $L_{Ar} = 0.1$ dB, we can immediately determine $n = 5$ by inspecting the design data, that is, Ω_s and L_{As} listed in Table 3.3. This also shows that the elliptic-function design is superior to both the Butterworth and Chebyshev designs for this type of specification.

3.2.4 Gaussian Lowpass Prototype Filters

The filter networks shown in Figure 3.10 can also serve as the Gaussian lowpass prototype filters, since the Gaussian filters are the all-pole filters, as are the Butterworth or Chebyshev filters. The element values of the Gaussian prototype filters are normally obtained by network synthesis [3,4]. For convenience, some element values, which are most commonly used for design of this type filter, are listed in Table 3.4, together with two useful design parameters. The first one is the value of Ω, denoted by $\Omega_{1\%}$, for which the group delay has fallen off by 1% from its value at $\Omega = 0$. Along with this parameter is the insertion loss at $\Omega_{1\%}$, denoted by $L_{\Omega_{1\%}}$ in dB. Not listed in the table is that for the $n = 1$ Gaussian lowpass prototype, which is actually identical to the first-order Butterworth lowpass prototype given in Table 3.1.

It can be observed from the tabulated element values that even with the equal terminations ($g_0 = g_{n+1} = 1$), the Gaussian filters ($n \geq 2$) are structurally asymmetrical. It is noteworthy that the higher-order ($n \geq 5$) Gaussian filters extend the flat group-delay property into the frequency range where the insertion loss has been exceeded by 3 dB. If we define a 3-dB bandwidth as the passband and require that the group delay is flat within 1% over the passband, the five-pole ($n = 5$) Gaussian prototype would be the best choice for the design with the minimum number of elements. This is because the four-pole Gaussian prototype filter only covers 91% of the 3-dB bandwidth within 1% group-delay flatness.

TABLE 3.3 Element Values for Elliptic-Function Lowpass Prototype Filters ($g_0 = g_{n+1} = 1.0$, $\Omega_c = 1$, $L_{Ar} = 0.1$ dB)

n	Ω_s	L_{As} dB	g_1	g_2	g_2'	g_3	g_4	g_4'	g_5	g_6	g_6'	g_7
3	1.4493	13.5698	0.7427	0.7096	0.5412	0.7427						
	1.6949	18.8571	0.8333	0.8439	0.3252	0.8333						
	2.0000	24.0012	0.8949	0.9375	0.2070	0.8949						
	2.5000	30.5161	0.9471	1.0173	0.1205	0.9471						
4	1.2000	12.0856	0.3714	0.5664	1.0929	1.1194	0.9244					
	1.2425	14.1259	0.4282	0.6437	0.8902	1.1445	0.9289					
	1.2977	16.5343	0.4877	0.7284	0.7155	1.1728	0.9322					
	1.3962	20.3012	0.5675	0.8467	0.5261	1.2138	0.9345					
	1.5000	23.7378	0.6282	0.9401	0.4073	1.2471	0.9352					
	1.7090	29.5343	0.7094	1.0688	0.2730	1.2943	0.9348					
	2.0000	36.0438	0.7755	1.1765	0.1796	1.3347	0.9352					
5	1.0500	13.8785	0.7081	0.7663	0.7357	1.1276	0.2014	4.3812	0.0499			
	1.1000	20.0291	0.8130	0.9242	0.4934	1.2245	0.3719	2.1350	0.2913			
	1.1494	24.5451	0.8726	1.0084	0.3845	1.3097	0.4991	1.4450	0.4302			
	1.2000	28.3031	0.9144	1.0652	0.3163	1.3820	0.6013	1.0933	0.5297			
	1.2500	31.4911	0.9448	1.1060	0.2694	1.4415	0.6829	0.8827	0.6040			
	1.2987	34.2484	0.9681	1.1366	0.2352	1.4904	0.7489	0.7426	0.6615			
	1.4085	39.5947	1.0058	1.1862	0.1816	1.5771	0.8638	0.5436	0.7578			
	1.6129	47.5698	1.0481	1.2416	0.1244	1.6843	1.0031	0.3540	0.8692			
	1.8182	54.0215	1.0730	1.2741	0.0919	1.7522	1.0903	0.2550	0.9367			
	2.000	58.9117	1.0876	1.2932	0.0732	1.7939	1.1433	0.2004	0.9772			

(*Continued*)

TABLE 3.3 Element Values for Elliptic-Function Lowpass Prototype Filters ($g_0 = g_{n+1} = 1.0$, $\Omega_c = 1$, $L_{Ar} = 0.1$ dB) (Continued)

n	Ω_s	L_{As} dB	g_1	g_2	g_2'	g_3	g_4	g_4'	g_5	g_6	g_6'	g_7
6	1.0500	18.6757	0.4418	0.7165	0.9091	0.8314	0.3627	2.4468	0.8046	0.9986		
	1.1000	26.2370	0.5763	0.8880	0.6128	0.9730	0.5906	1.3567	0.9431	1.0138		
	1.1580	32.4132	0.6549	1.0036	0.4597	1.0923	0.7731	0.9284	1.0406	1.0214		
	1.2503	39.9773	0.7422	1.1189	0.3313	1.2276	0.9746	0.6260	1.1413	1.0273		
	1.3024	43.4113	0.7751	1.1631	0.2870	1.2832	1.0565	0.5315	1.1809	1.0293		
	1.3955	48.9251	0.8289	1.2243	0.2294	1.3634	1.1739	0.4148	1.2366	1.0316		
	1.5962	58.4199	0.8821	1.3085	0.1565	1.4792	1.3421	0.2757	1.3148	1.0342		
	1.7032	62.7525	0.9115	1.3383	0.1321	1.5216	1.4036	0.2310	1.3429	1.0350		
	1.7927	66.0190	0.9258	1.3583	0.1162	1.5505	1.4453	0.2022	1.3619	1.0355		
	1.8915	69.3063	0.9316	1.3765	0.1019	1.5771	1.4837	0.1767	1.3794	1.0358		
7	1.0500	30.5062	0.9194	1.0766	0.3422	1.0962	0.4052	2.2085	0.8434	0.5034	2.2085	0.4110
	1.1000	39.3517	0.9882	1.1673	0.2437	1.2774	0.5972	1.3568	1.0403	0.6788	1.3568	0.5828
	1.1494	45.6916	1.0252	1.2157	0.1940	1.5811	0.9939	0.5816	1.2382	0.5243	0.5816	0.4369
	1.2500	55.4327	1.0683	1.2724	0.1382	1.7059	1.1340	0.4093	1.4104	0.7127	0.4093	0.6164
	1.2987	59.2932	1.0818	1.2902	0.1211	1.7478	1.1805	0.3578	1.4738	0.7804	0.3578	0.6759
	1.4085	66.7795	1.1034	1.3189	0.0940	1.8177	1.2583	0.2770	1.5856	0.8983	0.2770	0.7755
	1.5000	72.1183	1.1159	1.3355	0.0786	1.7569	1.1517	0.3716	1.6383	1.1250	0.3716	0.9559
	1.6129	77.9449	1.1272	1.3506	0.0647	1.8985	1.3485	0.1903	1.7235	1.0417	0.1903	0.8913
	1.6949	81.7567	1.1336	1.3590	0.0570	1.9206	1.3734	0.1675	1.7628	1.0823	0.1675	0.9231
	1.8182	86.9778	1.1411	1.3690	0.0479	1.9472	1.4033	0.1408	1.8107	1.1316	0.1408	0.9616

TABLE 3.4 Element Values for Gaussian Lowpass Prototype Filters ($g_0 = g_{n+1} = 1.0$, $\Omega_c = 1$)

n	$\Omega_{1\%}$	$L_{\Omega_{1\%}}$ dB	g_1	g_2	g_3	g_4	g_5	g_6	g_7	g_8	g_9	g_{10}
2	0.5627	0.4794	1.5774	0.4226								
3	1.2052	1.3365	1.2550	0.5528	0.1922							
4	1.9314	2.4746	1.0598	0.5116	0.3181	0.1104						
5	2.7090	3.8156	0.9303	0.4577	0.3312	0.2090	0.0718					
6	3.5245	5.3197	0.8377	0.4116	0.3158	0.2364	0.1480	0.0505				
7	4.3575	6.9168	0.7677	0.3744	0.2944	0.2378	0.1778	0.1104	0.0375			
8	5.2175	8.6391	0.7125	0.3446	0.2735	0.2297	0.1867	0.1387	0.0855	0.0289		
9	6.0685	10.349	0.6678	0.3203	0.2547	0.2184	0.1859	0.1506	0.1111	0.0682	0.0230	
10	6.9495	12.188	0.6305	0.3002	0.2384	0.2066	0.1808	0.1539	0.1240	0.0911	0.0557	0.0187

3.2.5 All-Pass Lowpass Prototype Filters

The basic network unit for realizing all-pass, lowpass prototype filters is a lattice structure, as shown in Figure 3.12a, where there is a conventional abbreviated representation on the right. This lattice is not only symmetrical with respect to the two ports, but also balanced with respect to ground. By inspection, the normalized two-port Z parameters of the network are

$$z_{11} = z_{22} = \frac{z_b + z_a}{2}$$
$$z_{12} = z_{21} = \frac{z_b - z_a}{2}$$
(3.31)

which are readily converted to the scattering parameters, as described in Chapter 2.

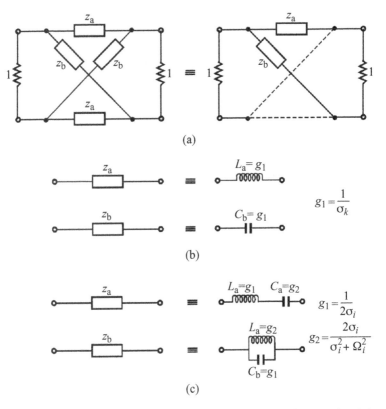

FIGURE 3.12 Lowpass prototype filters for all-pass filters: (**a**) basic network unit in a lattice structure; (**b**) the network elements for C-type, all-pass, lowpass prototype; (**c**) the network elements for D-type, all-pass, lowpass prototype.

For single-section C-type, all-pass, lowpass prototype, the network elements, as indicated in Figure 3.12b, are

$$z_a = j\Omega L_a = j\Omega g_1$$
$$z_b = \frac{1}{j\Omega C_a} = \frac{1}{j\Omega g_1} \qquad (3.32)$$
$$g_1 = \frac{1}{\sigma_k}$$

where $\sigma_k > 0$ is the design parameter that will control the group-delay characteristics, as shown in Figure 3.8. Since a C-type section is the first-order all-pass network, there is actually only one lowpass prototype element g_1, which will represent either the inductance of an inductor in a series arm or the capacitance of a capacitor in a cross arm.

The network elements for a single-section D-type, all-pass, lowpass prototype, as shown in Figure 3.12c, are given by

$$z_a = j\Omega L_a + \frac{1}{j\Omega C_a}, \qquad \frac{1}{z_b} = j\Omega C_b + \frac{1}{j\Omega L_b}$$
$$L_a = C_b = g_1 = \frac{1}{2\sigma_i}, \qquad C_a = L_b = g_2 = \frac{2\sigma_i}{\sigma_i^2 + \Omega_i^2} \qquad (3.33)$$

where $\sigma_i > 0$ and $\Omega_i > 0$ are the two design parameters that will shape the group-delay response, as illustrated in Figure 3.9. Since a D-type section is the second-order all-pass network, there are actually two lowpass prototype elements, namely g_1 and g_2, which will represent both the inductance of an inductor and the capacitance of a capacitor, depending on the locations of these reactive elements, as indicated in Figure 3.12c.

Higher-order all-pass prototype filters can be constructed by a chain connection of several C- and D-type sections. The composite-delay curves are then built up by adding their individual delay contributions to obtain the overall delay characteristics.

3.3 FREQUENCY AND ELEMENT TRANSFORMATIONS

Thus far, we have only considered the lowpass prototype filters, which have a normalized source resistance/conductance $g_0 = 1$ and a cutoff frequency of $\Omega_c = 1$. To obtain frequency characteristics and element values for practical filters, based on the lowpass prototype, one may apply frequency and element transformations, which will be addressed in this section.

The frequency transformation, which is also referred to as frequency mapping, is required to map, for example, a Chebyshev response in the lowpass prototype frequency domain Ω to that in the frequency domain ω in which a practical filter response such as lowpass, highpass, bandpass, and bandstop are expressed. The frequency transformation will have an effect on all the reactive elements accordingly, but no effect on the resistive elements.

In addition to the frequency mapping, impedance scaling is also required to accomplish the element transformation. The impedance scaling will remove the $g_0 = 1$ normalization and adjusts the filter to work for any value of the source impedance denoted by Z_0. For our formulation, it is convenient to define an impedance scaling factor γ_0 as

$$\gamma_0 = \begin{cases} Z_0/g_0 & \text{for } g_0 \text{ being the resistance} \\ g_0/Y_0 & \text{for } g_0 \text{ being the conductance} \end{cases} \quad (3.34)$$

where $Y_0 = 1/Z_0$ is the source admittance. In principle, applying the impedance scaling upon a filter network in such a way that

$$\begin{aligned} L &\to \gamma_0 L \\ C &\to C/\gamma_0 \\ R &\to \gamma_0 R \\ G &\to G/\gamma_0 \end{aligned} \quad (3.35)$$

has no effect on the response shape.

Let g be the generic term for the lowpass prototype elements in the element transformation to be discussed. Because it is independent of the frequency transformation, the following resistive-element transformation holds for any type of filter:

$$\begin{aligned} R &= \gamma_0 g & \text{for } g \text{ representing the resistance} \\ G &= \frac{g}{\gamma_0} & \text{for } g \text{ representing the conductance} \end{aligned} \quad (3.36)$$

3.3.1 Lowpass Transformation

The frequency transformation from a lowpass prototype to a practical lowpass filter having a cutoff frequency ω_c in the angular frequency axis ω is simply given by

$$\Omega = \left(\frac{\Omega_c}{\omega_c}\right)\omega \quad (3.37)$$

Applying Eq. (3.37), together with the impedance scaling described above, yields the element transformation:

$$\begin{aligned} L &= \left(\frac{\Omega_c}{\omega_c}\right)\gamma_0 g & \text{for } g \text{ representing the inductance} \\ C &= \left(\frac{\Omega_c}{\omega_c}\right)\frac{g}{\gamma_0} & \text{for } g \text{ representing the capacitance} \end{aligned} \quad (3.38)$$

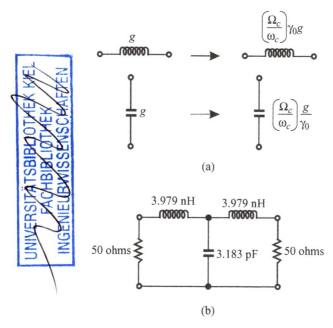

FIGURE 3.13 Lowpass prototype to lowpass transformation: (**a**) basic element transformation; (**b**) a practical lowpass filter based on the transformation.

which is shown in Figure 3.13a. To demonstrate the use of the element transformation, let us consider a design of practical lowpass filter with a cutoff frequency $f_c = 2$ GHz and a source impedance $Z_0 = 50\ \Omega$. A three-pole Butterworth lowpass prototype with the structure of Figure 3.10b is chosen for this example, which gives $g_0 = g_4 = 1.0$ mhos, $g_1 = g_3 = 1.0$ H, and $g_2 = 2.0$ F for $\Omega_c = 1.0$ rad/s, from Table 3.1. The impedance scaling factor is $\gamma_0 = 50$, according to Eq. (3.34). The angular cutoff frequency is $\omega_c = 2\pi \times 2 \times 10^9$ rad/s. Applying Eq. (3.38), we find $L_1 = L_3 = 3.979$ nH and $C_2 = 3.183$ pF. The resultant lowpass filter is illustrated in Figure 3.13b.

3.3.2 Highpass Transformation

For highpass filters with a cutoff frequency ω_c in the ω-axis, the frequency transformation is

$$\Omega = -\frac{\omega_c \Omega_c}{\omega} \quad (3.39)$$

Applying this frequency transformation to a reactive element g in the lowpass prototype leads to

$$j\Omega g \rightarrow \frac{\omega_c \Omega_c g}{j\omega}$$

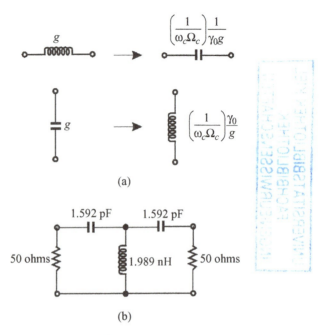

FIGURE 3.14 Lowpass prototype to highpass transformation: (a) basic element transformation; (b) a practical highpass filter based on the transformation.

It is then obvious that an inductive/capacitive element g in the lowpass prototype will be inversely transformed to a capacitive/inductive element in the highpass filter. With impedance scaling, the element transformation is given by

$$C = \left(\frac{1}{\omega_c \Omega_c}\right) \frac{1}{\gamma_0 g} \quad \text{for } g \text{ representing the inductance}$$
$$L = \left(\frac{1}{\omega_c \Omega_c}\right) \frac{\gamma_0}{g} \quad \text{for } g \text{ representing the capacitance} \qquad (3.40)$$

This type of element transformation is shown in Figure 3.14a. Figure 3.14b demonstrates a practical highpass filter with a cutoff frequency at 2 GHz and 50-Ω terminals, which is obtained from the transformation of the three-pole Butterworth lowpass prototype given above.

3.3.3 Bandpass Transformation

Assume that a lowpass prototype response is to be transformed to a bandpass response having a passband $\omega_2 - \omega_1$, where ω_1 and ω_2 indicate the passband-edge angular frequency. The required frequency transformation is

$$\Omega = \frac{\Omega_c}{FBW}\left(\frac{\omega}{\omega_0} - \frac{\omega_0}{\omega}\right) \quad (3.41a)$$

with

$$FBW = \frac{\omega_2 - \omega_1}{\omega_0} \quad (3.41b)$$

$$\omega_0 = \sqrt{\omega_1 \omega_2}$$

where ω_0 denotes the center angular frequency and FBW is defined as the fractional bandwidth. If we apply this frequency transformation to a reactive element g of the lowpass prototype, we have

$$j\Omega g \rightarrow j\omega \frac{\Omega_c g}{FBW\omega_0} + \frac{1}{j\omega}\frac{\Omega_c \omega_0 g}{FBW}$$

which implies that an inductive/capacitive element g in the lowpass prototype will transform to a series/parallel LC resonant circuit in the bandpass filter. The elements for the series LC resonator in the bandpass filter are

$$L_s = \left(\frac{\Omega_c}{FBW\omega_0}\right)\gamma_0 g$$
$$C_s = \left(\frac{FBW}{\omega_0 \Omega_c}\right)\frac{1}{\gamma_0 g} \quad \text{for } g \text{ representing the inductance} \quad (3.42a)$$

where the impedance scaling has been taken into account as well. Similarly, the elements for the parallel LC resonator in the bandpass filter are

$$C_p = \left(\frac{\Omega_c}{FBW\omega_0}\right)\frac{g}{\gamma_0}$$
$$L_p = \left(\frac{FBW}{\omega_0 \Omega_c}\right)\frac{\gamma_0}{g} \quad \text{for } g \text{ representing the capacitance} \quad (3.42b)$$

It should be noted that $\omega_0 L_s = 1/(\omega_0 C_s)$ and $\omega_0 L_p = 1/(\omega_0 C_p)$ hold in Eq. (3.42). The element transformation, in this case, is shown in Figure 3.15a. With the same three-pole Butterworth lowpass prototype as that used previously in Section 3.3.1, Figure 3.15b illustrates a bandpass having a passband from 1 to 2 GHz obtained using the element transformation.

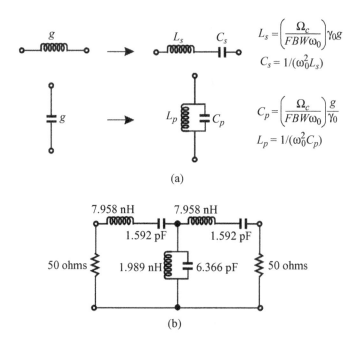

FIGURE 3.15 Lowpass prototype to bandpass transformation: (a) basic element transformation; (b) a practical bandpass filter based on the transformation.

3.3.4 Bandstop Transformation

The frequency transformation from lowpass prototype to bandstop is achieved by the frequency mapping

$$\Omega = \frac{\Omega_c FBW}{(\omega_0/\omega - \omega/\omega_0)} \tag{3.43a}$$

$$\omega_0 = \sqrt{\omega_1 \omega_2}$$
$$FBW = \frac{\omega_2 - \omega_1}{\omega_0} \tag{3.43b}$$

where $\omega_2 - \omega_1$ is the bandwidth. This form of the transformation is opposite to the bandpass transformation in that an inductive/capacitive element g in the lowpass prototype will transform to a parallel/series LC resonant circuit in the bandstop filter. The elements for the LC resonators transformed to the bandstop filter are

$$C_p = \left(\frac{1}{FBW\omega_0 \Omega_c}\right) \frac{1}{\gamma_0 g}$$
$$L_p = \left(\frac{\Omega_c FBW}{\omega_0}\right) \gamma_0 g \qquad \text{for } g \text{ representing the inductance} \tag{3.44a}$$

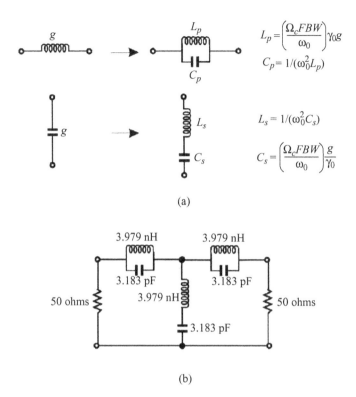

FIGURE 3.16 Lowpass prototype to bandstop transformation: (a) basic element transformation; (b) a practical bandstop filter based on the transformation.

$$L_s = \left(\frac{1}{FBW\omega_0\Omega_c}\right)\frac{\gamma_0}{g}$$
$$C_s = \left(\frac{\Omega_c FBW}{\omega_0}\right)\frac{g}{\gamma_0}$$ for g representing the capacitance (3.44b)

It is also true in Eq. (3.44) that $\omega_0 L_p = 1/(\omega_0 C_p)$ and $\omega_0 L_s = 1/(\omega_0 C_s)$. The element transformation of this type is shown in Figure 3.16a. An example of its application for designing a practical bandstop filter, with a bandwidth of 1 to 2 GHz, is demonstrated in Figure 3.16b, which is based on the three-pole Butterworth lowpass prototype, as described previously.

3.4 IMMITTANCE INVERTERS

3.4.1 Definition of Immittance, Impedance, and Admittance Inverters

Immittance inverters are either impedance or admittance inverters. An idealized impedance inverter is a two-port network that has a unique property at all frequencies, that is, if it is terminated in an impedance Z_2 on one port, the impedance Z_1 seen looking in at the other port is

$$Z_1 = \frac{K^2}{Z_2} \tag{3.45}$$

where K is real and defined as the characteristic impedance of the inverter. As can be seen, if Z_2 is inductive/conductive, Z_1 will become conductive/inductive, and, hence, the inverter has a phase shift of $\pm 90°$ or an odd multiple thereof. Impedance inverters are also known as K inverters. The $ABCD$ matrix of ideal impedance inverters may generally be expressed as

$$\begin{bmatrix} A & B \\ C & D \end{bmatrix} = \begin{bmatrix} 0 & \mp jK \\ \pm \frac{1}{jK} & 0 \end{bmatrix} \tag{3.46}$$

Likewise, an ideal admittance inverter is a two-port network, which exhibits such a property at all frequencys that if an admittance Y_2 is connected at one port, the admittance Y_1 seen into the other port is

$$Y_1 = \frac{J^2}{Y_2} \tag{3.47}$$

where J is real and called the characteristic admittance of the inverter. Similarly, the admittance inverter has a phase shift of $\pm 90°$ or an odd multiple thereof. Admittance inverters are also referred to as J inverters. In general, ideal admittance inverters have the $ABCD$ matrix

$$\begin{bmatrix} A & B \\ C & D \end{bmatrix} = \begin{bmatrix} 0 & \pm \frac{1}{jJ} \\ \mp jJ & 0 \end{bmatrix} \tag{3.48}$$

3.4.2 Filters with Immittance Inverters

It can be shown by network analysis that a series inductance with an inverter on each side looks like a shunt capacitance from its exterior terminals, as indicated in Figure 3.17a. Likewise, a shunt capacitance with an inverter on each side looks likes a series inductance from its external terminals, as demonstrated in Figure 3.17b. Also, as indicated, inverters have the ability to shift impedance or admittance levels, depending on the choice of K or J parameters. Making use of these properties enable us to convert a filter circuit to an equivalent form that would be more convenient for implementation with microwave structures.

For example, the two common lowpass prototype structures in Figure 3.10 may be converted into the forms shown in Figure 3.18, where the g_i values are the original prototype element values, as defined earlier. The new element values, such as Z_0, Z_{n+1}, L_{ai}, Y_0, Y_{n+1}, and C_{ai} may be chosen arbitrarily and the filter response will be identical to that of the original prototype, provided that the immittance inverter

IMMITTANCE INVERTERS   57

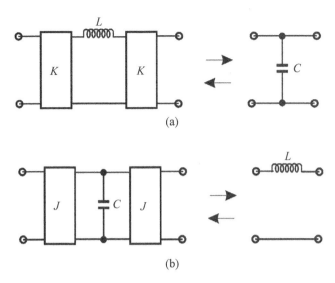

FIGURE 3.17 (a) Immittance inverters used to convert a shunt capacitance into an equivalent circuit with series inductance. (b) Immittance inverters used to convert a series inductance into an equivalent circuit with shunt capacitance.

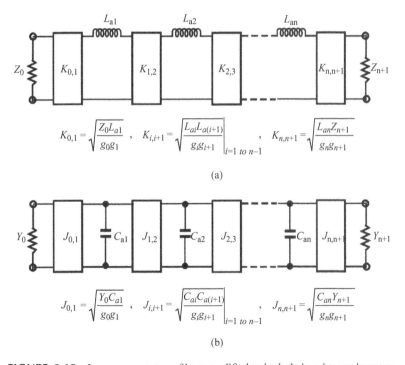

FIGURE 3.18 Lowpass prototype filters modified to include immittance inverters.

parameters $K_{i,i+1}$ and $J_{i,i+1}$ are specified as indicated by the equations in Figure 3.18. These equations can be derived by expanding the input immittances of the original prototype networks and the equivalent ones in continued fractions and by equating corresponding terms.

Since, ideally, immittance inverter parameters are frequency invariant, the lowpass filter networks in Figure 3.18 can easily be transformed to other types of filter by applying the element transformations similar to those described in the previous section. For instance, Figure 3.19 illustrates two bandpass filters using immittance inverters. In the case of Figure 3.19a, only series resonators are involved, whereas the filter in Figure 3.19b consists of only shunt-parallel resonators. The element transformations from Figures 3.18a to 3.19a are obtained as follows. Since the source impedances are assumed, the same in the both filters as indicated, no impedance scaling is required and the scaling factor is $\gamma_0 = 1$. Now, viewing L_{ai} as inductive

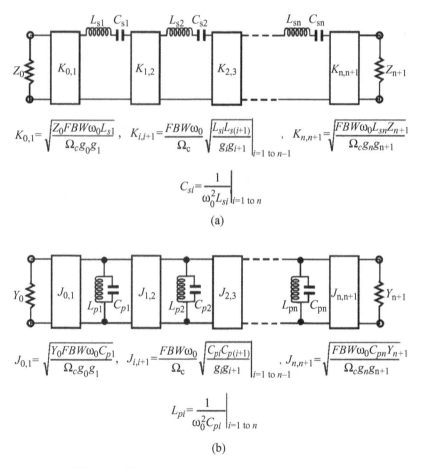

FIGURE 3.19 Bandpass filters using immittance inverters.

g in Figure 3.15a, and transforming the inductors of the lowpass filter to the series resonators of the bandpass filter, we obtain

$$L_{si} = \left(\frac{\Omega_c}{FBW\omega_0}\right) L_{ai}$$

$$C_{si} = \frac{1}{\omega_0^2 L_{si}}$$

As mentioned above, the K parameters must remain unchanged with respect to the frequency transformation. Replacing L_{ai} in the equations in Figure 3.18a with $L_{ai} = (FBW\omega_0/\Omega_c)L_{si}$ yields the equations in Figure 3.19a. Similarly, the transformations and equations in Figure 3.19b can be obtained on a dual basis.

Two important generalizations, shown in Figure 3.20, are obtained by replacing the lumped LC resonators by distributed circuits [10], which can be microwave cavities, microstrip resonators, or any other suitable resonant structures. In the ideal case,

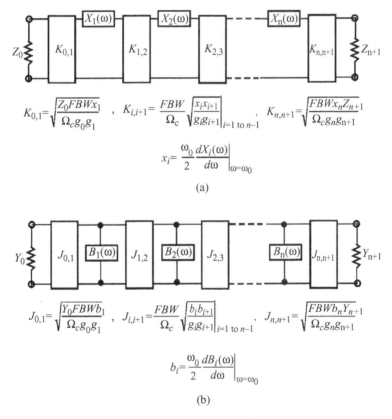

FIGURE 3.20 Generalized bandpass filters (including distributed elements) using immittance inverters.

the reactances or susceptances of the distributed circuits (not restricted to bandpass filters) should equal those of the lumped resonators at all frequencies. In practice, they approximate the reactances or susceptances of the lumped resonators only near resonance. Nevertheless, this is sufficient for narrow-band filters. For convenience, the distributed resonator reactance/susceptance and reactance/susceptance slope are made equal to their corresponding lumped-resonator values at band center. For this, two quantities, called the reactance- and susceptance-slope parameter, respectively, are introduced. The reactance-slope parameter for resonators having zero reactance at center frequency ω_0 is defined by

$$x = \frac{\omega_0}{2} \frac{dX(\omega)}{d\omega}\bigg|_{\omega=\omega_0} \tag{3.49}$$

where $X(\omega)$ is the reactance of the distributed resonator. In the dual case, the susceptance slope parameter for resonators having zero susceptance at center frequency ω_0 is defined by

$$b = \frac{\omega_0}{2} \frac{dB(\omega)}{d\omega}\bigg|_{\omega=\omega_0} \tag{3.50}$$

where $B(\omega)$ is the susceptance of the distributed resonator. It can be shown that the reactance-slope parameter of a lumped LC series resonator is $\omega_0 L$, and the susceptance-slope parameters of a lumped LC parallel resonator is $\omega_0 C$. Thus, replacing $\omega_0 L_{si}$ and $\omega_0 C_{pi}$ in the equations in Figure 3.19 with the general terms x_i and b_i, as defined by Eqs. (3.49) and (3.50), respectively, results in the equations indicated in Figure 3.20.

3.4.3 Practical Realization of Immittance Inverters

One of the simplest forms of inverters is a quarter-wavelength of transmission line. It can easily be shown that such a line has a $ABCD$ matrix of the form given in Eq. (3.46) with $K = Z_c$ Ω, where Z_c is the characteristic impedance of the line. Therefore, it will obey the basic impedance inverter definition of Eq. (3.45). Of course, a quarter-wavelength of line can be also used as an admittance inverter with $J = Y_c$, where $Y_c = 1/Z_c$ is the characteristic admittance of the line. Although its inverter properties are relatively narrow-band in nature, a quarter-wavelength line can be used satisfactorily as an immittance inverter in narrow-band filters.

Besides a quarter-wavelength line, there are numerous other circuits that operate as immittance inverters. All necessarily produce a phase shift of some odd multiple of $\pm 90°$ and many work over a much wider bandwidth than does a quarter-wavelength line. Figure 3.21 shows four typical lumped-element immittance inverters. While the inverters in Figure 3.21a and b are of interest for use as K inverters, those shown in Figure 3.21c and d are of interest for use as J inverters. This is simply because

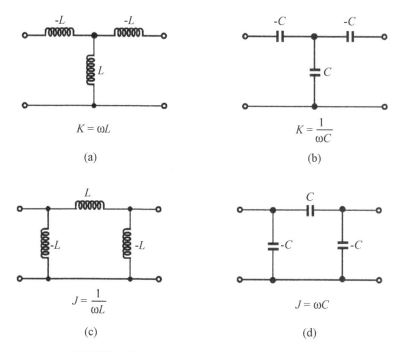

FIGURE 3.21 Lumped-element immittance inverters.

of the consideration that the negative elements of the inverters could conveniently be absorbed into adjacent elements in practical filters. Otherwise, any one of these inverters can be treated as either K or J inverter. It can be shown that the inverters in Figure 3.21a and -d have a phase shift (the phase of S_{21}) of $+90°$, while those in Figure 3.21b and c have a phase shift of $-90°$. This is why the "$\pm$" and "$\mp$" signs appear in the $ABCD$ matrix expressions of immittance inverters.

Another type of practical immittance inverter is a circuit mixed with lumped and transmission-line elements, as shown in Figure 3.22, where Z_0 and Y_0 are the characteristic impedance and admittance of the line, and ϕ denotes the total electrical length of the line. In practice, the line of positive or negative electrical length can be added to or subtracted from adjacent lines of the same characteristic impedance. Numerous other circuit networks may be constructed to operate as immittance inverters as long as their $ABCD$ matrices are of the form as that defined in Eqs. (3.46) or (3.48) in the frequency band of operation.

In reality, the J and K parameters of practical immittance inverters are frequency dependent; they can only approximate an ideal immittance, for which a constant J and K parameter is required, over a certain frequency range. The limited bandwidth of the practical immittance inverters limits how faithfully the desired transfer function is reproduced as the desired filter bandwidth is increased. Therefore, filters designed using immittance inverter theory are best applied to narrow-band filters.

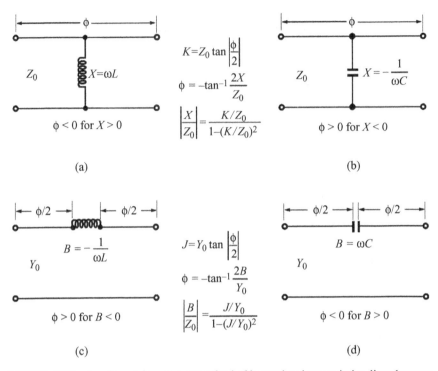

FIGURE 3.22 Immittance inverters comprised of lumped and transmission-line elements.

3.5 RICHARDS' TRANSFORMATION AND KURODA IDENTITIES

3.5.1 Richards' Transformation

Distributed transmission-line elements are of importance for designing practical microwave filters. A commonly used approach to the design of a practical distributed filter is to seek some approximate equivalence between lumped and distributed elements. Such equivalence can be established by applying Richards' transformation [12]. Richards showed that distributed networks, comprised of commensurate-length (equal electrical length) transmission lines and lumped resistors, could be treated in analysis or synthesis as lumped-element LCR networks under the transformation

$$t = \tanh \frac{lp}{v_p} \qquad (3.51)$$

where $p = \sigma + j\omega$ is the usual complex frequency variable, and l/v_p is the ratio of the length of the basic commensurate transmission-line element to the phase velocity of the wave in such a line element. t is a new complex frequency variable, also known as Richards' variable. The new complex plane, where t is defined, is called

RICHARDS' TRANSFORMATION AND KURODA IDENTITIES 63

the t-plane. Equation (3.51) is referred to as Richards' transformation. For lossless passive networks $p = j\omega$ and the Richards' variable is simply expressed by

$$t = j \tan \theta \qquad (3.52)$$

where

$$\theta = \frac{\omega}{v_p} l = \text{the electrical length} \qquad (3.53)$$

Assuming that the phase velocity v_p is independent of frequency, which is true for TEM transmission lines, the electrical length is then proportional to frequency and may be expressed as

$$\theta = \theta_0 \omega / \omega_0$$

where θ_0 is the electrical length at a reference frequency ω_0. It is convenient for discussion to let ω_0 be the radian frequency at which all line lengths are quarter-wave long with $\theta_0 = \pi/2$ and to let $\Omega = \tan \theta$, so that

$$\Omega = \tan\left(\frac{\pi}{2} \frac{\omega}{\omega_0}\right) \qquad (3.54)$$

This frequency mapping is illustrated in Figure 3.23a. As ω varies between 0 and ω_0, Ω varies between 0 and ∞, and the mapping from ω to Ω is not one to one but periodic, corresponding to the periodic nature of the distributed network.

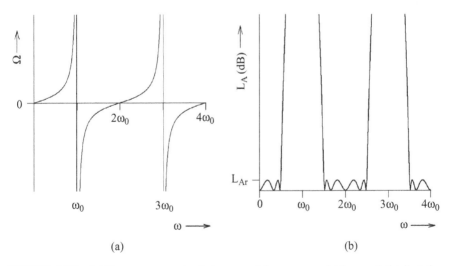

FIGURE 3.23 (a) Frequency mapping between real frequency variable ω and distributed frequency variable Ω. (b) Chebyshev lowpass response using the Richards' transformation.

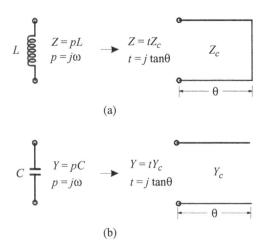

FIGURE 3.24 Lumped- and distributed-element correspondence under Richards' transformation.

The periodic frequency response of the distributed filter network is demonstrated in Figure 3.23b, which is obtained by applying the Richards' transformation of Eq. (3.54) to the Chebyshev lowpass prototype transfer function of Eq. (3.9), showing that the response repeats in frequency intervals of $2\omega_0$. It is interesting to notice that the response in Figure 3.23b may also be seen as a distributed bandstop filter response centered at ω_0. Therefore, a lowpass response in the p-plane may be mapped into either the lowpass or the bandstop one in the t-plane, depending on the design objective. Similarly, it can be shown that a highpass response in the p-plane may be transformed as either the highpass or the bandpass one in the t-plane.

Under Richards' transformation, a close correspondence exists between lumped inductors and capacitors in the p-plane and short- and open-circuited transmission lines in the t-plane. As a one-port inductive element with an impedance $Z = pL$, a lumped inductor corresponds to a short-circuited line element (stub) with an input impedance $Z = tZ_c = jZ_c \tan\theta$, where Z_c is the characteristic impedance of the line. Likewise, a lumped capacitor with an admittance $Y = pC$ corresponds to an open-circuited stub of input admittance $Y = tY_c = jY_c \tan\theta$ and characteristic admittance Y_c. These correspondences are illustrated in Figure 3.24 and, as a consequence, the short- and open-circuited line elements are sometimes referred to as the t-plane inductor and capacitor, respectively, and use the corresponding lumped-element symbols as well.

Another important distributed element, which has no lumped-element counterpart, is a two-port network consisting of a commensurate-length line. A transmission line of characteristic impedance Z_u has a ABCD matrix

$$\begin{bmatrix} A & B \\ C & D \end{bmatrix} = \begin{bmatrix} \cos\theta & jZ_u \sin\theta \\ j\sin\theta/Z_u & \cos\theta \end{bmatrix} \quad (3.55a)$$

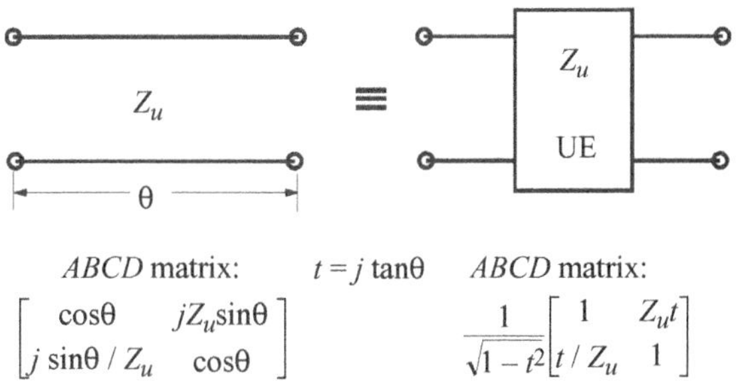

$$Z_u \quad \theta \quad \equiv \quad \boxed{\begin{array}{c} Z_u \\ UE \end{array}}$$

ABCD matrix: $\quad t = j\tan\theta \quad$ ABCD matrix:

$$\begin{bmatrix} \cos\theta & jZ_u\sin\theta \\ j\sin\theta/Z_u & \cos\theta \end{bmatrix} \qquad \frac{1}{\sqrt{1-t^2}}\begin{bmatrix} 1 & Z_u t \\ t/Z_u & 1 \end{bmatrix}$$

FIGURE 3.25 Unit element (UE).

which, in terms of Richards' variable, becomes

$$\begin{bmatrix} A & B \\ C & D \end{bmatrix} = \frac{1}{\sqrt{1-t^2}}\begin{bmatrix} 1 & Z_u t \\ t/Z_u & 1 \end{bmatrix} \qquad (3.55b)$$

This line element is referred to as a unit element, hereafter as UE, and its symbol is illustrated in Figure 3.25. It is interesting to note that the unit element has a half-order transmission zero at $t = \pm 1$. Unit elements are usually employed to separate the circuit elements in distributed filters that are otherwise located at the same physical point. We will demonstrate later (see Chapter 6) that unit element can be used in the filter design either as redundant or nonredundant elements. The former do not have any effect on the filter selectivity, but the latter can improve it.

3.5.2 Kuroda Identities

In designing transmission line filters, various network identities may be desirable to obtain filter networks that are electrically equivalent, but that differ in form or in element values. Such transformations not only provide designers with flexibility, but also are essential, in many cases, to obtain networks that are physically realizable with physical dimensions. The Kuroda identities [13], shown in Figure 3.26, form a basis to achieve such transformations, where the commensurate line elements with the same electrical length θ are assumed for each identity. The first two Kuroda identities interchange a unit element with a shunt open- or a series short-circuited stub, and a unit element with a series short- or a shunt open-circuited stub. The other two Kuroda identities, involving the ideal transformers, interchange stubs of the same kind. The Kuroda identities may be deduced by comparing the *ABCD* matrices of the corresponding networks in Figure 3.26.

3.5.3 Coupled-Line Equivalent Circuits

A pair of coupled transmission lines with the imposed terminal conditions, such as open- or short-circuited at any two of its four ports, is an important type of two-port

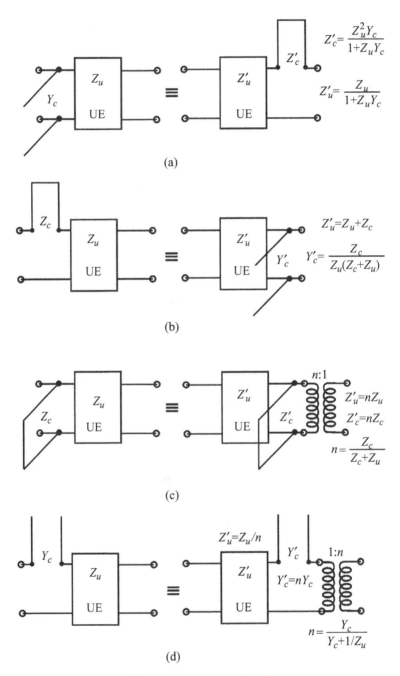

FIGURE 3.26 Kuroda identities.

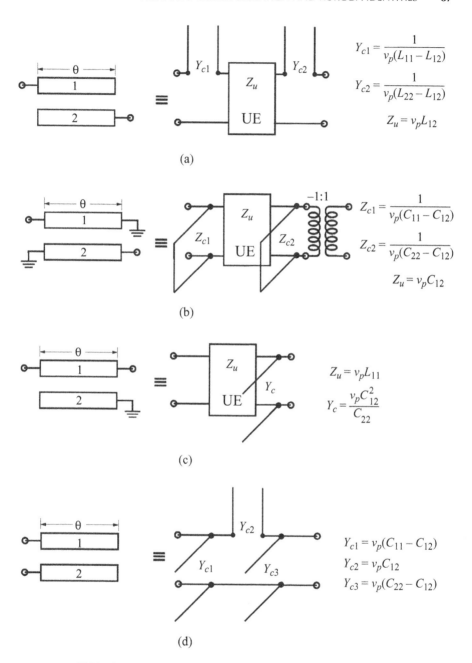

FIGURE 3.27 Equivalent circuits for coupled transmission lines.

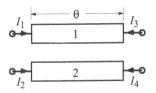

FIGURE 3.28 General coupled transmission-line network.

network in filter designs. Figure 3.27 illustrates some typical networks of this type with their equivalent circuits in the t-plane. These equivalent circuits may be derived from the general coupled-line network in Figure 3.28 by utilizing its general four-port voltage-current relationships:

$$\begin{bmatrix} I_1 \\ I_2 \\ I_3 \\ I_4 \end{bmatrix} = \frac{v_p}{t} \begin{bmatrix} [C] & -\sqrt{1-t^2}[C] \\ -\sqrt{1-t^2}[C] & [C] \end{bmatrix} \cdot \begin{bmatrix} V_1 \\ V_2 \\ V_3 \\ V_4 \end{bmatrix} \quad (3.56)$$

where I_k are the port currents, as indicated, and V_k are the port voltages with respect to a common ground (not shown).

$$[C] = \begin{bmatrix} C_{11} & -C_{12} \\ -C_{12} & C_{22} \end{bmatrix}$$

C_{11} and C_{22} are the self-capacitance per unit length of lines 1 and 2, respectively, and C_{12} is the mutual capacitance per unit length. Note that $C_{11} = C_{22}$ if the coupled lines are symmetrical.

Alternatively, the formulation with the impedance matrix may be used. This gives

$$\begin{bmatrix} V_1 \\ V_2 \\ V_3 \\ V_4 \end{bmatrix} = \frac{v_p}{t} \begin{bmatrix} [L] & \sqrt{1-t^2}[L] \\ \sqrt{1-t^2}[L] & [L] \end{bmatrix} \cdot \begin{bmatrix} I_1 \\ I_2 \\ I_3 \\ I_4 \end{bmatrix} \quad (3.57)$$

where

$$[L] = \begin{bmatrix} L_{11} & L_{12} \\ L_{12} & L_{22} \end{bmatrix}$$

In this case, L_{11} and L_{22} are the self-inductance per unit length of lines 1 and 2, respectively, and L_{12} is the mutual inductance per unit length. If the coupled lines are symmetrical, then $L_{11} = L_{22}$. It may be remarked that $[L]$ and $[C]$ together satisfy

$$[L] \cdot [C] = [C] \cdot [L] = [U]/v_p^2 \quad (3.58)$$

where $[U]$ denotes the identity matrix.

3.6 DISSIPATION AND UNLOADED QUALITY FACTOR

So far, we have only considered filters comprised of lossless elements, except for resistive terminations. However, in reality, any practical microwave filter will have lossy elements with finite unloaded quality factors in association with power dissipation in these elements. Such parasitic dissipation may frequently lead to substantial differences between the response of the filter actually realized and that of the ideal one designed with lossless elements. It is thus desirable to estimate the effects of dissipation on insertion loss characteristics.

3.6.1 Unloaded Quality Factors of Lossy Reactive Elements

In general, the losses in an inductor are conventionally represented by a resistance R connected in series with a pure inductance L, as indicated in Figure 3.29a. The unloaded quality factor Q_u of the lossy inductor is defined by

$$Q_u = \frac{\omega L}{R} \tag{3.59}$$

In a similar fashion, a lossy capacitor may have an equivalent circuit, as shown in Figure 3.29b, where G is a conductance connected in parallel with a pure capacitance C. The Q_u of the lossy capacitor is defined by

$$Q_u = \frac{\omega C}{G} \tag{3.60}$$

Note that in the above definitions, ω denotes some particular frequency at which the Q_u is measured. For a lowpass or a highpass filter, ω is usually the cutoff frequency; while for a bandpass or bandstop filter, ω is the center frequency.

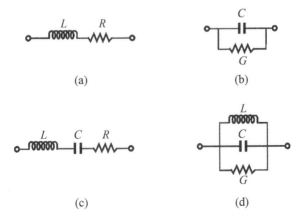

FIGURE 3.29 Circuit representations of lossy reactive elements and resonators.

70 BASIC CONCEPTS AND THEORIES OF FILTERS

For lossy resonators, they are most conveniently represented by the equivalent circuits shown in Figure 3.29c and d. The unloaded quality factors of these two equivalent resonant circuits have the same forms as those defined in Eq. (3.59) and (3.60), respectively, but in this case ω is normally the resonant frequency, namely, $\omega = 1/\sqrt{LC}$.

As can be seen for lossless reactive elements, $R \to 0$ and $G \to 0$ so that $Q_u \to \infty$. However, in practice, the Q_u will be finite because the inherent losses of the microwave components.

3.6.2 Dissipation Effects on Lowpass and Highpass Filters

Assuming that the unloaded quality factors of all reactive elements in a filter are known, determined theoretically or experimentally, we can find R and G for the lossy reactive elements from Eqs. (3.59) and (3.60). The dissipation effects on the filter insertion loss response can easily estimated by analysis of the whole-filter equivalent circuit, including the dissipative elements R and G.

As an example, let us consider the lowpass filter designed previously in Figure 3.13, which has a Butterworth response with a cutoff frequency at $f_c = 2$ GHz. To take into account the finite unloaded quality factors of the reactive elements, the filter circuit becomes that of Figure 3.30a. For simplicity, we have also assumed that the unloaded quality factors of all the reactive elements are equal, denoted by Q_u, although, in reality, they can be different. Figure 3.30b shows typical effects of the finite Q_u on the insertion loss response of the filter. The values of the Q_u are given at the cutoff frequency of this lowpass filter. Two effects are obvious:

1. A shift of insertion loss by a constant amount determined by the additional loss at zero frequency
2. A gradual rounding off the insertion loss curve at the passband edge, resulting in diminished width of the passband and, hence, in reduced selectivity

The two effects are simultaneous and the second one becomes more important for smaller Q_u, in this case, $Q_u < 100$.

When a filter has been designed from a lowpass prototype filter, it is convenient to relate the microwave filter element Q_u values to dissipative elements in the prototype filter and then determine the effects of the dissipation based on the prototype filter. Cohn [14] has presented such a simple formula for estimating the effects of dissipation loss of ladder-type of lowpass filters at $\omega = 0$, which may be expressed in the form [10]

$$\Delta L_{A0} = 4.343 \sum_{i=1}^{n} \frac{\Omega_c}{Q_{ui}} g_i \text{ dB} \qquad (3.61)$$

where ΔL_{A0} is the dB increase in insertion loss at $\omega = 0$, Ω_c and g_i are the cutoff frequency and element values of the lowpass prototypes, as discussed previously in

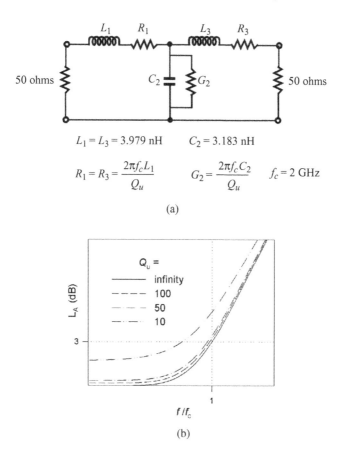

FIGURE 3.30 (a) Lowpass filter circuit including lossy reactive elements. (b) Dissipation effects on the insertion loss characteristic of the lowpass filter.

this chapter, and Q_{ui} are the unloaded quality factors of microwave elements corresponding to g_i, which are given at the cutoff ω_c of the practical lowpass filters. This formula does not require that the dissipation be uniform (equal Q_{ui}), and generally gives very good results if the terminations are equal or at least not very greatly different. As an application of this formula, consider the same example given above. The lowpass filter has used an $n = 3$ Butterworth lowpass prototype with $g_1 = g_3 = 1$, $g_2 = 2$ and $\Omega_c = 1$. As assumed in the above example that $Q_{u1} = Q_{u2} = Q_{u3} = Q_u$, we then have $\Delta L_{A0} = 0.174$ dB for $Q_u = 100$ and $\Delta L_{A0} = 1.737$ dB for $Q_u = 10$, according to Eq. (3.61), which are in excellent agreement with the results obtained by the network analysis.

The effects of dissipation in the lossy reactive elements of a highpass filter are analogous because of inverse frequency transformation used to generate such filters from lowpass prototypes. For highpass filters designed from lowpass prototypes, ΔL_{A0} on the inverse frequency scale now relates to the increased insertion loss as $\omega \to \infty$.

3.6.3 Dissipation Effects on Bandpass and Bandstop Filters

To discuss the dissipation effects on bandpass filters, consider the one depicted in Figure 3.15b, which has been designed previously for a Butterworth response with a passband from 1 to 2 GHz. Assume the equal Q_u for all the resonators, although, in principle, they may be unequal. The resultant filter circuit, including the dissipative elements for the resonators, is illustrated in Figure 3.31a; the insertion loss response is plotted in Figure 3.31b for different values of the Q_u. These Q values are supposed to be evaluated at the resonant frequency, which, in this case, is the center frequency of the filter. It should be mentioned that not only does the passband insertion loss increase and the selectivity becomes worse as the Q_u is decreased, but it also can be

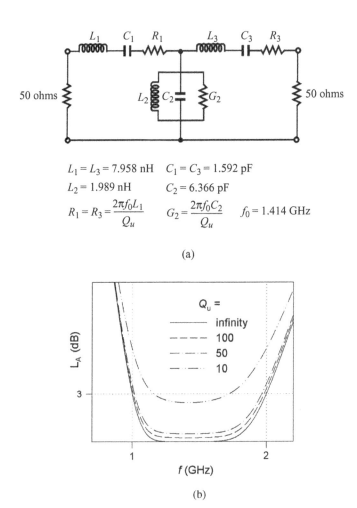

FIGURE 3.31 (a) Bandpass filter circuit including lossy resonators. (b) Dissipation effects on the insertion loss characteristic of the bandpass filter.

shown that for a given Q_u, the same tendencies occur as the fractional bandwidth of filter is reduced.

Similarly, it is convenient to use some closed-form expression to estimate the effects of the dissipation on bandpass filters that are designed from the lowpass prototypes. The formula in Eq. (3.61) may be modified for the bandpass filters

$$\Delta L'_{A0} = 4.343 \sum_{i=1}^{n} \frac{\Omega_c}{FBW Q_{ui}} g_i \text{ dB} \qquad (3.62)$$

Here $\Delta L'_{A0}$ is the dB increase in insertion loss at the center frequency of the filter and Q_{ui} are the unloaded quality factors of microwave resonators corresponding to g_i, which are evaluated at the center frequency of the filter. Consider the same bandpass filter example given above, which has used a $n = 3$ Butterworth lowpass prototype with $g_1 = g_3 = 1$, $g_2 = 2$, and $\Omega_c = 1$. The bandpass filter has a fractional bandwidth $FBW = 0.707$ and a center frequency of $f_0 = 1.414$ GHz. Again, we assume that $Q_{u1} = Q_{u2} = Q_{u3} = Q_u$. Substituting these data into Eq. (3.62) yields $\Delta L'_{A0} = 0.246$ dB for $Q_u = 100$ and $\Delta L'_{A0} = 2.457$ dB for $Q_u = 10$, which are almost the same as those obtained by the network analysis, as can be seen from Figure 3.31b. The expression of Eq. (3.62) indicates that the midband insertion loss of a bandpass filter is inversely proportional to the fractional bandwidth for given finite Q_u of resonators, as mentioned earlier.

An analogous demonstration can be given for bandstop filters consisting of lossy elements. The effects of parasitic dissipation in the filter elements are normally more serious in the stop- than in the passband. The stop band usually has one or more attenuation poles where, if the filter had no dissipation loss, the attenuation would be infinite. In practice, however, dissipation loss in the resonators will prevent the attenuation from going to infinity and, in some cases, may reduce the maximum stopband attenuation to an unacceptable low level.

In addition to using the network analysis, if the bandstop filters are designed, based on the lowpass prototypes, which have a ladder network structure, as described previously in this chapter, the maximum stopband attenuation $L_{A\max}$ in dB may be estimated using the simple formula [10]

$$L_{A\max} = 20 \log \left(\prod_{i=1}^{n} \Omega_c FBW Q_{ui} g_i \right) - 10 \log \left(\frac{4}{g_0 g_{n+1}} \right) \text{ dB} \qquad (3.63)$$

where Q_{ui} are the unloaded quality factors of microwave resonators evaluated at the center (midband) frequency of bandstop filters, and FBW is the fractional bandwidth as defined in Eq. (3.43). As an example, let us suppose that a bandstop filter is designed with a fractional stopband bandwidth of $FBW = 0.201$ (referred to the 3-dB points) based on a $n = 3$ Butterworth lowpass prototype with $g_0 = g_4 = 1, g_1 = g_3 = 1$, $g_2 = 2$, and $\Omega_c = 1$. Let us assume further that the microwave resonators have equal unloaded quality factors of 50, namely, $Q_{u1} = Q_{u2} = Q_{u3} = 50$. Using Eq. (3.63) to calculate the maximum stopband attenuation results in $L_{A\max} = 60.131$ dB. In comparison, using the network analysis gives $L_{A\max} = 61.857$ dB.

Dissipation loss in the resonators will also round off the attenuation characteristic of a bandstop filter. One obvious effect is to increase the attenuation around the band-edge frequencies ω_1 and ω_2, as defined in Eq. (3.43). For instance, the above Butterworth bandstop filter would have a desired 3-dB attenuation at the band edges if the Q_u were infinite for all resonators; however, the band-edge attenuation will, in fact, be larger, because of the finite Q_u. The increase in loss because of dissipation at the band-edge frequencies may be estimated by use of the formula [10]

$$\Delta L_{A\text{band-edge}} = 8.686 \sum_{i=1}^{n} \frac{\Omega_c g_i}{FBW Q_{ui}} \quad (3.64)$$

This formula represents only a rough estimate. It usually results in a better estimate for higher Q_{ui}, but a larger error when the Q_{ui} are lower.

REFERENCES

[1] G. C. Temes and S. K. Mitra, *Modern Filter Theory and Design*, John Wiley & Sons, New York, 1973.

[2] J. D. Rhodes, *Theory of Electrical Filters*, John Wiley & Sons, New York, 1976.

[3] J. Helszajn, *Synthesis of Lumped Element, Distributed and Planar Filters*, McGraw-Hill, London, 1990.

[4] L. Weinberg, *Network Analysis and Synthesis*, McGraw-Hill, New York, 1962.

[5] A. Papoulis, *The Fourier Integral and Its Applications*, McGraw-Hill, New York, 1962.

[6] W. Cauer, *Synthesis of Linear Communications Networks*, McGraw-Hill, New York, 1958.

[7] W. E. Thomson, Delay network having maximally flat frequency characteristics, *Proc. IEE*, **96**, 1949, 487–490.

[8] S. Darlington, Synthesis of reactance-four-poles which produce prescribed insertion loss characteristics, *J. Math. Phys.* **30**, 1939, 257–353.

[9] R. Saal and E. Ulbrich, On the design of filters by synthesis, *IRE Trans.* **CT-5**, 1958, 284–327.

[10] G. Mattaei, L. Young, and E. M. T. Jones, *Microwave Filters, Impedance-Matching Networks, and Coupling Structures*, Artech House, Norwood, MA, 1980.

[11] R. Saal, *Der Entwurf von Filtern mit Hilfe des Kataloges Normierter Tiefpasse*, Telefunkn GmbH, Backnang, Germany, 1961.

[12] P. I. Richards, Resistor-transmission-line circuits, *Proc. IRE.* **36**, 1948, 217–220.

[13] H. Ozaki and J. Ishii, Synthesis of a class of strip-line filters, *IRE Trans. Circuit Theory*, **CT-5**, 1958, 104–109.

[14] S. B. Cohn, Dissipation loss in multiple-coupled resonator filters, *Proc. IRE*, **47**, 1959, 1342–1348.

CHAPTER FOUR

Transmission Lines and Components

In this chapter, basic concepts and design equations for microstrip lines, coupled microstrip lines, discontinuities, and components useful for design of filters are briefly described. Another two types of planar transmission lines, that is, coplanar waveguide (CPW) and slotlines are also included in this chapter as there is an increasing interest in developing planar filters using mixed planar transmission-line components. Although comprehensive treatments of these topics can be found in the open literature, they are summarized here for easy reference.

4.1 MICROSTRIP LINES

4.1.1 Microstrip Structure

The general structure of a microstrip is illustrated in Figure 4.1. A conducting strip (microstrip line) with a width W and a thickness t is on the top of a dielectric substrate that has a relative dielectric constant ε_r and a thickness h, and the bottom of the substrate is a ground (conducting) plane.

4.1.2 Waves in Microstrip

The fields in the microstrip extend within two media — air above and dielectric below — so that the structure is inhomogeneous. Because of this inhomogeneous nature, the microstrip does not support a pure TEM wave. This is because a pure TEM wave has only transverse components and its propagation velocity depends only on the material properties, namely, the permittivity ε and the permeability μ.

Microstrip Filters for RF/Microwave Applications by Jia-Sheng Hong
Copyright © 2011 John Wiley & Sons, Inc.

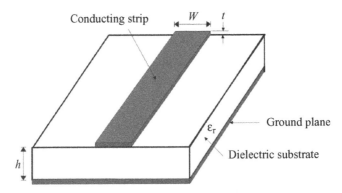

FIGURE 4.1 General microstrip structure.

However, with the presence of the two guided-wave media (the dielectric substrate and the air), the waves in a microstrip line will have no vanished longitudinal components of electric and magnetic fields and their propagation velocities will depend not only on the material properties, but also on the physical dimensions of the microstrip.

4.1.3 Quasi-TEM Approximation

When the longitudinal components of the fields for the dominant mode of a microstrip line remain very much smaller than the transverse components, they may be neglected. In this case, the dominant mode then behaves like a TEM mode and the TEM transmission-line theory is applicable for the microstrip line as well. This is called the quasi-TEM approximation and is valid over most of the operating frequency ranges of microstrip.

4.1.4 Effective Dielectric Constant and Characteristic Impedance

In the quasi-TEM approximation, a homogeneous dielectric material with an effective dielectric permittivity replaces the inhomogeneous dielectric air media of microstrip. Transmission characteristics of microstrip are described by two parameters, namely, the effective dielectric constant ε_{re} and characteristic impedance Z_c, which may then be obtained by quasistatic analysis [1]. In quasistatic analysis, the fundamental mode of wave propagation in a microstrip is assumed to be pure TEM. The above two parameters of microstrip are then determined from the values of two capacitances as follows

$$\varepsilon_{re} = C_d/C_a$$
$$Z_c = \frac{1}{c\sqrt{C_a C_d}} \tag{4.1}$$

MICROSTRIP LINES   77

in which C_d is the capacitance per unit length with the dielectric substrate present, C_a is the capacitance per unit length with the dielectric substrate replaced by air, and c is the velocity of electromagnetic waves in free space ($c \approx 3.0 \times 10^8$ m/s).

For very thin conductors (that is, $t \to 0$), the closed-form expressions that provide an accuracy better than 1% are given [2] as follows:

For $W/h \leq 1$:

$$\varepsilon_{re} = \frac{\varepsilon_r + 1}{2} + \frac{\varepsilon_r - 1}{2} \left\{ \left(1 + 12\frac{h}{W}\right)^{-0.5} + 0.04\left(1 - \frac{W}{h}\right)^2 \right\} \quad (4.2a)$$

$$Z_c = \frac{\eta}{2\pi \sqrt{\varepsilon_{re}}} \ln\left(\frac{8h}{W} + 0.25\frac{W}{h}\right) \quad (4.2b)$$

where $\eta = 120\pi$ Ω is the wave impedance in free space.

For $W/h \geq 1$:

$$\varepsilon_{re} = \frac{\varepsilon_r + 1}{2} + \frac{\varepsilon_r - 1}{2} \left(1 + 12\frac{h}{W}\right)^{-0.5} \quad (4.3a)$$

$$Z_c = \frac{\eta}{\sqrt{\varepsilon_{re}}} \left\{ \frac{W}{h} + 1.393 + 0.677 \ln\left(\frac{W}{h} + 1.444\right) \right\}^{-1} \quad (4.3b)$$

Hammerstad and Jensen [3] report more accurate expressions for the effective dielectric constant and characteristic impedance:

$$\varepsilon_{re} = \frac{\varepsilon_r + 1}{2} + \frac{\varepsilon_r - 1}{2}\left(1 + \frac{10}{u}\right)^{-ab} \quad (4.4)$$

where $u = W/h$, and

$$a = 1 + \frac{1}{49} \ln\left(\frac{u^4 + \left(\frac{u}{52}\right)^2}{u^4 + 0.432}\right) + \frac{1}{18.7} \ln\left[1 + \left(\frac{u}{18.1}\right)^3\right]$$

$$b = 0.564 \left(\frac{\varepsilon_r - 0.9}{\varepsilon_r + 3}\right)^{0.053}$$

The accuracy of this model is better than 0.2% for $\varepsilon_r \leq 128$ and $0.01 \leq u \leq 100$. The more accurate expression for the characteristic impedance is

$$Z_c = \frac{\eta}{2\pi \sqrt{\varepsilon_{re}}} \ln\left[\frac{F}{u} + \sqrt{1 + \left(\frac{2}{u}\right)^2}\right]. \quad (4.5)$$

where $u = W/h$, $\eta = 120\pi \Omega$, and

$$F = 6 + (2\pi - 6)\exp\left[-\left(\frac{30.666}{u}\right)^{0.7528}\right]$$

The accuracy for $Z_c\sqrt{\varepsilon_{re}}$ is better than 0.01% for $u \leq 1$ and 0.03% for $u \leq 1000$.

4.1.5 Guided Wavelength, Propagation Constant, Phase Velocity, and Electrical Length

Once the effective dielectric constant of a microstrip is determined, the guided wavelength of the quasi-TEM mode of microstrip is given by

$$\lambda_g = \frac{\lambda_0}{\sqrt{\varepsilon_{re}}} \quad (4.6a)$$

where λ_0 is the free space wavelength at operation frequency f. More conveniently, where the frequency is given in gigahertz (GHz), the guided wavelength can be evaluated directly in millimeters as follows:

$$\lambda_g = \frac{300}{f(\text{GHz})\sqrt{\varepsilon_{re}}} \text{mm} \quad (4.6b)$$

The associated propagation constant β and phase velocity v_p can be determined by

$$\beta = \frac{2\pi}{\lambda_g} \quad (4.7)$$

$$v_p = \frac{\omega}{\beta} = \frac{c}{\sqrt{\varepsilon_{re}}} \quad (4.8)$$

where c is the velocity of light ($c \approx 3.0 \times 10^8$ m/s) in free space.

The electrical length θ for a given physical length l of the microstrip is defined by

$$\theta = \beta l \quad (4.9)$$

Therefore, $\theta = \pi/2$ when $l = \lambda_g/4$, and $\theta = \pi$ when $l = \lambda_g/2$. These so-called quarter-wavelength and half-wavelength microstrip lines are important for design of microstrip filters.

4.1.6 Synthesis of W/h

Approximate expressions for W/h in terms of Z_c and ε_r, derived by Wheeler [4] and Hammerstad [2], are available

for $W/h \leq 2$

$$\frac{W}{h} = \frac{8\exp(A)}{\exp(2A) - 2} \tag{4.10}$$

with

$$A = \frac{Z_c}{60}\left\{\frac{\varepsilon_r + 1}{2}\right\}^{0.5} + \frac{\varepsilon_r - 1}{\varepsilon_r + 1}\left\{0.23 + \frac{0.11}{\varepsilon_r}\right\}$$

and for $W/h \geq 2$

$$\frac{W}{h} = \frac{2}{\pi}\left\{(B-1) - \ln(2B-1) + \frac{\varepsilon_r - 1}{2\varepsilon_r}\left[\ln(B-1) + 0.39 - \frac{0.61}{\varepsilon_r}\right]\right\} \tag{4.11}$$

with

$$B = \frac{60\pi^2}{Z_c\sqrt{\varepsilon_r}}$$

These expressions also provide accuracy better than 1%. If more accurate values are needed, an iterative or optimization process, based on the more accurate analysis models described previously, can be employed.

4.1.7 Effect of Strip Thickness

Thus far we have not considered the effect of conducting strip thickness t (as referred to in Figure 4.1). The thickness t is usually very small when the microstrip line is realized by conducting thin films; therefore, its effect may quite often be neglected. Nevertheless, its effect on the characteristic impedance and effective dielectric constant may be included [5]:

For $W/h \leq 1$:

$$Z_c(t) = \frac{\eta}{2\pi\sqrt{\varepsilon_{re}}}\ln\left\{\frac{8}{W_e(t)/h} + 0.25\frac{W_e(t)}{h}\right\} \tag{4.12a}$$

For $W/h \geq 1$:

$$Z_c(t) = \frac{\eta}{\sqrt{\varepsilon_{re}}} \left\{ \frac{W_e(t)}{h} + 1.393 + 0.667 \ln\left(\frac{W_e(t)}{h} + 1.444\right) \right\}^{-1} \quad (4.12b)$$

where

$$\frac{W_e(t)}{h} = \begin{cases} \dfrac{W}{h} + \dfrac{1.25}{\pi}\dfrac{t}{h}\left(1 + \ln\dfrac{4\pi W}{t}\right) & (W/h \leq 0.5\pi) \\ \dfrac{W}{h} + \dfrac{1.25}{\pi}\dfrac{t}{h}\left(1 + \ln\dfrac{2h}{t}\right) & (W/h \geq 0.5\pi) \end{cases} \quad (4.13a)$$

$$\varepsilon_{re}(t) = \varepsilon_{re} - \frac{\varepsilon_r - 1}{4.6}\frac{t/h}{\sqrt{W/h}} \quad (4.13b)$$

In the above expressions, ε_{re} is the effective dielectric constant for $t = 0$. It can be observed that the effect of strip thickness on both the characteristic impedance and effective dielectric constant is insignificant for small values of t/h. However, the effect of strip thickness is significant on conductor loss of the microstrip line.

4.1.8 Dispersion in Microstrip

Generally speaking, there is dispersion in microstrips; namely, its phase velocity is not a constant but depends on frequency. It follows that its effective dielectric constant ε_{re} is a function of frequency and can, in general, be defined as frequency-dependent effective dielectric constant $\varepsilon_{re}(f)$. The previous expressions for ε_{re} are obtained based on the quasi-TEM or quasistatic approximation and, therefore, are rigorous only at DC. At low microwave frequencies, these expressions provide a good approximation. To take into account the effect of dispersion, the formula of $\varepsilon_{re}(f)$ reported in Kobayashi [6] may be used and is given as follows:

$$\varepsilon_{re}(f) = \varepsilon_r - \frac{\varepsilon_r - \varepsilon_{re}}{1 + (f/f_{50})^m} \quad (4.14)$$

where

$$f_{50} = \frac{f_{TM_0}}{0.75 + \left(0.75 - 0.332\varepsilon_r^{-1.73}\right) W/h} \quad (4.15a)$$

$$f_{TM_0} = \frac{c}{2\pi h\sqrt{\varepsilon_r - \varepsilon_{re}}} \tan^{-1}\left(\varepsilon_r\sqrt{\frac{\varepsilon_{re} - 1}{\varepsilon_r - \varepsilon_{re}}}\right) \quad (4.15b)$$

$$m = m_0 m_c \leq 2.32 \tag{4.16a}$$

$$m_0 = 1 + \frac{1}{1+\sqrt{W/h}} + 0.32\left(\frac{1}{1+\sqrt{W/h}}\right)^3 \tag{4.16b}$$

$$m_c = \begin{cases} 1 + \dfrac{1.4}{1+W/h}\left\{0.15 - 0.235\exp\left(\dfrac{-0.45f}{f_{50}}\right)\right\} & \text{For } W/h \leq 0.7 \\ 1 & \text{For } W/h \geq 0.7 \end{cases} \tag{4.16c}$$

where c is the velocity of light in free space and, whenever the product $m_0 m_c$ is greater than 2.32, the parameter m is chosen equal to 2.32. The dispersion model shows that the $\varepsilon_{re}(f)$ increases with frequency and $\varepsilon_{re}(f) \to \varepsilon_r$ as $f \to \infty$. The accuracy is estimated to be within 0.6% for $0.1 \leq W/h \leq 10$, $1 \leq \varepsilon_r \leq 128$, and for any value of h/λ_0.

The effect of dispersion on the characteristic impedance may be estimated [3] by

$$Z_c(f) = Z_c \frac{\varepsilon_{re}(f)-1}{\varepsilon_{re}-1} \sqrt{\frac{\varepsilon_{re}}{\varepsilon_{re}(f)}} \tag{4.17}$$

where Z_c is the quasistatic value of characteristic impedance obtained earlier.

4.1.9 Microstrip Losses

The loss components of a single microstrip line include conductor, dielectric, and radiation losses, while the magnetic loss plays a role only for magnetic substrates, such as ferrites [8,9]. The propagation constant on a lossy transmission line is complex; namely, $\gamma = \alpha + j\beta$, where the real part α in nepers per unit length is the attenuation constant, which is the sum of the attenuation constants arising from each effect. In practice, one may prefer to express α in decibels (dB) per unit length, which can be related by

$$\alpha(\text{dB/unit length}) = (20\log_{10} e)\,\alpha(\text{nepers/unit length})$$
$$\approx 8.686\alpha(\text{nepers/unit length})$$

A simple expression for the estimation of the attenuation produced by the conductor loss is given [9] by

$$\alpha_c = \frac{8.686 R_s}{Z_c W}\,\text{dB/unit length} \tag{4.18}$$

where Z_c is the characteristic impedance of the microstrip of the width W and R_s represents the surface resistance in ohms per square for the strip conductor and ground plane. For a conductor

$$R_s = \sqrt{\frac{\omega \mu_0}{2\sigma}}$$

where σ is the conductivity, μ_0 is the permeability of free space, and ω is the angular frequency. The surface resistance of superconductors is expressed differently; this will be addressed in Chapter 11. Strictly speaking, the simple expression of Eq. (4.18) is only valid for large strip widths, because it assumes that the current distribution across the microstrip is uniform and, therefore, it would overestimate the conductor loss for narrower microstrip lines. Nevertheless, it may be found to be accurate enough in many practical situations, because of extraneous sources of loss, such as conductor surface roughness.

The attenuation because of the dielectric loss in microstrip can be determined [8,9] by

$$\alpha_d = 8.686\pi \left(\frac{\varepsilon_{re} - 1}{\varepsilon_r - 1} \right) \frac{\varepsilon_r}{\varepsilon_{re}} \frac{\tan \delta}{\lambda_g} \quad \text{dB/unit length} \quad (4.19)$$

where $\tan \delta$ denotes the loss tangent of the dielectric substrate.

Since the microstrip is a semi-open structure, any radiation is either free to propagate away or to induce currents on the metallic enclosure, causing radiation loss or the so-called housing loss.

4.1.10 Effect of Enclosure

A metallic enclosure is normally required for most microstrip circuit applications, such as filters. The presence of conducting top and side walls will affect both the characteristic impedance and the effective dielectric constant. Closed formulas are available [1] for a microstrip shielded with a conducting top cover (without side walls), which show how both the parameters are modified in comparison with the unshielded ones given previously. In practice, a rule of thumb may be applied in the filter design to reduce the effect of enclosure: the height up to the cover should be more than eight times and the distance to walls more than five times the substrate thickness. For more accurate design, the effect of enclosure, including the housing loss, can be taken into account by using full-wave EM simulation.

4.1.11 Surface Waves and Higher-Order Modes

A surface wave is a propagating mode guided by the air−dielectric surface for a dielectric substrate on the conductor ground plane, even without the upper conductor strip. Although the lowest surface wave mode can propagate at any frequency (it has

COUPLED LINES   83

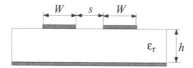

FIGURE 4.2   Cross section of coupled microstrip lines.

no cutoff), its coupling to the quasi-TEM mode of microstrip only becomes significant at the frequency

$$f_s = \frac{c \tan^{-1} \varepsilon_r}{\sqrt{2\pi} h \sqrt{\varepsilon_r - 1}} \tag{4.20}$$

at which the phase velocities of the two modes are close [10].

The excitation of higher-order modes in a microstrip can be avoided by operating it below the cutoff frequency of the first higher-order mode, which is given by Vendelin [10]

$$f_c = \frac{c}{\sqrt{\varepsilon_r} (2W + 0.8h)} \tag{4.21}$$

In practice, the lowest value (the worst case) of the two frequencies given by Eqs. (4.20) and (4.21) is taken as the upper limit of operating frequency of a microstrip line.

4.2   COUPLED LINES

Coupled microstrip lines are widely used for implementing microstrip filters. Figure 4.2 illustrates the cross section of a pair of coupled microstrip lines under consideration in this section, where the two microstrip lines of width W are in the parallel- or edge-coupled configuration with a separation s. This coupled line structure supports two quasi-TEM modes, that is, the even and the odd mode, as shown in Figure 4.3. For an even-mode excitation, both microstrip lines have the same voltage potentials or

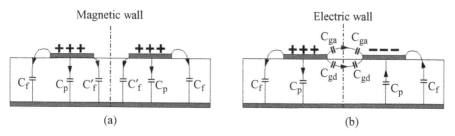

FIGURE 4.3   Quasi-TEM modes of a pair of coupled microstrip lines. (**a**) Even mode; (**b**) odd mode.

84 TRANSMISSION LINES AND COMPONENTS

carry the same sign charges, say, the positive ones, resulting in a magnetic wall at the symmetry plane, as shown in Figure 4.3a. In the case when an odd mode is excited, both microstrip lines have the opposite voltage potentials or carry the opposite sign charges, so that the symmetric plane is an electric wall, as indicated in Figure 4.3b. In general, these two modes will be excited at the same time. However, they propagate with different phase velocities because they are not pure TEM modes, which mean that they experience different permittivities. Therefore, the coupled microstrip lines are characterized by the characteristic impedances, as well as the effective dielectric constants for the two modes.

4.2.1 Even- and Odd-Mode Capacitances

In a static approach similar to the single microstrip, the even- and odd-mode characteristic impedances and effective dielectric constants of the coupled microstrip lines may be obtained in terms of the even- and odd-mode capacitances, denoted by C_e and C_o. As shown in Figure 4.3, the even- and odd-mode capacitances C_e and C_o may be expressed [11] as

$$C_e = C_p + C_f + C'_f \tag{4.22}$$

$$C_o = C_p + C_f + C_{gd} + C_{ga} \tag{4.23}$$

In these expressions, C_p denotes the parallel-plate capacitance between the strip and the ground plane and, hence, is simply given by

$$C_p = \varepsilon_o \varepsilon_r W/h \tag{4.24}$$

C_f is the fringe capacitance as if for an uncoupled single microstrip line and is evaluated by

$$2C_f = \sqrt{\varepsilon_{re}}/(cZ_c) - C_p \tag{4.25}$$

The term C'_f accounts for the modification of fringe capacitance C_f of a single line because of the presence of another line. An empirical expressions for C'_f is given below

$$C'_f = \frac{C_f}{1 + A\,(h/s)\tanh(8s/h)} \tag{4.26}$$

where

$$A = \exp\left[-0.1\exp(2.33 - 2.53W/h)\right]$$

For the odd-mode C_{ga} and C_{gd} represent, respectively, the fringe capacitances for the air and dielectric regions across the coupling gap. The capacitance C_{gd} may be found from the corresponding coupled stripline geometry, with the spacing between the ground planes given by $2h$. A closed-form expression for C_{gd} is

$$C_{gd} = \frac{\varepsilon_0 \varepsilon_r}{\pi} \ln\left(\coth\left(\frac{\pi}{4}\frac{s}{h}\right)\right) + 0.65 C_f \left(\frac{0.02\sqrt{\varepsilon_r}}{s/h} + 1 - \frac{1}{\varepsilon_r^2}\right) \quad (4.27)$$

The capacitance C_{ga} can be modified from the capacitance of the corresponding coplanar strips and expressed in terms of a ratio of two elliptic functions

$$C_{ga} = \varepsilon_o \frac{K(k')}{K(k)} \quad (4.28a)$$

where

$$k = \frac{s/h}{s/h + 2W/h} \quad (4.28b)$$
$$k' = \sqrt{1 - k^2}$$

and the ratio of the elliptic functions is given by

$$\frac{K(k')}{K(k)} = \begin{cases} \dfrac{1}{\pi} \ln\left(2\dfrac{1+\sqrt{k'}}{1-\sqrt{k'}}\right) & \text{for } 0 \le k^2 \le 0.5 \\ \dfrac{\pi}{\ln\left(2\dfrac{1+\sqrt{k}}{1-\sqrt{k}}\right)} & \text{for } 0.5 \le k^2 \le 1 \end{cases} \quad (4.28c)$$

The capacitances obtained by using above design equations [11] are found to be accurate to within 3% over the ranges $0.2 \le W/h \le 2$, $0.05 \le s/h \le 2$, and $\varepsilon_r \ge 1$.

4.2.2 Even- and Odd-Mode Characteristic Impedances and Effective Dielectric Constants

The even- and odd-mode characteristic impedances Z_{ce} and Z_{co} can be obtained from the capacitances. This yields

$$Z_{ce} = \left(c\sqrt{C_e^a C_e}\right)^{-1} \quad (4.29)$$

$$Z_{co} = \left(c\sqrt{C_o^a C_o}\right)^{-1} \quad (4.30)$$

where C_e^a and C_o^a are even- and odd-mode capacitances for the coupled microstrip line configuration with air as a dielectric.

86 TRANSMISSION LINES AND COMPONENTS

Effective dielectric constants ε_{re}^e and ε_{re}^o for even and odd modes, respectively, can be obtained from C_e and C_o by using the relations

$$\varepsilon_{re}^e = C_e/C_e^a \tag{4.31}$$

and

$$\varepsilon_{re}^o = C_o/C_o^a \tag{4.32}$$

4.2.3 More Accurate Design Equations

More accurate closed-form expressions for the effective dielectric constants and the characteristic impedances of coupled microstrip are available [12]. For a static approximation, namely, without considering dispersion, these are given as follows:

$$\varepsilon_{re}^e = \frac{\varepsilon_r + 1}{2} + \frac{\varepsilon_r - 1}{2}\left(1 + \frac{10}{v}\right)^{-a_e b_e} \tag{4.33}$$

with

$$v = \frac{u\left(20 + g^2\right)}{10 + g^2} + g\exp(-g)$$

$$a_e = 1 + \frac{1}{49}\ln\left[\frac{v^4 + (v/52)^2}{v^4 + 0.432}\right] + \frac{1}{18.7}\ln\left[1 + \left(\frac{v}{18.1}\right)^3\right]$$

$$b_e = 0.564\left(\frac{\varepsilon_r - 0.9}{\varepsilon_r + 3}\right)^{0.053}$$

where $u = W/h$ and $g = s/h$. The error in ε_{re}^e is within 0.7% over the ranges of $0.1 \leq u \leq 10$, $0.1 \leq g \leq 10$, and $1 \leq \varepsilon_r \leq 18$.

$$\varepsilon_{re}^o = \varepsilon_{re} + [0.5(\varepsilon_r + 1) - \varepsilon_{re} + a_o]\exp\left(-c_o g^{d_o}\right) \tag{4.34}$$

with

$$a_o = 0.7287[\varepsilon_{re} - 0.5(\varepsilon_r + 1)]\left[1 - \exp(-0.179u)\right]$$

$$b_o = \frac{0.747\varepsilon_r}{0.15 + \varepsilon_r}$$

$$c_o = b_o - (b_o - 0.207)\exp(-0.414u)$$

$$d_o = 0.593 + 0.694\exp(-0.562u)$$

where ε_{re} is the static effective dielectric constant of single microstrip of width W as discussed previously. The error in ε_{re}^o is stated to be on the order of 0.5%.

The even- and odd-mode characteristic impedances given by the following closed-form expressions are accurate to within 0.6% over the ranges $0.1 \leq u \leq 10$, $0.1 \leq g \leq 10$, and $1 \leq \varepsilon_r \leq 18$.

$$Z_{ce} = Z_c \frac{\sqrt{\varepsilon_{re}/\varepsilon_{re}^e}}{1 - Q_4 \sqrt{\varepsilon_{re}} \cdot Z_c/377} \tag{4.35}$$

where Z_c is the characteristic impedance of single microstrip of width W, and

$$Q_1 = 0.8695 u^{0.194}$$

$$Q_2 = 1 + 0.7519g + 0.189g^{2.31}$$

$$Q_3 = 0.1975 + \left[16.6 + \left(\frac{8.4}{g}\right)^6\right]^{-0.387} + \frac{1}{241} \ln\left[\frac{g^{10}}{1 + (g/3.4)^{10}}\right]$$

$$Q_4 = \frac{2Q_1}{Q_2} \cdot \frac{1}{u^{Q_3} \exp(-g) + [2 - \exp(-g)] u^{-Q_3}}$$

$$Z_{co} = Z_c \frac{\sqrt{\varepsilon_{re}/\varepsilon_{re}^o}}{1 - Q_{10} \sqrt{\varepsilon_{re}} \cdot Z_c/377} \tag{4.36}$$

with

$$Q_5 = 1.794 + 1.14 \ln\left[1 + \frac{0.638}{g + 0.517g^{2.43}}\right]$$

$$Q_6 = 0.2305 + \frac{1}{281.3} \ln\left[\frac{g^{10}}{1 + (g/5.8)^{10}}\right] + \frac{1}{5.1} \ln\left(1 + 0.598g^{1.154}\right)$$

$$Q_7 = \frac{10 + 190g^2}{1 + 82.3g^3}$$

$$Q_8 = \exp\left[-6.5 - 0.95 \ln(g) - (g/0.15)^5\right]$$

$$Q_9 = \ln(Q_7) \cdot (Q_8 + 1/16.5)$$

$$Q_{10} = Q_4 - \frac{Q_5}{Q_2} \exp\left[\frac{Q_6 \ln(u)}{u^{Q_9}}\right]$$

Closed-form expressions for characteristic impedance and effective dielectric constants, as given above, may also be used to obtain accurate values of capacitances for the even and odd modes from the relationships defined in Eqs. (4.29–4.32). The formulations that include the effect of dispersion can be found in [12].

4.3 DISCONTINUITIES AND COMPONENTS

In this section, some typical microstrip discontinuities and components that are often encountered in microstrip filter designs are described.

4.3.1 Microstrip Discontinuities

Microstrip discontinuities commonly encountered in the layout of practical filters include steps, open-ends, bends, gaps, and junctions. Figure 4.4 illustrates some typical structures and their equivalent circuits. Generally speaking, the effects of discontinuities can be more accurately modeled and taken into account in the filter designs with full-wave electromagnetic (EM) simulations, which will be addressed later on. Nevertheless, closed-form expressions for equivalent circuit models of these discontinuities are still useful whenever they are appropriate. These expressions are used in many circuit analysis programs. There are numerous closed-form expressions for microstrip discontinuities available in open literature [1,13–16]; for convenience, some typical ones are given as follows.

4.3.1.1 Steps in Width
For a symmetrical step, the capacitance and inductances of the equivalent circuit indicated in Figure 4.4a may be approximated by the following formulation [1]

$$C = 0.00137h \frac{\sqrt{\varepsilon_{re1}}}{Z_{c1}} \left(1 - \frac{W_2}{W_1}\right)\left(\frac{\varepsilon_{re1} + 0.3}{\varepsilon_{re1} - 0.258}\right)\left(\frac{W_1/h + 0.264}{W_1/h + 0.8}\right) \text{ (pF)} \quad (4.37)$$

$$L_1 = \frac{L_{w1}}{L_{w1} + L_{w2}} L, \quad L_2 = \frac{L_{w2}}{L_{w1} + L_{w2}} L \quad (4.38)$$

with

$$L_{wi} = Z_{ci}\sqrt{\varepsilon_{rei}}/c$$

$$L = 0.000987h \left(1 - \frac{Z_{c1}}{Z_{c2}}\sqrt{\frac{\varepsilon_{re1}}{\varepsilon_{re2}}}\right)^2 \text{ (nH)}$$

where L_{wi} for $i = 1, 2$ are the inductances per unit length of the appropriate microstrips, having widths W_1 and W_2, respectively. While Z_{ci} and ε_{rei} denote the

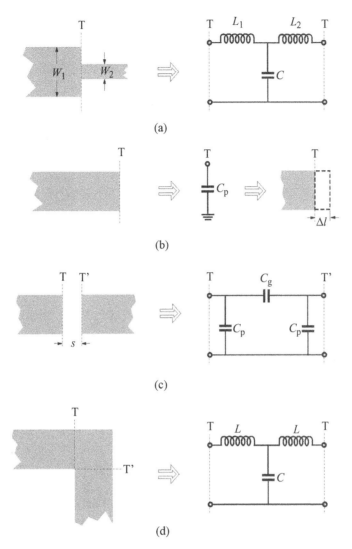

FIGURE 4.4 Microstrip discontinuities: (a) step; (b) open-end; (c) gap; (d) bend.

characteristic impedance and effective dielectric constant corresponding to width W_i, c is the light velocity in free space and h is the substrate thickness in micrometers.

4.3.1.2 Open Ends At the open end of a microstrip line with a width of W, the fields do not stop abruptly, but extend slightly further because of the effect of fringing field. This effect can be modeled either with an equivalent shunt capacitance C_p or with an equivalent length of transmission line Δl, as shown in Figure 4.4b. The equivalent length is usually more convenient for filter design. The relation between the two equivalent parameters may be found [13] by

$$\Delta l = \frac{c Z_c C_p}{\sqrt{\varepsilon_{re}}} \tag{4.39}$$

where c is the light velocity in free space. A closed-form expression for $\Delta l/h$ is given [14] by

$$\frac{\Delta l}{h} = \frac{\xi_1 \xi_3 \xi_5}{\xi_4} \tag{4.40}$$

where

$$\xi_1 = 0.434907 \frac{\varepsilon_{re}^{0.81} + 0.26 (W/h)^{0.8544} + 0.236}{\varepsilon_{re}^{0.81} - 0.189 (W/h)^{0.8544} + 0.87}$$

$$\xi_2 = 1 + \frac{(W/h)^{0.371}}{2.358 \varepsilon_r + 1}$$

$$\xi_3 = 1 + \frac{0.5274 \tan^{-1} \left[0.084 (W/h)^{1.9413/\xi_2} \right]}{\varepsilon_{re}^{0.9236}}$$

$$\xi_4 = 1 + 0.037 \tan^{-1} \left[0.067 (W/h)^{1.456} \right] \cdot \{6 - 5 \exp \left[0.036 (1 - \varepsilon_r) \right] \}$$

$$\xi_5 = 1 - 0.218 \exp(-7.5 W/h)$$

The accuracy is better than 0.2% for the range of $0.01 \leq w/h \leq 100$ and $\varepsilon_r \leq 128$.

4.3.1.3 Gaps A microstrip gap can be represented by an equivalent circuit, as shown in Figure 4.4c. The shunt and series capacitances C_p and C_g may be determined [1] by

$$\begin{aligned} C_p &= 0.5 C_e \\ C_g &= 0.5 C_o - 0.25 C_e \end{aligned} \tag{4.41}$$

where

$$\frac{C_o}{W} (\text{pF/m}) = \left(\frac{\varepsilon_r}{9.6} \right)^{0.8} \left(\frac{s}{W} \right)^{m_o} \exp(k_o)$$

$$\frac{C_e}{W} (\text{pF/m}) = 12 \left(\frac{\varepsilon_r}{9.6} \right)^{0.9} \left(\frac{s}{W} \right)^{m_e} \exp(k_e)$$

with

$$m_o = \frac{W}{h}(0.619 \log(W/h) - 0.3853)$$
$$k_o = 4.26 - 1.453 \log(W/h)$$

for $0.1 \leq s/W \leq 1.0$

$$m_e = 0.8675$$
$$k_e = 2.043 \left(\frac{W}{h}\right)^{0.12}$$

for $0.1 \leq s/W \leq 0.3$

$$m_e = \frac{1.565}{(W/h)^{0.16}} - 1$$
$$k_e = 1.97 - \frac{0.03}{W/h}$$

for $0.3 \leq s/W \leq 1.0$

The accuracy of these expressions is within 7% for $0.5 \leq W/h \leq 2$ and $2.5 \leq \varepsilon_r \leq 15$.

4.3.1.4 Bends Right-angle bends of microstrips may be modeled by an equivalent T-network, as shown in Figure 4.4d. Gupta et al. [1] have given closed-form expressions for evaluation of capacitance and inductance:

$$\frac{C}{W}(\text{pF/m}) = \begin{cases} \dfrac{(14\varepsilon_r + 12.5)W/h - (1.83\varepsilon_r - 2.25)}{\sqrt{W/h}} + \dfrac{0.02\varepsilon_r}{W/h} & \text{for } W/h < 1 \\ (9.5\varepsilon_r + 1.25)W/h + 5.2\varepsilon_r + 7.0 & \text{for } W/h \geq 1 \end{cases}$$

(4.42a)

$$\frac{L}{h}(\text{nH/m}) = 100\left\{4\sqrt{\frac{W}{h}} - 4.21\right\}$$

(4.42b)

The accuracy on the capacitance is quoted as within 5% over the ranges of $2.5 \leq \varepsilon_r \leq 15$ and $0.1 \leq W/h \leq 5$. The accuracy on the inductance is about 3% for $0.5 \leq W/h \leq 2.0$.

4.3.2 Microstrip Components

Microstrip components, which are often encountered in microstrip filter designs, may include lumped inductors and capacitors, quasilumped elements (namely, short-line sections and stubs), and resonators. In most cases, the resonators are the distributed elements, such as quarter-wavelength and half-wavelength line resonators. The choice of individual components may mainly depend on the types of filters, the fabrication

92 TRANSMISSION LINES AND COMPONENTS

techniques, the acceptable losses or Q factors, the power handling, and the operating frequency. These components are briefly described as follows.

4.3.2.1 Lumped Inductors and Capacitors Some typical configurations of planar microwave lumped inductors and capacitors are shown in Figures 4.5 and 4.6. These components may be categorized as the elements whose physical dimensions are much smaller than the free space wavelength λ_0 of highest operating frequency, say smaller than $0.1\lambda_0$ [18,19]. Thus, they have the advantage of small size, low cost, and wide-band characteristics, but have lower Q and power handling than disturbed elements. Because of a considerable size reduction, lumped elements are normally attractive for the realization of monolithic microwave integrated circuits (MMICs). The applications of lumped elements can be extended to millimeter wave with the emerging fabrication techniques, such as the micromachining technique [20].

As illustrated in Figure 4.5, the high-impedance, straight-line section is the simplest form of inductor, used for low inductance values (typically up to 3 nH), while the spiral inductor (circular or rectangular) can provide higher inductance values, typically up to 10 nH. The innermost turn of the spiral inductor can be connected to

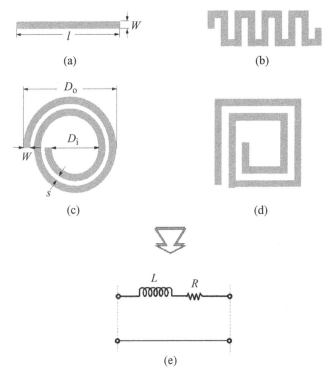

FIGURE 4.5 Lumped-element inductors: (**a**) high-impedance line; (**b**) meander line; (**c**) circular spiral; (**d**) square spiral; (**e**) their ideal circuit representation.

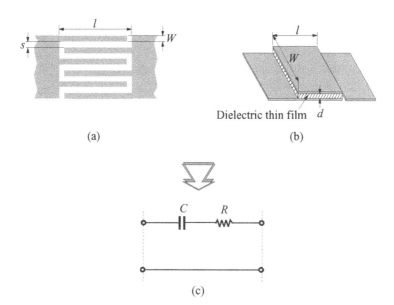

FIGURE 4.6 Lumped-element capacitors: (**a**) interdigital capacitor; (**b**) MIM capacitor; (**c**) ideal circuit representation.

outside circuit through a dielectric-spaced underpass or using a wire-bond air–bridge crossover.

In Figure 4.6, the interdigital capacitor is more suitable for applications where low values of capacitance (less than 1.0 pF) are required. The metal-insulator-metal (MIM) capacitor, constructed by using a thin layer of a low-loss dielectric (typically 0.5-μm thick) between two metal plates, is used to achieve higher values, for example, as high as 30 pF in small areas. The metal plates should be thicker than three skin depths to minimize conductor losses. The top plate is generally connected to other circuitry by using an air bridge that provides higher breakdown voltages.

Bear in mind that to function well as a lumped element at microwave frequencies, the total line length of a lumped inductor or an overall size of a lumped capacitor in whatever form must be a small fraction of a wavelength. Unfortunately, this condition is not often satisfied. Moreover, there are other parasitics that make it difficult to realize a truly lumped element. For instance, there always exists shunt capacitance to ground when a lumped inductor is realized in microstrip; this capacitance can become important enough to significantly affect the performance of the inductor. Therefore, to accurately characterize lumped elements over the entire operation frequency band, while taking into account all parasitics and other effects, usually necessitates the use of full-wave EM simulations. Nevertheless, some basic design equations described below may be found useful for initial designs.

4.3.2.1.1 Design of Inductors Approximate design equations are available for inductances and associated resistances of various types of inductors [1,21]. Let W, t, and l represent the width, thickness, and length of the conductor, respectively. The conductor thickness t should be greater than three skin depths. In the case of spirals, n denotes the number of turns and s is the spacing between the turns. Also, let R_s denote the surface resistance of the conductor in ohms per square.

For the straight-line inductor:

$$L(\text{nH}) = 2 \times 10^{-4} l \left[\ln\left(\frac{l}{W+t}\right) + 1.193 + 0.2235 \frac{W+t}{l} \right] \cdot K_g \quad \text{for } l \text{ in } \mu\text{m} \tag{4.43a}$$

$$R = \frac{R_s l}{2(W+t)} \cdot \left[1.4 + 0.217 \ln\left(\frac{W}{5t}\right) \right] \quad \text{for } 5 < \frac{W}{t} < 100 \tag{4.43b}$$

For the circular spiral inductor:

$$L(\text{nH}) = 0.03937 \frac{a^2 n^2}{8a + 11c} \cdot K_g \quad \text{for } a \text{ in } \mu\text{m} \tag{4.44a}$$

$$a = \frac{D_o + D_i}{4} \qquad c = \frac{D_o - D_i}{2}$$

$$R = 1.5 \frac{\pi a n R_s}{W} \tag{4.44b}$$

The design of a loop inductor may be obtained from a single-turn ($n = 1$) spiral inductor. It may be noticed that the inductance of one single turn is less (because of the proximity effect) than the inductance of a straight line with the same length and width.

In the inductance expressions, K_g is a correction factor to take into account the effect of a ground plane, which tends to decrease the inductance value as the ground plane is brought nearer. To a first-order approximation, the following closed-form expression for K_g may be used

$$K_g = 0.57 - 0.145 \ln \frac{W}{h} \quad \text{for } \frac{W}{h} > 0.05 \tag{4.45}$$

where h is the substrate thickness. The unloaded Q of an inductor may be calculated from

$$Q = \frac{\omega L}{R} \tag{4.46}$$

4.3.2.1.2 Design of Capacitors Letting the finger width W equal the space s to achieve maximum capacitance density, and assuming that the substrate thickness h is much larger than the finger width, a very simple closed-form expression [22] for

DISCONTINUITIES AND COMPONENTS   95

estimation of capacitance of the interdigital capacitor may be given by

$$C(\text{pF}) = 3.937 \times 10^{-5} l (\varepsilon_r + 1)[0.11(n-3) + 0.252] \quad \text{for } l \text{ in } \mu\text{m} \quad (4.47a)$$

where n is the number of fingers and ε_r is the relative dielectric constant of the substrate. The Q factor corresponding to conductor losses is given by

$$Q_c = \frac{1}{\omega C R} \quad \text{for } R = \frac{4}{3} \frac{R_s l}{W n} \quad (4.47b)$$

The dielectric Q factor is approximately $Q_d = 1/\tan \delta$, where $\tan \delta$ is the dielectric loss tangent. The total Q factor is then found from

$$\frac{1}{Q} = \frac{1}{Q_c} + \frac{1}{Q_d} \quad (4.48)$$

The capacitance of a MIM capacitor is very close to a simple parallel plate value:

$$C = \frac{\varepsilon (W \times l)}{d} \quad (4.49a)$$

where $(W \times l)$ is the area of the metal plates that are separated by a dielectric thin film with a thickness d and a dielectric constant ε. The conductor Q_c is

$$Q_c = \frac{1}{\omega C R} \quad \text{for } R = \frac{R_s l}{W} \quad (4.49b)$$

Similarly, the total Q can be determined from Eq. (4.48).

4.3.2.2 Quasi-Lumped Elements Microstrip line short sections and stubs, whose physical lengths are smaller than a quarter of guided wavelength λ_g at which they operate, are the most common components for approximate microwave realization of lumped elements in microstrip filter structures, and are termed quasilumped elements. They may also be regarded as lumped elements if their dimensions are even smaller, say smaller than $\lambda_g/8$. Some important microstrip quasilumped elements are discussed in this section.

4.3.2.2.1 High- and Low-Impedance Short-Line Sections In Figure 4.7, a short length of high-impedance (Z_c) lossless line terminated at both ends by relatively low impedance (Z_0) is represented by a π-equivalent circuit. For a propagation constant $\beta = 2\pi/\lambda_g$ of the short line, the circuit parameters are given by

$$x = Z_c \sin\left(\frac{2\pi}{\lambda_g} l\right) \quad \text{and} \quad \frac{B}{2} = \frac{1}{Z_c} \tan\left(\frac{\pi}{\lambda_g} l\right) \quad (4.50)$$

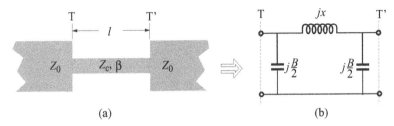

FIGURE 4.7 High-impedance short-line element.

which can be obtained by equating the *ABCD* parameters of the two circuits. If $l < \lambda_g/8$, then

$$x \approx Z_c \left(\frac{2\pi}{\lambda_g} l\right) \quad \text{and} \quad \frac{B}{2} \approx \frac{1}{Z_c}\left(\frac{\pi}{\lambda_g} l\right) \quad (4.51)$$

It can further be shown that for $Z_c \gg Z_0$, the effect of the shunt susceptances may be neglected and this short-line section has an effect equivalent to that of a series inductance having a value of $L = Z_c l/v_p$, where $v_p = \omega/\beta$ is the phase velocity of propagation along the short line.

For the dual case shown in Figure 4.8, a short length of low-impedance (Z_c) lossless line terminated at either end by relatively high impedance (Z_0) is represented by a T-equivalent circuit with the circuit parameters

$$B = \frac{1}{Z_c}\sin\left(\frac{2\pi}{\lambda_g}l\right) \quad \text{and} \quad \frac{x}{2} = Z_c \tan\left(\frac{\pi}{\lambda_g}l\right) \quad (4.52)$$

For $l < \lambda_g/8$ the values of the circuit parameters can be approximated by

$$B \approx \frac{1}{Z_c}\left(\frac{2\pi}{\lambda_g}l\right) \quad \text{and} \quad \frac{x}{2} \approx Z_c\left(\frac{\pi}{\lambda_g}l\right) \quad (4.53)$$

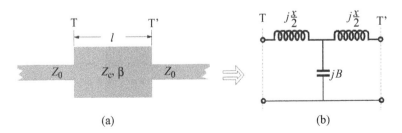

FIGURE 4.8 Low-impedance short-line element.

Similarly, if $Z_c \ll Z_0$ the effect of the series reactances may be neglected, and this short-line section has an effect equivalent to that of a shunt capacitance $C = l/(Z_c v_p)$.

To evaluate the quality factor Q of these short-line elements, losses may be included by considering a lossy transmission line with a complex propagation constant $\gamma = \alpha + j\beta$. The total equivalent series resistance associated with the series reactance is then approximated by $R \approx Z_c \alpha l$, whereas the total equivalent shunt conductance associated with the shunt susceptance is $G \approx \alpha l/Z_c$. Since $Q_Z = x/R$ for a lossy reactance element and $Q_Y = B/G$ for a lossy susceptance element, it can be shown that the total Q factor ($1/Q = 1/Q_Z + 1/Q_Y$) of the short-line elements is estimated by

$$Q = \frac{\beta}{2\alpha} \tag{4.54}$$

where β is in radians per unit length and α is in nepers per unit length.

4.3.2.2.2 Open- and Short-Circuited Stubs We will now demonstrate that a short open-circuited stub of lossless microstrip line can be equivalent to a shunt capacitor and that a similar short-circuited stub can be equivalent to a shunt inductor, as indicated in Figure 4.9.

According to the transmission-line theory, the input admittance of an open-circuited transmission line having a characteristic admittance $Y_c = 1/Z_c$ and

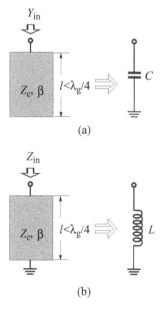

FIGURE 4.9 Short stub elements: (**a**) open-circuited stub; (**b**) short-circuited stub.

propagation constant $\beta = 2\pi/\lambda_g$ is given by

$$Y_{in} = jY_c \tan\left(\frac{2\pi}{\lambda_g}l\right) \qquad (4.55)$$

where l is the length of the stub. If $l < \lambda_g/4$ this input admittance is capacitive. If the stub is even shorter, say $l < \lambda_g/8$, the input admittance may be approximated by

$$Y_{in} \approx jY_c\left(\frac{2\pi}{\lambda_g}l\right) = j\omega\left(\frac{Y_c l}{v_p}\right) \qquad (4.56)$$

where v_p is the phase velocity of propagation in the stub. It is now clearer that such a short open-circuited stub is equivalent to a shunt capacitance $C = Y_c l/v_p$.

For the dual case, the input impedance of a similar short-circuited transmission line is given by

$$Z_{in} = jZ_c \tan\left(\frac{2\pi}{\lambda_g}l\right) \qquad (4.57)$$

This input impedance is inductive for $l < \lambda_g/4$. If $l < \lambda_g/8$, an approximation of the input impedance is

$$Z_{in} \approx jZ_c\left(\frac{2\pi}{\lambda_g}l\right) = j\omega\left(\frac{Z_c l}{v_p}\right) \qquad (4.58)$$

Such a short section of the short-circuited stub functions, therefore, as a shunt lumped-element inductance $L = Z_c l/v_p$.

4.3.2.3 Resonators A microstrip resonator is any structure that is able to contain at least one oscillating electromagnetic field. There are numerous forms of microstrip resonator. In general, microstrip resonators for filter designs may be classified as lumped-element or quasilumped-element resonators and distributed-line resonators or patch resonators. Some typical configurations of these resonators are illustrated in Figure 4.10.

Lumped-element or quasilumped-element resonators, formed by the lumped or quasilumped inductors and capacitors as shown in Figure 4.10a and b, will obviously resonate at $\omega_0 = 1/\sqrt{LC}$. However, they may resonate at some higher frequencies at which their sizes are no longer much smaller than a wavelength and, thus, by definition, are no longer lumped or quasilumped elements.

The distributed-line resonators shown in Figure 4.10c and d may be termed as the quarter-wavelength resonators, since they are $\lambda_{g0}/4$ long, where λ_{g0} is the guided wavelength at the fundamental resonant frequency f_0. They can also resonate at other higher frequencies when $f \approx (2n-1)f_0$ for $n = 2, 3, \ldots$. Another typical distributed-line resonator is the half-wavelength resonator, as shown in Figure 4.10e,

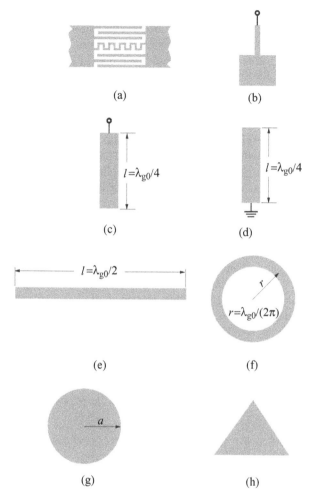

FIGURE 4.10 Some typical microstrip resonators: (**a**) lumped-element resonator; (**b**) quasilumped-element resonator; (**c**) $\lambda_{g0}/4$ line resonator (shunt-series -resonance); (**d**) $\lambda_{g0}/4$ line resonator (shunt-parallel resonance); (**e**) $\lambda_{g0}/2$ line resonator; (**f**) ring resonator; (**g**) circular patch resonator; (**h**) triangular patch resonator.

which is $\lambda_{g0}/2$ long at its fundamental resonant frequency, and can also resonate at $f \approx nf_0$ for $n = 2, 3, \ldots$. It will be demonstrated later when we discuss filter designs that this type of line resonator can be shaped into many different configurations for filter implementations, such as the open-loop resonator [23].

The ring resonator shown in Figure 4.10f is another type of distributed-line resonator [24], where r is the median radius of the ring. The ring will resonate at its fundamental frequency f_0 when its median circumference $2\pi r \approx \lambda_{g0}$. The higher resonant modes occur at $f \approx nf_0$ for $n = 2, 3, \ldots$. Because of its symmetrical geometry, a resonance can occur in either of two orthogonal coordinates. This type of line resonator, therefore, has a distinct feature that can support a pair of degenerate modes that have the same resonant frequencies, but orthogonal field distributions. This feature can be utilized to design dual-mode filters (see Chapter 10 for details). Similarly, it is possible to construct this type of line resonator into different configurations, such as square and meander loops [25,26].

Patch resonators are of interest for the design of microstrip filters, in order to increase the power-handling capability [27,28]. An associated advantage of microstrip patch resonators is their lower conductor losses as compared with narrow microstrip line resonators. Although patch resonators tend to have a stronger radiation, they are normally enclosed in a metal housing for filter applications so that the radiation loss can be minimized. Patch resonators usually are a larger size; however, this would not be a problem for the application in which the power handling or low loss has a higher priority. The size may not be an issue at all for the filters operating at very high frequencies. Depending on the applications, patches may take different shapes, such as circular in Figure 4.10g and triangular in Figure 4.10h. These microstrip patch resonators can be analyzed as waveguide cavities with magnetic walls on the sides. The fields within the cavities can be expanded by the TM^z_{nm0} modes, where z is perpendicular to the ground plane. For instance, the fields for each of the cavity modes in a circular microstrip patch (disk) resonator may be expressed in a cylindrical coordinate system (ϕ, r, z) [29] as

$$
\begin{aligned}
E_z &= A_{nm} J_n(K_{nm} r/a) \cos(n\phi) \\
H_r &= \frac{j}{\omega\mu} \left(\frac{1}{r} \frac{\partial E_z}{\partial \phi} \right) = -\frac{j}{\omega\mu} \frac{n}{r} A_{nm} J_n(K_{nm} r/a) \sin(n\phi) \\
H_\phi &= \frac{-j}{\omega\mu} \left(\frac{\partial E_z}{\partial r} \right) = \frac{-j}{\omega\mu} \frac{K_{nm}}{a} A_{nm} J'_n(K_{nm} r/a) \cos(n\phi) \\
E_\phi &= E_r = H_z = 0
\end{aligned}
\qquad (4.59)
$$

where A_{nm} is a reference amplitude, J_n is the Bessel function of the first kind of order n, J'_n is the derivative of J_n, and K_{nm} is the mth zero of J'_n. For $m = 1$ the zeros are

$$
K_{n1} = \begin{cases} 3.83171 & n = 0 \\ 1.84118 & n = 1 \\ 3.05424 & n = 2 \\ 4.20119 & n = 3 \end{cases}
\qquad (4.60)
$$

The resonant frequencies of these modes are given by

$$
f_{nm0} = \frac{K_{nm} c}{2\pi a_e \sqrt{\varepsilon_r}} \quad \text{for } a_e = a\sqrt{1 + \frac{2h}{\pi a}\left[\ln\left(\frac{\pi a}{2h}\right) + 1.7726\right]}
\qquad (4.61)
$$

where a_e is an effective radius that takes into account the fringing fields ($a_e \approx a$ for $h \ll a$) [30]. Thus, the lowest order or fundamental mode is the TM_{110}^z mode. A circular patch operating at this mode is a dual-mode resonator as well. Another interesting mode is the TM_{010}^z mode, which does not have current along the edge. This mode has been used to design superconducting filters having higher power-handling capability [31].

4.3.3 Loss Considerations for Microstrip Resonators

In many practical filter designs, it is desirable to evaluate the achievable unloaded quality factor Q_u of microstrip resonators. For example, this will serve as a justification for whether or not the required insertion loss of a bandpass filter can be met. A very general definition of Q_u that is applicable to any resonator is

$$Q_u = \omega \frac{\text{Time-average energy stored in resonator}}{\text{Average power lost in resonator}} \qquad (4.62)$$

The losses in a microstrip resonator arise because of a number of mechanisms. The most important are usually the losses associated with the conductor, dielectric substrate, and radiation. The total unloaded quality factor can be found by adding these losses together, resulting in

$$\frac{1}{Q_u} = \frac{1}{Q_c} + \frac{1}{Q_d} + \frac{1}{Q_r} \qquad (4.63)$$

where Q_c, Q_d, and Q_r are the conductor, dielectric, and radiation quality factors, respectively. It should be mentioned that for filter applications, microstrip resonators are usually shielded in package housings, and, in this case, the radiation Q may be replaced by a quality factor associated with the housing loss. Calculations of these quality factors are nontrivial because they require knowledge of electromagnetic field distributions that depend on the geometry of microstrip resonator, the size of housing, and the boundary conditions imposed. With the advent of modern design tools based on full-wave EM simulations, these parameters can be modeled. Nevertheless, the following considerations are useful as guide lines for filter designs.

The conductor Q of a microstrip line resonator can be evaluated by [32,33]

$$Q_c = \frac{\pi}{\alpha_c \lambda_g} \qquad (4.64)$$

where α_c is the conductor attenuation constant in nepers per unit length and λ_g is the guided wavelength of the microstrip line. Assuming a uniform field between the microstrip and its ground plane, the Q_c may be approximated by

$$Q_c \approx \pi \left(\frac{h}{\lambda}\right)\left(\frac{\eta}{R_s}\right) \qquad (4.65)$$

where h is the substrate thickness, λ and η ($\approx 377\Omega$) are the wavelength and wave impedance in free space, and R_s the surface resistance of conductor sheets. It is commonly accepted that the R_s of normal conductors varies as $\sqrt{f}$, with the frequency f, while the R_s of superconductors varies as f^2. Since $\lambda = c/f$, the Q_c for normal conductors will actually be proportional to $\sqrt{f}$, whereas the Q_c for superconductors is inversely proportional to frequency. Using a thick substrate can increase the Q_c, but care must be taken because this increases the radiation and unwanted couplings as well.

The dielectric loss can be taken into account in terms of a complex permittivity of the dielectric substrate $\varepsilon = \varepsilon' - j\varepsilon''$, with the negative imaginary part denoting energy loss [34]. Hence, the loss in a dielectric substrate may be attributed to an effective conductivity $\omega\varepsilon''$. It can be shown that

$$Q_d \geq \frac{\varepsilon'}{\varepsilon''} = \frac{1}{\tan\delta} \qquad (4.66)$$

where $\tan\delta$ is the dielectric loss tangent. Alternatively, the dielectric Q may be evaluated by

$$Q_d = \frac{\pi}{\alpha_d \lambda_g} \qquad (4.67)$$

where α_d is the dielectric attenuation constant in nepers per unit length.

If a microstrip resonator is not enclosed, it will then radiate. A radiation quality factor can, in general, be defined as

$$Q_r = \omega \frac{\text{Time-average energy stored in resonator}}{\text{Average power radiated}} \qquad (4.68)$$

However, as mentioned above, for filter applications, microstrip resonators are normally shielded in a conductor housing. In this case, the so-called housing quality factor should be considered instead of the radiation Q. A general definition of the housing Q is

$$Q_h = \omega \frac{\text{Time-average energy stored in resonator}}{\text{Average power lost in housing walls}} \qquad (4.69)$$

The power loss of the housing because of nonperfect conducting walls at resonance, may be expressed as

$$P_h = \frac{R_h}{2} \int |\underline{n} \times \underline{H}|^2 \, dS \qquad (4.70)$$

Here R_h is the surface resistance of the housing walls, $\underline{n}$ is the unit normal to the housing walls, and $\underline{H}$ is the magnetic field at resonance. The housing walls can be gold-plated to a thickness that is thicker than several skin depths. The fields intercepted by the housing walls, which are normally located within the near field region of the resonator, are weaker because they actually decay very fast, proportionally to $1/r^3$ (static field) or $1/r^2$ (induced field), where r is the distance from the microstrip resonator. Therefore, the housing quality factor Q_h, which is inversely proportional to P_h, can generally reach a higher value with a larger housing size. Alternatively, one could optimize the shape of a resonator or use lumped resonators to confine the fields in the substrate to obtain a higher housing Q, which will, however, reduce Q_c.

4.4 OTHER TYPES OF MICROSTRIP LINES

There are several derivatives of microstrip lines that can be used as alternative structures for microstrip filter implementations. These include suspended and inverted microstrip lines, multilayered microstrips, thin film microstrips (TFM), and valley microstrips [1]. In addition, there are many other types of transmission lines that are of interest to filter designs.

Suspended and inverted microstrip lines, as shown in Figure 4.11, provide a higher Q (500–1500 for normal conductor) than the conventional microstrip lines. They are normally enclosed for filter applications [35], as indicated in Figure 4.11c. Although they can be used for realizing any types of filters, the wide range of impedance values achievable makes them particularly suitable for lowpass and highpass filters, which can be cascaded together to form broadband bandpass filters and multiplexers. By using very thin dielectric substrates of low dielectric constant, the dielectric loss can be minimized. This makes these media attractive for developing filters, such as micromachined filters for millimeter-wave applications.

The closed-form expression in Eq. (4.5) may be employed to find the characteristic impedance with $u = W/(h_1 + h_2)$ for suspended microstrip in Figure 4.11a, and

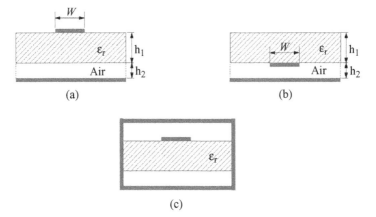

FIGURE 4.11 Other types of microstrip lines: (**a**) suspended microstrip line; (**b**) inverted microstrip line; (**c**) suspended or inverted microstrip line enclosed.

$u = W/h_2$ for the inverted microstrip in Figure 4.11b, and the effective dielectric constants for the both structures can be obtained from Eq. [36]

$$\sqrt{\varepsilon_{re}} = \left[1 + \frac{h_1}{h_2}\left(a - b\ln\frac{W}{h_2}\right)\left(\frac{1}{\sqrt{\varepsilon_r}} - 1\right)\right]^{-1} \quad \text{for suspended microstrip}$$

(4.71a)

where

$$a = (0.8621 - 0.1251\ln(h_1/h_2))^4$$
$$b = (0.4986 - 0.1397\ln(h_1/h_2))^4$$

(4.71b)

$$\sqrt{\varepsilon_{re}} = 1 + \frac{h_1}{h_2}\left(a - b\ln\frac{W}{h_2}\right)(\sqrt{\varepsilon_r} - 1) \quad \text{for inverted microstrip} \quad (4.72a)$$

where

$$a = [0.5173 - 0.1515\ln(h_1/h_2)]^2$$
$$b = [0.3092 - 0.1047\ln(h_1/h_2)]^2$$

(4.72b)

The accuracy of Eqs. (4.71a) and (4.72a) is within 1% over the ranges $1 \le W/h_2 \le 8$, $0.2 \le h_1/h_2 \le 1$, and $\varepsilon_r \le 6$.

4.5 COPLANAR WAVEGUIDE (CPW)

The coplanar waveguide (CPW), as shown in Figure 4.12, was first proposed by Wen [37]. It consists of a metal strip with a width W and two adjacent ground planes printed on a dielectric substrate that has a relative dielectric constant ε_r and a thickness h. The gap between the strip and ground is denoted by G. Similar to the microstrip, the CPW is not a pure TEM wave-transmission line because a small longitudinal component of the electromagnetic field exists at the air–dielectric interface. An obvious advantage of the CPW arises from the fact that only a single-layer metal process is required and the connecting of a component (active or passive) in shunt configuration is much easier, as via-hole grounding is not needed to reach the ground plane.

With a quasistatic approximation of the dominant mode (TEM) of wave propagation in a CPW, the effective dielectric constant ε_{re} and characteristic impedance Z_c can also be obtained from the formulation of Eq. (4.1). For the CPW shown in

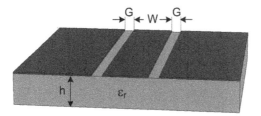

FIGURE 4.12 Coplanar waveguide (CPW).

Figure 4.12, assuming that the ground planes are much wider than the center strip and the two gaps, the two transmission-line parameters, that is, ε_{re} and Z_c, are given [37,38] by

$$\varepsilon_{re} = 1 + \left(\frac{\varepsilon_r - 1}{2}\right)\left[\frac{K(k_2)}{K'(k_2)}\right]\left[\frac{K'(k_1)}{K(k_1)}\right] \quad (4.73a)$$

$$Z_c = \left(\frac{30\pi}{\sqrt{\varepsilon_{re}}}\right)\left[\frac{K'(k_1)}{K(k_1)}\right] \quad \text{(ohm)} \quad (4.73b)$$

where

$K(k)$ = complete elliptic integral of the first kind
$K'(k) = K(k')$
$k' = \sqrt{1 - k^2}$

and the ratio of the elliptic functions can be calculated by

$$\frac{K(k)}{K'(k)} = \begin{cases} \dfrac{\pi}{\ln\left(2\dfrac{1+\sqrt{k'}}{1-\sqrt{k'}}\right)} & \text{for } 0 \leq k \leq 0.707 \\[2ex] \dfrac{1}{\pi}\ln\left(2\dfrac{1+\sqrt{k}}{1-\sqrt{k}}\right) & \text{for } 0.707 \leq k \leq 1 \end{cases} \quad (4.74)$$

Note that k represents a generic argument for either k_1 or k_2. Referring to Figure 4.12, let

$$a = \frac{W}{2}, \quad b = G + a \quad (4.75)$$

The two arguments for the elliptic integrals in Eq. (4.73) are obtained as

$$k_1 = \frac{a}{b} = \frac{W}{W+2G} \quad (4.76a)$$

$$k_2 = \frac{\sinh\left(\frac{\pi a}{2h}\right)}{\sinh\left(\frac{\pi b}{2h}\right)} \quad (4.76b)$$

The values of Z_c and ε_{re} computed using the above formulations are plotted in Figure 4.13. The line impedance is found to decrease as the ratio of the strip width to gap (W/G) increases, and is less dependent on the ratio of W/h or the substrate thickness. The effective dielectric constant increases against W/G, but the change is only more significant for a larger ratio of W/h or a thinner substrate.

In a practical application, there are other factors that can also affect the line impedance and effective dielectric constant. For example, a practical CPW has finite width ground planes, and the truncation of lateral ground planes can cause the line impedance to increase and effective dielectric constant to slightly decrease [39]. It can also support another quasi-TEM mode whose characteristic impedance could be

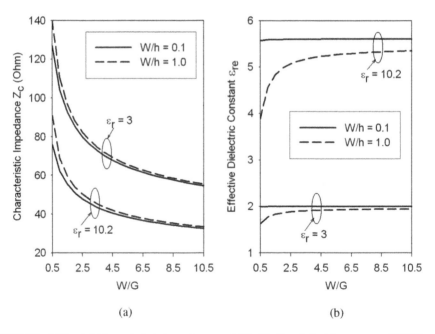

FIGURE 4.13 (a) Characteristic impedance and (b) effective dielectric constant of CPW as a function of the ratio of W/G with the ratio of W/h and the relative dielectric constant ε_r as parameters.

comparable to the desired mode. To suppress this mode, lateral ground planes are kept at the same potential by means of properly spaced conductive air bridges [39]. Another practical application is to introduce a conductor backing to the CPW structure in Figure 4.12 resulting in a so-called conductor-backed CPW that looks like a mixed coplanar-microstrip structure [39]. Therefore, the conductor-backed CPW also supports the parasitic microstrip mode. The microstrip behavior becomes dominant when the substrate is thin and the slots are wide. The conductor backing considerably reduces the line impedance. However, it can improve both the mechanical strength and the power-handling capability. Moreover, it allows easy implementation of mixed CPW-microstrip circuits.

4.6 SLOTLINES

Another planar transmission line is the slotline [40], whose basic structure is shown in Figure 4.14. It consists of a dielectric substrate with a narrow slot etched in the metallization on one side of the substrate. The width of the slot is denoted by W and the substrate has a thickness h and a relative dielectric constant ε_r. The other side of the substrate is without any metallization. Slotlines can be included in microstrip or CPW circuits by etching the slotline circuit on the ground plane for microstrip or CPW circuits. This type of hybrid combination allows flexibility in the design of planar filters.

A voltage difference exists between the slot edges. The electric field extends across the slot; the magnetic field is perpendicular to the slot. Because the voltage occurs across the slot, the configuration is especially convenient for connecting shunt elements, such as diodes, resistors, and capacitors. In a slotline, the wave propagates along the slot with the major electric field component oriented across the slot in the plane of metallization on the dielectric substrate. On the other hand, a slotline differs from a waveguide in that it has no cutoff frequency. Propagation along the slot occurs at all frequencies down to $f = 0$, where, if the metal-coated substrate is assumed infinite in length and width. For slotline to be practical as a transmission line, radiation must be minimized. This is accomplished through the use of a high permittivity substrate, which causes the slot-mode wavelength λ_g to be small compared to free-space wavelength λ_0, and thereby results in the fields being closely confined to the slot with negligible radiation loss [40].

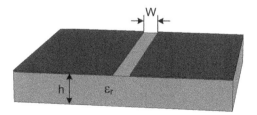

FIGURE 4.14 Slotline configuration.

108 TRANSMISSION LINES AND COMPONENTS

The basic electrical parameters of slot line are the effective dielectric constant ε_{re}, for $\lambda_g/\lambda_0 = 1/\sqrt{\varepsilon_{re}}$, and the characteristic impedance Z_c. Because of the non-TEM nature of the slotline mode, these parameters are not constant, but vary with frequency. A semiempirical expression for computing ε_{re} is given [42] by

$$\varepsilon_{re} = 1 + \left(\frac{\varepsilon_r - 1}{2}\right) \left[\frac{K(k'_b)}{K(k_b)}\right] \left[\frac{K(k_a)}{K(k'_a)}\right] \tag{4.77}$$

where

$$k_a = \sqrt{\frac{2a}{1+a}} \qquad k'_a = \sqrt{1 - k_a^2}$$

$$k_b = \sqrt{\frac{2b}{1+b}} \qquad k'_b = \sqrt{1 - k_b^2}$$

$$a = \tanh\left\{\left(\frac{\pi}{2}\right) \cdot \left(\frac{W}{h}\right) \cdot \left[1 + \frac{0.0133}{\varepsilon_r + 2} \cdot \left(\frac{\lambda_0}{h}\right)^2\right]^{-1}\right\}$$

$$b = \tanh\left[\left(\frac{\pi}{2}\right) \cdot \left(\frac{W}{h}\right)\right]$$

The ratio of the elliptic functions in Eq. (4.77) can be calculated using the formulation of Eq. (4.28c). It is reported Svacina [41] that Eq. (4.77) gives good results for the following range of parameters:

$$\begin{aligned} 9.7 &\leq \varepsilon_r \leq 20 \\ 0.02 &\leq W/h \leq 1.0 \\ 0.01 &\leq h/\lambda_0 \leq \frac{0.25}{\sqrt{\varepsilon_r - 1}} \end{aligned} \tag{4.78}$$

where the expression on the right-hand side of the last term gives the cutoff value for the TE_0 surface-wave mode of the slotline. In Svacina [41], a semiempirical expression is also proposed for estimating the characteristic impedance as

$$Z_c = \frac{60\pi}{\sqrt{\varepsilon_{re}}} \left[\frac{K(k_a)}{K(k'_a)}\right] \quad \text{(ohm)} \tag{4.79}$$

Note that because of the non-TEM nature of the slotline mode, the definition of characteristic impedance is somewhat arbitrary [42]. Figure 4.15 plots some computed results using Eqs. (4.77) and (4.79), showing that the slotline is a more dispersive

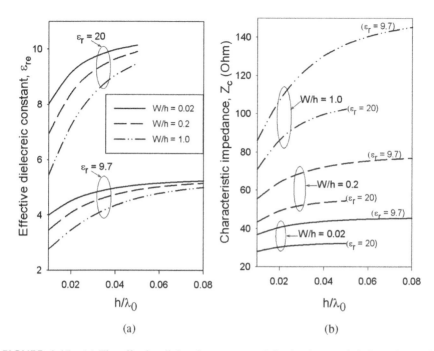

FIGURE 4.15 (a) The effective dielectric constant and (b) the characteristic impedance of slotline as a function of the ratio of h/λ_0 with the ratio of W/h and the relative dielectric constant ε_r as parameters.

transmission line than the microstrip or CPW. Closed-form expressions for slotline wavelength and impedance and for a wider range of parameters can be found in Garg and Gupta [43] and Janaswasny and Schaubert [44].

REFERENCES

[1] K. C. Gupta, R. Garg, I. Bahl, and P. Bhartis, *Microstrip Lines and Slotlines*, second edition, Artech House, Boston, MA, 1996.
[2] E. O. Hammerstard, Equations for microstrip circuit design. In: Proceedings of the European Microwave Conference; Hamburg, Germany, 1975, pp. 268–272.
[3] E. O. Hammerstad and O. Jensen, Accurate models for microstrip computer-aided design. *IEEE MTT-S Dig.* 1980, pp. 407–409.
[4] H. Wheeler, Transmission line properties of parallel strips separated by a dielectric sheet. *IEEE Trans.* **MTT-13**, 1965, 172–185.
[5] I. J. Bahl and R. Garg, Simple and accurate formulas for microstrip with finite strip thickness. *Proc. IEEE* **65**, 1977, 1611–1612.
[6] M. Kobayashi, A dispersion formula satisfying recent requirements in microstrip CAD. *IEEE Trans.* **MTT-36**, 1988, 1246–1250.

[7] I. J. Bahl, Easy and exact methods for shielded microstrip design. *Microwaves* **17**, 1978, 61–62.

[8] E. J. Denlinger, Losses of microstrip lines. *IEEE Trans.* **MTT-28**, 1980, 513–522.

[9] R. A. Pucel, D. J. Masse, and C. P. Hartwig, Losses in microstrip. *IEEE Trans.* **MTT-16**, 1968, 342–350. Correction in *IEEE Trans.* **MTT-16**, 1968, 1064.

[10] G. D. Vendelin, Limitations on strip line Q. *Microwave J.* **13**, 1970, 63–69.

[11] R. Garg and I. J. Bahl, Characteristics of coupled microstriplines. *IEEE Trans.* **MTT-27**, 1979, 700–705. Corrections in *IEEE Trans.* **MTT-28**, 1980, 272.

[12] M. Kirschning and R. H. Jansen Accurate wide-range design equations for parallel coupled microstrip lines. *IEEE Trans.* **MTT-32**, 1984, 83–90. Corrections in *IEEE Trans.* **MTT-33**, 1985, 288.

[13] T. Edwards, *Foundations for Microstrip Circuit Design*, second edition, John Wiley & Sons, Chichester, U.K., 1991.

[14] M. Kirschning, R. H. Jansen, and N. H. L. Koster, Accurate model for open end effect of microstrip lines. *Electron. Lett.* **17**, 1981, 123–125.

[15] B. Easter, The equivalent circuit of some microstrip discontinuities. *IEEE Trans.* **MTT-23**, 1975, 655–660.

[16] J.-S. Hong and J.-M. Shi, Modeling microstrip step discontinuities by the transmission matrix. *Electron. Lett.* **23**(13), 1987, 678–679.

[17] K. Rambabu and M. Ramesh, Models describe microstrip circuit characteristics. *Microwave RF* **38**, 1999, 64–74.

[18] R. A. Pucel, Design considerations for monolithic microwave circuits. *IEEE Trans.* **MTT-29**, 1981, 513–534.

[19] E. Pettenpaul, H. Kapusta, A. Weisgerber, H. Mampe, J. Luginsland, and I. Wolff, CAD models of lumped elements on GaAs up to 18 GHz. *IEEE Trans.* **MTT-36**, 1988, 294–304.

[20] C.-Y. Chi and G. M. Rebeiz, Planar microwave and millimeter-wave lumped elements and coupled-line filters using micro-machining techniques. *IEEE Trans.* **MTT-43**, 1995, 730–738.

[21] M. Caulton, S. P. Knight, and D. A. Daly, Hybrid integrated lumped-element microwave amplifiers. *IEEE Trans.* **MTT-16**, 1968, 397–404.

[22] G. D. Alley, Interdigital capacitors and their application to lumped-element microwave integrated circuits. *IEEE Trans.* **MTT-18**, 1970, 1028–1033.

[23] J.-S. Hong and M. J. Lancaster, Couplings of microstrip square open-loop resonators for cross-coupled planar microwave filters. *IEEE Trans.* **MTT-44**, 1996, 2099–2109.

[24] I. Wolff and N. Knoppink, Microstrip ring resonator and dispersion measurement on microstrip lines. *Electron. Lett.* **7**, 1971, 779–781.

[25] J.-S. Hong and M. J. Lancaster, Bandpass characteristics of new dual-mode microstrip square loop resonators. *Electron. Lett.* **11**, 1995, 891–892.

[26] J.-S. Hong and M. J. Lancaster, Microstrip bandpass filter using degenerate modes of a novel meander loop resonator. *IEEE Microwave Guided Wave Lett.* **5**, 1995, 371–372.

[27] R. R. Monsour, B. Jolley, Shen Ye, F. S. Thomoson, and V. Dokas, On the power handling capability of high temperature superconductive filters. *IEEE Trans.* **MTT-44**, July 1996, 1322–1338.

[28] J.-S. Hong and M. J. Lancaster, Microstrip triangular patch resonator filters. *IEEE MTT-S Dig.* (1), 2000, 331–334.

[29] J. Watkins, Circular resonant structures in microstrip. *Electron. Lett.* **5**, 1969, 524–525.

[30] I. Wolff and N. Knoppin, Rectangular and circular microstrip disk capacitors and resonators. *IEEE Trans.* **MTT-22**, 1974, 857–864.

[31] H. Chaloupka, M. Jeck, B. Gurzinski, and S. Kolesov, Superconducting planar disk resonators and filters with high power handling capability. *Electron. Lett.* **32**, 1996, 1735–1737.

[32] E. Belohoubek and E. Denlinger, Loss considerations for microstrip resonators. *IEEE Trans.* **MTT-23**, 1975, 522–526.

[33] A. Gopinath, Maximum Q-factor of microstrip resonators. *IEEE Trans.* **MTT-29**, 1981, 128–131.

[34] R. E. Collin, Foundations for Microwave Engineering, second edition, McGraw-Hill, New York, 1992.

[35] J.-S. Hong, J.-M. Shi, and L. Sun, Exact computation of generalized scattering matrix of suspended microstrip step discontinuities. *Electron. Lett.* **25**(5), 1989, 335–336.

[36] P. Pramanick and P. Bhartia, Computer-aided design models for millimeter-wave finlines and suspend-substrate microstrip lines. *IEEE Trans.* **MTT-33**, 1985, 1429–1435.

[37] C. P. Wen, Coplanar waveguide: a surface strip transmission line suitable for nonreciprocal gyromagnetic device applications. *IEEE Trans.* **MTT-7**, 1969, 1087–1090.

[38] C. Veyres and V. F. Hanna, Extension of the application of conformal mapping techniques to coplanar lines with finite dimensions. *Int. J. Electron.* **48**, 1980, 47–56.

[39] G. Ghione and C. U. Naldi, Coplanar waveguide for MMIC applications: effect of upper shielding, conductor backing, finite extent ground planes, and line-to-line coupling. *IEEE Trans.* **MTT-35**, 1987, 260–267.

[40] S. B. Cohn, Slot line on a dielectric substrate. *IEEE Trans Microwave Theory Tech.* **MTT-17**, 1969, 768–778.

[41] J. Svacina, Dispersion characteristics of multilayered slotlines—a simple approach. *IEEE Trans.* **MTT-47**, 1999, 1826–1829.

[42] E. A. Mariani, C. P. Heinzman, J. P. Agrios, and S. B. Cohn, Slotline characteristics. *IEEE Trans. Microwave Theory Tech.* **MTT-17**, 1969, 1091–1096.

[43] R. Garg and K. C. Gupta, Expressions for wavelength and impedance of a slotline. *IEEE Trans.* **MTT-24**, 1976, 532.

[44] R. Janaswamy and D. H. Schaubert, Characteristic impedance of a wide slotline on low-permittivity substrates. *IEEE Trans. Microwave Theory Tech.* **MTT-34**, 1986, 900–902.

CHAPTER FIVE

Lowpass and Bandpass Filters

Conventional microstrip lowpass and bandpass filters such as stepped-impedance filters, open-stub filters, semi-lumped element filters, end- and parallel-coupled half-wavelength resonator filters, hairpin-line filters, interdigital and combline filters, pseudocombline filters, and stub-line filters are widely used in many RF/microwave applications. This chapter presents examples of the illustrative designs of these filters.

5.1 LOWPASS FILTERS

In general, design of microstrip lowpass filters involves two main steps. The first is to select an appropriate lowpass prototype, such as one as described in Chapter 3. The choice of the type of response, including passband ripple and the number of reactive elements, will depend on the required specifications. The element values of the lowpass prototype filter, which are usually normalized to make a source impedance $g_0 = 1$ and a cutoff frequency $\Omega_c = 1.0$, are then transformed to the L–C elements for the desired cutoff frequency and source impedance, which is normally 50 Ω for microstrip filters. Having obtained a suitable lumped-element filter design, the next main step in the design of microstrip lowpass filters is to find an appropriate microstrip realization that approximates the lumped-element filter. In this section, we concentrate on the second step. Several microstrip lowpass filters will be described.

5.1.1 Stepped-Impedance L-C Ladder-Type Lowpass Filters

Figure 5.1a shows a general structure of the stepped-impedance lowpass microstrip filter, which use a cascaded structure of alternating high- and low-impedance transmission lines. Since these are much shorter than the associated guided wavelength, they act as semilumped elements. The high-impedance lines act as series inductors

---

Microstrip Filters for RF/Microwave Applications by Jia-Sheng Hong
Copyright © 2011 John Wiley & Sons, Inc.

LOWPASS FILTERS

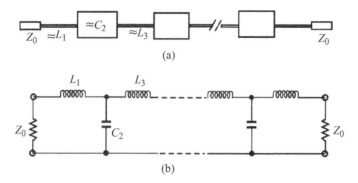

FIGURE 5.1 (a) General structure of the stepped-impedance lowpass microstrip filters. (b) L–C ladder-type of lowpass filters to be approximated.

and the low-impedance lines act as shunt capacitors. Therefore, this filter structure directly resembles the L-C ladder-type of lowpass filters shown in Figure 5.1b.

Some *a priori* design information must be provided about the microstrip lines, because expressions for inductance and capacitance depend upon both the characteristic impedance and length. It would be practical to initially fix the characteristic impedances of high- and low-impedance lines by considering

- $Z_{0C} < Z_0 < Z_{0L}$, where Z_{0C} and Z_{0L} denote the characteristic impedances of the low- and high-impedance lines, respectively, and Z_0 is the source impedance, which is usually 50 Ω for microstrip filters.
- A lower Z_{0C} results in a better approximation of a lumped-element capacitor, but the resulting line width W_C must not allow any transverse resonance to occur at operation frequencies.
- A higher Z_{0L} leads to a better approximation of a lumped-element inductor, but Z_{0L} must not be so high that its fabrication becomes inordinately difficult as a narrow line, or that its current-carrying capability becomes a limitation.

In order to illustrate the design procedure for this type of filter, a three-pole lowpass filter is described.

The specifications for the filter under consideration are

Cutoff frequency $f_c = 1$ GHz
Passband ripple 0.1 dB (or return loss ≤ -16.42 dB)
Source/load impedance $Z_0 = 50$ Ω

A lowpass prototype with Chebyshev response is chosen, whose element values are

$$g_0 = g_4 = 1$$
$$g_1 = g_3 = 1.0316$$
$$g_2 = 1.1474$$

for the normalized cutoff $\Omega_c = 1.0$. Using the element transformations described in Chapter 3, we have

$$L_1 = L_3 = \left(\frac{Z_0}{g_0}\right)\left(\frac{\Omega_c}{2\pi f_c}\right) g_1 = 8.209 \times 10^{-9} \text{ H}$$

$$C_2 = \left(\frac{g_0}{Z_0}\right)\left(\frac{\Omega_c}{2\pi f_c}\right) g_2 = 3.652 \times 10^{-12} \text{ F}$$

(5.1)

The filter is fabricated on a substrate with a relative dielectric constant of 10.8 and a thickness of 1.27 mm. Following the above-mentioned considerations, the characteristic impedances of the high- and low-impedance lines are chosen as $Z_{0L} = 93 \, \Omega$ and $Z_{0C} = 24 \, \Omega$. The relevant design parameters of microstrip lines, which are determined using the formulas given in Chapter 4, are listed in Table 5.1, where the guided-wavelengths are calculated at the cutoff frequency $f_c = 1.0$ GHz.

Initially, the physical lengths of the high- and low-impedance lines may be found by

$$l_L = \frac{\lambda_{gL}}{2\pi} \sin^{-1}\left(\frac{\omega_c L}{Z_{0L}}\right)$$

$$l_C = \frac{\lambda_{gC}}{2\pi} \sin^{-1}(\omega_c C Z_{0C})$$

(5.2)

which gives $l_L = 11.04$ mm and $l_C = 9.75$ mm for this example. The results of Eq. (5.2) do not take into account the series reactance of the low-impedance line and shunt susceptance of the high-impedance lines. To include these effects, the lengths of the high- and low-impedance lines should be adjusted to satisfy

$$\omega_c L = Z_{0L} \sin\left(\frac{2\pi l_L}{\lambda_{gL}}\right) + Z_{0C} \tan\left(\frac{\pi l_C}{\lambda_{gC}}\right)$$

$$\omega_c C = \frac{1}{Z_{0C}} \sin\left(\frac{2\pi l_C}{\lambda_{gC}}\right) + 2 \times \frac{1}{Z_{0L}} \tan\left(\frac{\pi l_L}{\lambda_{gL}}\right)$$

(5.3)

where L and C are the required element values of lumped inductors and capacitor given above. This set of equations is solved for l_L and l_C, resulting in $l_L = 9.81$ mm and $l_C = 7.11$ mm.

A layout of this designed microstrip filter is illustrated in Figure 5.2a, and its performance obtained by full-wave EM simulation, which is plotted in Figure 5.2b.

TABLE 5.1 Design Parameters of Microstrip Lines for a Stepped-Impedance Lowpass Filter

Characteristic impedance (Ω)	$Z_{0C} = 24$	$Z_0 = 50$	$Z_{0L} = 93$
Guided-wavelengths (mm)	$\lambda_{gC} = 105$	$\lambda_{g0} = 112$	$\lambda_{gL} = 118$
Microstrip line width (mm)	$W_C = 4.0$	$W_0 = 1.1$	$W_L = 0.2$

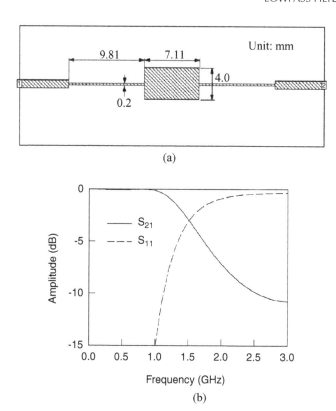

FIGURE 5.2 (a) Layout of a three-pole stepped-impedance microstrip lowpass filter on a substrate with a relative dielectric constant of 10.8 and a thickness of 1.27 mm. (b) Full-wave EM simulated performance of the filter.

5.1.2 L-C Ladder-Type of Lowpass Filters Using Open-Circuited Stubs

The previous stepped-impedance lowpass filter realizes the shunt capacitors of the lowpass prototype as low-impedance lines in the transmission path. An alternative realization of a shunt capacitor is to use an open-circuited stub subject to

$$\omega C = \frac{1}{Z_0} \tan\left(\frac{2\pi}{\lambda_g}l\right) \quad \text{for } l < \lambda_g/4 \tag{5.4}$$

where the term on the left-hand side is the susceptance of shunt capacitor, while the term on the right-hand side represents the input susceptance of open-circuited stub, which has characteristic impedance Z_0 and a physical length l that is smaller than one-quarter of the guided-wavelength λ_g.

The following example will demonstrate how to accomplish this type of microstrip lowpass filter. For comparison, the same prototype filter and the substrate for the previous design example of stepped-impedance microstrip lowpass filter is employed.

The same high-impedance ($Z_{0L} = 93\,\Omega$) lines are also used for the series inductors, while the open-circuited stub will have the same low characteristic impedance as $Z_{0C} = 24\,\Omega$. Thus, the design parameters of the microstrip lines listed in Table 5.1 are valid for this design example.

To produce the lumped L–C elements, the physical lengths of the high-impedance lines and the open-circuited stub are initially determined by

$$l_L = \frac{\lambda_{gL}}{2\pi} \sin^{-1}\left(\frac{\omega_c L}{Z_{0L}}\right) = 11.04\,\text{mm}$$

$$l_C = \frac{\lambda_{gC}}{2\pi} \tan^{-1}(\omega_c C Z_{0C}) = 8.41\,\text{mm}$$

To compensate for the unwanted susceptance resulting from the two adjacent high-impedance lines, the initial l_C should be changed to satisfy

$$\omega_c C = \frac{1}{Z_{0C}} \tan\left(\frac{2\pi l_C}{\lambda_{gC}}\right) + 2 \times \frac{1}{Z_{0L}} \tan\left(\frac{\pi l_L}{\lambda_{gL}}\right) \tag{5.5}$$

which is solved for l_C and results in $l_C = 6.28$ mm. Furthermore, the open-end effect of the open-circuited stub must also be taken into account. According to the discussions in Chapter 4, in this case, a length of $\Delta l = 0.5$ mm should be compensated for. Therefore, the final dimension of the open-circuited stub is $l_C = 6.28 - 0.5 = 5.78$ mm.

The layout and EM-simulated performance of the designed filter are given in Figure 5.3. Comparing to the filter response to that in Figure 5.2, both filters show a very similar filtering characteristic in the given frequency range, which is expected, as they are designed based on the same prototype filter. However, one should bear in mind that the two filters have different realizations that only approximate the lumped elements of the prototype in the vicinity of the cutoff frequency and, hence, their wideband frequency responses can be different, as shown in Figure 5.4. The filter using an open-circuited stub exhibits a better stopband characteristic with an attenuation peak at about 5.6 GHz. This is because at this frequency, the open-circuited stub is about one-quarter guided wavelength so as to almost short out a transmission and cause the attenuation peak.

To obtain a shaper rate of cutoff, a higher degree of filter can be designed in the same way. Figure 5.5a is a seven-pole, lumped-element lowpass filter with its microstrip implementation illustrated in Figure 5.5b. The four open-circuited stubs, which have the same line width W_C, are used to approximate the shunt capacitors; the three narrow microstrip lines of width W_L are for approximation of the series inductors. The lowpass filter is designed to have a Chebyshev response with a passband ripple of 0.1 dB and a cutoff frequency at 1.0 GHz. The lumped element values in Figure 5.5a are then given by

$$Z_0 = 50\,\Omega \qquad C_1 = C_7 = 3.7596\,\text{pF}$$
$$L_2 = L_6 = 11.322\,\text{nH} \qquad C_3 = C_5 = 6.6737\,\text{pF}$$
$$L_4 = 12.52\,\text{nH}$$

LOWPASS FILTERS 117

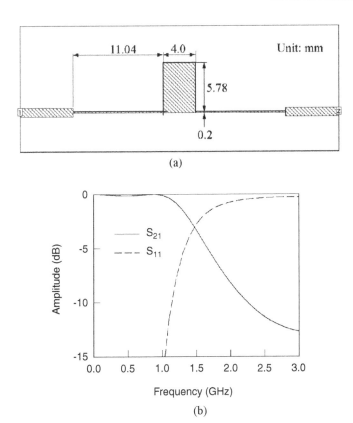

FIGURE 5.3 (a) Layout of a three-pole microstrip lowpass filter using open-circuited stubs on a substrate with a relative dielectric constant of 10.8 and a thickness of 1.27 mm. (b) Full-wave EM simulated performance of the filter.

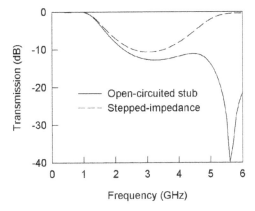

FIGURE 5.4 Comparison of wide-band frequency responses of the filters in Figures 5.2a and 5.3a.

118 LOWPASS AND BANDPASS FILTERS

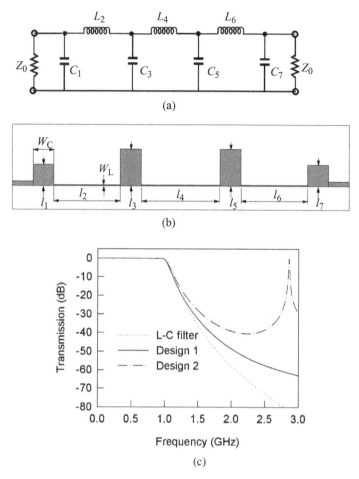

FIGURE 5.5 (a) A seven-pole lumped-element lowpass filter. (b) Microstrip realization. (c) Comparison of filter performance for the lumped-element design and the two microstrip designs given in Table 5.2.

The microstrip filter design uses a substrate having a relative dielectric constant $\varepsilon_r = 10.8$ and a thickness $h = 1.27$ mm. To demonstrate that the microstrip realization in Figure 5.5b can only approximate the ideal lumped-element filter in Figure 5.5a, two microstrip filter designs that use different characteristic impedances for the high-impedance lines are presented in Table 5.2. The first design (design 1) uses high-impedance lines that have a characteristic impedance $Z_{0L} = 110 \ \Omega$ and a line width $W_L = 0.1$ mm on the substrate used. The second design (design 2) uses a characteristic impedance $Z_{0L} = 93 \ \Omega$ and a line width $W_L = 0.2$ mm. The performance of these two microstrip filters is shown in Figure 5.5c, as compared to that of the lumped-element filter. As can be seen, the two microstrip filters not only behave differently from the lumped-element one, but also differently from each other. The main difference lies in the stopband behaviors. The microstrip filter (design 1)

LOWPASS FILTERS

TABLE 5.2 Two Microstrip Lowpass Filter Designs with Open-Circuited Stubs

Substrate ($\varepsilon_r = 10.8$, $h = 1.27$ mm) $W_C = 5$ mm	$l_1 = l_7$ (mm)	$l_2 = l_6$ (mm)	$l_3 = l_5$ (mm)	l_4 (mm)
Design 1 ($W_L = 0.1$ mm)	5.86	13.32	9.54	15.09
Design 2 ($W_L = 0.2$ mm)	5.39	16.36	8.67	18.93

that uses the narrower inductive lines ($W_L = 0.1$ mm) has a better matched stopband performance. This is because the use of the inductive lines with the higher characteristic impedance and the shorter lengths (referring to Table 5.2) achieves a better approximation of the lumped inductors. The other microstrip filter (design 2), with the wider inductive lines ($W_L = 0.2$ mm), exhibits an unwanted transmission peak at 2.86 GHz, which is because of its longer inductive lines being about half-wavelength and resonating at about this frequency.

5.1.3 Semilumped Lowpass Filters Having Finite-Frequency Attenuation Poles

The previous two types of microstrip lowpass filter represent the lowpass prototype filters having their frequencies of infinite attenuation at $f = \infty$. In order to obtain an even sharper rate of cutoff for a given number of reactive elements, it is desirable to use filter structures giving infinite attenuation at finite frequencies. A prototype of this type may have an elliptic-function response, as discussed in Chapter 3. Figure 5.6a shows an elliptic-function lowpass filter that has two series-resonant branches connected in a shunt that shorts out transmission at their resonant frequencies, and thus gives two finite-frequency attenuation poles. Note that at $f = \infty$ these two branches have no effect and the inductances L_1, L_3, and L_5 block transmission by having infinite series

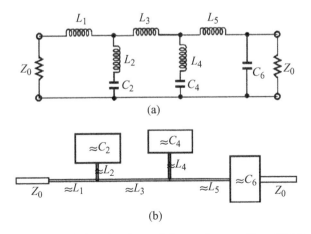

FIGURE 5.6 (a) An elliptic-function, lumped-element lowpass filter. (b) Microstrip realization of the elliptic-function lowpass filter.

reactance, whereas the capacitance C_6 shorts out transmission by having infinite shunt susceptance.

A microstrip filter structure that can approximately represent such a filtering characteristic is illustrated in Figure 5.6b, which is much the same as that for the stripline realization in Mattaei et al. [1]. Similar to the stepped-impedance microstrip filters described in Section 5.1, the lumped L-C elements in Figure 5.6a are approximated using short lengths of high- and low-impedance lines, the actual dimensions of the lines of which are determined in a similar way to that discussed previously. A design example is described below.

The element values for elliptic function lowpass prototype filters may be obtained from Table 3.3 or from Saal and Umbrich [2] and Saal [3]. For this example, we use the lowpass prototype element values

$$g_0 = g_7 = 1.000 \qquad g_{L4} = g'_4 = 0.7413$$
$$g_{L1} = g_1 = 0.8214 \qquad g_{C4} = g_4 = 0.9077$$
$$g_{L2} = g'_2 = 0.3892 \qquad g_{L5} = g_5 = 1.1170$$
$$g_{C2} = g_2 = 1.0840 \qquad g_{C6} = g_6 = 1.1360$$
$$g_{L3} = g_3 = 1.1880$$

where we use g_{Li} and g_{Ci} to denote the inductive and capacitive elements, respectively. This prototype filter has a passband ripple $L_{Ar} = 0.18$ dB and a minimum stopband attenuation $L_{As} = 38.1$ dB at $\Omega_s = 1.194$ for the cutoff $\Omega_c = 1.0$ [2]. The microstrip filter is designed to have a cutoff frequency $f_c = 1.0$ GHz and input/output terminal impedance $Z_0 = 50\ \Omega$. Therefore, the L-C element values, which are scaled to Z_0 and f_c, can be determined by

$$L_i = \frac{1}{2\pi f_c} Z_0 g_{Li}$$
$$C_i = \frac{1}{2\pi f_c} \frac{1}{Z_0} g_{Ci} \tag{5.6}$$

This yields

$$L_1 = 6.53649\text{ nH} \qquad L_2 = 3.09716\text{ nH}$$
$$L_3 = 9.45380\text{ nH} \qquad C_2 = 3.45048\text{ pF}$$
$$L_5 = 8.88880\text{ nH} \qquad L_4 = 5.89908\text{ nH} \tag{5.7}$$
$$C_6 = 3.61600\text{ pF} \qquad C_4 = 2.88930\text{ pF}$$

The two finite-frequency attenuation poles occur at

$$f_{p1} = \frac{1}{2\pi\sqrt{L_4 C_4}} = 1.219\text{ GHz}$$
$$f_{p2} = \frac{1}{2\pi\sqrt{L_2 C_2}} = 1.540\text{ GHz} \tag{5.8}$$

LOWPASS FILTERS

TABLE 5.3 Microstrip Design Parameters for an Elliptic-Function Lowpass Filter

Characteristic impedance (Ω)	$Z_{0C} = 14$	$Z_0 = 50$	$Z_{0L} = 93$
Microstrip line width (mm)	$W_C = 8.0$	$W_0 = 1.1$	$W_L = 0.2$
Guided wavelength (mm) at f_c	$\lambda_{gC}(f_c) = 101$	$\lambda_{g0} = 112$	$\lambda_{gL}(f_c) = 118$
Guided wavelength (mm) at f_{p1}	$\lambda_{gC}(f_{p1}) = 83$		$\lambda_{gL}(f_{p1}) = 97$
Guided wavelength (mm) at f_{p2}	$\lambda_{gC}(f_{p2}) = 66$		$\lambda_{gL}(f_{p2}) = 77$

For microstrip representation, a substrate with a relative dielectric constant of 10.8 and a thickness of 1.27 mm is assumed. All inductors will be realized using high-impedance lines with characteristic impedance $Z_{0L} = 93\ \Omega$, whereas the all capacitors are realized using low-impedance lines with characteristic impedance $Z_{0C} = 14\ \Omega$. Table 5.3 lists all relevant microstrip design parameters calculated using the microstrip design equations presented in the Chapter 4.

Initial physical lengths of the high- and low-impedance lines for accomplishing the desired L–C elements can be determined according to the design equations

$$l_{Li} = \frac{\lambda_{gL}(f_c)}{2\pi} \sin^{-1}\left(2\pi f_c \frac{L_i}{Z_{0L}}\right)$$

$$l_{Ci} = \frac{\lambda_{gC}(f_c)}{2\pi} \sin^{-1}(2\pi f_c Z_{0C} C_i)$$

(5.9)

Substituting the corresponding parameters from (5.7) and Table 5.3 results in

$$l_{L1} = 8.59 \quad l_{L2} = 3.96$$
$$l_{L3} = 13.01 \quad l_{C2} = 4.96$$
$$l_{L5} = 12.10 \quad l_{L4} = 7.70$$
$$l_{C6} = 5.20 \quad l_{C4} = 4.13$$

where the all dimensions are in millimeters. To achieve a more accurate design, compensations are required for some unwanted reactance/susceptance and microstrip discontinuities.

To compensate for the unwanted reactance and susceptance presented at the junction of the microstrip line elements for L_5 and C_6, the lengths l_{L5} and l_{C6} may be corrected by solving a pair of equations

$$2\pi f_c L_5 = Z_{0L} \sin\left(\frac{2\pi l_{L5}}{\lambda_{gL}(f_c)}\right) + Z_{0C} \tan\left(\frac{\pi l_{C6}}{\lambda_{gC}(f_c)}\right)$$

$$2\pi f_c C_6 = \frac{1}{Z_{0C}} \sin\left(\frac{2\pi l_{C6}}{\lambda_{gC}(f_c)}\right) + \frac{1}{Z_{0L}} \tan\left(\frac{\pi l_{L5}}{\lambda_{gL}(f_c)}\right)$$

(5.10)

which yields $l_{L5} = 11.62$ mm and $l_{C6} = 4.39$ mm.

The compensation for the unwanted reactance/susceptance at the junction of the inductive line elements for L_1, L_2, and L_3, as well as at the junction of the line elements for L_2 and C_2, may be achieved by correcting l_{L2} and l_{C2} while keeping l_{L1} and l_{L3} unchanged so that

$$\frac{1}{(2\pi f L_2) - 1/(2\pi f C_2)} = B_2(f) + \Delta B_{123}(f) \quad \text{for } f = f_c \text{ and } f_{p2} \quad (5.11)$$

where the term on the left-hand side is the desired susceptance of the series-resonant branch formed by L_2 and C_2 and, on the right-hand side, $B_2(f)$, denoting a "compensated" susceptance formed by the line elements for L_2 and C_2, is given by

$$B_2(f) = \frac{1}{Z_{0L} \sin\left(\frac{2\pi l_{L2}}{\lambda_{gL}(f)}\right) + Z_{0C} \tan\left(\frac{\pi l_{C2}}{\lambda_{gC}(f)}\right) - \frac{1}{\frac{1}{Z_{0C}} \sin\left(\frac{2\pi l_{C2}}{\lambda_{gC}(f)}\right) + \frac{1}{Z_{0L}} \tan\left(\frac{\pi l_{L2}}{\lambda_{gL}(f)}\right)}}$$

The unwanted ΔB_{123} represents the total equivalent susceptance because of the three inductive line elements; it is evaluated by

$$\Delta B_{123}(f) = \frac{1}{Z_{0L}} \tan\left(\frac{\pi l_{L1}}{\lambda_{gL}(f)}\right) + \frac{1}{Z_{0L}} \tan\left(\frac{\pi l_{L2}}{\lambda_{gL}(f)}\right) + \frac{1}{Z_{0L}} \tan\left(\frac{\pi l_{L3}}{\lambda_{gL}(f)}\right)$$

Note that the Eq. (5.11) is solved at the cutoff frequency f_c and the desired attenuation pole frequency f_{p2} for l_{L2} and l_{C2}. The solutions are found to be $l_{L2} = 2.98$ mm and $l_{C2} = 5.61$ mm.

The compensation for the unwanted reactance/susceptance at the junction of the inductive-line elements for L_3, L_4, and L_5, as well as at the junction of the line elements for L_4 and C_4, can be done in the same way as the above. This results in the corrected lengths $l_{L4} = 6.49$ mm and $l_{C4} = 4.24$ mm.

To correct for the fringing capacitance at the ends of the line elements for C_2 and C_4, the open-end effect is calculated using the equations presented in Chapter 4 and found to be $\Delta l = 0.54$ mm. We need to subtract Δl from the above-determined l_{C2} and l_{C4}, which gives $l_{C2} = 5.61 - 0.54 = 5.07$ mm and $l_{C4} = 4.24 - 0.54 = 3.70$ mm.

The layout of the microstrip filter with the final design dimensions is given in Figure 5.7a. The design is verified by full-wave EM simulation; the simulated frequency response of this microstrip filter is illustrated in Figure 5.7b, showing the two attenuation poles near the cutoff frequency, which result in a sharp rate of cutoff. It is also shown that a spurious transmission peak occurs at about 2.81 GHz. This unwanted transmission peak could be moved away up to a higher frequency if higher characteristic impedance could be used for the inductive lines.

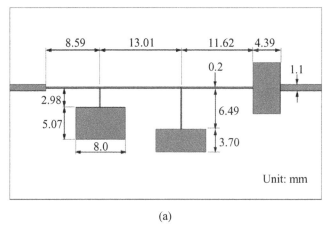

(a)

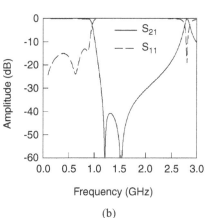

(b)

FIGURE 5.7 (a) Layout of the designed microstrip elliptic-function lowpass filter on a substrate with a relative dielectric constant of 10.8 and a thickness of 1.27 mm. (b) Full-wave EM simulated performance of the filter.

5.2 BANDPASS FILTERS

5.2.1 End-Coupled Half-Wavelength Resonator Filters

The general configuration of an end-coupled microstrip bandpass filter is illustrated in Figure 5.8, where each open-end microstrip resonator is approximately a half-guided wavelength long at the midband frequency f_0 of the bandpass filter. The coupling from one resonator to the other is through the gap between the two adjacent open ends and, hence, is capacitive. In this case, the gap can be represented by the inverters, which are of the form in Figure 3.22(d). These J inverters tend to reflect high impedance levels to the ends of each of the half-wavelength resonators and it can be shown that this causes the resonators to exhibit a shunt-type resonance [1].

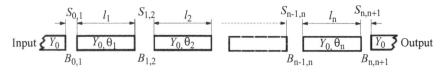

FIGURE 5.8 General configuration of end-coupled microstrip bandpass filter.

Thus, the filter under consideration operates like the shunt-resonator type of filter whose general design equations are given as follows:

$$\frac{J_{01}}{Y_0} = \sqrt{\frac{\pi}{2} \frac{FBW}{g_0 g_1}} \tag{5.12a}$$

$$\frac{J_{j,j+1}}{Y_0} = \frac{\pi FBW}{2} \frac{1}{\sqrt{g_j g_{j+1}}}, j = 1 \text{ to } n-1 \tag{5.12b}$$

$$\frac{J_{n,n+1}}{Y_0} = \sqrt{\frac{\pi FBW}{2 g_n g_{n+1}}} \tag{5.12c}$$

where $g_0, g_1 \ldots g_n$ are the element of a ladder-type lowpass prototype with a normalized cutoff $\Omega_c = 1$, and FBW is the fractional bandwidth of bandpass filter as defined in Chapter 3. The $J_{j,j+1}$ are the characteristic admittances of J inverters and Y_0 is the characteristic admittance of the microstrip line.

Assuming the capacitive gaps act as perfect, series-capacitance discontinuities of susceptance $B_{j,j+1}$, as in Figure 3.22d

$$\frac{B_{j,j+1}}{Y_0} = \frac{\frac{J_{j,j+1}}{Y_0}}{1 - \left(\frac{J_{j,j+1}}{Y_0}\right)^2} \tag{5.13}$$

and

$$\theta_j = \pi - \frac{1}{2}\left[\tan^{-1}\left(\frac{2B_{j-1,j}}{Y_o}\right) + \tan^{-1}\left(\frac{2B_{j,j+1}}{Y_o}\right)\right] \quad \text{radians} \tag{5.14}$$

where the $B_{j,j+1}$ and θ_j are evaluated at f_0. Note that the second term on the right-hand side of Eq. (5.14) indicates the absorption of the negative electrical lengths of the J inverters associated with the jth half-wavelength resonator.

By referring to the equivalent circuit of microstrip gap in Figure 4.4c, the coupling gaps $s_{j,j+1}$ of the microstrip end-coupled resonator filter can be determined as to obtain the series capacitances that satisfy

$$C_g^{j,j+1} = \frac{B_{j,j+1}}{\omega_0} \tag{5.15}$$

where $\omega_0 = 2\pi f_0$ is the angular frequency at the midband. The physical lengths of resonators are given by

$$l_j = \frac{\lambda_{g0}}{2\pi}\theta_j - \Delta l_j^{e1} - \Delta l_j^{e2} \tag{5.16}$$

where $\Delta l_j^{e1,e2}$ are the effective lengths of the shunt capacitances on the both ends of resonator j. Because the shunt capacitances $C_p^{j,j+1}$ are associated with the series capacitances $C_g^{j,j+1}$, as defined in the equivalent circuit of microstrip gap, they are also determined once $C_g^{j,j+1}$ in Eq. (5.15) are solved for the required coupling gaps. The effective lengths can then found by

$$\Delta l_j^{e1} = \frac{\omega_0 C_p^{j-1,j}}{Y_0} \frac{\lambda_{g0}}{2\pi}$$

$$\Delta l_j^{e2} = \frac{\omega_0 C_p^{j,j+1}}{Y_0} \frac{\lambda_{g0}}{2\pi} \tag{5.17}$$

Design Example

As an example, a microstrip end-coupled bandpass filter is designed to have a fractional bandwidth $FBW = 0.028$ or 2.8% at the midband frequency $f_0 = 6$ GHz. A three-pole ($n = 3$) Chebyshev lowpass prototype with 0.1-dB passband ripple is chosen, whose element values are $g_0 = g_4 = 1.0$, $g_1 = g_3 = 1.0316$, and $g_2 = 1.1474$. From Eq. (5.12) we have

$$\frac{J_{01}}{Y_0} = \frac{J_{3,4}}{Y_0} = \sqrt{\frac{\pi}{2} \times \frac{0.028}{1.0 \times 1.0316}} = 0.2065$$

$$\frac{J_{1,2}}{Y_0} = \frac{J_{2,3}}{Y_0} = \frac{\pi \times 0.028}{2} \frac{1}{\sqrt{1.0316 \times 1.1474}} = 0.0404$$

The susceptances associated with the J inverters are calculated from Eq. (5.13)

$$\frac{B_{0,1}}{Y_0} = \frac{B_{3,4}}{Y_0} = \frac{0.2065}{1 - (0.2065)^2} = 0.2157$$

$$\frac{B_{1,2}}{Y_0} = \frac{B_{2,3}}{Y_0} = \frac{0.0404}{1 - (0.0404)^2} = 0.0405$$

The electrical lengths of the half-wavelength resonators after absorbing the negative electrical lengths attributed to the J inverters are determined by Eq. (5.14)

$$\theta_1 = \theta_3 = \pi - \frac{1}{2}\left[\tan^{-1}(2 \times 0.2157) + \tan^{-1}(2 \times 0.0405)\right] = 2.8976 \text{ radians}$$

$$\theta_2 = \pi - \frac{1}{2}\left[\tan^{-1}(2 \times 0.0405) + \tan^{-1}(2 \times 0.0405)\right] = 3.0608 \text{ radians} \tag{5.18}$$

Using Eq. (5.15) we obtain the coupling capacitances

$$C_g^{0,1} = C_g^{3,4} = 0.11443 \, \text{pF}$$
$$C_g^{1,2} = C_g^{2,3} = 0.021483 \, \text{pF}$$
(5.19)

For microstrip implementation, we use a substrate with a relative dielectric constant $\varepsilon_r = 10.8$ and a thickness $h = 1.27$ mm. The line width for microstrip half-wavelength resonators is also chosen as $W = 1.1$ mm, which gives characteristic impedance $Z_0 = 50 \, \Omega$ on the substrate. To determine the other physical dimensions of the microstrip filter, such as the coupling gaps, we need to find the desired coupling capacitances $C_g^{j,j+1}$ given in Eq. (5.19) in terms of gap dimensions. To do so, we might have used the closed-form expressions for microstrip gap given in Chapter 4. However, the dimensions of the coupling gaps for the filter seem to be outside the parameter range available for these closed-form expressions. This very often will be the case when we design this type of microstrip filter. We will describe next how to utilize the EM simulation (see Chapter 8) to complete the filter design of this type.

In principle, any EM simulator can simulate the two-port network parameters of a microstrip gap without restricting any of its physical parameters, such as the substrate, the line width, or the dimension of the gap. Figure 5.9 shows a layout of a microstrip gap for EM simulation, where arrows indicate the reference planes for de-embedding to obtain the two-port parameters of the microstrip gap. Assume that the two-port parameters obtained by the EM simulation are the Y parameters given by

$$[Y] = \begin{bmatrix} Y_{11} & Y_{12} \\ Y_{21} & Y_{22} \end{bmatrix}$$

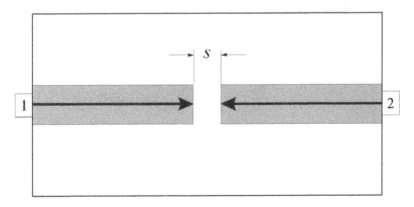

FIGURE 5.9 Layout of a microstrip gap for EM simulation.

TABLE 5.4 Characterization of Microstrip Gaps with Line Width $W = 1.1$ mm on the Substrate with $\varepsilon_r = 10.8$ and $h = 1.27$ mm

s(mm)	$Y_{11} = Y_{22}$ (mhos) at 6 GHz	$Y_{12} = Y_{21}$ (mhos) at 6 GHz	C_g (pF)	C_p (pF)
0.05	j0.0045977	$-j$0.004434	0.11762	0.00434
0.1	j0.0039240	$-j$0.003604	0.09560	0.00849
0.2	j0.0032933	$-j$0.0026908	0.07138	0.01598
0.5	j0.0026874	$-j$0.0014229	0.03774	0.03354
0.8	j0.0025310	$-j$0.00081105	0.02151	0.04562
1.0	j0.0024953	$-j$0.00055585	0.01474	0.05145
1.5	j0.0024808	$-j$0.0001876	0.00498	0.06083

The capacitances C_g and C_p appear in the equivalent π-network as shown in Figure 4.4c may be determined on a narrow-band basis by

$$C_g = -\frac{\text{Im}(Y_{21})}{\omega_0}$$
$$C_p = \frac{\text{Im}(Y_{11} + Y_{21})}{\omega_0} \quad (5.20)$$

where ω_0 is the filter midband angular frequency used in the simulation and $\text{Im}(x)$ denotes the imaginary part of x. If the microstrip gap simulated is lossless, the real parts of the Y parameters are actually zero.

For this filter design example, the simulated Y parameters at 6 GHz and the extracted capacitances based on Eq. (5.20) are listed in Table 5.4 compared to the microstrip gaps. Interpolating the data in Table 5.4, we can determine the dimensions $s_{j,j+1}$ of the microstrip gaps that produce the desired capacitances given in Eq. (5.19). The results of this are

$$s_{0,1} = s_{3,4} = 0.057 \text{ mm}$$
$$s_{1,2} = s_{2,3} = 0.801 \text{ mm}$$

Also, by interpolation, the shunt capacitances associated with these gaps are found to be

$$C_p^{0,1} = C_p^{3,4} = 0.0049 \text{ pF}$$
$$C_p^{1,2} = C_p^{2,3} = 0.0457 \text{ pF}$$

At the midband frequency, $f_0 = 6$ GHz, the guided-wavelength of the microstrip line resonators is $\lambda_{g0} = 18.27$ mm. The effective lengths of the shunt capacitances are calculated using Eq. (5.17)

$$\Delta l_1^{e1} = \Delta l_3^{e2} = \frac{2\pi \times 6 \times 10^9 \times 0.0049 \times 10^{-12}}{(1/50)} \frac{18.27}{2\pi} = 0.0269 \text{ mm}$$

$$\Delta l_1^{e2} = \Delta l_3^{e1} = \frac{2\pi \times 6 \times 10^9 \times 0.0457 \times 10^{-12}}{(1/50)} \frac{18.27}{2\pi} = 0.2505 \text{ mm}$$

$$\Delta l_2^{e1} = \Delta l_2^{e2} = \Delta l_1^{e2}$$

Finally, the physical lengths of the resonators are found by substituting the above effective lengths and the electrical lengths θ_j determined in Eqs. (5.18) into (5.16). This results in

$$l_1 = l_3 = \frac{18.27}{2\pi} \times 2.8976 - 0.0269 - 0.2505 = 8.148 \text{ mm}$$

$$l_2 = \frac{18.27}{2\pi} \times 3.0608 - 0.2505 - 0.2505 = 8.399 \text{ mm}$$

The design of the filter is completed, and the layout of the filter is given in Figure 5.10a with all the determined dimensions. Figure 5.10b shows the EM simulated performance of the filter.

5.2.2 Parallel-Coupled Half-Wavelength Resonator Filters

Figure 5.11 illustrates a general structure of parallel-coupled (or edge-coupled) microstrip bandpass filters, which use half-wavelength line resonators. They are positioned so that adjacent resonators are parallel to each other along one-half of their length. This parallel arrangement gives relatively large coupling for a given spacing between resonators and, thus, this filter structure is particularly convenient for constructing filters having a wider bandwidth as compared to the structure for the end-coupled microstrip filters described in the last section. The design equations [1] for this type of filter are given by

$$\frac{J_{01}}{Y_0} = \sqrt{\frac{\pi \, FBW}{2 \, g_0 g_1}} \quad (5.21\text{a})$$

$$\frac{J_{j,j+1}}{Y_0} = \frac{\pi \, FBW}{2} \frac{1}{\sqrt{g_j g_{j+1}}} \quad j = 1 \text{ to } n-1 \quad (5.21\text{b})$$

$$\frac{J_{n,n+1}}{Y_0} = \sqrt{\frac{\pi \, FBW}{2 g_n g_{n+1}}} \quad (5.21\text{c})$$

where $g_0, g_1 \ldots g_n$ are the element of a ladder-type lowpass prototype with a normalized cutoff $\Omega_c = 1$, and FBW is the fractional bandwidth of bandpass filter, as defined in Chapter 3. $J_{j,j+1}$ are the characteristic admittances of J inverters and Y_0 is the

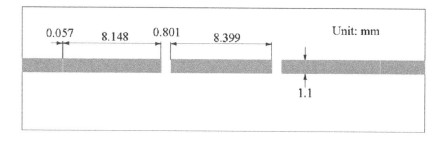

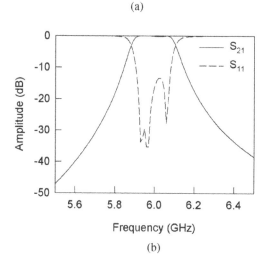

FIGURE 5.10 (a) Layout of the three-pole microstrip, end-coupled half-wavelength resonator filter on a substrate with a relative dielectric constant of 10.8 and a thickness of 1.27 mm. (b) Full-wave EM simulated frequency response of the filter.

characteristic admittance of the terminating lines. One might note that Eq. (5.21) is actually the same as Eq. (5.12). The reason for this is because the both types of filter can have the same lowpass network representation. However, the implementation will be different. To realize the J inverters obtained above, the even- and odd-mode characteristic impedances of the coupled microstrip line resonators are determined by

$$(Z_{0e})_{j,j+1} = \frac{1}{Y_0}\left[1 + \frac{J_{j,j+1}}{Y_0} + \left(\frac{J_{j,j+1}}{Y_0}\right)^2\right] \text{ for } j = 0 \text{ to } n \quad (5.22a)$$

$$(Z_{0o})_{j,j+1} = \frac{1}{Y_0}\left[1 - \frac{J_{j,j+1}}{Y_0} + \left(\frac{J_{j,j+1}}{Y_0}\right)^2\right] \text{ for } j = 0 \text{ to } n \quad (5.22b)$$

The use of the design equations and the implementation of microstrip filter of this type are best illustrated by use of an example. Let us consider a design of five-pole ($n = 5$) microstrip bandpass filter that has a fractional bandwidth $FBW = 0.15$ at a

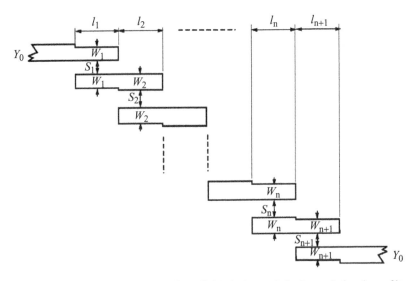

FIGURE 5.11 General structure of parallel (edge)-coupled microstrip bandpass filter.

midband frequency $f_0 = 10$ GHz. Suppose a Chebyshev prototype with a 0.1-dB ripple is to be used in the design. The desired $n = 5$ prototype parameters are

$$g_0 = g_6 = 1.0 \qquad g_1 = g_5 = 1.1468$$
$$g_2 = g_4 = 1.3712 \qquad g_3 = 1.9750$$

The calculations using Eqs. (5.21) and (5.22) yield the design parameters, one-half of which are listed in Table 5.5, because of symmetry of the filter, where the even- and odd-mode impedances are calculated for $Y_0 = 1/Z_0$ and $Z_0 = 50$ Ω.

The next step of the filter design is to find the dimensions of coupled microstrip lines that exhibit the desired even- and odd-mode impedances. For instance, referring to Figure 5.11, W_1 and s_1 are determined such that the resultant even- and odd-mode impedances match $(Z_{0e})_{0,1}$ and $(Z_{0o})_{0,1}$. Assume that the microstrip filter is constructed on a substrate with a relative dielectric constant of 10.2 and thickness of 0.635 mm. Using the design equations for coupled microstrip lines given in Chapter 4, the width and spacing for each pair of quarter-wavelength coupled sections are found;

TABLE 5.5 Circuit Design Parameters of the Five-Pole Parallel-Coupled Half-Wavelength Resonator Filter

j	$J_{j,j+1}/Y_0$	$(Z_{0e})_{j,j+1}$	$(Z_{0o})_{j,j+1}$
0	0.4533	82.9367	37.6092
1	0.1879	61.1600	42.3705
2	0.1432	58.1839	43.8661

BANDPASS FILTERS

TABLE 5.6 Microstrip Design Parameters of the Five-Pole Parallel-Coupled Half-Wavelength Resonator Filter

j	W_j (mm)	s_j (mm)	$(\varepsilon_{re})_j$	$(\varepsilon_{ro})_j$
1 and 6	0.385	0.161	6.5465	5.7422
2 and 5	0.575	0.540	6.7605	6.0273
3 and 4	0.595	0.730	6.7807	6.1260

they are listed in Table 5.6, together with the effective dielectric constants of even mode and odd mode.

The actual length of each coupled line section are then determined by

$$l_j = \frac{\lambda_0}{4\left(\sqrt{(\varepsilon_{re})_j \times (\varepsilon_{ro})_j}\right)^{1/2}} - \Delta l_j \tag{5.23}$$

where Δl_j is the equivalent length of microstrip open end, as discussed in Chapter 4. The final filter layout with all the determined dimensions is illustrated in Figure 5.12a. The EM simulated frequency responses of the filter are plotted in Figure 5.12b.

5.2.3 Hairpin-Line Bandpass Filters

Hairpin-line bandpass filters are compact structures. They may conceptually be obtained by folding the resonators of parallel-coupled, half-wavelength resonator filters, which are discussed in the previous section, into a "U" shape. This type of "U-"shape resonator is the so-called hairpin resonator. Consequently, the same design equations for the parallel-coupled, half-wavelength resonator filters may be used [4]. However, to fold the resonators, it is necessary to take into account the reduction of the coupled-line lengths, which reduces the coupling between resonators. If the two arms of each hairpin resonator are also closely spaced, they function as a pair of coupled lines, which can also have an effect on the coupling. To design this type of filter more accurately, a design approach employing full-wave EM simulation will be described.

For this design example, a microstrip hairpin bandpass filter is designed to have a fractional bandwidth 20% or $FBW = 0.2$ at a midband frequency $f_0 = 2$ GHz. A five-pole ($n = 5$) Chebyshev lowpass prototype with a passband ripple of 0.1 dB is chosen. The lowpass prototype parameters, given for a normalized lowpass cutoff frequency $\Omega_c = 1$, are $g_0 = g_6 = 1.0$, $g_1 = g_5 = 1.1468$, $g_2 = g_4 = 1.3712$, and $g_3 = 1.9750$. Having obtained the lowpass parameters, the bandpass design parameters can be calculated by

$$Q_{e1} = \frac{g_0 g_1}{FBW}, \quad Q_{en} = \frac{g_n g_{n+1}}{FBW}$$
$$M_{i,i+1} = \frac{FBW}{\sqrt{g_i g_{i+1}}} \quad \text{for } i = 1 \text{ to } n-1 \tag{5.24}$$

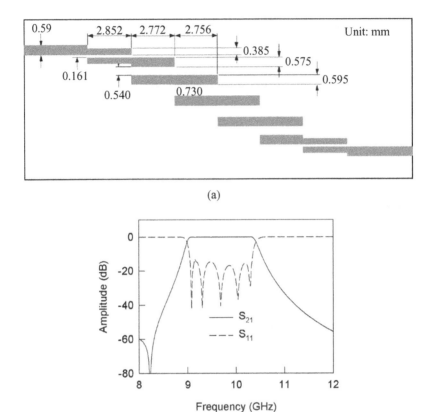

FIGURE 5.12 (a) Layout of a five-pole microstrip bandpass filter designed on a substrate with a relative dielectric constant of 10.2 and a thickness of 0.635 mm. (b) Frequency responses of the filter obtained by full-wave EM simulations.

where Q_{e1} and Q_{en} are the external quality factors of the resonators at the input and output and $M_{i,i+1}$ are the coupling coefficients between the adjacent resonators (referring to Chapter 7).

For this design example, we have

$$Q_{e1} = Q_{e5} = 5.734$$
$$M_{1,2} = M_{4,5} = 0.160 \tag{5.25}$$
$$M_{2,3} = M_{3,4} = 0.122$$

We use a commercial substrate (RT/D 6006) with a relative dielectric constant of 6.15 and a thickness of 1.27 mm for microstrip representation. Using a parameter-extraction technique described in Chapter 7, we then carry out full-wave EM

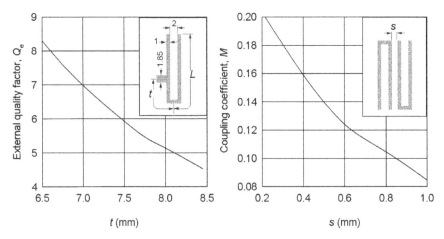

FIGURE 5.13 Design curves obtained by full-wave EM simulations for design of a hairpin-line microstrip bandpass filter. (a) External quality factor. (b) Coupling coefficient.

simulations to extract the external Q and coupling coefficient M against the physical dimensions. Two design curves obtained in this way are plotted in Figure 5.13. It should be noted that the hairpin resonators used have a line width of 1 mm and a separation of 2 mm between the two arms, as indicated by a small drawing inserted in Figure 5.13a. Another dimension of the resonator as indicated by L is about $\lambda_{g0}/4$ long with λ_{g0} the guided wavelength at the midband frequency and, in this case, $L = 20.4$ mm. The filter is designed to have tapped line input and output. The tapped line is chosen to have characteristic impedance that matches a terminating impedance $Z_0 = 50$ Ω. Hence, the tapped line is 1.85 mm wide on the substrate. Also in Figure 5.13a, the tapping location is denoted by t, and the design curve gives the value of external quality factor, Q_e, as a function of t. In Figure 5.13b, the value of coupling coefficient M is given against the coupling spacing (denoted by s) between two adjacent hairpin resonators with the opposite orientations as shown. The required external Q and coupling coefficients as designed in Eq. (5.25) can be read off the two design curves above and the filter designed.

The layout of the final filter design with all the determined dimensions is illustrated in Figure 5.14a. The filter is quite compact with a substrate size of 31.2 by 30 mm. The input and output resonators are slightly shortened to compensate for the effect of the tapping line and the adjacent coupled resonator. The EM simulated performance of the filter is shown in Figure 5.14b.

An experimental hairpin filter of this type has been demonstrated [5], where a design equation is proposed for estimating the tapping point t as

$$t = \frac{2L}{\pi} \sin^{-1}\left(\sqrt{\frac{\pi}{2} \frac{Z_0/Z_r}{Q_e}}\right) \quad (5.26)$$

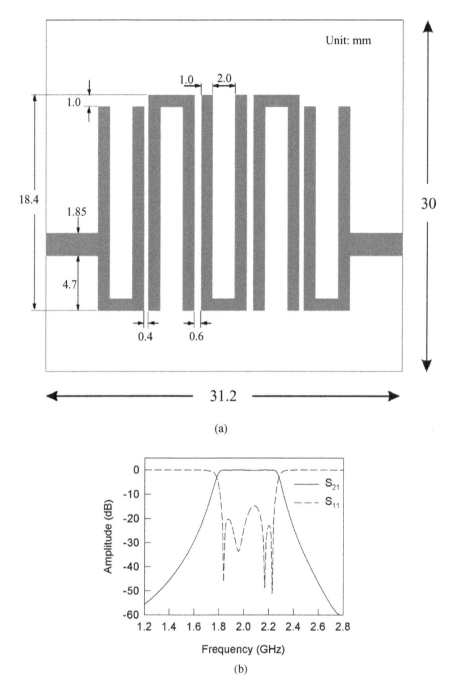

FIGURE 5.14 (a) Layout of a five-pole, hairpin-line microstrip bandpass filter on a 1.27-mm thick substrate with a relative dielectric constant of 6.15. (b) Full-wave simulated performance of the filter.

in which, Z_r is the characteristic impedance of the hairpin line, Z_0 is the terminating impedance, and L is about $\lambda_{g0}/4$ long, as mentioned above. This design equation ignores the effect of discontinuity at the tapped point, as well as the effect of coupling between the two folded arms. Nevertheless, it gives a good estimation. For instance, in the filter design example above, the hairpin line is 1.0-mm wide, which results in $Z_r = 68.3$ Ω on the substrate used. Recall that $L = 20.4$ mm, $Z_0 = 50$ Ω, and the required $Q_e = 5.734$. Substituting them into Eq. (5.26) yields $t = 6.03$ mm, which is close to the t of 7.625 mm found from the EM simulation above.

5.2.4 Interdigital Bandpass Filters

Figure 5.15 shows a type of interdigital bandpass filter commonly used for microstrip implementation. The filter configuration, as shown, consists of an array of n TEM-mode or quasi-TEM-mode transmission-line resonators, each of which has an electrical length of 90° at the midband frequency and is short-circuited at one end and open-circuited at the other end with alternative orientation. In general, the physical dimensions of the line elements or the resonators can be different, as indicated by the lengths $l_1, l_2 \cdots l_n$ and the widths $W_1, W_2 \cdots W_n$. Coupling is achieved by the fields fringing between adjacent resonators separated by spacing $s_{i,i+1}$ for $i = 1 \cdots n-1$. The filter input and output use tapped lines with a characteristic admittance Y_t, which may be set to equal the source/load characteristic admittance Y_0. An electrical length θ_t, measured away from the short-circuited end of the input/output resonator,

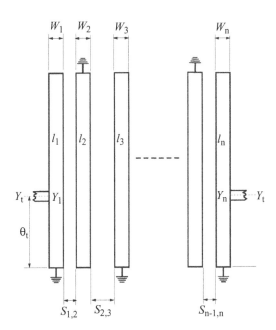

FIGURE 5.15 General configuration of interdigital bandpass filter.

indicates the tapping position, where $Y_1 = Y_n$ denotes the single microstrip characteristic impedance of the input/output resonator.

This type of microstrip bandpass filter is compact, but requires use of grounding microstrip resonators, which is usually accomplished with via-holes. However, because the resonators are quarter-wavelength long using the grounding, the second passband of filter is centered at about three times the midband frequency of the desired first pass band and there is no possibility of any spurious response in between. For the filters with parallel-coupled, half-wavelength resonators, described in the previous section, a spurious pass band at around twice the midband frequency is almost always excited.

Original theory and design procedure for interdigital bandpass filters with coupled-line input/output (I/O) can be found [6]. Explicit design equations based on Caspi and Adelman [7] for the type of bandpass filter with tapped-line I/O in Figure 5.15 are given by

$$\theta = \frac{\pi}{2}\left(1 - \frac{FBW}{2}\right), \quad Y = \frac{Y_1}{\tan\theta}$$

$$J_{i,i+1} = \frac{Y}{\sqrt{g_i g_{i+1}}} \quad \text{for } i = 1 \text{ to } n-1$$

$$Y_{i,i+1} = J_{i,i+1}\sin\theta \quad \text{for } i = 1 \text{ to } n-1$$

$$C_1 = \frac{Y_1 - Y_{1,2}}{v}, \quad C_n = \frac{Y_1 - Y_{n-1,n}}{v}$$

$$C_i = \frac{Y_1 - Y_{i-1,i} - Y_{i,i+1}}{v} \quad \text{for } i = 2 \text{ to } n-1$$

$$C_{i,i+1} = \frac{Y_{i,i+1}}{v} \quad \text{for } i = 1 \text{ to } n-1$$

$$Y_t = Y_1 - \frac{Y_{1,2}^2}{Y_1}$$

$$\theta_t = \frac{\sin^{-1}\left(\sqrt{\dfrac{Y\sin^2\theta}{Y_0 g_0 g_1}}\right)}{1 - \dfrac{FBW}{2}} \tag{5.27}$$

$$C_t = \frac{\cos\theta_t \sin^3\theta_t}{\omega_0 Y_t \left(\dfrac{1}{Y_0^2} + \dfrac{\cos^2\theta_t \sin^2\theta_t}{Y_t^2}\right)}$$

where FBW is the fractional bandwidth and g_i represents the element values of a ladder-type lowpass prototype filter with a normalized cutoff frequency at $\Omega_c = 1$. $C_i (i = 1 \text{ to } n)$ are the self-capacitances per unit length for the line elements, whereas $C_{i,i+1}(i = 1 \text{ to } n - 1)$ are the mutual capacitances per unit length between adjacent line elements. Note that v denotes the wave-phase velocity in the medium of propagation. The physical dimensions of the line elements may then be found from the required self- and mutual capacitances. C_t is the capacitance to be loaded to the input and output resonators in order to compensate for resonant frequency shift because of the effect of the tapped input and output.

It may also be desirable to use the even- and odd-mode impedances for filter designs. The self- and mutual capacitances per unit length of a pair of parallel-coupled lines denoted by a and b may be related to the line characteristic admittances and impedances [6] by

$$Y_{0e}^a = vC_a, \qquad Y_{0o}^a = v(C_a + 2C_{ab})$$
$$Y_{0e}^b = vC_b, \qquad Y_{0o}^b = v(C_b + 2C_{ab})$$
$$Z_{0o}^a = \frac{C_b}{vF}, \qquad Z_{0e}^a = \frac{C_b + 2C_{ab}}{vF} \qquad (5.28)$$
$$Z_{0o}^b = \frac{C_a}{vF}, \qquad Z_{0e}^b = \frac{C_a + 2C_{ab}}{vF}$$
$$F = C_a C_b + C_{ab}(C_a + C_b)$$

In order to obtain the desired even- and odd-mode impedances, the coupled lines in association with adjacent coupled resonators will, in general, have different line widths, resulting in pairs of asymmetric coupled lines. The two modes, which are also termed "c" and "π" modes [8], as corresponding to the even and odd modes in the symmetric case, have different characteristic impedances, as can be seen from Eq. (5.28). Using Eq. (5.28) directly may cause some difficulty for filter designs. For instance, if $C_1 \neq C_2 \neq C_3$ and $C_{1,2} \neq C_{2,3}$ obtained from Eq. (5.27), the line element 2 may have two values for the even-mode impedance and two values for the odd-mode impedance when it is related to the line elements 1 and 3, respectively. An approximate design approach has been reported in [9] to overcome this difficulty with the following design equations

$$Z_{0e1,2} = \frac{1}{Y_1 - Y_{1,2}}, \qquad Z_{0o1,2} = \frac{1}{Y_1 + Y_{1,2}}$$
$$Z_{0ei,i+1} = \frac{1}{2Y_1 - 1/Z_{0ei-1,i} - Y_{i,i+1} - Y_{i-1,i}} \quad \text{for } i = 2 \text{ to } n-2$$
$$Z_{0oi,i+1} = \frac{1}{2Y_{i,i+1} + 1/Z_{0ei,i+1}} \quad \text{for } i = 2 \text{ to } n-2 \qquad (5.29)$$
$$Z_{0en-1,n} = \frac{1}{Y_1 - Y_{n-1,n}}, \qquad Z_{0on-1,n} = \frac{1}{Y_1 + Y_{n-1,n}}$$

where $Z_{0ei,i+1}$ and $Z_{0oi,i+1}$ are the even- and odd-mode impedances of coupled lines associated with resonators i and $i+1$ and all the admittance parameters are those given in Eq. (5.27).

If we allow the use of asymmetrical coupled lines for a filter design, each of the even-mode impedances in Eq. (5.29) may be seen as an average of the two c-mode

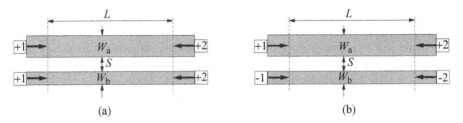

FIGURE 5.16 Microstrip layouts for full-wave EM simulations to extract even- and odd-mode impedances. (**a**) Even-mode excitation. (**b**) Odd-mode excitation.

impedances for adjacent coupled lines. Similarly, each of the odd-mode impedances may be seen as an average of the two associated π-mode impedances. A technique, which will extract such even- and odd-mode impedances for the filter design, is discussed below.

The technique to be discussed enable one to take advantage of full-wave electromagnetic (EM) simulation for filter design, which is available in many CAD tools, as described in Chapter 8. To obtain the average even-mode characteristic impedance of a pair of asymmetric coupled lines, we may impose the two asymmetric coupled lines to have an even-mode excitation, as illustrated in Figure 5.16a, where the ports with the same number are electrically connected in parallel. Similarly, to obtain the average odd-mode impedance, we may impose the two lines to have an odd-mode excitation, as shown in Figure 5.16b, where the ports with the same number, but opposite signs, indicate the odd-mode excitation (equal magnitude and opposite polarization), and are electrically connected in series. In either case, from the results of EM simulation, a set of two-port S parameters for the mode of interest can be found in the form

$$\begin{aligned} S_{11} &= S_{22} = |S_{11}| e^{j\phi_{11}} \\ S_{12} &= S_{21} = |S_{21}| e^{j\phi_{21}} \end{aligned} \quad (5.30)$$

We can then extract an effective dielectric constant for the mode under consideration by

$$\varepsilon_{re} = \left(\frac{\phi_{21}}{2\pi} \frac{\lambda_0}{L} \right)^2 \quad (5.31)$$

where ϕ_{21} is the phase in radian, λ_0 is the wavelength in free space at the frequency used for the simulation, and L is the line length between the two reference planes, where the S parameters are de-embedded. Theoretically, the L can be set to any length; however, practically it may be set to be about quarter-wavelength to obtain more accurate numerical data for the parameter extraction. It will be shown later that the extracted relative dielectric constants for the both modes are useful for microstrip filter design. We can also extract a characteristic impedance

$$Z_c = \text{Re}\left\{\frac{Z_{in} - Z_0 + \sqrt{(Z_{in} - Z_0)^2 - 4Z_0 Z_{in} \tan^2 \phi_{21}}}{j2\tan\phi_{21}}\right\} \quad (5.32)$$

with

$$Z_{in} = Z_0 \frac{1 + S_{11}}{1 - S_{11}} \quad (5.33)$$

and Z_0 is the port terminal resistance. Some commercial EM simulators such as *em* [12] can automatically extract ε_{re} and Z_c. For the even-mode excitation, the average even-mode impedance is then found by

$$Z_{0e} = 2Z_c \quad (5.34)$$

While, in the case of the odd-mode excitation, the average odd-mode impedance is determined by

$$Z_{0o} = Z_c/2 \quad (5.35)$$

Design Example with Asymmetric Coupled Lines

To demonstrate how to design the microstrip interdigital bandpass filter, a design example is detailed below. For this example, the design is worked out using an $n = 5$ Chebyshev lowpass prototype with a passband ripple 0.1 dB. The prototype parameters are

$$g_0 = g_6 = 1.0 \quad g_1 = g_5 = 1.1468$$
$$g_2 = g_4 = 1.3712 \quad g_3 = 1.9750$$

The bandpass filter is designed for a fractional bandwidth $FBW = 0.5$ centered at the midband frequency $f_0 = 2.0$ GHz. Table 5.7 lists the bandpass design parameters obtained by using the design equations given in Eqs. (5.27) and (5.29). The characteristic admittance Y_1 is chosen so that the characteristic impedance of the tapped lines $Z_t = 1/Y_t$ is equal to 50 Ω.

A commercial dielectric substrate (RT/D 6006) with a relative dielectric constant of 6.15 and a thickness of 1.27 mm is chosen for the filter design. Using the technique described above, some design data of asymmetric coupled microstrip lines on the substrate are extracted from the results of EM simulations and are given in Table 5.8, where W_a and W_b are the widths of two coupled microstrip lines and s is the spacing between them. Referring to the design parameters in Table 5.7, one can find that the extracted even- and odd-mode impedances for $W_a = 2.2$ mm, $W_b = 1.1$ mm, and $s = 0.2$ mm match to the desired $Z_{0e1,2} = Z_{0e4,5}$ and $Z_{0o1,2} = Z_{0o4,5}$; the extracted

TABLE 5.7 Circuit Design Parameters of the Five-Pole Interdigital Bandpass Filter with Asymmetric Coupled Lines

i	$Z_{0ei,i+1}$	$Z_{0oi,i+1}$
1	65.34	34.78
2	59.16	36.83
3	59.16	36.83
4	65.34	34.78

$Y_1 = 1/45.4$ mhos
$Y_t = 1/50$ mhos
$\theta_t = 0.82939$ radians
$C_t = 3.45731 \times 10^{-13}$ F

even- and odd-mode impedances for $W_a = 2.6$ mm, $W_b = 1.1$ mm, and $s = 0.3$ mm match to the desired $Z_{0e2,3} = Z_{0e3,4}$ and $Z_{0o2,3} = Z_{0o3,4}$. Therefore, these two sets of dimensions will form the basis of physical design parameters of the filter, namely, $W_1 = W_5 = 2.2$ mm, $W_2 = W_4 = 1.1$ mm, $W_3 = 2.6$ mm, $s_{1,2} = s_{4,5} = 0.2$ mm, and $s_{2,3} = s_{3,4} = 0.3$ mm. It should be noted that choosing the line width for the input and output resonators is also restricted to have a single-line characteristic admittance $Y_1 = 1/45.4$, as specified in Table 5.7.

Next, we need to decide the lengths of microstrip interdigital resonators. Basically, they can be found by

$$l_i = \lambda_{g0i}/4 - \Delta l_i \tag{5.36}$$

where λ_{g0i} is the guided wavelength and Δl_i is the equivalent line length of microstrip open end associated with resonator i. Since microstrip is not a pure TEM-mode

TABLE 5.8 Microstrip Design Parameters of the Five-Pole Interdigital Bandpass Filter with Asymmetric Coupled Lines

W_a (mm)	W_b (mm)	s (mm)	Z_{0e} (Ω)	Z_{0o} (Ω)	ε_{re}^e	ε_{re}^o
2.2	1.2	0.2	63.6	34.15	4.68	3.76
2.2	1.1	0.2	64.92	34.83	4.66	3.75
2.0	1.1	0.5	74.7	41.82	4.70	3.80
2.4	1.1	0.5	59.8	40.44	4.73	3.82
2.6	1.1	0.5	57.6	39.86	4.75	3.83
2.8	1.1	0.5	55.6	39.33	4.76	3.84
2.8	1.1	0.4	56.4	37.85	4.76	3.82
2.8	1.1	0.3	57.2	36.01	4.75	3.80
2.7	1.1	0.3	58.26	36.22	4.75	3.79
2.6	1.1	0.3	59.36	36.43	4.74	3.79

BANDPASS FILTERS   141

transmission line, there will be unequal guided wavelengths for the even-mode and odd-mode as evidence of unequal effective dielectric constants for the both modes given in Table 5.8. Hence, the λ_{g0i} may be seen as an average value given by

$$\lambda_{g0i} = \lambda_0 \left(\sqrt{\varepsilon_{rei}^e \varepsilon_{rei}^o} \right)^{-1/2} \tag{5.37}$$

where λ_0 is the wavelength in free space at the midband frequency of the filter. The Δl_i can be determined using the design equation for microstrip open-end presented in Chapter 4.

Recall that there is a capacitance C_t that needs to be loaded to the input and output resonators. This capacitive loading may be achieved by an open-circuit stub, namely, an extension in length of the resonators. Let Δl_C denote the length extension, which may be found by

$$\Delta l_C = \frac{\lambda_{g01}}{2\pi} \tan^{-1} \left(\frac{2\pi f_0 C_t}{Y_1} \right) \tag{5.38}$$

Therefore, the length for the input and output resonators are actually determined by

$$l_1 = l_n = \lambda_{g01}/4 - \Delta l_1 + \Delta l_C \tag{5.39}$$

Finally, the physical length l_t measured from the input/output resonator ground to the tap point is calculated by

$$l_t = \frac{\theta_t}{2\pi} \lambda_{g01} \tag{5.40}$$

All determined physical design parameters including the feed-line width W_t for this filter are summarized in Table 5.9.

Figure 5.17a shows the layout of the designed microstrip interdigital bandpass filter. The filter frequency responses obtained using EM simulation are plotted in Figure 5.17b. The performance of the designed filter is excellent, except for a slight shift in the midband frequency, which is lower than 2 GHz. The frequency shift may be because of the effect of via-holes [10] and should easily be corrected by slightly shortening the resonator lengths. It can be shown that the wide-band response of this filter exhibits a transmission zero near the twice the designed midband frequency, whereas the second pass band is centered at about three times the designed midband frequency, as expected for this type of filter.

TABLE 5.9 Filter Dimensions (mm) on Substrate with $\varepsilon_r = 6.15$ and $h = 1.27$ mm

$W_1 = W_5 = 2.2$	$W_2 = W_4 = 1.1$	$W_3 = 2.6$	$s_{1,2} = s_{4,5} = 0.2$	$W_t = 1.85$
$l_1 = l_5 = 20.14$	$l_2 = l_4 = 17.85$	$l_3 = 17.72$	$s_{2,3} = s_{3,4} = 0.3$	$l_t = 9.68$

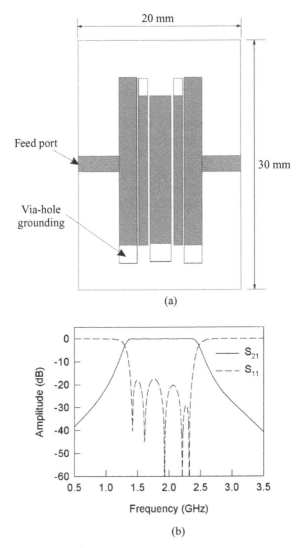

FIGURE 5.17 (a) Layout of a five-pole, microstrip interdigital bandpass filter using asymmetrical coupled lines. The dimensions are given in Table 5.9 as referring to Figure 5.15. (b) Full-wave EM simulated performance of the filter.

Design Example with Symmetric Coupled Lines

It may also be desirable to design interdigital bandpass filters with symmetric coupled lines. This means that all resonators of an interdigital filter in Figure 5.15 will have the same line widths. There are two advantages arising from this configuration. The one advantage is that more design equations and data on symmetric coupled lines are

TABLE 5.10 Circuit Design Parameters of the Five-Pole Interdigital Bandpass Filter with Symmetric Coupled Lines

i	$Z_{0ei,i+1}$	$Z_{0oi,i+1}$
1	68.18	31.94
2	60.07	34.09
3	60.07	34.09
4	68.18	31.94

$Y_1 = 1/43.5$ mhos
$Y_t = 1/50$ mhos
$\theta_t = 0.98609$ rad
$C_t = 4.20655 \times 10^{-13}$ F

available for the filter design; second, the unloaded quality factor of each resonator will be much the same.

However, a difficulty arises because it is generally not possible to realize arbitrary even- and odd-mode impedances with a fixed-line width. To make such a filter design possible, a technique is presented here demonstrating a filter design.

For this demonstration, the design uses the same lowpass prototype filter element values for the above design example. This design technique requires a fractional bandwidth larger than the specified one in order to achieve the desirable passband bandwidth. Recall that the specified fractional bandwidth is 50%. In this design, we assume a fractional bandwidth of 60% or $FBW = 0.6$ to calculate design parameters according to Eqs. (5.27) and (5.29). The results are listed in Table 5.10.

The same substrate ($\varepsilon_r = 6.15$ and a thickness of 1.27 mm) for the above design example is also used for this design. On this substrate, the line width for a characteristic impedance $Z_1 = 1/Y_1 = 43.5\ \Omega$ is found to be 2.39 mm by using microstrip design equations given in Chapter 4. The line width for all coupled lines is then fixed by $W_1 = W_2 = \cdots = W_5 = 2.39$ mm. As mentioned above, with the fixed same line width, it is almost impossible to obtain the desired $Z_{0ei,i+1}$ and $Z_{0oi,i+1}$ by adjusting the spacing $s_{i,i+1}$ alone. Therefore, instead, for matching to the desired $Z_{0ei,i+1}$ and $Z_{0oi,i+1}$, the spacing $s_{i,i+1}$ are adjusted for matching to

$$k_{i,i+1} = \frac{Z_{0ei,i+1} - Z_{0oi,i+1}}{Z_{0ei,i+1} + Z_{0oi,i+1}} \quad (5.41)$$

In this way, the all spacing can be determined. Since we are now dealing with symmetric coupled microstrip lines, the design equations described in Chapter 4 can be utilized to find $s_{i,i+1}$. The other physical design parameters, such as the lengths of the interdigital resonators, can be determined by using Eqs. (5.36–5.39). All the physical design parameters determined for this filter are summarized in Table 5.11.

TABLE 5.11 Filter Dimensions (mm) with Symmetric Coupled Microstrip Lines on Substrate with $\varepsilon_r = 6.15$ and $h = 1.27$ mm

$W_1 = W_2 = W_3 = W_4 = W_5 = 2.39$		$s_{1,2} = s_{4,5} = 0.13$	$W_t = 1.85$
$l_1 = l_5 = 20.45$	$l_2 = l_4 = 17.91$ $\quad l_3 = 17.83$	$s_{2,3} = s_{3,4} = 0.36$	$l_t = 11.5$

Figure 5.18 shows the layout of the designed microstrip interdigital bandpass filter with the same line width for the all resonators. The EM simulated frequency response of this filter shows a very similar performance as compared with that in Figure 5.17b.

It should be emphasized that the actual bandwidth of filter resulting from this design technique is different from that used for the design. The difference between the two could be reduced when the actual bandwidth becomes small.

5.2.5 Combline Filters

As shown in Figure 5.19, the combline bandpass filter is comprised of an array of coupled resonators. The resonators consist of line elements 1 to n, which are short-circuited at one end, with a lumped capacitance C_{Li} loaded between the other end of each resonator line element and ground. The input and output of filter are through coupled-line elements 0 and $n + 1$, which are not resonators. With the lumped

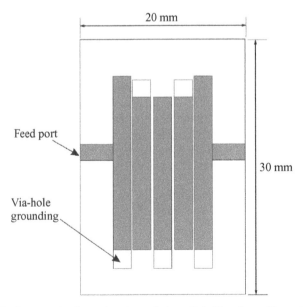

FIGURE 5.18 Layout of a five-pole microstrip interdigital bandpass filter using symmetrical coupled lines. The dimensions are given in Table 5.11 as referring to Figure 5.15.

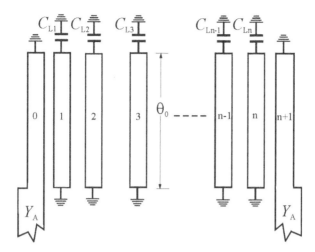

FIGURE 5.19 General structure of combline bandpass filter.

capacitors present, the resonator lines will be less than $\lambda_{g0}/4$ long at resonance, where λ_{g0} is the guided wavelength in the medium of propagation at the midband frequency of filter. It is interesting that if the capacitors were not present, the resonator lines would be a full $\lambda_{g0}/4$ long at resonance, and the filter structure in Figure 5.19 would have no pass band at all when the line elements are constructed from a pure TEM-mode transmission line such as stripline. This is because, in this case, the magnetic and electric couplings totally cancel out each other.

The larger the loading capacitances C_{Li}, the shorter the resonator lines, which results in a more compact filter structure with a wider stopband between the first (desired) and the second passband (unwanted). For instance, if the resonator lines are $\lambda_{g0}/8$ long at the primary passband, the second passband will be centered somewhat over four times the midband frequency of the first passband. In practice, the minimum resonator line length could be limited by the decrease of the unloaded quality factor of resonator and a requirement for heavy capacitive loading. The lumped capacitors may offer a convenient means for filter tuning, which may be required particularly for narrow-band filters.

The type of filter in Figure 5.19 can be designed from a chosen ladder type of lowpass prototype with the lowpass parameters g_i for $i = 0$ to $n+1$. A design procedure as described [1] starts with choosing the resonator susceptance slope parameters b_i

$$\frac{b_i}{Y_A} = \frac{Y_{ai}}{Y_A}\left(\frac{\cot\theta_0 + \theta_0 \csc^2\theta_0}{2}\right) \quad \text{for } i = 1 \text{ to } n \qquad (5.42a)$$

where Y_A is the terminating line admittance, θ_0 is the midband electrical length of the resonators, Y_{ai} is interpreted physically as the admittance of line with the adjacent lines $i-1$ and $i+1$ grounded. The choice of Y_{ai} fixes the admittance level within the filter and can influence the unloaded quality factors of the resonators. For a specified fractional bandwidth FBW, calculate

$$\frac{J_{0,1}}{Y_A} = \sqrt{\frac{FBW\frac{b_1}{Y_A}}{g_0 g_1}}, \quad \frac{J_{n,n+1}}{Y_A} = \sqrt{\frac{FBW\frac{b_n}{Y_A}}{g_n g_{n+1}}} \quad (5.42b)$$

$$\frac{J_{i,i+1}}{Y_A} = FBW\sqrt{\frac{(b_i/Y_A)(b_{i+1}/Y_A)}{g_i g_{i+1}}} \quad \text{for } i = 1 \text{ to } n-1$$

where the lowpass element values g_i are given for a normalized cut-off frequency $\Omega_c = 1$. The design procedure leads to the determination of the lumped capacitances, as well as the self- and mutual capacitances per unit length of the distributed line elements.

The lumped capacitances C_{Li} are:

$$C_{Li} = Y_A \left(\frac{Y_{ai}}{Y_A}\right) \frac{\cot \theta_0}{\omega_0} \quad \text{for } i = 1 \text{ to } n \quad (5.43)$$

where ω_0 is the angular frequency at the midband.

The self-capacitances C_i are given by

$$\frac{C_0}{\varepsilon_0 \sqrt{\varepsilon_{re}}} = \eta_0 Y_A \left(1 - \frac{J_{0,1}}{Y_A}\right), \quad \frac{C_{n+1}}{\varepsilon_0 \sqrt{\varepsilon_{re}}} = \eta_0 Y_A \left(1 - \frac{J_{n,n+1}}{Y_A}\right)$$

$$\frac{C_1}{\varepsilon_0 \sqrt{\varepsilon_{re}}} = \eta_0 Y_A \left(\frac{Y_{a1}}{Y_A} - 1 + \left(\frac{J_{0,1}}{Y_A}\right)^2 - \frac{J_{1,2}}{Y_A} \tan \theta_0\right) + \frac{C_0}{\varepsilon_0 \sqrt{\varepsilon_{re}}}$$

$$\frac{C_n}{\varepsilon_0 \sqrt{\varepsilon_{re}}} = \eta_0 Y_A \left(\frac{Y_{an}}{Y_A} - 1 + \left(\frac{J_{n,n+1}}{Y_A}\right)^2 - \frac{J_{n-1,n}}{Y_A} \tan \theta_0\right) + \frac{C_{n+1}}{\varepsilon_0 \sqrt{\varepsilon_{re}}}$$

$$\frac{C_i}{\varepsilon_0 \sqrt{\varepsilon_{re}}} = \eta_0 Y_A \left(\frac{Y_{ai}}{Y_A} - \frac{J_{i-1,i}}{Y_A} \tan \theta_0 - \frac{J_{i,i+1}}{Y_A} \tan \theta_0\right) \quad \text{for } i = 2 \text{ to } n-2$$

(5.44a)

and the mutual capacitances $C_{i,i+1}$ are

$$\frac{C_{0,1}}{\varepsilon_0 \sqrt{\varepsilon_{re}}} = \eta_0 Y_A - \frac{C_0}{\varepsilon_0 \sqrt{\varepsilon_{re}}}, \quad \frac{C_{n,n+1}}{\varepsilon_0 \sqrt{\varepsilon_{re}}} = \eta_0 Y_A - \frac{C_{n+1}}{\varepsilon_0 \sqrt{\varepsilon_{re}}}$$

$$\frac{C_{i,i+1}}{\varepsilon_0 \sqrt{\varepsilon_{re}}} = \eta_0 Y_A \frac{J_{i,i+1}}{Y_A} \tan \theta_0$$

(5.44b)

Note that in Eqs. (5.44a) and (5.44b) the self- and mutual capacitances are normalized with the free-space permittivity ε_0 and scaled by $\sqrt{\varepsilon_{re}}$ with ε_{re} denoting the relative effective dielectric constant of the line elements and $\eta_0 = 120\pi \; \Omega$ is the wave impedance in the free space. It should be mentioned that for a pure TEM transmission line, such as stripline, ε_{re} is simply the relative dielectric constant ε_r of the medium of propagation, but for a quasi-TEM transmission line such as microstrip, ε_{re} will also depend on the geometry of the transmission line, as discussed in Chapter 4.

Having obtained the $C_i/(\varepsilon_0\sqrt{\varepsilon_{re}})$ and $C_{i,i+1}/(\varepsilon_0\sqrt{\varepsilon_{re}})$, the filter design is then turned to find the dimensions of the coupled lines, namely the widths and spacing that give the required self- and mutual capacitances on the specified transmission-line medium. A method for finding dimensions of coupled lines on microstrip in terms of the self- and mutual capacitances can be found [13].

Instead of working on the self- and mutual capacitances, an alternative design approach is to determine the dimensions in terms of another set of design parameters consisting of external quality factors and coupling coefficients. The theory of this approach is discussed in Chapter 7; here we should be only concerned with its application for the design of combline filters. These design parameters are given by

$$Q_{e1} = \frac{b_1}{J_{0,1}^2/Y_A} = \frac{g_0 g_1}{FBW}, \qquad Q_{en} = \frac{b_n}{J_{n,n+1}^2/Y_A} = \frac{g_n g_{n+1}}{FBW}$$

$$M_{i,i+1} = \frac{J_{i,i+1}}{\sqrt{b_i b_{i+1}}} = \frac{FBW}{\sqrt{g_i g_{i+1}}} \qquad \text{for } i = 1 \text{ to } n-1$$

(5.45)

where Q_{e1} and Q_{en} are the external quality factors of the resonators at the input and output and $M_{i,i+1}$ are the coupling coefficients between the adjacent resonators. The required dimensions can easily be found by using full-wave EM simulation, as described in Chapter 7. A design example of microstrip combline bandpass filter is detailed next.

Design Example

Assume that a five-pole Chebyshev lowpass prototype filter with 0.1-dB passband ripple has been chosen for the bandpass filter design. The lowpass prototype parameters are $g_0 = g_6 = 1.0$, $g_1 = g_5 = 1.1468$, $g_2 = g_4 = 1.3712$, and $g_3 = 1.9750$. The bandpass filter is designed to have a fractional bandwidth $FBW = 0.1$ at a midband frequency $f_0 = 2$ GHz. From Eq. (5.45) we obtain

$$Q_{e1} = Q_{e5} = 11.468$$
$$M_{1,2} = M_{4,5} = 0.07975 \qquad (5.46)$$
$$M_{2,3} = M_{3,4} = 0.06077$$

As mentioned above, the combline resonators cannot be $\lambda_{g0}/4$ long if they are realized with TEM transmission lines. However, this is not necessarily the case for a microstrip combline filter, because the microstrip is not a pure TEM transmission line. For demonstration, we also let the microstrip resonators be $\lambda_{g0}/4$ long and require no capacitive loading for this filter design. The microstrip filter is designed

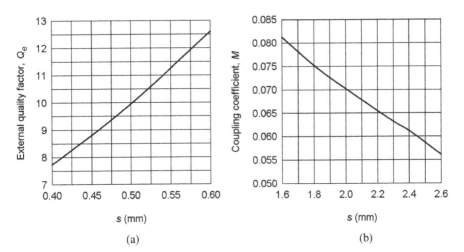

FIGURE 5.20 Design curves obtained by full-wave EM simulations for designing the microstrip combline filter. (**a**) External quality factor. (**b**) Coupling coefficient.

on a commercial substrate (RT/D 6010) with a relative dielectric constant of 10.8 and a thickness of 1.27 mm. With the EM simulation we are able to work directly on the dimensions of the filters. Therefore, we first fix a line width $W = 0.8$ mm (although other line widths may be chosen) for the all line elements except for the terminating lines. The terminating lines are 1.1 mm wide, which matches to a 50-Ω terminating impedance. Using the parameter extraction technique described in Chapter 7, the design curves for external quality factor and coupling coefficient against spacing s can be obtained; these are plotted in Figure 5.20. The required dimensions of spacing for the design parameters given in Eq. (5.46) can then be found from these two design curves. For instance, a spacing $s = 0.56$ mm is identified from Figure 5.20a for the required external quality of 11.468. The final filter design with the all determined physical dimensions is shown in Figure 5.21a. It should be noted that the lengths of the adjacent resonators are slightly different because the phase velocities are different. This would not be the case for the pure TEM line resonators. The effect of the via-hole grounding on frequency shift has been also taken into account for determining the lengths. Figure 5.21b shows the EM simulated filter performance. It is interesting to note that there is an attenuation pole near the high edge of the passband, resulting in a higher selectivity on that side. This attenuation pole is likely because of cross couplings between the nonadjacent resonators. It is also shown that the second passband of the filter is centered at about 6 GHz, which is threes times the midband frequency, as expected, because the $\lambda_{g0}/4$ resonators are used without any lumped capacitor.

Combline filters can also be designed with tapped-line input and output (I/O). The design equations, which correspond to those in Eqs. (5.42) – (5.44), can be found in [7]. Of course, without employing these design equations, the tapped combline filter can be designed by using the alternative design approach based on the external Q

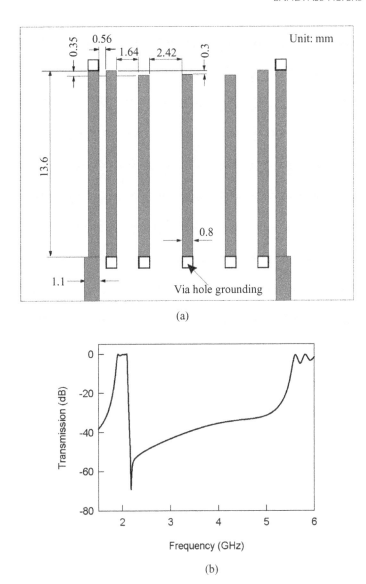

FIGURE 5.21 (a) Layout of the designed microstrip combline bandpass filter on a 1.27-mm thick substrate with a relative dielectric constant of 10.8. (b) Full-wave EM simulated performance of the filter.

and coupling coefficients, as just described. For a larger bandwidth, the spacing for the coupled-line I/O of the structure in Figure 5.19 can be very tight and physical realization becomes impractical, but the same filter can still be realized by tapping. A microstrip combline filter of this type with a fractional bandwidth 40% at a center frequency of 10 GHz has been reported [14].

5.2.6 Pseudocombline Filters

Figure 5.22 shows a so-called pseudocombline bandpass filter that is comprised of an array of coupled resonators. The resonators consist of line elements 1 to n, which are open-circuited at one end, with a lumped capacitance C_{Li} loaded between the other end of each resonator line element and ground. The filter uses the tapped lines as the input and output. With the lumped capacitors present, the resonator lines will be less than $\lambda_{g0}/2$ long at resonance, where λ_{g0} is the guided wavelength in the medium of propagation at the midband frequency of filter.

This type of filter may be conceptually obtained from a combline filter shown in Figure 5.23a and an array of $\lambda_{g0}/4$ open-circuited stubs in Figure 5.23b. It should be noticed that at the midband frequency f_0, each open end of the $\lambda_{g0}/4$ open-circuited stubs reflects an electric shorted-circuit on the other end, resulting in a virtual grounding. Therefore, the physical groundings of the resonators in Figure 5.23a may be removed and replaced with the virtual groundings, produced by the array of the $\lambda_{g0}/4$ open-circuited stubs, if they are jointed together. This connection leads to the filter structure of Figure 5.22. However, such a resultant pseudocombline filter can only be electrically equivalent to the combline filter in the vicinity of the midband frequency; its stopband behavior is different from that of the combline filter.

Similar to the combline filter, if the capacitors were not present, the resonator lines would be a full $\lambda_{g0}/2$ long at resonance and the filter structure in Figure 5.22 would have no passband at all when it is realized in stripline, because of a total cancellation of electric and magnetic couplings. For microstrip realization, this would not be the case [15]. However, if the removal of all lumped capacitors is desired for stripline

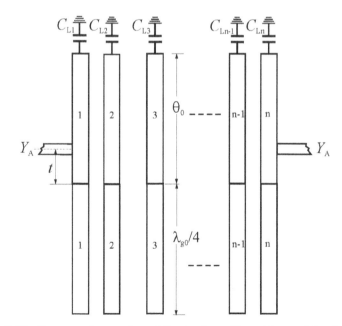

FIGURE 5.22 Structure of a pseudocombline bandpass filter with tapped-line input and output.

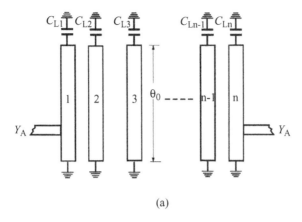

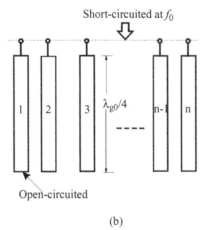

FIGURE 5.23 (a) A tapped combline filter. (b) An array of quarter-wavelength open-circuited stubs.

realization, a staggered stripline resonator array may be used to restore a passband [16]. In that case, each $\lambda_{g0}/2$ resonator is staggered with its neighbors; coupling between the resonators is then introduced. The more the resonators are staggered, the more coupling will result.

The type of filter in Figure 5.22 can be designed with a set of bandpass design parameters consisting of external quality factors and coupling coefficients. For a given fractional bandwidth FBW and a chosen ladder type of lowpass prototype (such as a Chebyshev one) having the lowpass parameters g_i, which are given for a normalized lowpass cutoff frequency $\Omega_c = 1$; the bandpass design parameters are calculated by

$$Q_{e1} = \frac{g_0 g_1}{FBW}, \quad Q_{en} = \frac{g_n g_{n+1}}{FBW}$$

$$M_{i,i+1} = \frac{FBW}{\sqrt{g_i g_{i+1}}} \quad \text{for } i = 1 \text{ to } n-1 \tag{5.47}$$

where Q_{e1} and Q_{en} are the external quality factors of the resonators at the input and output and $M_{i,i+1}$ are the coupling coefficients between the adjacent resonators. The required dimensions can be found by using full-wave EM simulation (see Chapter 7 for details). In what follows, we will design a microstrip pseudocombline bandpass filter that uses $\lambda_{g0}/2$ resonators without any capacitive loading.

Design Example

For this design, we work out with a five-pole Chebyshev lowpass prototype filter with 0.1-dB passband ripple. The lowpass prototype parameters are $g_0 = g_6 = 1.0$, $g_1 = g_5 = 1.1468$, $g_2 = g_4 = 1.3712$, and $g_3 = 1.9750$. The bandpass filter is designed to have a 15% fractional bandwidth or $FBW = 0.15$ at a midband frequency $f_0 = 2$ GHz. From Eq. (5.47) we obtain

$$Q_{e1} = Q_{e5} = 7.645$$
$$M_{1,2} = M_{4,5} = 0.11962 \qquad (5.48)$$
$$M_{2,3} = M_{3,4} = 0.09115$$

A commercial substrate (RT/D 6010) with a relative dielectric constant of 10.8 and a thickness of 1.27 mm is chosen for the microstrip filter design. We fix a line width $W = 0.8$ mm for the all half-wavelength resonators. The tapped lines are 1.1 mm wide, which matches to 50-Ω terminating impedance. Using the parameter extraction technique described in Chapter 7, the design curves for external quality factor and coupling coefficient are obtained, as shown in Figure 5.24. The required physical dimensions for the design parameters given in Eq. (5.48) can then be found from these two design curves. For instance, a tapping position (referring to Figure 5.22)

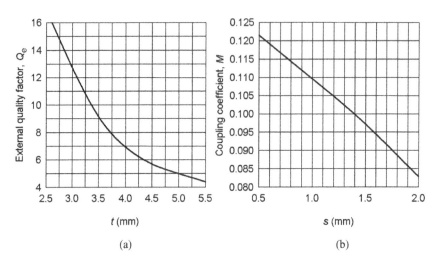

FIGURE 5.24 Design curves for the microstrip pseudocombline filter. (**a**) External quality factor. (**b**) Coupling coefficients.

$t = 3.8$ mm can be identified from Figure 5.24a for the required external quality of 7.645. The final filter design with the all determined physical dimensions is shown in Figure 5.25a. It should be noted that the lengths of the adjacent resonators are slightly different because of the difference in the phase velocities. Figure 5.25b shows the EM simulated filter performance. Similar to the combline filter discussed in Section 5.2.5, there is an attenuation pole near the high edge of the passband, resulting in a higher selective on that side. This attenuation pole is likely to be because of cross couplings between the nonadjacent resonators. However, the second pass band of the pseudocombline filter is centered at about 4 GHz, which is only twice the midband frequency. This is expected because the $\lambda_{g0}/2$ resonators are used without any lumped capacitor loading.

5.2.7 Stub Bandpass Filters

5.2.7.1 Filters with $\lambda_{g0}/4$ Short-Circuited Stubs Bandpass filters can be designed to have a form in Figure 5.26, which is comprised of shunt short-circuited stubs that are $\lambda_{g0}/4$ long with connecting lines that are also $\lambda_{g0}/4$ long, where λ_{g0} is the guided wavelength in the medium of propagation at the midband frequency f_0. For a given filter degree n, the stub bandpass filter characteristics will then depend on the characteristic admittances of the stub lines denoted by Y_i ($i = 1$ to n) and the characteristic admittances of the connecting lines denoted by $Y_{i,i+1}$ ($i = 1$ to $n - 1$). The design equations for determining these characteristic admittances described in Matthaei et al. [1] are given by

$$\theta = \frac{\pi}{2}\left(1 - \frac{FBW}{2}\right)$$

$$h = 2$$

$$\frac{J_{1,2}}{Y_0} = g_0 \sqrt{\frac{hg_1}{g_2}}, \quad \frac{J_{n-1,n}}{Y_0} = g_0 \sqrt{\frac{hg_1 g_{n+1}}{g_0 g_{n-1}}}$$

$$\frac{J_{i,i+1}}{Y_0} = \frac{hg_0 g_1}{\sqrt{g_i g_{i+1}}} \quad \text{for } i = 2 \text{ to } n - 2$$

$$N_{i,i+1} = \sqrt{\left(\frac{J_{i,i+1}}{Y_0}\right)^2 + \left(\frac{hg_0 g_1 \tan\theta}{2}\right)^2} \quad \text{for } i = 1 \text{ to } n - 1$$

$$Y_1 = g_0 Y_0 \left(1 - \frac{h}{2}\right) g_1 \tan\theta + Y_0 \left(N_{1,2} - \frac{J_{1,2}}{Y_0}\right)$$

$$Y_n = Y_0 \left(g_n g_{n+1} - g_0 g_1 \frac{h}{2}\right) \tan\theta + Y_0 \left(N_{n-1,n} - \frac{J_{n-1,n}}{Y_0}\right)$$

$$Y_i = Y_0 \left(N_{i-1,i} + N_{i,i+1} - \frac{J_{i-1,i}}{Y_0} - \frac{J_{i,i+1}}{Y_0}\right) \quad \text{for } i = 2 \text{ to } n - 1$$

$$Y_{i,i+1} = Y_0 \left(\frac{J_{i,i+1}}{Y_0}\right) \quad \text{for } i = 1 \text{ to } n - 1 \qquad (5.49)$$

154 LOWPASS AND BANDPASS FILTERS

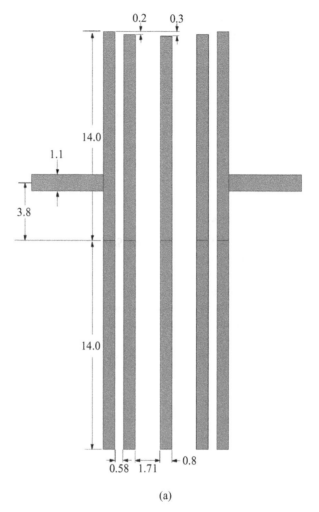

(a)

FIGURE 5.25 (a) Layout of the designed microstrip pseudocombline bandpass filter on the substrate with a relative dielectric constant of 10.8 and thickness of 1.27 mm. (b) Full-wave EM simulated performance of the filter.

where g_i is the element value of a ladder-type lowpass prototype filter, such as a Chebyshev, given for a normalized cutoff $\Omega_c = 1.0$. h is a dimensionless constant, which may be assigned to another value so as to give a convenient admittance level in the interior of the filter.

Design Example

To demonstrate how to design this type of microstrip filter, let us start with a five-pole ($n = 5$) Chebyshev lowpass prototype with a 0.1-dB passband ripple. The prototype

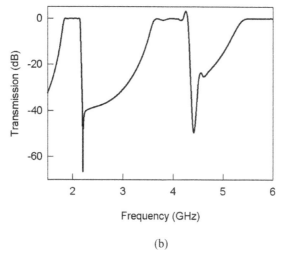

(b)

FIGURE 5.25 (*Continued*)

parameters are

$$g_0 = g_6 = 1.0 \qquad g_1 = g_5 = 1.1468$$
$$g_2 = g_4 = 1.3712 \qquad g_3 = 1.9750$$

The bandpass filter is designed to have a fractional bandwidth $FBW = 0.5$ at a midband frequency $f_0 = 2$ GHz. A 50-Ω terminal line impedance is chosen, which gives

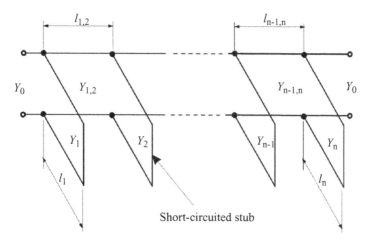

FIGURE 5.26 Transmission line bandpass filter with quarter-wavelength short-circuited stubs.

TABLE 5.12 Circuit Design Parameters of a Five-Pole Stub Bandpass Filter with $\lambda_{g0}/4$ Short-Circuited Stubs

i	Y_i (mhos)	$Y_{i,i+1}$ (mhos)
1	0.03525	0.02587
2	0.06937	0.02787
3	0.06824	0.02787
4	0.06937	0.02587
5	0.03525	

$Y_0 = 1/50$ mhos. The computed design parameters using Eq. (5.49) are summarized in Table 5.12.

For the microstrip filter design, we use a dielectric substrate with a relative dielectric constant of 10.2 and a thickness of 0.635 mm. Using the microstrip design equations described in Chapter 4, the widths and guided quarter-wavelengths associated with the characteristic admittances in Table 5.12 can be found and are listed in Table 5.13.

Figure 5.27a shows the layout of the designed microstrip filter and Figure 5.27b plots the filter frequency responses obtained by full-wave EM simulations. In general, the performance is seen to be in good agreement with the design objective. As can also be seen, the filter has a second passband centered by $3f_0$, but exhibits an attenuation pole at $2f_0$, which are typical stopband characteristics of this type of filter. Filters of this type are candidates for use primarily as wide-band filters, because if narrow-band filters are designed in this form, their stubs will have unreasonable low impedance levels.

5.2.7.2 Filters with $\lambda_{g0}/2$ Open-Circuited Stubs
If the $\lambda_{g0}/4$ short-circuited stubs in Figure 5.26 are replaced with $\lambda_{g0}/2$ open-circuited stubs, as shown in Figure 5.28, where l_{ia} and l_{ib} are both $\lambda_{g0}/4$, long associated with the transmission lines of characteristic admittances Y_{ia} and Y_{ib}, the resultant filter will have similar passband characteristics but quite different stopband characteristics. If $Y_{ib} = Y_{ia}$ for

TABLE 5.13 Microstrip Design Parameters of a Five-Pole Stub Bandpass Filter with $\lambda_{g0}/4$ Short-Circuited Stubs

i	W_i (mm)	$\lambda_{g0i}/4$ (mm)	$W_{i,i+1}$ (mm)	$\lambda_{g0i,i+1}/4$ (mm)
1	1.61	13.67	0.97	14.03
2	4.00	13.07	1.10	13.97
3	3.93	13.03	1.10	13.97
4	4.00	13.07	0.97	14.03
5	1.61	13.67		

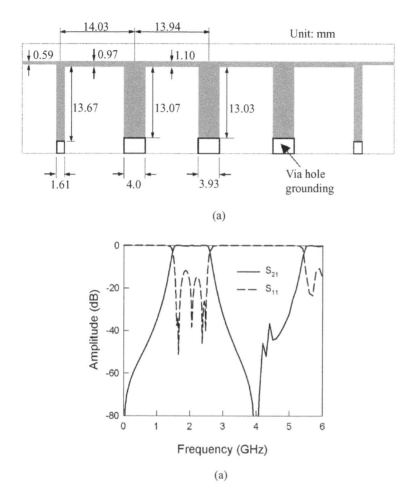

FIGURE 5.27 (a) Layout of the designed microstrip bandpass filter with quarter-wavelength short-circuited stubs on a 0.635-mm thick substrate with a relative dielectric constant of 10.2. (b) Full-wave EM simulated performance of the filter.

each $\lambda_{g0}/2$ stub, then the stopband will have attenuation poles at the frequencies $f_0/2$ and $3f_0/2$. If $Y_{ib} = \alpha Y_{ia}$ with α a constant, then the attenuation poles can be made to occur at frequencies other than $f_0/2$ and $3f_0/2$. This type of filter will have additional passbands in the vicinity of $f = 0$ and $f = 2f_0$, and at other corresponding periodic frequencies.

It has been pointed out [1] that this type of filter can be readily designed by a modified use of the design equations in Eq. (5.49). The design is carried out first to give a filter in the form in Figure 5.26 with the desired passband characteristics and bandwidth. Y_i is then replaced, as shown in Figure 5.28, by a shunt, half-wavelength,

158 LOWPASS AND BANDPASS FILTERS

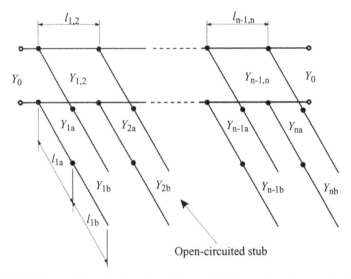

FIGURE 5.28 Transmission line bandpass filter with half-wavelength open-circuited stubs.

open-circuited stub having an inner quarter-wavelength portion with a characteristic admittance

$$Y_{ia} = \frac{Y_i(\alpha_i \tan^2 \theta - 1)}{(\alpha_i + 1)\tan^2 \theta} \tag{5.50}$$

and an outer quarter-wavelength portion with a characteristic admittance

$$Y_{ib} = \alpha_i Y_{ia} \tag{5.51}$$

where θ has been defined in Eq. (5.49); the parameter α_i is given by

$$\alpha_i = \cot^2\left(\frac{\pi f_{zi}}{2 f_0}\right) \quad \text{for } f_{zi} < f_1 \tag{5.52}$$

where f_1 is the low band-edge frequency of the passband, f_{zi} is a frequency at which the shunt open-circuited stub presents a short circuit to the main line, and causes a transmission zero or attenuation pole. Although using the same f_{zi} for all the stubs should give the best passband response, it may be permissible to stagger the f_{zi} points of the stubs slightly to achieve broader regions of high rejection. The modified design equations of Eqs. (5.50–5.52) are constrained to yield half-wavelength, open-circuited stubs that have exactly the same susceptances at the band-edge frequency f_1 as did the quarter-wavelength short-circuited stubs that they replace; both kinds of stubs have zero admittance at the midband frequency f_0.

BANDPASS FILTERS   159

TABLE 5.14 Microstrip Design Parameters of a Five-Pole Stub Bandpass Filter with $\lambda_{g0}/2$ Open-Circuited Stubs

i	$Y_{ai} = Y_{bi}$ (mhos)	W_i (mm)	$\lambda_{g0i}/4$ (mm)	$Y_{i,i+1}$ (mhos)	$W_{i,i+1}$ (mm)	$\lambda_{g0i,i+1}/4$ (mm)
1	0.01460	0.28	14.73	0.02587	0.97	14.03
2	0.02873	1.16	13.91	0.02787	1.10	13.97
3	0.02826	1.13	13.92	0.02787	1.10	13.97
4	0.02873	1.16	13.91	0.02587	0.97	14.03
5	0.01460	0.28	14.73			

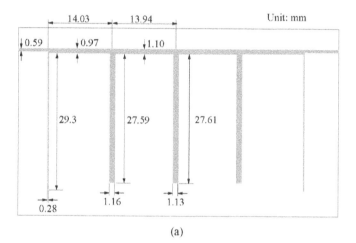

(a)

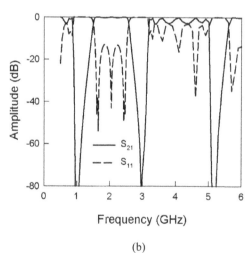

(b)

FIGURE 5.29 (a) Layout of the designed microstrip bandpass filter with half-wavelength open-circuited stubs on a 0.635-mm thick substrate with a relative dielectric constant of 10.2. (b) Full-wave EM simulated performance of the filter.

Design Example

For a demonstration, a microstrip filter of this type is designed using a five-pole ($n = 5$) Chebyshev lowpass prototype with a 0.1-dB passband ripple and a fractional bandwidth $FBW = 0.5$ at a midband frequency $f_0 = 2$ GHz. Assuming 50-Ω terminal line impedance, the initial design parameters obtained using Eq. (5.49) are the same as those listed in Table 5.12, Choosing $f_{zi} = 1.0$ GHz for all the stubs, then gives $\alpha_i = 1$ from Eq. (5.52). Using Eqs. (5.51) and (5.52) yields the characteristic admittances of the open-circuited stubs. The design results for this example are summarized in Table 5.14, including the associated microstrip widths and guided quarter-wavelengths at f_0 on the substrate, having a relative dielectric constant of 10.2 and a thickness of 0.635 mm.

The final layout of the designed microstrip filter is illustrated in Figure 5.29a, where the effect of microstrip open end on each stub has been taken into account so that each stub is slightly shorter than $\lambda_{g0}/2$. The EM simulated frequency responses of the filter are plotted in Figure 5.29b. Note that the passband response of this filter is almost the same as prescribed and that an attenuation pole does occur at 1 GHz. On the both sides of the desired passband there also are spurious passbands in the vicinity of $f = 0$ and $f = 2f_0$, which are to be expected.

Filters of this type should be particularly useful where the spurious passbands around $f = 0$ and $f = 2f_0$ are not objectionable and where there is a relatively narrow band of signals near the desired passband to be rejected. They are also practical for bandwidths narrower than those of filters of the form in Figure 5.26.

REFERENCES

[1] G. Matthaei, L.Young, and E. M. T. Jones, *Microwave Filters, Impedance-Matching Networks, and Coupling Structures*, Artech House, Norwood, MA, 1980.

[2] R. Saal and E. Ulbrich, On the design of filters by synthesis, *IRE Trans.* **CT-5**, 1958, 284–327.

[3] R. Saal, *Der Entwurf von Filtern mit Hilfe des Kataloges Normierter Tiefpasse*, Telefunkn GmbH, Backnang (Germany), 1961.

[4] E. G. Cristal and S. Frankel, Design of hairpin-line and hybrid haripin-parallel-coupled-line filters, *IEEE MTT-S Dig.* 1971, 12–13.

[5] J. S. Wong, Microstrip tapped-line filter design, *IEEE Trans.* **MTT-27**, 1, 1979, 44–50.

[6] G. L. Matthaei, Interdigital band-pass filters, *IEEE Trans.* **MTT-10**, 1962, 479–492.

[7] S. Caspi and J. Adelman, Design of combline and interdigital filters with tapped-lone input. *IEEE Trans.* **MTT-36**, 1988, 759–763.

[8] V. K. Tripathi, Asymmetric coupled transmission lines in an inhomogeneous medium, *IEEE Trans.* **MTT-23**, 1975, 734–739.

[9] C. Dening, Using microwave CAD programs to analyze microstrip interdigital filters, *Microwave J.* 1989, 147–152.

[10] D. G. Swanson, Grounding microstrip lines with via holes, *IEEE* **MTT-40**, 1992, 1719–1721.

[11] A. B. Dalby, Interdigital microstrip circuit parameters using empirical formulas and simplified model, *IEEE Trans.* **MTT-27**, 1979, 744–752.

[12] *em User's Manual*, Sonnet Software Inc., Liverpool, New York, 1993.

[13] T. A. Millgan, Dimensions of microstrip coupled lines and interdigital structures. *IEEE Trans.* **MTT**, 1977, 405–410.

[14] C.-K. C. Tzuang and W.-T. Lo, Printed-circuit realization of a tapped combline bandpass filter, *IEEE MTT-S Dig.* 1990, 131–134.

[15] D. Zhang, G.-C. Liang, C. F. Shih, R. S. Withers, M. E. Johansson, and A. Dela Cruz, Compact forward-coupled superconducting microstrip filters for cellular communication, Applied Superconductivity Conference, Oct. 16–21, 1994, Boston, MA.

[16] G. L. Matthaei and G. L. Hey-Shipton, Novel staggered resonator array superconducting 2.3-GHz bandpass filter. *IEEE Trans.* **MTT-41**, 1993, 2345–2352.

CHAPTER SIX

Highpass and Bandstop Filters

In this chapter, we will discuss some typical microstrip highpass and bandstop filters. These include quasilumped element and optimum distributed highpass filters, narrow- and wide-band bandstop filters, as well as filters for RF chokes. Design equations, tables, and examples are presented for easy reference.

6.1 HIGHPASS FILTERS

6.1.1 Quasilumped Highpass Filters

Highpass filters constructed from quasilumped elements may be desirable for many applications, provided that these elements can achieve good approximation of desired lumped elements over an entire operating frequency band. Care should be taken when designing this type of filter, because as the size of any quasilumped element becomes comparable with the wavelength of an operating frequency, it no longer behaves as a lumped element.

The simplest form of a highpass filter may just consist of a series capacitor, which is often applied for direct current or dc block. For more selective highpass filters, more elements are required. This type of highpass filter can be easily designed based on a lumped-element lowpass prototype, such as one shown in Figure 6.1a, where g_i denotes the element values normalized by a terminating impedance Z_0 and obtained at a lowpass cutoff frequency Ω_c. Following the discussions in the Chapter 3, if we apply the frequency mapping

$$\Omega = -\frac{\omega_c \Omega_c}{\omega} \qquad (6.1)$$

Microstrip Filters for RF/Microwave Applications by Jia-Sheng Hong
Copyright © 2011 John Wiley & Sons, Inc.

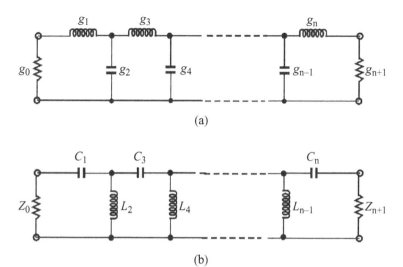

FIGURE 6.1 (a) A lowpass prototype filter; (b) highpass filter transformed from the lowpass prototype.

Ω and ω are the angular frequency variables of the lowpass and highpass filters respectively, while ω_c is the cutoff frequency of the highpass filter. Any series inductive element in the lowpass prototype filter is transformed to a series capacitive element in the highpass filter, with a capacitance

$$C_i = \frac{1}{Z_0 \omega_c \Omega_c g_i} \tag{6.2}$$

Likewise, any shunt-capacitive element in the lowpass prototype is transformed to a shunt-inductive element in the highpass filter, with an inductance

$$L_i = \frac{Z_0}{\omega_c \Omega_c g_i} \tag{6.3}$$

Figure 6.1b illustrates such a lumped-element highpass filter resulting from the transformations.

In order to demonstrate the technique for designing a quasilumped element highpass filter in microstrip, we will consider the design of a three-pole highpass microstrip filter with 0.1-dB passband ripple and a cutoff frequency $f_c = 1.5$ GHz ($\omega_c = 2\pi f_c$). The normalized element values of a corresponding Chebyshev lowpass prototype filter are $g_0 = g_4 = 1.0$, $g_1 = g_3 = 1.0316$, and $g_2 = 1.1474$ for $\Omega_c = 1$. The highpass filter will operate between 50-Ω terminations so that $Z_0 = 50$ Ω. Using design Eqs. (6.2) and (6.3), we find

$$C_1 = C_3 = \frac{1}{Z_0 \omega_c \Omega_c g_1} = \frac{1}{50 \times 2\pi \times 1.5 \times 10^9 \times 1.0316} = 2.0571 \times 10^{-12} \text{ F}$$

$$L_2 = \frac{Z_0}{\omega_c \Omega_c g_2} = \frac{50}{2\pi \times 1.5 \times 10^9 \times 1.1474} = 4.6236 \times 10^{-9} \text{ H}$$

A possible realization of such a highpass filter in microstrip, using quasilumped elements, is shown in Figure 6.2a. It is seen here that the series capacitors for C_1 and C_3 are realized by two identical interdigital capacitors, while the shunt inductor for L_2 is realized by a short-circuited stub. The microstrip highpass filter is designed on a commercial substrate (RT/D 5880) with a relative dielectric constant of 2.2 and a thickness of 1.57 mm. In determining the dimensions of the interdigital capacitors, such as the finger width, length, and space, as well as the number of the fingers, the closed-form design formulation for interdigital capacitor discussed in the Chapter 4 may be used. Alternatively, full-wave EM simulations can be performed to extract the two-port admittance parameters of an interdigital capacitor for different dimensions. The desired dimensions are found such that the extracted admittance parameter $Y_{12} = Y_{21}$ at the cutoff frequency f_c is equal to $-j\omega_c C_1$. The interdigital capacitor determined by this approach is comprised of 10 fingers, each of which is 10 mm long and 0.3 mm wide, with a 0.2 mm space with respect to the adjacent ones. The dimensions of the short-circuited stub, namely, the width W and length l, can be estimated from

$$jZ_c \tan\left(\frac{2\pi}{\lambda_{gc}}l\right) = j\omega_c L_2 \tag{6.4}$$

where Z_c is the characteristic impedance of the stub line, λ_{gc} is its guided wavelength at the cutoff frequency f_c, and both depend on the line width W on a substrate. One might recognize that the term on the left-hand side of Eq. (6.4) is the input impedance of a short-circuited transmission line. With a line width $W = 2.0$ mm on the given substrate, it is found by using the microstrip design equations in Chapter 4 that $Z_c = 84.619$ Ω and $\lambda_{gc} = 149.66$ mm. Therefore, $l = 11.327$ mm is obtained from Eq. (6.4), which is equivalent to an electrical length of 27.25° at 1.5 GHz. Although the short-circuited stub of this length has a reactance matching that of the ideal inductor at the cutoff frequency, it will have about 36% higher reactance than the idealized lumped-element design at 3 GHz. Generally speaking, to achieve a good approximation of a lumped-element inductor over a wide frequency band, it is essential to keep the length of a short-circuited stub as short as possible. This would normally occur for a narrower line with higher characteristic impedance, which, however, is restricted by fabrication tolerance and power-handling capability.

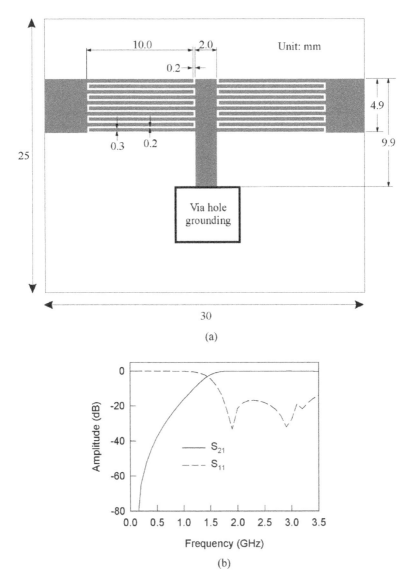

FIGURE 6.2 (a) A quasilumped highpass filter in microstrip on a substrate with a relative dielectric constant of 2.2 and a thickness of 1.57 mm. (b) EM-simulated performance of the quasilumped highpass filter.

The final dimensions of the designed microstrip highpass filter, as shown in Figure 6.2a, were determined by EM simulation of the whole filter, taking into account the effects of discontinues and parasitic parameters. The EM-simulated performance of the final filter is plotted in Figure 6.2b. It should be mentioned that the interdigital capacitors start to resonate at about 3.7 GHz, which limits the usable bandwidth.

Reducing the size of the interdigital capacitors or replacing them with appropriate microwave chip or beam-lead capacitors can lead to an increase in the bandwidth.

6.1.2 Optimum Distributed Highpass Filters

Highpass filters can also be constructed from distributed elements, such as commensurate (equal electrical length) transmission-line elements. Since any commensurate network exhibits periodic frequency response, the wide-band bandpass stub filters discussed in Chapter 5 may be used as pseudohighpass filters as well, particularly for wide-band applications, but they may not be optimum ones. This is because the unit elements (connecting lines) in those filters are redundant and their filtering properties are not fully utilized. For this reason, we will discuss another type of distributed highpass filter [1] in this section.

The type of filter is shown in Figure 6.3a, which consists of a cascade of shunt short-circuited stubs of electrical length θ_c at some specified frequency f_c (usually the cutoff frequency of high pass), separated by connecting lines (unit elements) of electrical length $2\theta_c$. Although the filter consists of only n stubs, it has an insertion function of degree $2n - 1$ in frequency so that its highpass response has $2n - 1$ ripples. This compares with n ripples for an n-stub bandpass (pseudo highpass) filter discussed in Chapter 5. Therefore, the stub filter of Figure 6.3a will have a fast rate of cutoff and may be argued to be optimum in this sense. Figure 6.3b illustrates the typical transmission characteristics of this type of filter, where f is the frequency variable and θ is the electrical length, which is proportional to f, that is,

$$\theta = \theta_c \frac{f}{f_c} \tag{6.5}$$

For high-pass applications, the filter has a primary passband from θ_c to $\pi - \theta_c$ with a cutoff at θ_c. The harmonic passbands occur periodically, centered at $\theta = 3\pi/2$, $5\pi/2, \ldots$, and separated by attenuation poles located at $\theta = \pi, 2\pi, \ldots$. The filtering characteristics of the network in Figure 6.3a can be described by a transfer (insertion) function

$$|S_{21}(\theta)|^2 = \frac{1}{1 + \varepsilon^2 F_N^2(\theta)} \tag{6.6}$$

where ε is the passband ripple constant, θ is the electrical length as defined in Eq. (6.5), and F_N is the filtering function given by

$$F_N(\theta) = \frac{\left(1 + \sqrt{1 - x_c^2}\right) T_{2n-1}\left(\frac{x}{x_c}\right) - \left(1 - \sqrt{1 - x_c^2}\right) T_{2n-3}\left(\frac{x}{x_c}\right)}{2\cos\left(\frac{\pi}{2} - \theta\right)} \tag{6.7}$$

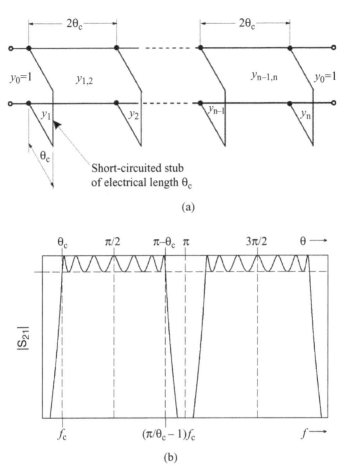

FIGURE 6.3 (a) Optimum distributed highpass filter; (b) typical filtering characteristics of the optimum distributed highpass filter.

where n is the number of short-circuited stubs,

$$x = \sin\left(\frac{\pi}{2} - \theta\right), \qquad x_c = \sin\left(\frac{\pi}{2} - \theta_c\right) \tag{6.8}$$

and $T_n(x) = \cos(n \cos^{-1} x)$ is the Chebyshev function of the first kind of degree n.

Theoretically, this type of highpass filter can have an extremely wide primary passband as θ_c becomes very small. However, this may require unreasonably high impedance levels for short-circuited stubs. Nevertheless, practical stub filter designs will meet many wide-band applications. Table 6.1 tabulates some typical element values of the network in Figure 6.3a for practical design of optimum highpass filters with two −six stubs and a passband ripple of 0.1 dB for $\theta_c = 25°$, $30°$, and $35°$. Note that the tabulated elements are the normalized characteristic admittances of transmission line elements and, for given terminating impedance Z_0, the associated characteristic line impedances are determined by

$$Z_i = Z_0/y_i$$
$$Z_{i,i+1} = Z_0/y_{i,i+1}$$
(6.9)

Design Example

To demonstrate how to design this type of filter, let us consider the design of an optimum distributed highpass filter having a cutoff frequency $f_c = 1.5$ GHz and a 0.1-dB ripple passband up to 6.5 GHz. Referring to Figure 6.3b, the electrical length θ_c can be found from

$$\left(\frac{\pi}{\theta_c} - 1\right) f_c = 6.5$$

This gives $\theta_c = 0.589$ radians or $\theta_c = 33.75°$. Assume that the filter is designed with six shorted-circuited stubs. From Table 6.1, we could choose the element values for $n = 6$ and $\theta_c = 30°$, which will gives a wider passband up to 7.5 GHz, because the smaller the electrical length at the cutoff frequency, the wider the passband. Alternatively, we can find the element values for $\theta_c = 33.75°$ by interpolation

TABLE 6.1 Element Values of Optimum Distributed Highpass Filters with 0.1-dB Ripple

n	θ_c	$y_1\ y_n$	$y_{1,2}\ y_{n-1,n}$	$y_2\ y_{n-1}$	$y_{2,3}\ y_{n-2,n-1}$	$y_3\ y_{n-2}$	$y_{3,4}$
2	25°	0.15436	1.13482				
	30°	0.22070	1.11597				
	35°	0.30755	1.08967				
3	25°	0.19690	1.12075	0.18176			
	30°	0.28620	1.09220	0.30726			
	35°	0.40104	1.05378	0.48294			
4	25°	0.22441	1.11113	0.23732	1.10361		
	30°	0.32300	1.07842	0.39443	1.06488		
	35°	0.44670	1.03622	0.60527	1.01536		
5	25°	0.24068	1.10540	0.27110	1.09317	0.29659	
	30°	0.34252	1.07119	0.43985	1.05095	0.48284	
	35°	0.46895	1.02790	0.66089	0.99884	0.72424	
6	25°	0.25038	1.10199	0.29073	1.08725	0.33031	1.08302
	30°	0.35346	1.06720	0.46383	1.04395	0.52615	1.03794
	35°	0.48096	1.02354	0.68833	0.99126	0.77546	0.98381

from the element values presented in the table. As an illustration, for $n = 6$ and $\theta_c = 33.75°$, the element value y_1 is calculated as follows:

$$y_1 = 0.35346 + \frac{(0.48096 - 0.35346)}{5} \times 3.75 = 0.44909$$

In a similar way, the rest of element values are found to be

$$y_{1,2} = 1.03446, \; y_2 = 0.63221, \; y_{2,3} = 1.00443, \; y_3 = 0.71313, \; y_{3,4} = 0.99734$$

These interpolated element values are well within 1% of directly synthesized element values. The filter is supposed to be doubly terminated by $Z_0 = 50$ Ω. Using Eq. (6.9), the characteristic impedances for the line elements are $Z_1 = Z_6 = 111.3$ Ω, $Z_2 = Z_5 = 79.1$ Ω, $Z_3 = Z_4 = 70.1$ Ω, $Z_{1,2} = Z_{5,6} = 48.3$ Ω, $Z_{2,3} = Z_{4,5} = 49.8$ Ω, and $Z_{3,4} = 50.1$ Ω.

The filter is represented in a microstrip on a substrate with a relative dielectric constant of 2.2 and a thickness of 1.57 mm. The initial dimensions of the filter can be easily estimated by using the microstrip design equations discussed in Chapter 4 for realizing these characteristic impedances and the required electrical lengths at the cutoff frequency, namely, $\theta_c = 33.75°$ for all the stubs and $2\theta_c = 67.5°$ for all the connecting lines. The final filter design with all the determined dimensions is shown in Figure 6.4a, where the final dimensions have taken into account the effects of discontinues, and have been slightly modified to allow all the connecting lines to have a 50-Ω characteristic impedance. The design is verified by full-wave EM simulation. Figure 6.4b is the simulated performance of the filter. We can see that the filter frequency responses does show eleven or $2n - 1$ ripples in the designed passband, as would be expected for this type of optimum highpass filter with only $n = 6$ stubs.

6.2 BANDSTOP FILTERS

6.2.1 Narrow-Band Bandstop Filters

Figure 6.5 shows two typical configurations for TEM or quasi-TEM narrow-band bandstop filters. In Figure 6.5a, a main transmission line is electrically coupled to half-wavelength resonators, while in Figure 6.5b, a main transmission line is magnetically coupled to half-wavelength resonators in a hairpin shape. In either case, the resonators are spaced a quarter-guided wavelength apart. If desired, the half-wavelength, open-circuited resonators may be replaced with short-circuited, quarter-wavelength resonators having one short-circuited end.

A simple and general approach for design of narrow-band bandstop filters is based on reactance/susceptance slope parameters of the resonators. To employ a lowpass prototype, for bandstop filter design, such as those discussed in Chapter 3,

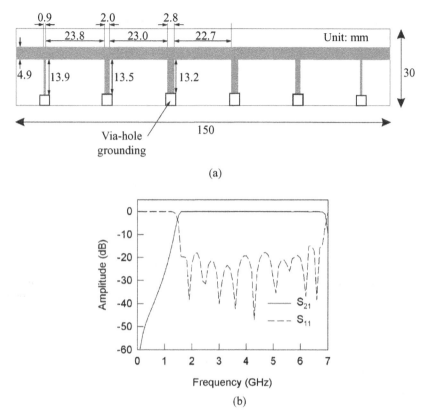

FIGURE 6.4 (a) A microstrip optimum highpass filter on a substrate with a relative dielectric constant of 2.2 and a thickness of 1.57 mm. (b) EM-simulated performance of the microstrip optimum highpass filter.

the transition from lowpass to bandstop characteristics can be effected by frequency mapping

$$\Omega = \frac{\Omega_c FBW}{(\omega/\omega_0 - \omega_0/\omega)}$$
$$\omega_0 = \sqrt{\omega_1 \omega_2} \qquad (6.10)$$
$$FBW = \frac{\omega_2 - \omega_1}{\omega_0}$$

where Ω is the normalized frequency variable of a lowpass prototype, Ω_c is its cutoff, and ω_0 and FBW are the midband frequency and fractional bandwidth of the bandstop filter. The band-edge frequencies ω_1 and ω_2 are indicated in Figure 6.6. Two equivalent circuits for the bandstop filters of Figure 6.5 can then be obtained as depicted in Figure 6.7, where Z_0 and Y_0 denote the terminating impedance and

BANDSTOP FILTERS 171

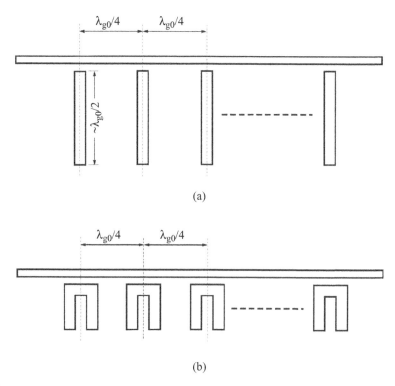

FIGURE 6.5 TEM or quasi-TEM narrow-band bandstop with (**a**) electric and (**b**) magnetic couplings.

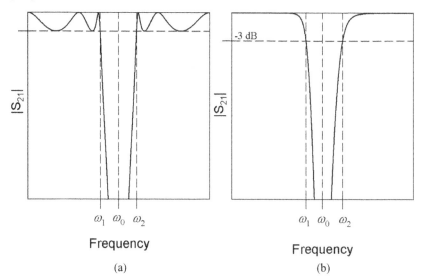

FIGURE 6.6 Bandstop filter characteristics defining midband frequency and bandedge frequencies. (**a**) Chebyshev characteristic; (**b**) Butterworth characteristic.

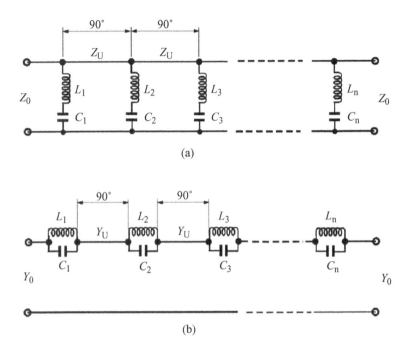

FIGURE 6.7 Equivalent circuits of bandstop filters with (a) shunt-series resonant branches and (b) series-parallel resonant branches.

admittance, Z_U and Y_U are the characteristic impedance and admittance of immittance inverters, and all the circuit parameters including inductances L_i and capacitances C_i can be defined in terms of lowpass prototype elements [2]. For the circuit in Figure 6.7a:

$$\left(\frac{Z_U}{Z_0}\right)^2 = \frac{1}{g_0 g_{n+1}}$$

$$x_i = \omega_0 L_i = \frac{1}{\omega_0 C_i} = Z_0 \left(\frac{Z_U}{Z_0}\right)^2 \frac{g_0}{g_i \Omega_c FBW} \quad \text{for } i = 1 \text{ to } n \quad (6.11a)$$

where g_i are the element values of lowpass prototype and x_i are the reactance slope parameters of shunt-series resonators. For series branches in Figure 6.7b:

$$\left(\frac{Y_U}{Y_0}\right)^2 = \frac{1}{g_0 g_{n+1}}$$

$$b_i = \omega_0 C_i = \frac{1}{\omega_0 L_i} = Y_0 \left(\frac{Y_U}{Y_0}\right)^2 \frac{g_0}{g_i \Omega_c FBW} \quad \text{for } i = 1 \text{ to } n \quad (6.11b)$$

where b_i are the susceptance slope parameters of series-parallel resonators.

It is obvious that for a chosen lowpass prototype, with known element values, the desired reactance/susceptance slope parameters can easily be determined using Eq. (6.11). The next step is to design microwave bandstop resonators, such as those in Figure 6.5, so as to have prescribed slope parameters. A practical and general technique that allows one to extract slope parameters of microwave bandstop resonators using EM simulations or experiments is discussed next.

Consider a two-port network with a single-shunt branch of $Z = j\omega L + 1/(j\omega C)$, such as one in Figure 6.7a. The shunt-branch resonates at $\omega_0 = 1/\sqrt{LC}$ and has a reactance slope $x = \omega_0 L$. The transmission parameter for this two-port network terminated with Z_0 is given by

$$S_{21} = \frac{1}{1 + \dfrac{Z_0}{2Z}} \tag{6.12}$$

Let $\omega = \omega_0 + \Delta\omega$. In narrow-band cases, $\Delta\omega << \omega_0$, and thus the shunt impedance may be approximated by

$$Z \approx j\omega_0 L \left(\frac{2\Delta\omega}{\omega_0}\right) \tag{6.13}$$

in which the approximation $(\omega/\omega_0 - \omega_0/\omega) \approx 2\Delta\omega/\omega_0$ has been made. By substitution, we can obtain from Eq. (6.12)

$$|S_{21}| = \frac{1}{\sqrt{1 + \left[\dfrac{1}{4(x/Z_0)}\dfrac{\omega_0}{\Delta\omega}\right]^2}} \tag{6.14}$$

This is at resonance when $\omega = \omega_0$ or $\Delta\omega = 0$, $|S_{21}| = 0$ because the resonant shunt branch shorts out the transmission and causes an attenuation pole. When the frequency shifts such that

$$\frac{1}{4(x/Z_0)}\frac{\omega_0}{\Delta\omega_\pm} = \pm 1 \tag{6.15}$$

the value of $|S_{21}|$ has risen to 0.707 or -3 dB. From Eq. (6.15), a 3-dB bandwidth can be defined by

$$\Delta\omega_{3dB} = \Delta\omega_+ - \Delta\omega_- = \frac{\omega_0}{2(x/Z_0)} \tag{6.16}$$

and, thus,

$$\left(\frac{x}{Z_0}\right) = \frac{\omega_0}{2\Delta\omega_{3dB}} = \frac{f_0}{2\Delta f_{3dB}} \quad (6.17)$$

This equation is very useful, because it relates the normalized reactance slope parameter to the frequency response of a microwave bandstop resonator, and the latter can quite easily be obtained by EM simulation or measurement. It should be mentioned that if another attenuation bandwidth, other than the 3-dB bandwidth is desirable for extracting the normalized reactance slope parameter, the relationship between the desired attenuation bandwidth and the normalized reactance slope parameter can be derived in the steps similar to Eqs. (6.15–6.17).

Similarly, to derive the formulation for extracting susceptance slope parameter, let us consider a two-port network with a single series branch of $Y = j\omega C + 1/(j\omega L)$, such as one in Figure 6.7b. The series branch has a parallel-resonant frequency $\omega_0 = 1/\sqrt{LC}$ and a susceptance-slope parameter $b = \omega_0 C$. The transmission parameter for this two-port network terminated with Y_0 is given by

$$S_{21} = \frac{1}{1 + \dfrac{Y_0}{2Y}} \quad (6.18)$$

For narrow-band applications, the amplitude of Eq. (6.18) may be approximated by

$$|S_{21}| = \frac{1}{\sqrt{1 + \left[\dfrac{1}{4(b/Y_0)}\dfrac{\omega_0}{\Delta\omega}\right]^2}} \quad (6.19)$$

where $(\omega/\omega_0 - \omega_0/\omega) \approx 2\Delta\omega/\omega_0$ using $\omega = \omega_0 + \Delta\omega$. This is at resonance when $\omega = \omega_0$ or $\Delta\omega = 0$, $|S_{21}| = 0$, because the resonant series branch blocks out the transmission and causes an attenuation pole. The attenuation will then be reduced or the value of $|S_{21}|$ will rise when the frequency shifts away from ω_0. When the frequency shifts such that

$$\frac{1}{4(b/Y_0)}\frac{\omega_0}{\Delta\omega_\pm} = \pm 1 \quad (6.20)$$

the value of $|S_{21}|$ will reach to 0.707 or −3 dB. From Eq. (6.20) a 3-dB bandwidth can be defined by

$$\Delta\omega_{3dB} = \Delta\omega_+ - \Delta\omega_- = \frac{\omega_0}{2(b/Y_0)} \quad (6.21)$$

Therefore, we have

$$\left(\frac{b}{Y_0}\right) = \frac{\omega_0}{2\Delta\omega_{3dB}} = \frac{f_0}{2\Delta f_{3dB}} \tag{6.22}$$

Similar to Eq. (6.17), Eq. (6.22) is useful for extraction of the normalized susceptance slope parameter of a microwave bandstop resonator from its frequency response by either EM simulation or measurement.

One might notice that for a given lowpass prototype, the prescribed normalized slope parameters obtained from Eq. (6.11a) and (6.11b) have the same value for the ith resonator, referred to in Figs. 6.7a and b. Equations (6.17) and (6.22) have the exact formulation for extracting the normalized slope parameters. From these two facts, an important and useful conclusion can be drawn: when we design a narrow-band bandstop filter based on the normalized slope parameters, either sets of design equations, namely, Eqs. (6.11a) and (6.17) or Eqs. (6.11b) and (6.22), can be used, regardless of actual structures of microwave bandstop resonators and regardless of whether the couplings are electric or magnetic or mixed.

Design Example

A five-pole ($n = 5$) Chebyshev lowpass prototype with a passband ripple of 0.1 dB is chosen for designing a microstrip bandstop filter, as shown in Figure 6.8. The microstrip bandstop filter uses L-shape resonators coupled both electrically and magnetically to the main line. The desired band-edge frequencies to equal-ripple (0.1-dB) points are $f_1 = 3.3$ and $f_2 = 3.5$ GHz. Hence, the midband frequency of the stopband is $f_0 = 3.3985$ GHz and the fractional bandwidth is $FBW = 0.0588$, according to Eq. (6.10). The element values of the chosen lowpass prototype for $\Omega_c = 1$ are

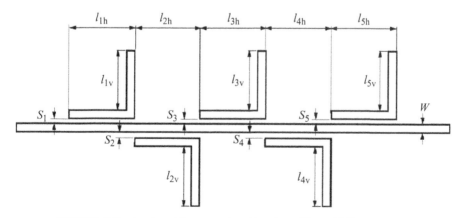

FIGURE 6.8 A microstrip narrow-band bandstop filter with L resonators.

$g_0 = g_6 = 1.0$, $g_1 = g_5 = 1.1468$, $g_2 = g_4 = 1.3712$, and $g_3 = 1.9750$. Using Eq. (6.11a) to calculate the desired design parameters yields

$$Z_U = Z_0 \qquad \frac{x_1}{Z_0} = \frac{x_5}{Z_0} = 14.8170$$

$$\frac{x_3}{Z_0} = 8.6038 \qquad \frac{x_2}{Z_0} = \frac{x_4}{Z_0} = 12.3924$$

where $Z_0 = 50\,\Omega$. For microstrip design, a commercial substrate (RT/D 6010) with a relative dielectric constant of 10.8 and thickness of 1.27 mm is used. The 50-Ω main line has a width $W = 1.1$ mm. For simplicity, the line width for the L-resonators is fixed to be 1.1 mm as well. The resonator lengths are made the same to be half-guided wavelength at f_0 with $l_h = 8.9$ mm and $l_v = 7.9$ mm. Frequency responses of a single L-resonator coupled to the main line for different coupling spacing s are then simulated using a full-wave EM simulator. The normalized reactance slope parameters are then extracted according to Eq. (6.17). The typical simulated frequency response and extracted normalized reactance slope parameters are plotted in Figure 6.9, from which the desired coupling spacings are determined to be $s_1 = s_5 = 0.292$ mm, $s_2 = s_4 = 0.221$ mm, and $s_3 = 0.102$ mm. Figure 6.10a is a photograph of the fabricated microstrip bandstop filter. The measured and simulated performances of the filter are illustrated in Figure 6.10b, showing a good agreement between the two. It should be remarked that the measured filter was enclosed in a copper housing to reduce radiation losses, otherwise the stopband attenuation around the midband would be

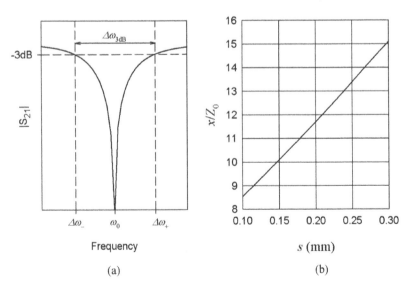

FIGURE 6.9 (a) Typical simulated frequency response of a single microstrip L resonator coupled to a main transmission line. (b) Extracted normalized reactance-slope parameters against coupling spacings.

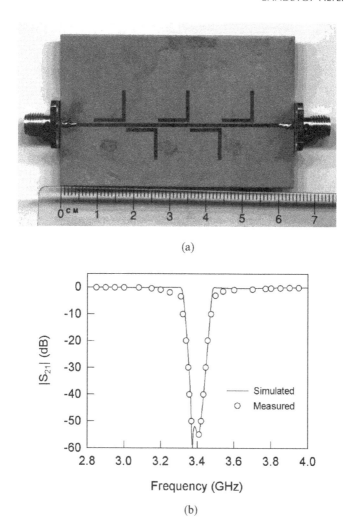

FIGURE 6.10 (a) Photograph of the fabricated microstrip bandstop filter on a substrate with a relative dielectric constant of 10.8 and a thickness of 1.27 mm. (b) Measured and simulated performances of the filter.

degraded. Frequency tuning is also normally required for narrow-band bandstop filters to compensate for fabrication tolerances. In this case, the length l_v for each resonator could be slightly trimmed.

6.2.2 Bandstop Filters with Open-Circuited Stubs

Figure 6.11a is a transmission line network of bandstop filter with open-circuited stubs, where the shunt quarter-wavelength, open-circuited stubs are separated by unit

178 HIGHPASS AND BANDSTOP FILTERS

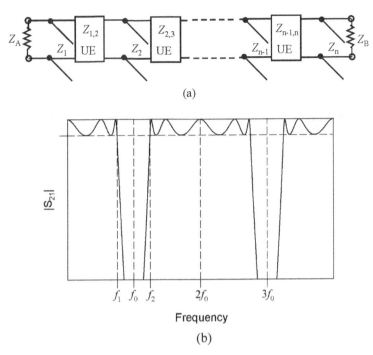

FIGURE 6.11 Bandstop filter with open-circuited stubs. (a) Transmission-line network representation. (b) Frequency characteristic defining midband frequency f_0 and band edge frequencies f_1 and f_2 ($f_1 < f_0 < f_2$).

elements (connecting lines) that are quarter-wavelength long at the mid-stopband frequency. The filtering characteristic of the filter then entirely depends on design of characteristic impedances Z_i for the open-circuited stubs and characteristic impedances $Z_{i,i+1}$ for the unit elements, as well as two terminating impedances Z_A and Z_B. Theoretically, this type of filter can be designed to have any stopband width. However, in practice, the impedance of the open-circuited stubs becomes unreasonably high if the stopband width is very narrow. Therefore, this type of bandstop filter is more suitable for realization of wide-band bandstop filters.

This type of bandstop filter may be designed using a design procedure [3]. The design procedure starts with a chosen ladder-type lowpass prototype (that is, with Chebyshev response). It then uses a frequency mapping

$$\Omega = \Omega_c \alpha \tan\left(\frac{\pi}{2} \frac{f}{f_0}\right)$$
$$\alpha = \cot\left(\frac{\pi}{2}\left(1 - \frac{FBW}{2}\right)\right) \quad (6.23)$$

where Ω and Ω_c are the normalized frequency variable and the cutoff frequency of a lowpass prototype filter, f and f_0 are the frequency variable and the midband frequency of the corresponding bandstop filter, and FBW is the fractional bandwidth of the bandstop filter defined by

$$FBW = \frac{f_2 - f_1}{f_0} \text{ with } f_0 = \frac{f_1 + f_2}{2} \quad (6.24)$$

f_1 and f_2 are frequency points in the bandstop response as indicated in Figure 6.11b. It should be mentioned that the bandstop filter of this type have spurious stop bands periodically centered at frequencies that are odd multiples of f_0. At these frequencies, the shunt open-circuited stubs in the filter of Figure 6.11a are odd multiple of $\lambda_{g0}/4$ long, with λ_{g0} being the guided wavelength at frequency f_0, so that they short out the main line and cause spurious stop bands. Note that the frequency mapping in Eq. (6.23) actually involves the Richards' transformation. Therefore, under the mapping of Eq. (6.23), the shunt (capacitive) elements of lowpass prototype become shunt (open-circuited) stubs of the mapped bandstop filter, while the series (inductive) elements become series (short-circuited) stubs. The series short-circuited stubs are then removed by utilizing Kuroda's identities to obtain the desired transmission line bandstop filter of Figure 6.11a. To demonstrate this design procedure, let us consider a six-pole ($n = 6$) ladder-type lowpass prototype in Figure 6.12a, where g_i ($i = 0$ to $n + 1$) are the normalized lowpass elements. Applying Eq. (6.23) to the shunt capacitors and series inductors of this prototype filter results in a transmission-line filter of Figure 6.12b, where y'_i, which are the normalized characteristic admittances of the shunt open-circuited stubs, are given by

$$y'_i = \Omega_c \alpha g_i \quad \text{for } i = 2, 4, \text{ and } 6 \quad (6.25a)$$

while z'_i, which are the normalized characteristic impedances of the series shorted-circuited stubs, and are given by

$$z'_i = \Omega_c \alpha g_i \quad \text{for } i = 1, 3, \text{ and } 5 \quad (6.25b)$$

All the stubs are $\lambda_{g0}/4$ long, with λ_{g0} being the guided wavelength at frequency f_0.

In order to remove the three series short-circuited stubs, three unit elements of normalized characteristic impedance z'_A are inserted on the left after the normalized terminating impedance z'_A and, similarly, two unit elements of normalized characteristic impedance z'_B are inserted on the right before the terminating impedance z'_B, as shown in Figure 6.12c. Since the characteristic impedance of the inserted unit elements on each side match that of the termination on the same side, the inserted unit elements have no effect on the amplitude characteristic of the filter, but just add

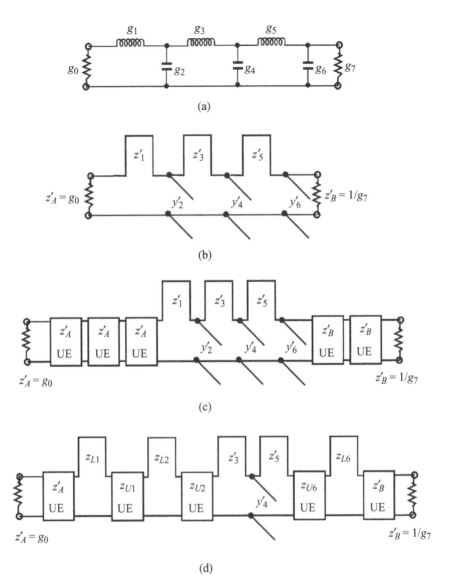

FIGURE 6.12 Primary steps in the transformation of a lowpass prototype filter into a bandstop transmission-line filter. (a) Prototype lowpass filter ($n = 6$). (b) After applying frequency mapping. (c) After adding $n - 1$ matched unit elements from the two terminals. (d) After applying a sequence of Kuroda's identities.

BANDSTOP FILTERS 181

some phase shift. However, if we apply Kuroda's identities (as described in Chapter 3) in a sequence with the following equivalent parameter transformations

$$y_{C0} = \frac{z'_1}{z'_A(z'_A + z'_1)}, \qquad z_{U0} = z'_A + z'_1 \qquad (6.26a)$$

$$z_{L6} = \frac{z'^2_B y'_6}{1 + z'_B y'_6}, \qquad z_{U6} = \frac{1 + z'_B}{1 + z'_B y'_6} \qquad (6.26b)$$

$$z_{L2} = \frac{z^2_{U0} y'_2}{1 + z_{U0} y'_2}, \qquad z_{U2} = \frac{z_{U0}}{1 + z_{U0} y'_2} \qquad (6.26c)$$

$$z_{L1} = \frac{z'^2_A y_{C0}}{1 + z'_A y_{C0}}, \qquad z_{U1} = \frac{z'_A}{1 + z'_A y_{C0}} \qquad (6.26d)$$

we obtain an equivalent network as shown in Figure 6.12d. Applying Kuroda's identity simultaneously to stub z'_3 and unit element z_{U2}, to stub z_{L2} and unit element z_{U1}, to stub z_{L1} and unit element z'_A, to stub z'_5 and unit element z_{U6}, and to stub z_{L6} and unit element z'_B in Figure 6.12d then yields a desired network shown in Figure 6.11a, with the following design parameters

$$Z_A = Z_0 z'_A, \qquad Z_B = Z_0 z'_B$$

$$Z_1 = Z_0 z'_A \frac{z'_A + z_{L1}}{z_{L1}}, \qquad Z_{1,2} = Z_0(z'_A + z_{L1})$$

$$Z_2 = Z_0 z_{U1} \frac{z_{U1} + z_{L2}}{z_{L2}}, \qquad Z_{2,3} = Z_0(z_{U1} + z_{L2})$$

$$Z_3 = Z_0 z_{U2} \frac{z_{U2} + z'_3}{z'_3}, \qquad Z_{3,4} = Z_0(z_{U2} + z'_3) \qquad (6.27)$$

$$Z_4 = Z_0/y'_4$$

$$Z_5 = Z_0 z_{U6} \frac{z_{U6} + z'_5}{z'_5}, \qquad Z_{4,5} = Z_0(z_{U6} + z'_5)$$

$$Z_6 = Z_0 z'_B \frac{z'_B + z_{L6}}{z_{L6}}, \qquad Z_{5,6} = Z_0(z'_B + z_{L6})$$

where Z_0 is a reference impedance, which is usually chosen as 50 Ω for microstrip filter designs.

Design equations for other filter orders can be derived in the similar way. For convenience, the design equations for first- to fifth-order bandstop filters of this type available in [3] are given as follows for $\Omega_c = 1$, $Z_A = Z_0 g_0$ and the bandwidth parameter α that is defined in Eq. (6.23).

For $n = 1$:

$$Z_1 = \frac{Z_A}{\alpha g_0 g_1}; \quad Z_B = \frac{Z_A g_2}{g_0} \qquad (6.28)$$

For $n = 2$:

$$Z_1 = Z_A\left(1 + \frac{1}{\alpha g_0 g_1}\right); \quad Z_{1,2} = Z_A(1 + \alpha g_0 g_1)$$
$$Z_2 = \frac{Z_A g_0}{\alpha g_2}; \quad Z_B = Z_A g_0 g_3 \qquad (6.29)$$

For $n = 3$:
Z_1, $Z_{1,2}$, and Z_2 are the same formulas as case $n = 2$

$$Z_3 = \frac{Z_A g_0}{g_4}\left(1 + \frac{1}{\alpha g_3 g_4}\right); \quad Z_{2,3} = \frac{Z_A g_0}{g_4}(1 + \alpha g_3 g_4)$$
$$Z_B = \frac{Z_A g_0}{g_4} \qquad (6.30)$$

For $n = 4$:

$$Z_1 = Z_A\left(2 + \frac{1}{\alpha g_0 g_1}\right); \qquad Z_{1,2} = Z_A\left(\frac{1 + 2\alpha g_0 g_1}{1 + \alpha g_0 g_1}\right)$$
$$Z_2 = Z_A\left(\frac{1}{1 + \alpha g_0 g_1} + \frac{g_0}{\alpha g_2(1 + \alpha g_0 g_1)^2}\right); \quad Z_{2,3} = \frac{Z_A}{g_0}\left(\alpha g_2 + \frac{g_0}{1 + \alpha g_0 g_1}\right)$$
$$Z_3 = \frac{Z_A}{\alpha g_0 g_3}; \qquad Z_{3,4} = \frac{Z_A}{g_0 g_5}(1 + \alpha g_4 g_5)$$
$$Z_4 = \frac{Z_A}{g_0 g_5}\left(1 + \frac{1}{\alpha g_4 g_5}\right); \qquad Z_B = \frac{Z_A}{g_0 g_5}$$

$$(6.31)$$

For $n = 5$:
Z_1, $Z_{1,2}$, Z_2, $Z_{2,3}$, Z_3 are the same formulas as case $n = 4$

$$Z_4 = \frac{Z_A}{g_0}\left(\frac{1}{1 + \alpha g_5 g_6} + \frac{g_6}{\alpha g_4(1 + \alpha g_5 g_6)^2}\right); \quad Z_{3,4} = \frac{Z_A}{g_0}\left(\alpha g_4 + \frac{g_6}{1 + \alpha g_5 g_6}\right)$$
$$Z_5 = \frac{Z_A g_6}{g_0}\left(2 + \frac{1}{\alpha g_5 g_6}\right); \qquad Z_{4,5} = \frac{Z_A g_6}{g_0}\left(\frac{1 + 2\alpha g_5 g_6}{1 + \alpha g_5 g_6}\right)$$
$$Z_B = \frac{Z_A g_6}{g_0}$$

$$(6.32)$$

The design parameters for $n = 6$ can be found from Eqs. (6.25–6.27).

BANDSTOP FILTERS

Design Example

In this example, a microstrip stopband filter of this type is designed based on a three-pole ($n = 3$) Chebyshev lowpass prototype with 0.05-dB passband ripple. The element values of the lowpass prototype are $g_0 = g_4 = 1.0$, $g_1 = g_3 = 0.8794$, and $g_2 = 1.1132$. The bandstop filter is designed to have a fractional bandwidth $FBW = 1.0$ at a midband frequency $f_0 = 2.5$ GHz for the band-edge frequencies $f_1 = 1.25$ GHz and $f_2 = 3.75$ GHz, as defined in Figure 6.11b. Thus, $\alpha = 1.0$ from Eq. (6.23). Using the design equations for $n = 3$ given above and $Z_0 = 50\ \Omega$, we obtain

$$Z_A = Z_B = 50\ \Omega$$
$$Z_1 = Z_3 = 106.8544\ \Omega, \quad Z_{1,2} = Z_{2,3} = 93.9712\ \Omega$$
$$Z_2 = 44.9169\ \Omega$$

A commercial substrate (RT/D 6006) having a relative dielectric constant of 6.15 and thickness of 1.27 mm is chosen for this microstrip filter representation. Using the microstrip design equations given in Chapter 4, the microstrip widths and guided quarter-wavelengths at 2.5 GHz associated with above-required characteristic impedances can be found, as can the physical dimensions of the filter. However, the open-end and T-junction effects should also be taken into account for determining the final filter dimensions. The layout of the final filter design with all the determined dimensions is illustrated in Figure 6.13a. The design is verified by the EM simulation. The EM-simulated performance of the filter is shown in Figure 6.13b.

6.2.3 Optimum Bandstop Filters

With the design method discussed in the last section, the unit elements of the bandstop filter in Figure 6.11a are redundant, and their filtering properties are not utilized, so that in this sense, the resultant bandstop filter is not an optimum one. It has been pointed out [4,5] that for wide-band bandstop filters of the type in Figure 6.11a, the unit elements can be made nearly as effective as the open-circuited stubs. Therefore, by incorporating the unit elements in the design, significantly steeper attenuation characteristics can be obtained for the same number of stubs than is possible for filters designed with redundant unit elements. A specified filter characteristic can also be met with a more compact configuration using fewer stubs, if the filter is designed by an optimum method.

To design optimum bandstop filters with n stubs, the network in Figure 6.11a is synthesized based on an optimum transfer function

$$|S_{21}(f)|^2 = \frac{1}{1 + \varepsilon^2 F_N^2(f)} \qquad (6.33)$$

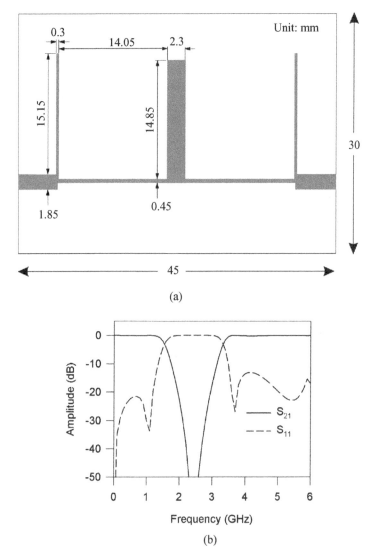

FIGURE 6.13 (a) Layout of a wide-band microstrip bandstop filter on a substrate with a relative dielectric constant of 6.15 and a thickness of 1.27 mm. (b) Full-wave EM simulated frequency responses of the filter.

where ε is the passband ripple constant and F_N is the filtering function give by

$$F_N(f) = T_n\left(\frac{t}{t_c}\right) T_{n-1}\left(\frac{t\sqrt{1-t_c^2}}{t_c\sqrt{1-t^2}}\right) - U_n\left(\frac{t}{t_c}\right) U_{n-1}\left(\frac{t\sqrt{1-t_c^2}}{t_c\sqrt{1-t^2}}\right) \quad (6.34)$$

in which t is the Richards' transform variable

$$t = j \tan\left(\frac{\pi}{2}\frac{f}{f_0}\right) \tag{6.35}$$

and

$$t_c = j \tan\left(\frac{\pi}{4}(2 - FBW)\right) \tag{6.36}$$

where f_0 is the midband frequency of bandstop filter and FBW is the fractional bandwidth as defined in the last section. $T_n(x)$ and $U_n(x)$ are the Chebyshev functions of the first and second kinds of order n:

$$\begin{aligned}T_n(x) &= \cos(n\cos^{-1}x)\\ U_n(x) &= \sin(n\cos^{-1}x)\end{aligned} \tag{6.37}$$

For convenience, element values of the network in Figure 6.11a, for design of optimum bandstop filters with two to six stubs and a passband return loss level of -20 dB, are tabulated in Tables 6.2–6.6 for bandwidths between 30% and 150%. Note that the tabulated elements are the normalized admittances and, for a given reference impedance Z_0, the impedances are determined by

$$\begin{aligned}Z_A &= Z_B = Z_0\\ Z_i &= Z_0/g_i\\ Z_{i,i+1} &= Z_0/J_{i,i+1}\end{aligned} \tag{6.38}$$

TABLE 6.2 Element Values of Optimum Bandstop Filters for $n = 2$ and $\varepsilon = 0.1005$

FBW	$g_1 = g_2$	$J_{1,2}$
0.3	0.16989	0.98190
0.4	0.23418	0.93880
0.5	0.30386	0.89442
0.6	0.38017	0.84857
0.7	0.46470	0.80106
0.8	0.55955	0.75173
0.9	0.66750	0.70042
1.0	0.79244	0.64700
1.1	0.93992	0.59137
1.2	1.11821	0.53346
1.3	1.34030	0.47324
1.4	1.62774	0.41077
1.5	2.01930	0.34615

TABLE 6.3 Element Values of Optimum Bandstop Filters for $n = 3$ and $\varepsilon = 0.1005$

FBW	$g_1 = g_3$	g_2	$J_{1,2} = J_{2,3}$
0.3	0.16318	0.26768	0.97734
0.4	0.23016	0.38061	0.92975
0.5	0.37754	0.63292	0.83956
0.6	0.46895	0.79494	0.78565
0.7	0.56896	0.97488	0.73139
0.8	0.67986	1.17702	0.67677
0.9	0.80477	1.40708	0.62180
1.0	0.94806	1.67311	0.56648
1.1	1.11601	1.98667	0.51082
1.2	1.15215	2.06604	0.49407
1.3	1.37952	2.49473	0.43430
1.4	1.67476	3.05136	0.37349
1.5	2.07059	3.79862	0.31262

Design Example

An optimum microstrip bandstop filter with three open-circuited stubs ($n = 3$) and a fractional bandwidth $FBW = 1.0$ at a midband frequency $f_0 = 2.5$ GHz will be designed. Assume a passband return loss of -20 dB, which corresponds to a ripple constant $\varepsilon = 0.1005$. From Table 6.3, we obtain the normalized element values $g_1 = g_3 = 0.94806$, $g_2 = 1.67311$, and $J_{1,2} = J_{2,3} = 0.56648$. The filter is designed

TABLE 6.4 Element Values of Optimum Bandstop Filters for $n = 4$ and $\varepsilon = 0.1005$

FBW	$g_1 = g_4$	$g_2 = g_3$	$J_{1,2} = J_{3,4}$	$J_{2,3}$
0.3	0.23069	0.40393	0.93372	0.91337
0.4	0.31457	0.55651	0.87752	0.85157
0.5	0.40366	0.72118	0.82172	0.79093
0.6	0.49941	0.90054	0.76623	0.73145
0.7	0.60366	1.09802	0.71101	0.67313
0.8	0.71884	1.31815	0.65598	0.61597
0.9	0.79436	1.46655	0.62025	0.57951
1.0	0.99642	1.85355	0.54634	0.50503
1.1	1.10390	2.06672	0.50871	0.46793
1.2	1.37861	2.59505	0.43702	0.39831
1.3	1.55326	2.94111	0.39654	0.35972
1.4	1.97310	3.74861	0.32781	0.29526
1.5	2.43047	4.63442	0.27321	0.24488

TABLE 6.5 Element Values of Optimum Bandstop Filters for $n = 5$ and $\varepsilon = 0.1005$

FBW	$g_1 = g_5$	$g_2 = g_4$	g_3	$J_{1,2} = J_{4,5}$	$J_{2,3} = J_{3,4}$
0.3	0.23850	0.42437	0.45444	0.92798	0.90213
0.4	0.32455	0.58273	0.62307	0.87068	0.83818
0.5	0.41542	0.75293	0.80324	0.81413	0.77611
0.6	0.51385	0.93717	0.99711	0.75705	0.71472
0.7	0.62178	1.14215	1.21166	0.70221	0.65683
0.8	0.74624	1.38212	1.46326	0.64418	0.59700
0.9	0.87071	1.62166	1.71464	0.59299	0.54475
1.0	1.01167	1.89960	2.00417	0.54092	0.49299
1.1	1.20308	2.26455	2.38104	0.48338	0.43663
1.2	1.40157	2.66645	2.79799	0.42804	0.38305
1.3	1.68069	3.19873	3.36796	0.37569	0.33651
1.4	2.00690	3.84473	4.02293	0.32252	0.28580
1.5	2.47075	4.74882	4.96115	0.26871	0.23694

to match 50-Ω terminations. Therefore, $Z_0 = 50\ \Omega$, and from Eq. (6.38) we determine the electrical design parameters for the filter network representation in Figure 6.11a

$$Z_A = Z_B = 50\ \Omega$$
$$Z_1 = Z_3 = 52.74\ \Omega$$
$$Z_2 = 29.88\ \Omega$$
$$Z_{1,2} = Z_{2,3} = 88.26\ \Omega$$

For microstrip design, we use a commercial substrate (RT/D 6006) with a relative dielectric constant of 6.15 and a thickness of 1.27 mm. Using the microstrip

TABLE 6.6 Element Values of Optimum Bandstop Filters for $n = 6$ and $\varepsilon = 0.1005$

FBW	$g_1 = g_6$	$g_2 = g_5$	$g_3 = g_4$	$J_{1,2} = J_{5,6}$	$J_{2,3} = J_{4,5}$	$J_{3,4}$
0.3	0.24270	0.43420	0.47301	0.92496	0.89714	0.89192
0.4	0.32964	0.59521	0.64657	0.86715	0.83242	0.82616
0.5	0.42153	0.76772	0.83149	0.81025	0.76986	0.76287
0.6	0.51988	0.95454	1.03066	0.75411	0.70933	0.70191
0.7	0.62664	1.15929	1.24785	0.69860	0.65071	0.64311
0.8	0.74437	1.38676	1.48806	0.64361	0.59385	0.58629
0.9	0.87646	1.64341	1.75810	0.58904	0.53863	0.53128
1.0	1.02761	1.93826	2.06742	0.53481	0.48491	0.47793
1.1	1.20459	2.28427	2.42968	0.48084	0.43255	0.42606
1.2	1.41745	2.70085	2.86529	0.42707	0.38142	0.37550
1.3	1.68193	3.21846	3.40624	0.37346	0.33137	0.32610
1.4	2.02423	3.88780	4.10584	0.31997	0.28228	0.27769
1.5	2.49141	4.80007	5.05987	0.26655	0.23399	0.23014

TABLE 6.7 Microstrip Design Parameters for a Three-Pole Optimum Bandstop Filter

Line Impedance	W(mm)	$\lambda_{g0}/4$ (mm)
Z_1 and Z_3	1.7	14.32
Z_2	4.25	13.73
$Z_{1,2}$ and $Z_{2,3}$	0.528	14.89
Z_A and Z_B	1.85	

design equations given in Chapter 4, the microstrip line width W and guided quarter-wavelength $\lambda_{g0}/4$ at 2.5 GHz for each of the above determined line impedances are found and listed in Table 6.7.

The final microstrip filter design, which takes into account the open-end and T-junction effects, is shown in Figure 6.14a. The full-wave EM simulated performance of this designed filter is illustrated in Figure 6.14b.

It is interesting to compare the filter design given in the last section, since both designs used the same number of stubs ($n = 3$), the same fractional bandwidth and midband frequency, as well as the same substrate. The layouts of the two designed filters in Figures 6.13a and 6.14a are similar, but the optimum design has more uniform stub line widths with a ratio of $4.25/1.7 = 2.5$, whereas the redundant design has a stub-width ratio of $2.3/0.3 \approx 7.67$. For filters with more stubs or narrower bandwidths, the outer stubs in a redundant design can become extremely narrow and, hence, cause some practical realization difficulties. Comparing the both filter performances in Figures 6.13b and 6.14b, the optimum design demonstrates substantially improved performance with a steeper stop band response.

6.2.4 Bandstop Filters for RF Chokes

A bandstop filter with quarter-wavelength open-circuited stubs may be designed as a pseudolowpass filter. Such a bandstop filter would function efficiently in a bias network to choke off RF transmission over its stopband, while maintaining a perfect transmission for direct current.

Figure 6.15a shows a basic bias network incorporating a bandstop filter (from points A to B). This type of network is also referred to as a bias T, which is commonly used for feeding dc into active RF components in such a way that the RF behavior is not at all affected by the dc connection. Referring to Figure 6.15a, the connection of the bandstop filter should not affect the RF transmission from ports 1 to 2. In other words, the bandstop filter should produce an RF open circuit at the connecting point A. Since RF active components, such as amplifiers and oscillators, operate over a limited frequency band, it is normally only necessary for a dc bias circuit to choke off the RF over such an operation bandwidth. On this ground, to use a stub bandstop filter instead of a lowpass filter would be preferred, because the open-circuited stubs, which are quarter-wavelength long at the center frequency of the operation bandwidth, can more efficiently choke off the operating RF signals.

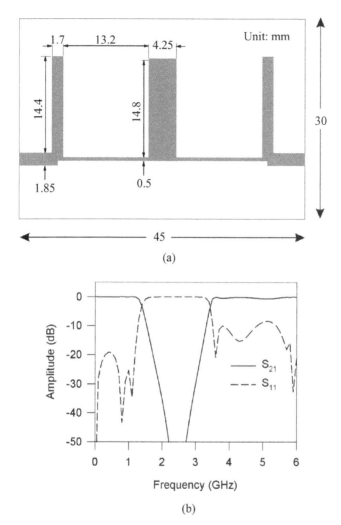

FIGURE 6.14 (a) Layout of a microstrip optimum bandstop filter with three open-circuited stubs on a substrate with a relative dielectric constant of 6.15 and thickness of 1.27 mm. (b) Full-wave EM simulated performance of the microstrip optimum bandstop filter.

In many situations, such a bandstop filter composed of a single section of an open-circuited stub and a quarter-wavelength connecting line, may be found to be adequate for the required RF choke. To obtain a wider stop band, more sections may be cascaded. Furthermore, the open-circuited stub may take a different form than a conventional straight line. For instance, the bandstop filter in Figure 6.15a consists of two radial stubs and two quarter- wavelength connecting lines between points A and B. The dc bias is applied to point B through another line of arbitrary length from port 3. The radial stub is another common component used in both hybrid and

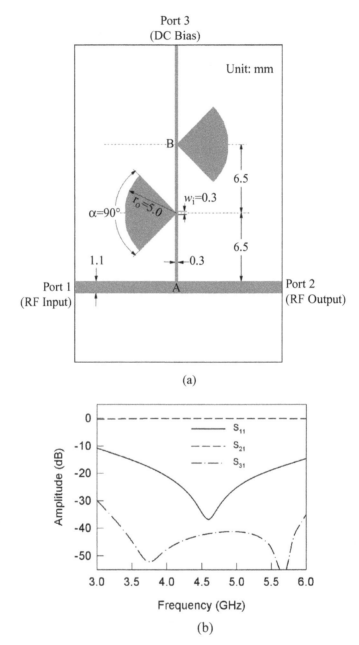

FIGURE 6.15 (a) A microstrip bias T incorporating a bandstop (pseudolowpass) filter for RF choke. (b) EM simulated performance of the microstrip bias T.

monolithic microstrip circuits to provide a low impedance level at a well-specified insertion point in a wide frequency band [7–9].

For the design of a bandstop filter of the type in Figure 6.15a, a radial stub, such as a conventional quarter-wavelength open-circuited stub is designed to short out the RF transmission, to cause an attenuation pole at a midband frequency of the stopband. The outer radius r_o of a radial stub will mainly decide the attenuation pole frequency, whereas the angle α mainly affects the bandwidth. It is obvious that a larger radius r_o results in a lower attenuation pole frequency. It can also be shown that as the angle α is increased, the reactance slope of input reactance of the radial stub is decreased and, as a consequence, the bandwidth of attenuation is increased. The input port width w_i (refer to Figure 6.15a) of a radial stub can have some effect on both the attenuation pole frequency and the bandwidth. For other fixed parameters, increasing the input port width w_i will, in general, increase the attenuation pole frequency and the bandwidth. In practice, w_i should be very small compared to the guided wavelength in order to excite only the dominant TEM radial mode and to have a well-defined point of low impedance at its input. Usually, w_i may be chosen to be the same as the width of connecting inductive line.

The connecting inductive line for each radial stub is $\lambda_{g0}/4$ long, where λ_{g0} is the guided wavelength of the inductive line at the midband frequency of the desired stopband for RF choke. A quarter-wavelength inductive line at the input of a bandstop filter is necessary for transforming an RF short circuit resulting from a radial stub into an RF open circuit at the input port of the filter, or at junction point A in a bias T circuit, as shown in Figure 6.15a. To achieve a wider stop band and better RF choke, the width of inductive lines should be chosen as narrow as possible; in other words, the characteristic impedance should be as high as possible. In practice, the possible narrow line width will be limited by fabrication tolerance and handling capability of dc current.

The dimensions given in Figure 6.15a are for a microstrip bias T on a RT/D 6010 substrate with a relative dielectric constant of 10.8 and a thickness of 1.27 mm. The bandstop filter is designed to provide an RF rejection better than 40 dB over an operating frequency band from 3.5 to 5.5 GHz. The EM simulated performance of the microstrip bias T is illustrated in Figure 6.15b, showing a good RF choke (S_{31}) over the desired frequency band.

REFERENCES

[1] R. Levy, A new class of distributed prototype filters with applications to mixed lumped/distributed component design. *IEEE Trans.* **MTT-18**, 1970, 1064–1071.

[2] G. Matthaei, L. Young, and E. M. T. Jones, *Microwave Filters, Impedance-Matching Networks, and Coupling Structures*, Artech House, Norwood, MA, 1980.

[3] B. M. Schiffman and G. L. Matthaei, Exact design of band-stop microwave filters, *IEEE Trans.* **MTT-12**, 1964, 6–15.

[4] M. C. Horton and R. J. Menzel, General theory and design of optimum quarter wave TEM filters, *IEEE Trans.* **MTT-13**, 1965, 316–327.

[5] M. C. Horton and R. J. Menzel, The effectiveness of component elements in commensurate linelength filters, *IEEE Trans.* **MTT-16**, 1968, 555–557.

[6] O. P. Cupta and R. J. Menzel, Design tables for a class of optimum microwave bandstop filters, *IEEE Trans.* **MTT-18**, 1970, 402–404.

[7] B. A. Syrett, A broad-band element for microstrip bias or tuning, *IEEE Trans.* **MTT-28**, 1980, 925–927.

[8] H. A. Atwater, Microstrip reactive circuit elements, *IEEE Trans.* **MTT-31**, 1983, 488–491.

[9] R. Sorrentino and L. Roselli, A new simple and accurate formula for microstrip radial stub, *IEEE Microwave Guided-Wave Lett.* **2** (12), 1992, 480–482.

CHAPTER SEVEN

Coupled-Resonator Circuits

Coupled-resonator circuits are of importance for design of RF/microwave filters, in particular, the narrow-band bandpass filters that play a significant role in many applications. There is a general technique for designing coupled-resonator filters in the sense that it can be applied to any type of resonator despite its physical structure. It has been applied to the design of waveguide filters [1,2], dielectric resonator filters [3], ceramic combline filters [4], microstrip filters [5–7], superconducting filters [8], and micromachined filters [9]. This design method is based on coupling coefficients of intercoupled resonators and the external quality factors of the input and output resonators. We actually saw some examples in Chapter 5 when we discussed the design of hairpin-resonator filters and combline filters. We will discuss more applications for designing various filters through the remainder of this book. Since this design technique is so useful and flexible, it would be desirable to have a extensive understanding of not only the approach, but also the theory behind it. For this purpose, this chapter will present a comprehensive treatment of relevant subjects.

The general coupling matrix is of importance for representing a wide range of coupled-resonator filter topologies. Section 7.1 shows how it can be formulated either from a set of loop equations or from a set of node equations. This leads to a very useful formula for analysis and synthesis of coupled-resonator filter circuits in terms of coupling coefficients and external quality factors. Section 7.2 considers general theory of couplings in order to establish the relationship between the coupling coefficient and the physical structure of synchronously or asynchronously tuned coupled resonators. Following this, a discussion of a general formulation for extracting coupling coefficients is given in Section 7.3. Formulations for extracting the external quality factors from frequency responses of the externally loaded input/output resonators are derived in Section 7.4. Some numerical examples are described in Section 7.5 to demonstrate how the formulations obtained can be applied to extract

---

Microstrip Filters for RF/Microwave Applications by Jia-Sheng Hong
Copyright © 2011 John Wiley & Sons, Inc.

coupling coefficients and external quality factors of microwave coupling structures from EM simulations. The final section of this chapter addresses a more advanced topic on general coupling matrix, including source and load.

7.1 GENERAL COUPLING MATRIX FOR COUPLED-RESONATOR FILTERS

7.1.1 Loop Equation Formulation

Shown in Figure 7.1a is an equivalent circuit of n-coupled resonators. L, C, and R denote the inductance, capacitance, and resistance, respectively, i represents the loop current, and e_s the voltage source. Using the voltage law, which is one of Kirchhoff's two circuit laws, and states that the algebraic sum of the voltage drops around any closed path in a network is zero, we can write down the loop equations for the circuit of Figure 7.1a

$$\left(R_1 + j\omega L_1 + \frac{1}{j\omega C_1}\right) i_1 - j\omega L_{12} i_2 \cdots - j\omega L_{1n} i_n = e_s$$
$$-j\omega L_{21} i_1 + \left(j\omega L_2 + \frac{1}{j\omega C_2}\right) i_2 \cdots - j\omega L_{2n} i_n = 0 \quad (7.1)$$
$$\vdots$$
$$-j\omega L_{n1} i_1 - j\omega L_{n2} i_2 \cdots + \left(R_n + j\omega L_n + \frac{1}{j\omega C_n}\right) i_n = 0$$

in which $L_{ij} = L_{ji}$ represents the mutual inductance between resonators i and j and the all loop currents are supposed to have the same direction, as shown in Figure 7.1a,

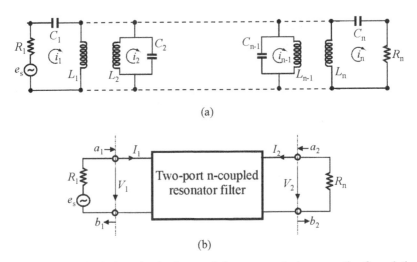

FIGURE 7.1 (a) Equivalent circuit of n-coupled resonators for loop-equation formulation. (b) Its network representation.

GENERAL COUPLING MATRIX FOR COUPLED-RESONATOR FILTERS   195

so that the voltage drops because of the mutual inductance have a negative sign. This set of equations can be represented in matrix form

$$\begin{bmatrix} R_1 + j\omega L_1 + \dfrac{1}{j\omega C_1} & -j\omega L_{12} & \cdots & -j\omega L_{1n} \\ -j\omega L_{21} & j\omega L_2 + \dfrac{1}{j\omega C_2} & \cdots & -j\omega L_{2n} \\ \vdots & \vdots & \vdots & \vdots \\ -j\omega L_{n1} & -j\omega L_{n2} & \cdots & R_n + j\omega L_n + \dfrac{1}{j\omega C_n} \end{bmatrix} \begin{bmatrix} i_1 \\ i_2 \\ \vdots \\ i_n \end{bmatrix} = \begin{bmatrix} e_s \\ 0 \\ \vdots \\ 0 \end{bmatrix}$$

(7.2)

or

$$[Z] \cdot [i] = [e]$$

where $[Z]$ is an $n \times n$ impedance matrix.

For simplicity, let us first consider a synchronously tuned filter. In this case, all resonators resonate at the same frequency, namely, the midband frequency of filter $\omega_0 = 1/\sqrt{LC}$, where $L = L_1 = L_2 = \cdots = L_n$ and $C = C_1 = C_2 = \cdots = C_n$. The impedance matrix in Eq. (7.2) may be expressed by

$$[Z] = \omega_0 L \cdot FBW \cdot [\bar{Z}] \qquad (7.3)$$

where $FBW = \Delta\omega/\omega_0$ is the fractional bandwidth of filter and $[\bar{Z}]$ is the normalized impedance matrix, which, in the case of synchronously tuned filter, is given by

$$[\bar{Z}] = \begin{bmatrix} \dfrac{R_1}{\omega_0 L \cdot FBW} + p & -j\dfrac{\omega}{\omega_0}\dfrac{L_{12}}{L} \cdot \dfrac{1}{FBW} & \cdots & -j\dfrac{\omega}{\omega_0}\dfrac{L_{1n}}{L} \cdot \dfrac{1}{FBW} \\ -j\dfrac{\omega}{\omega_0}\dfrac{L_{21}}{L} \cdot \dfrac{1}{FBW} & p & \cdots & -j\dfrac{\omega}{\omega_0}\dfrac{L_{2n}}{L} \cdot \dfrac{1}{FBW} \\ \vdots & \vdots & \vdots & \vdots \\ -j\dfrac{\omega}{\omega_0}\dfrac{L_{n1}}{L} \cdot \dfrac{1}{FBW} & -j\dfrac{\omega}{\omega_0}\dfrac{L_{n2}}{L} \cdot \dfrac{1}{FBW} & \cdots & \dfrac{R_n}{\omega_0 L \cdot FBW} + p \end{bmatrix}$$

(7.4)

with

$$p = j\dfrac{1}{FBW}\left(\dfrac{\omega}{\omega_0} - \dfrac{\omega_0}{\omega}\right)$$

the complex lowpass frequency variable. It should be noticed that

$$\dfrac{R_i}{\omega_0 L} = \dfrac{1}{Q_{ei}} \quad \text{for } i = 1, n \qquad (7.5)$$

196 COUPLED-RESONATOR CIRCUITS

Q_{e1} and Q_{en} are the external quality factors of the input and output resonators, respectively. Defining the coupling coefficient as

$$M_{ij} = \frac{L_{ij}}{L} \tag{7.6}$$

and assuming $\omega/\omega_0 \approx 1$ for a narrow-band approximation, we can simplify Eq. (7.4) as

$$[\bar{Z}] = \begin{bmatrix} \frac{1}{q_{e1}} + p & -jm_{12} & \cdots & -jm_{1n} \\ -jm_{21} & p & \cdots & -jm_{2n} \\ \vdots & \vdots & \vdots & \vdots \\ -jm_{n1} & -jm_{n2} & \cdots & \frac{1}{q_{en}} + p \end{bmatrix} \tag{7.7}$$

where q_{e1} and q_{en} are the scaled external quality factors

$$q_{ei} = Q_{ei} \cdot FBW \quad \text{for } i = 1, n \tag{7.8}$$

and m_{ij} denotes the so-called normalized coupling coefficient

$$m_{ij} = \frac{M_{ij}}{FBW} \tag{7.9}$$

A network representation of the circuit of Figure 7.1a is shown in Figure 7.1b, where V_1, V_2 and I_1, I_2 are the voltage and current variables at the filter ports and the wave variables are denoted by a_1, a_2, b_1, and b_2. By inspecting the circuit of Figure 7.1a and the network of Figure 7.1b, it can be identified that $I_1 = i_1$, $I_2 = -i_n$ and $V_1 = e_s - i_1 R_1$. Referring to Chapter 2, we have

$$a_1 = \frac{e_s}{2\sqrt{R_1}} \qquad b_1 = \frac{e_s - 2i_1 R_1}{2\sqrt{R_1}} \tag{7.10}$$

$$a_2 = 0 \qquad b_2 = i_n \sqrt{R_n}$$

and, hence,

$$S_{21} = \left.\frac{b_2}{a_1}\right|_{a_2=0} = \frac{2\sqrt{R_1 R_n} i_n}{e_s}$$

$$S_{11} = \left.\frac{b_1}{a_1}\right|_{a_2=0} = 1 - \frac{2R_1 i_1}{e_s} \tag{7.11}$$

Solving Eq. (7.2) for i_1 and i_n, we obtained

$$i_1 = \frac{e_s}{\omega_0 L \cdot FBW} [\bar{Z}]_{11}^{-1}$$
$$i_n = \frac{e_s}{\omega_0 L \cdot FBW} [\bar{Z}]_{n1}^{-1}$$
(7.12)

where $[\bar{Z}]_{ij}^{-1}$ denotes the ith row and jth column element of $[\bar{Z}]^{-1}$. Substituting Eq. (7.12) into Eq. (7.11) yields

$$S_{21} = \frac{2\sqrt{R_1 R_n}}{\omega_0 L \cdot FBW} [\bar{Z}]_{n1}^{-1}$$

$$S_{11} = 1 - \frac{2R_1}{\omega_0 L \cdot FBW} [\bar{Z}]_{11}^{-1}$$

Recalling the external quality factors defined in Eqs. (7.5) and (7.8), we have

$$S_{21} = 2\frac{1}{\sqrt{q_{e1} \cdot q_{en}}} [\bar{Z}]_{n1}^{-1}$$
$$S_{11} = 1 - \frac{2}{q_{e1}} [\bar{Z}]_{11}^{-1}$$
(7.13)

In the case that the coupled-resonator circuit of Figure 7.1 is asynchronously tuned, and the resonant frequency of each resonator, which may be different, is given by $\omega_{0i} = 1/\sqrt{L_i C_i}$, the coupling coefficient of asynchronously tuned filter is defined as

$$M_{ij} = \frac{L_{ij}}{\sqrt{L_i L_j}} \quad \text{for } i \neq j \qquad (7.14)$$

It can be shown that Eq. (7.7) becomes

$$[\bar{Z}] = \begin{bmatrix} \frac{1}{q_{e1}} + p - jm_{11} & -jm_{12} & \cdots & -jm_{1n} \\ -jm_{21} & p - jm_{22} & \cdots & -jm_{2n} \\ \vdots & \vdots & \vdots & \vdots \\ -jm_{n1} & -jm_{n2} & \cdots & \frac{1}{q_{en}} + p - jm_{nn} \end{bmatrix} \qquad (7.15)$$

The normalized impedance matrix of Eq. (7.15) is almost identical to Eq. (7.7), except that it has the extra entries m_{ii} to account for asynchronous tuning.

7.1.2 Node Equation Formulation

As can be seen, the coupling coefficients introduced in the above section are all based on mutual inductance and, hence, the associated couplings are magnetic couplings. What formulation of the coupling coefficients would result from a two-port n-coupled resonator filter with electric couplings? We may find the answer on the dual basis directly. However, let us consider the n-coupled-resonator circuit shown in Figure 7.2a, where v_i denotes the node voltage, G represents the conductance, and i_s is the source current. According to the current law, which is the other one of Kirchhoff's two circuit laws and states that the algebraic sum of the currents leaving a node in a network is zero, with a driving or external current of i_s the node equations for the circuit of Figure 7.2a are

$$\left(G_1 + j\omega C_1 + \frac{1}{j\omega L_1}\right)v_1 - j\omega C_{12}v_2 \cdots - j\omega C_{1n}v_n = i_s$$
$$-j\omega C_{21}v_1 + \left(j\omega C_2 + \frac{1}{j\omega L_2}\right)v_2 \cdots - j\omega C_{2n}v_n = 0 \quad (7.16)$$
$$\vdots$$
$$-j\omega C_{n1}v_1 - j\omega C_{n2}v_2 \cdots + \left(G_n + j\omega C_n + \frac{1}{j\omega L_n}\right)v_n = 0$$

where $C_{ij} = C_{ji}$ represents the mutual capacitance across resonators i and j. Note that all node voltages are determined with respect to the reference node (ground), so

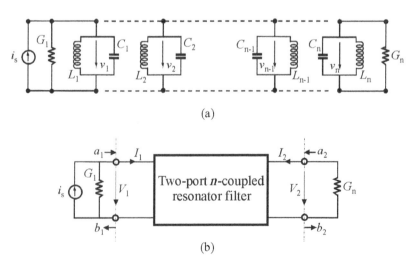

FIGURE 7.2 (a) Equivalent circuit of n-coupled resonators for node-equation formulation. (b) Its network representation.

that the currents resulting from the mutual capacitance have a negative sign. Arrange this set of equations in matrix form

$$\begin{bmatrix} G_1 + j\omega C_1 + \dfrac{1}{j\omega L_1} & -j\omega C_{12} & \cdots & -j\omega C_{1n} \\ -j\omega C_{21} & j\omega C_2 + \dfrac{1}{j\omega L_2} & \cdots & -j\omega C_{2n} \\ \vdots & \vdots & \vdots & \vdots \\ -j\omega C_{n1} & -j\omega C_{n2} & \cdots & G_n + j\omega C_n + \dfrac{1}{j\omega L_n} \end{bmatrix} \begin{bmatrix} v_1 \\ v_2 \\ \vdots \\ v_n \end{bmatrix} = \begin{bmatrix} i_s \\ 0 \\ \vdots \\ 0 \end{bmatrix}$$

(7.17)

or

$$[Y] \cdot [v] = [i]$$

in which $[Y]$ is an $n \times n$ admittance matrix.

Similarly, the admittance matrix in Eq. (7.17) may be expressed by

$$[Y] = \omega_0 C \cdot FBW \cdot [\bar{Y}] \tag{7.18}$$

where $\omega_0 = 1/\sqrt{LC}$ is the midband frequency of filter, $FBW = \Delta\omega/\omega_0$ is the fractional bandwidth, and $[\bar{Y}]$ is the normalized admittance matrix. In the case of synchronously tuned filter, $[\bar{Y}]$ is given by

$$[\bar{Y}] = \begin{bmatrix} \dfrac{G_1}{\omega_0 C \cdot FBW} + p & -j\dfrac{\omega}{\omega_0} \dfrac{C_{12}}{C} \cdot \dfrac{1}{FBW} & \cdots & -j\dfrac{\omega}{\omega_0} \dfrac{C_{1n}}{C} \cdot \dfrac{1}{FBW} \\ -j\dfrac{\omega}{\omega_0} \dfrac{C_{21}}{C} \cdot \dfrac{1}{FBW} & p & \cdots & -j\dfrac{\omega}{\omega_0} \dfrac{C_{2n}}{C} \cdot \dfrac{1}{FBW} \\ \vdots & \vdots & \vdots & \vdots \\ -j\dfrac{\omega}{\omega_0} \dfrac{C_{n1}}{C} \cdot \dfrac{1}{FBW} & -j\dfrac{\omega}{\omega_0} \dfrac{C_{n2}}{C} \cdot \dfrac{1}{FBW} & \cdots & \dfrac{G_n}{\omega_0 C \cdot FBW} + p \end{bmatrix}$$

(7.19)

where p is the complex lowpass frequency variable. Notice that

$$\dfrac{G_i}{\omega_0 C} = \dfrac{1}{Q_{ei}} \quad \text{for } i = 1, n \tag{7.20}$$

with Q_e being the external quality factor. Let us define the coupling coefficient

$$M_{ij} = \dfrac{C_{ij}}{C} \tag{7.21}$$

and assume $\omega/\omega_0 \approx 1$ for the narrow-band approximation. A simpler expression of Eq. (7.19) is obtained:

$$[\bar{Y}] = \begin{bmatrix} \dfrac{1}{q_{e1}} + p & -jm_{12} & \cdots & -jm_{1n} \\ -jm_{21} & p & \cdots & -jm_{2n} \\ \vdots & \vdots & \vdots & \vdots \\ -jm_{n1} & -jm_{n2} & \cdots & \dfrac{1}{q_{en}} + p \end{bmatrix} \qquad (7.22)$$

where q_{ei} and m_{ij} denote the scaled external quality factor and normalized coupling coefficient defined by Eqs. (7.8) and (7.9), respectively.

Similarly, it can be shown that if the coupled-resonator circuit of Figure 7.2a is asynchronously tuned, Eqs. (7.21) and (7.22) become

$$M_{ij} = \dfrac{C_{ij}}{\sqrt{C_i C_j}} \quad \text{for } i \neq j \qquad (7.23)$$

$$[\bar{Y}] = \begin{bmatrix} \dfrac{1}{q_{e1}} + p - jm_{11} & -jm_{12} & \cdots & -jm_{1n} \\ -jm_{21} & p - jm_{22} & \cdots & -jm_{2n} \\ \vdots & \vdots & \vdots & \vdots \\ -jm_{n1} & -jm_{n2} & \cdots & \dfrac{1}{q_{en}} + p - jm_{nn} \end{bmatrix} \qquad (7.24)$$

To derive the two-port S-parameters of coupled-resonator filter, the circuit of Figure 7.2a is represented by a two-port network of Figure 7.2b, where all the variables at the filter ports are the same as those in Figure 7.1b. In this case, $V_1 = v_1$, $V_2 = v_n$, and $I_1 = i_s - v_1 G_1$. We have

$$a_1 = \dfrac{i_s}{2\sqrt{G_1}} \qquad b_1 = \dfrac{2v_1 G_1 - i_s}{2\sqrt{G_1}}$$

$$a_2 = 0 \qquad b_2 = v_n \sqrt{G_n} \qquad (7.25)$$

$$S_{21} = \left.\dfrac{b_2}{a_1}\right|_{a_2=0} = \dfrac{2\sqrt{G_1 G_n} v_n}{i_s}$$

$$S_{11} = \left.\dfrac{b_1}{a_1}\right|_{a_2=0} = \dfrac{2v_1 G_1}{i_s} - 1 \qquad (7.26)$$

GENERAL COUPLING MATRIX FOR COUPLED-RESONATOR FILTERS 201

Finding the unknown node voltages v_1 and v_n from Eq. (7.17)

$$v_1 = \frac{i_s}{\omega_0 C \cdot FBW} [\bar{Y}]^{-1}_{11}$$
$$v_n = \frac{i_s}{\omega_0 C \cdot FBW} [\bar{Y}]^{-1}_{n1}$$
(7.27)

where $[\bar{Y}]^{-1}_{ij}$ denotes the ith row and jth column element of $[\bar{Y}]^{-1}$. Replacing the node voltages in Eq. (7.26) with those given by Eq. (7.27) results in

$$S_{21} = \frac{2\sqrt{G_1 G_n}}{\omega_0 C \cdot FBW} [\bar{Y}]^{-1}_{n1}$$
$$S_{11} = \frac{2 G_1}{\omega_0 C \cdot FBW} [\bar{Y}]^{-1}_{11} - 1$$
(7.28)

which can be simplified as

$$S_{21} = 2 \frac{1}{\sqrt{q_{e1} \cdot q_{en}}} [\bar{Y}]^{-1}_{n1}$$
$$S_{11} = \frac{2}{q_{e1}} [\bar{Y}]^{-1}_{11} - 1$$
(7.29)

7.1.3 General Coupling Matrix

In the foregoing formulations, the most notable is that the formulation of normalized impedance matrix $[\bar{Z}]$ is identical to that of normalized admittance matrix $[\bar{Y}]$. This is very important, because it implies that we could have a unified formulation for a n-coupled resonator filter regardless of whether the couplings are magnetic or electric or even the combination of both. Accordingly, Eqs. (7.13) and (7.29) may be incorporated into a general one:

$$S_{21} = 2 \frac{1}{\sqrt{q_{e1} \cdot q_{en}}} [A]^{-1}_{n1}$$
$$S_{11} = \pm \left(1 - \frac{2}{q_{e1}} [A]^{-1}_{11} \right)$$
(7.30)

with

$$[A] = [q] + p[U] - j[m]$$

where $[U]$ is the $n \times n$ unit or identity matrix, $[q]$ is an $n \times n$ matrix with all entries zero, except for $q_{11} = 1/q_{e1}$ and $q_{nn} = 1/q_{en}$, and $[m]$ is the so-called general coupling matrix, which is an $n \times n$ reciprocal matrix (that is, $m_{ij} = m_{ji}$) and is allowed to have nonzero diagonal entries m_{ii} for an asynchronously tuned filter.

For a given filtering characteristic of $S_{21}(p)$ and $S_{11}(p)$, the coupling matrix and the external quality factors may be obtained using the synthesis procedure developed by Atia et al. [10] and Cameron [11]. However, the elements of the coupling matrix $[m]$ that emerges from the synthesis procedure will, in general, all have nonzero values. The nonzero values will only occur in the diagonal elements of the coupling matrix for an asynchronously tuned filter. However, a nonzero entry everywhere else means that in the network that $[m]$ represents, couplings exist between every resonator and every other one. As this is clearly impractical, it is usually necessary to perform a similar transforms until a more convenient form for implementation is obtained. A more practical synthesis approach based on optimization will be presented in the next chapter.

7.2 GENERAL THEORY OF COUPLINGS

After determining the required coupling matrix for the desired filtering characteristic, a next important step for the filter design is to establish the relationship between the value of every required coupling coefficient and the physical structure of coupled resonators in order to find the physical dimensions of the filter for fabrication.

In general, the coupling coefficient of coupled RF/microwave resonators, which can be different in structure and can have different self-resonant frequencies (see Figure 7.3), may be defined on the basis of a ratio of coupled to stored energy [12], i.e.,

$$k = \frac{\iiint \varepsilon \underline{E}_1 \cdot \underline{E}_2 dv}{\sqrt{\iiint \varepsilon |\underline{E}_1|^2 dv \times \iiint \varepsilon |\underline{E}_2|^2 dv}} + \frac{\iiint \mu \underline{H}_1 \cdot \underline{H}_2 dv}{\sqrt{\iiint \mu |\underline{H}_1|^2 dv \times \iiint \mu |\underline{H}_2|^2 dv}}$$

(7.31)

where $\underline{E}$ and $\underline{H}$ represent the electric and magnetic field vectors, respectively; we now use the more traditional notation k instead of M for the coupling coefficient. Note that all fields are determined at resonance and the volume integrals are over entire effecting regions with permittivity of ε and permeability of μ. The first term on the right-hand side represents the electric coupling, while the second term represents the magnetic

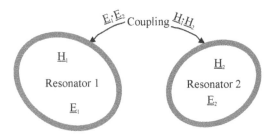

FIGURE 7.3 General coupled RF/microwave resonators where resonators 1 and 2 can be different in structure and have different resonant frequencies.

coupling. It should be remarked that the interaction of the coupled resonators is mathematically described by the dot operation of their space vector fields, which allows the coupling to have either positive or negative sign. A positive sign would imply that the coupling enhances the stored energy of uncoupled resonators, whereas a negative sign would indicate a reduction. Therefore, the electric and magnetic couplings could either have the same effect if they have the same sign, or have the opposite effect if their signs are opposite. Obviously, the direct evaluation of coupling coefficient from Eq. (7.31) requires the knowledge of the field distributions and performance of the space integrals. This is not an easy task unless analytical solutions of the fields exist.

On the other hand, it may be much easier by using full-wave EM simulation or experiment to find some characteristic frequencies that are associated with the coupling of coupled RF/microwave resonators. The coupling coefficient can then be determined against the physical structure of coupled resonators if the relationship between the coupling coefficient and the characteristic frequencies is established. In what follows, we derive the formulation of such relationships. Before processing further, it might be worth pointing out that although the following derivations are based on lumped-element circuit models, the outcomes are also valid for distributed-element coupled structures on a narrow-band basis.

7.2.1 Synchronously Tuned Coupled-Resonator Circuits

7.2.1.1 Electric Coupling

An equivalent lumped-element circuit model for electrically coupled RF/microwave resonators is given in Figure 7.4a, where L and C are the self-inductance and self-capacitance, so that $(LC)^{-1/2}$ equals the angular resonant frequency of uncoupled resonators, and C_m represents the mutual capacitance. As mentioned earlier, if the coupled structure is a distributed element, the lumped-element circuit equivalence is valid on a narrow-band basis, namely, near its resonance. The same comment is applicable for the other coupled structures discussed later. If we look into reference planes T_1-T_1' and T_2-T_2', we can see a two-port network that may be described by the following set of equations:

$$I_1 = j\omega C V_1 - j\omega C_m V_2$$
$$I_2 = j\omega C V_2 - j\omega C_m V_1$$
(7.32)

in which a sinusoidal waveform is assumed. It might be well to mention that Eq. (7.32) implies that the self-capacitance C is the capacitance seen in one resonant loop of Figure 7.4a when the capacitance in the adjacent loop is shorted out. Thus, the second terms on the R.H.S. of Eq. (7.32) are the induced currents resulting from the increasing voltage in resonant loops 2 and 1, respectively. From Eq. (7.32) four Y parameters

$$Y_{11} = Y_{22} = j\omega C$$
$$Y_{12} = Y_{21} = -j\omega C_m$$
(7.33)

can easily be found from definitions.

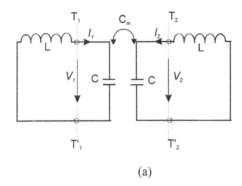

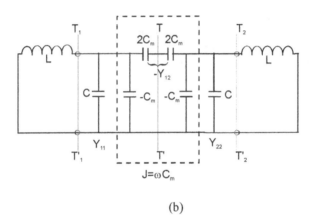

FIGURE 7.4 (a) Synchronously tuned coupled resonator circuit with electric coupling. (b) An alternative form of the equivalent circuit with an admittance inverter $J = \omega C_m$ to represent the coupling.

According to the network theory [13], an alternative form of the equivalent circuit in Figure 7.4a can be obtained, and is shown in Figure 7.4b. This form yields the same two-port parameters as those of the circuit of Figure 7.4a, but it is more convenient for our discussions. Actually, it can be shown that the electric coupling between the two resonant loops is represented by an admittance inverter $J = \omega C_m$. If the symmetry plane T-T′ in Figure 7.4b is replaced by an electric wall (or a short-circuit), the resultant circuit has a resonant frequency

$$f_e = \frac{1}{2\pi\sqrt{L(C + C_m)}} \tag{7.34}$$

This resonant frequency is lower than that of an uncoupled single resonator. A physical explanation is that the coupling effect enhances the capability to store charge of the single resonator when the electric wall is inserted in the symmetrical plane of the coupled structure. Similarly, replacing the symmetry plane in Figure 7.4b by a magnetic wall (or an open-circuit) results in a single resonant circuit having a resonant frequency

$$f_m = \frac{1}{2\pi\sqrt{L(C - C_m)}} \tag{7.35}$$

In this case, the coupling effect reduces the capability to store charge so that the resonant frequency is increased.

Equations (7.34) and (7.35) can be used to find the electric coupling coefficient k_E

$$k_E = \frac{f_m^2 - f_e^2}{f_m^2 + f_e^2} = \frac{C_m}{C} \tag{7.36}$$

which is not only identical to the definition of ratio of the coupled electric energy to the stored energy of uncoupled single resonator, but also consistent with the coupling coefficient defined by Eq. (7.21) for coupled-resonator filter.

7.2.1.2 Magnetic Coupling Shown in Figure 7.5a is an equivalent lumped-element circuit model for magnetically coupled resonator structures, where L and C are the self-inductance and self-capacitance, and L_m represents the mutual inductance. In this case, the coupling equations describing the two-port network at reference planes T_1-T_1' and T_2-T_2' are

$$\begin{aligned} V_1 &= j\omega L I_1 + j\omega L_m I_2 \\ V_2 &= j\omega L I_2 + j\omega L_m I_1 \end{aligned} \tag{7.37}$$

The equations in (7.37) also imply that the self-inductance L is the inductance seen in one resonant loop of Figure 7.5a when the adjacent loop is open-circuited. Thus, the second terms on the R.H.S. of Eq. (7.37) are the induced voltage resulting from the increasing current in loops 2 and 1, respectively. It should be noticed that the two loop currents in Figure 7.5a flow in the opposite directions, so that the voltage drops because of the mutual inductance have a positive sign. From Eq. (7.37) we can find four Z parameters

$$\begin{aligned} Z_{11} &= Z_{22} = j\omega L \\ Z_{12} &= Z_{21} = j\omega L_m \end{aligned} \tag{7.38}$$

Shown in Figure 7.5b is an alternative form of equivalent circuit having the same network parameters as those of Figure 7.5a. It can be shown that the magnetic coupling between the two resonant loops is represented by an impedance inverter

206 COUPLED-RESONATOR CIRCUITS

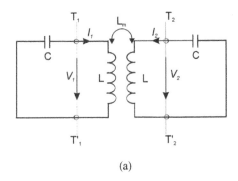

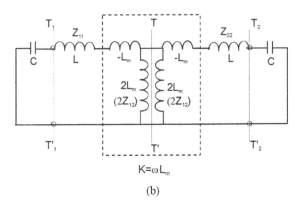

FIGURE 7.5 (a) Synchronously tuned coupled resonator circuit with magnetic coupling. (b) An alternative form of the equivalent circuit with an impedance inverter $K = \omega L_m$ to represent the coupling.

$K = \omega L_m$. If the symmetry plane T-T' in Figure 7.5b is replaced by an electric wall (or a short-circuit), the resultant single-resonant circuit has a resonant frequency

$$f_e = \frac{1}{2\pi \sqrt{(L - L_m)C}} \quad (7.39)$$

It can be shown that the increase in resonant frequency is because the coupling effect reducing the stored flux in the single resonator circuit when the electric wall is inserted in the symmetric plane. If a magnetic wall (or an open-circuit) replaces the symmetry plane in Figure 7.5b, the resultant single-resonant circuit has a resonant frequency

$$f_m = \frac{1}{2\pi \sqrt{(L + L_m)C}} \quad (7.40)$$

In this case, it turns out that the coupling effect increases the stored flux, so that the resonant frequency is shifted down.

Similarly, Eq. (7.39) and (7.40) can be used to find the magnetic coupling coefficient k_M

$$k_M = \frac{f_e^2 - f_m^2}{f_e^2 + f_m^2} = \frac{L_m}{L} \quad (7.41)$$

It should be emphasized that the magnetic coupling coefficient defined by Eq. (7.41) corresponds to the definition of ratio of the coupled magnetic energy to the stored energy of an uncoupled single resonator. It is also consistent with the definition given in Eq. (7.6) for coupled-resonator filter.

7.2.1.3 Mixed Coupling

For coupled-resonator structures, with both the electric and magnetic couplings, a network representation is given in Figure 7.6a. Notice that the Y parameters are the parameters of a two-port network located on the left side of reference plane T_1-T_1' and the right side of reference plane T_2'-T_2', while the Z parameters are the parameters of the other two-port network located on the right side of reference plane T_1-T_1' and the left side of reference plane T_2-T_2'. The Y and Z parameters are defined by

$$\begin{aligned} Y_{11} &= Y_{22} = j\omega C \\ Y_{12} &= Y_{21} = j\omega C_m' \end{aligned} \quad (7.42)$$

$$\begin{aligned} Z_{11} &= Z_{22} = j\omega L \\ Z_{12} &= Z_{21} = j\omega L_m' \end{aligned} \quad (7.43)$$

where C, L, C_m', and L_m' are the self-capacitance, the self-inductance, the mutual capacitance, and the mutual inductance of an associated equivalent lumped-element circuit shown in Figure 7.6b. One can also identify an impedance inverter $K = \omega L_m'$ and an admittance inverter $J = \omega C_m'$, which represent the magnetic coupling and the electric coupling, respectively.

By inserting an electric wall and a magnetic wall, respectively, into the symmetry plane of the equivalent circuit in Figure 7.6b, we obtain

$$f_e = \frac{1}{2\pi\sqrt{(L - L_m')(C - C_m')}} \quad (7.44)$$

$$f_m = \frac{1}{2\pi\sqrt{(L + L_m')(C + C_m')}} \quad (7.45)$$

As can be seen, both the magnetic and electric couplings have the same effect on the resonant frequency shifting.

From Eqs. (7.44) and (7.45), the mixed coupling coefficient k_X can be found to be

$$k_X = \frac{f_e^2 - f_m^2}{f_e^2 + f_m^2} = \frac{CL_m' + LC_m'}{LC + L_m'C_m'}. \quad (7.46)$$

208 COUPLED-RESONATOR CIRCUITS

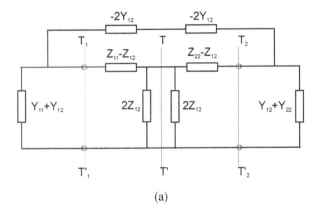

(a)

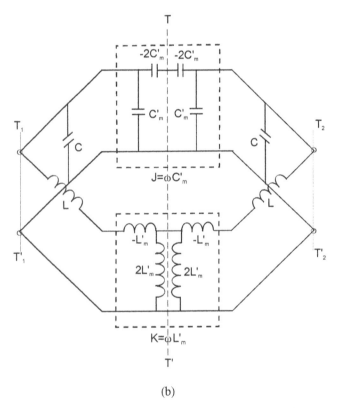

(b)

FIGURE 7.6 (a) Network representing of synchronously tuned coupled resonator circuit with mixed coupling. (b) An associated equivalent circuit with an impedance inverter $K = \omega L'_m$ and an admittance inverter $J = \omega C'_m$ to represent the magnetic coupling and electric coupling, respectively.

It is reasonable to assume that $L'_m C'_m \ll LC$, and thus Eq. (7.46) becomes

$$k_X \approx \frac{L'_m}{L} + \frac{C'_m}{C} = k'_M + k'_E \tag{7.47}$$

which clearly indicates that the mixed coupling results from the superposition of the magnetic and electric couplings. Care should be taken for the mixed coupling, because the superposition of both the magnetic and electric couplings can result in two opposite effects, either enhancing or canceling each other, as mentioned earlier. If we allow either the mutual inductance or the mutual capacitance in Figure 7.6b to change sign, we will find that the both couplings tend to cancel each other.

It should be remarked that for numerical computations, depending on the particular EM simulator used, as well as the coupling structure analyzed, it may sometimes be difficult to implement the electric wall, the magnetic wall, or even both in the simulation. This difficulty is more obvious for experiments. The difficulty can be removed easily by analyzing or measuring the whole coupling structure instead of the half and finding the natural resonant frequencies of two resonant peaks, observable from the resonant frequency response. It has been proved that the two natural resonant frequencies obtained in this way are f_e and f_m [5]. This can also be seen in the next section when we consider more general case, namely, the asynchronously tuned coupled-resonator circuits.

7.2.2 Asynchronously Tuned Coupled-Resonator Circuits

Asynchronously tuned narrow-band bandpass filters exhibit some attractive characteristics that may better meet the demanding requirements for rapid developments of mobile communications systems (see Chapter 9). In an asynchronously tuned filter, each of resonators may resonate at different frequencies. Hence, in order to achieve an accurate or first-pass filter design, it is essential to characterize couplings of coupled resonators whose self-resonant frequencies are different. This would seem more important for some filter technologies in which the post-tuning after fabrications are not convenient.

In general, two eigenfrequencies associated with the coupling between a pair of coupled resonators can be observed, despite whether or not the coupled resonators are synchronously or asynchronously tuned. In the last section, we have derived the formula for extracting coupling coefficients from these two eigenfrequencies for synchronously tuned resonators. However, if the coupled resonators are asynchronously tuned, a wrong result will occur if one attempts to extract the coupling coefficient by using the same formula derived for the synchronously tuned resonators. Therefore, other appropriate formulas should be sought. These will be derived below.

7.2.2.1 Electric Coupling In the case when we are only concerned with the electric coupling, an equivalent lumped-element circuit, as shown in Figure 7.7, may be employed to represent the coupled resonators. The two resonators may resonate at different frequencies of $\omega_{01} = (L_1 C_1)^{-1/2}$ and $\omega_{02} = (L_2 C_2)^{-1/2}$, respectively, and

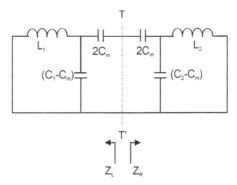

FIGURE 7.7 Asynchronously tuned coupled resonator circuits with electric coupling.

are coupled to each other electrically through mutual capacitance C_m. For natural resonance of the circuit of Figure 7.7, the condition is

$$Z_L = -Z_R \tag{7.48}$$

where Z_L and Z_R are the input impedances when we look into the left and the right of reference plane $T - T'$ of Figure 7.7. The resonant condition of Eq. (7.48) leads to an eigenequation

$$\frac{1}{j\omega C_m} + \frac{j\omega L_1}{1 - \omega^2 L_1 (C_1 - C_m)} + \frac{j\omega L_2}{1 - \omega^2 L_2 (C_2 - C_m)} = 0 \tag{7.49}$$

After some manipulations, Eq. (7.49) can be written as

$$\omega^4 \left(L_1 L_2 C_1 C_2 - L_1 L_2 C_m^2\right) - \omega^2 (L_1 C_1 + L_2 C_2) + 1 = 0 \tag{7.50}$$

We note that Eq. (7.50) is a biquadratic equation having four solutions or eigenvalues. Among those four, we are only interested in the two positive real ones that represent the resonant frequencies that are measurable, namely,

$$\omega_{1,2} = \sqrt{\frac{(L_1 C_1 + L_2 C_2) \pm \sqrt{(L_1 C_1 - L_2 C_2)^2 + 4 L_1 L_2 C_m^2}}{2 \left(L_1 L_2 C_1 C_2 - L_1 L_2 C_m^2\right)}} \tag{7.51}$$

The other two eigenvalues may be seen as their image frequencies. Define a parameter

$$K_E = \frac{\omega_2^2 - \omega_1^2}{\omega_2^2 + \omega_1^2} \tag{7.52}$$

where $\omega_2 > \omega_1$ is assumed. Since $\omega_{01} = (L_1 C_1)^{-1/2}$ and $\omega_{02} = (L_2 C_2)^{-1/2}$, we have by substitution

$$K_E^2 = \frac{C_m^2}{C_1 C_2} \frac{4}{\left(\dfrac{\omega_{02}}{\omega_{01}} + \dfrac{\omega_{01}}{\omega_{02}}\right)^2} + \left(\frac{\omega_{02}^2 - \omega_{01}^2}{\omega_{02}^2 + \omega_{01}^2}\right)^2 \qquad (7.53)$$

The electric coupling coefficient is defined

$$k_e = \frac{C_m}{\sqrt{C_1 C_2}} = \pm \frac{1}{2}\left(\frac{\omega_{02}}{\omega_{01}} + \frac{\omega_{01}}{\omega_{02}}\right)\sqrt{\left(\frac{\omega_2^2 - \omega_1^2}{\omega_2^2 + \omega_1^2}\right)^2 - \left(\frac{\omega_{02}^2 - \omega_{01}^2}{\omega_{02}^2 + \omega_{01}^2}\right)^2} \qquad (7.54)$$

in accordance with the ratio of the coupled electric to the average stored energy, where the positive sign should be chosen if a positive mutual capacitance C_m is defined.

7.2.2.2 Magnetic Coupling Shown in Figure 7.8 is a lumped-element circuit model of asynchronously tuned resonators that are coupled magnetically, denoted by mutual inductance L_m. The two resonant frequencies of uncoupled resonators are $\omega_{01} = (L_1 C_1)^{-1/2}$ and $\omega_{02} = (L_2 C_2)^{-1/2}$, respectively. The condition for natural resonance of the circuit of Figure 7.8 is

$$Y_L = -Y_R \qquad (7.55)$$

where Y_L and Y_R are the pair of admittances on the left and the right of reference plane $T - T'$. This resonant condition leads to

$$\frac{1}{j\omega L_m} + \frac{j\omega C_1}{1 - \omega^2 C_1(L_1 - L_m)} + \frac{j\omega C_2}{1 - \omega^2 C_2(L_2 - L_m)} = 0 \qquad (7.56)$$

The eigenequation (7.56) can be expanded as

$$\omega^4 (L_1 L_2 C_1 C_2 - C_1 C_2 L_m^2) - \omega^2 (L_1 C_1 + L_2 C_2) + 1 = 0 \qquad (7.57)$$

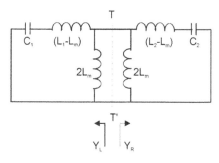

FIGURE 7.8 Asynchronously tuned coupled resonator circuits with magnetic coupling.

212 COUPLED-RESONATOR CIRCUITS

This biquadratic equation has four eigenvalues and the two positive real values of interest are

$$\omega_{1,2} = \sqrt{\frac{(L_1C_1 + L_2C_2) \pm \sqrt{(L_1C_1 - L_2C_2)^2 + 4C_1C_2L_m^2}}{2(L_1L_2C_1C_2 - C_1C_2L_m^2)}} \tag{7.58}$$

To extract the magnetic coupling coefficient, we define a parameter

$$K_M = \frac{\omega_2^2 - \omega_1^2}{\omega_2^2 + \omega_1^2} \tag{7.59}$$

Assume $\omega_2 > \omega_1$ in Eq. (7.59), and recall $\omega_{01} = (L_1C_1)^{-1/2}$ and $\omega_{02} = (L_2C_2)^{-1/2}$ so that

$$K_M^2 = \frac{L_m^2}{L_1L_2} \frac{4}{\left(\dfrac{\omega_{02}}{\omega_{01}} + \dfrac{\omega_{01}}{\omega_{02}}\right)^2} + \left(\frac{\omega_{02}^2 - \omega_{01}^2}{\omega_{02}^2 + \omega_{01}^2}\right)^2 \tag{7.60}$$

Defining the magnetic coupling coefficient as the ratio of the coupled magnetic energy to the average stored energy, we have

$$k_m = \frac{L_m}{\sqrt{L_1L_2}} = \pm\frac{1}{2}\left(\frac{\omega_{02}}{\omega_{01}} + \frac{\omega_{01}}{\omega_{02}}\right)\sqrt{\left(\frac{\omega_2^2 - \omega_1^2}{\omega_2^2 + \omega_1^2}\right)^2 - \left(\frac{\omega_{02}^2 - \omega_{01}^2}{\omega_{02}^2 + \omega_{01}^2}\right)^2} \tag{7.61}$$

The choice of a sign depends on the definition of the mutual inductance, which is normally allowed to be either positive or negative, corresponding to the same or opposite direction of the two loops currents.

7.2.2.3 Mixed Coupling In many coupled resonator structures, both electric and magnetic couplings exist. In this case, we may have a circuit model as depicted in Figure 7.9. It can be shown that the electric coupling is represented by an admittance inverter with $J = \omega C_m$ and the magnetic coupling is represented by an impedance inverter with $K = \omega L_m$. Note that the currents denoted by $I_1, I_2,$ and I_3 are the external currents flowing into the coupled resonator circuit. According to the circuit model of Figure 7.9, by assuming all internal currents flowing outward from each node, we can define a definite nodal admittance matrix with a reference at node "0"

$$\begin{bmatrix} I_1 \\ I_2 \\ I_3 \end{bmatrix} = \begin{bmatrix} y_{11} & y_{12} & y_{13} \\ y_{21} & y_{22} & y_{23} \\ y_{31} & y_{32} & y_{33} \end{bmatrix} \cdot \begin{bmatrix} V_1 \\ V_2 \\ V_3 \end{bmatrix} \tag{7.62}$$

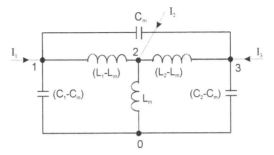

FIGURE 7.9 Asynchronously tuned coupled resonator circuits with the mixed electric coupling and magnetic coupling.

with

$$y_{11} = j\omega C_1 + \frac{1}{j\omega(L_1 - L_m)}$$

$$y_{12} = y_{21} = -\frac{1}{j\omega(L_1 - L_m)}$$

$$y_{13} = y_{31} = -j\omega C_m$$

$$y_{22} = \frac{1}{j\omega L_m} + \frac{1}{j\omega(L_1 - L_m)} + \frac{1}{j\omega(L_2 - L_m)}$$

$$y_{23} = y_{32} = -\frac{1}{j\omega(L_2 - L_m)}$$

$$y_{33} = j\omega C_2 + \frac{1}{j\omega(L_2 - L_m)}$$

For natural resonance, it implies that

$$\begin{bmatrix} V_1 \\ V_2 \\ V_3 \end{bmatrix} \neq \begin{bmatrix} 0 \\ 0 \\ 0 \end{bmatrix}, \quad \text{for} \quad \begin{bmatrix} I_1 \\ I_2 \\ I_3 \end{bmatrix} = \begin{bmatrix} 0 \\ 0 \\ 0 \end{bmatrix} \quad (7.63)$$

This requires that the determinant of admittance matrix to be zero, that is,

$$\begin{vmatrix} y_{11} & y_{12} & y_{13} \\ y_{21} & y_{22} & y_{23} \\ y_{31} & y_{32} & y_{33} \end{vmatrix} = 0 \quad (7.64)$$

After some manipulations, we can arrive at

$$\omega^4 \left(L_1 C_1 L_2 C_2 - L_m^2 C_1 C_2 - L_1 L_2 C_m^2 + L_m^2 C_m^2 \right) - \omega^2 \left(L_1 C_1 + L_2 C_2 - 2 L_m C_m \right) + 1 = 0$$

$$(7.65)$$

This biquadratic equation is the eigenequation for an asynchronously tuned coupled resonator circuit with the mixed coupling. One can immediately see that allowing either $L_m = 0$ or $C_m = 0$ in Eq. (7.65) reduces the equation to either Eq. (7.50) for the electric coupling or Eq. (7.57) for the magnetic coupling. There are four solutions of Eq. (7.65). However, only the two positive ones are of interest; they may be expressed as

$$\omega_1 = \sqrt{\frac{\Re_B - \Re_C}{\Re_A}}, \quad \omega_2 = \sqrt{\frac{\Re_B + \Re_C}{\Re_A}} \qquad (7.66)$$

with

$$\Re_A = 2\left(L_1 C_1 L_2 C_2 - L_m^2 C_1 C_2 - L_1 L_2 C_m^2 + L_m^2 C_m^2\right)$$
$$\Re_B = (L_1 C_1 + L_2 C_2 - 2 L_m C_m)$$
$$\Re_C = \sqrt{\Re_B^2 - 2\Re_A}$$

Define

$$K_X = \frac{\omega_2^2 - \omega_1^2}{\omega_2^2 + \omega_1^2} \qquad (7.67)$$

For narrow-band applications, we can assume that $(L_1 C_1 + L_2 C_2) \gg L_m C_m$ and $\frac{(L_1 C_1 + L_2 C_2)/2}{\sqrt{L_1 C_1 L_2 C_2}} \approx 1$; the latter actually represents a ratio of an arithmetic mean to a geometric mean of two resonant frequencies. Thus, we have

$$K_X^2 = \frac{4 L_1 C_1 L_2 C_2}{(L_1 C_1 + L_2 C_2)^2} k_x^2 + \frac{(L_1 C_1 - L_2 C_2)^2}{(L_1 C_1 + L_2 C_2)^2} \qquad (7.68)$$

in which

$$k_x^2 = \left(\frac{C_m^2}{C_1 C_2} + \frac{L_m^2}{L_1 L_2} - \frac{2 L_m C_m}{\sqrt{L_1 C_1 L_2 C_2}}\right)$$
$$= (k_e - k_m)^2 \qquad (7.69)$$

Now, it is clearer that k_x is the mixed coupling coefficient defined as

$$k_x = k_e - k_m = \pm\frac{1}{2}\left(\frac{\omega_{02}}{\omega_{01}} + \frac{\omega_{01}}{\omega_{02}}\right) \sqrt{\left(\frac{\omega_2^2 - \omega_1^2}{\omega_2^2 + \omega_1^2}\right)^2 - \left(\frac{\omega_{02}^2 - \omega_{01}^2}{\omega_{02}^2 + \omega_{01}^2}\right)^2} \qquad (7.70)$$

7.3 GENERAL FORMULATION FOR EXTRACTING COUPLING COEFFICIENT k

In the last section, we derived the formulas for extracting the electric, magnetic, and mixed coupling coefficients in terms of the characteristic frequencies of both synchronously and asynchronously tuned coupled resonators. It is interesting to note that the formulas of Eqs. (7.54), (7.61), and (7.70) are all the same. Therefore, we may use the universal formulation

$$k = \pm \frac{1}{2} \left(\frac{f_{02}}{f_{01}} + \frac{f_{01}}{f_{02}} \right) \sqrt{ \left(\frac{f_{p2}^2 - f_{p1}^2}{f_{p2}^2 + f_{p1}^2} \right)^2 - \left(\frac{f_{02}^2 - f_{01}^2}{f_{02}^2 + f_{01}^2} \right)^2 } \tag{7.71}$$

where $f_{0i} = \omega_{0i}/2\pi$ and $f_{pi} = \omega_i/2\pi$ for $i = 1,2$. The formulation of Eq. (7.71) can be used to extract the coupling coefficient of any two asynchronously tuned coupled resonators, regardless of whether the coupling is electric, magnetic, or mixed. Needless to say, the formulation is applicable for synchronously tuned coupled resonators as well and, in that case, it degenerates to

$$k = \pm \frac{f_{p2}^2 - f_{p1}^2}{f_{p2}^2 + f_{p1}^2} \tag{7.72}$$

Comparing Eq. (7.72) with Eqs. (7.36), (7.41), and (7.46), we notice that f_{p1} or f_{p2} corresponds to either f_e or f_m.

The sign of coupling may only be a matter for cross-coupled resonator filters (see Chapter 9). It should be borne in mind that the determination of the sign of the coupling coefficient is very dependent on the physical coupling structure of coupled resonators, which may, in general, be done by using Eq. (7.31). Nevertheless, for filter design, the meaning of positive or negative coupling is rather relative. This means that if we may refer to one particular coupling as the positive coupling and then the negative coupling would imply that its phase response is opposite to that of the positive coupling. The phase response of a coupling may be found from the S parameters of its associated coupling structure. Alternatively, the derivations in Section 7.2.1 have suggested another simple way to find whether the two coupling structures have the same signs or not. This can be done by applying either the electric or magnetic wall to find the f_e or f_m of both the coupling structures. If the frequency shifts of f_e or f_m with respect to their individual uncoupled resonant frequencies are in the same direction, the resultant coupling coefficients will have the same signs, if not the opposite signs. For instance, the electric and magnetic couplings discussed in Section 7.2.1 are said to have the opposite signs for their coupling coefficients. This is because, in the case of the electric coupling, the f_e of Eq. (7.34) is lower than the uncoupled resonant frequency, whereas in the case of the magnetic coupling, the f_e of Eq. (7.39) is higher than the uncoupled resonant frequency. Similarly, one should notice the opposite effects when referring to the f_m of Eq. (7.35) and the f_m of Eq. (7.40).

7.4 FORMULATION FOR EXTRACTING EXTERNAL QUALITY FACTOR Q_e

Two typical input/output (I/O) coupling structures for coupled microstrip resonator filters, namely the tapped-line and the coupled-line structures, are shown with the microstrip open- loop resonator, although other types of resonator may be used (see Figure 7.10). For the tapped-line coupling, usually a 50-Ω feed line is directly tapped on to the I/O resonator, and the coupling or the external quality factor is controlled by the tapping position t, as indicated in Figure 7.10a. For example, the smaller the t, the closer is the tapped line to a virtual grounding of the resonator, which results in a weaker coupling or a larger external quality factor. The coupling of the coupled line stricture in Figure 7.10b can be found from the coupling gap g and the line width w. Normally, a smaller gap and a narrower line result in a stronger I/O coupling or a smaller external quality factor of the resonator.

7.4.1 Singly Loaded Resonator

In order to extract the external quality factor from the frequency response of the I/O resonator, let us consider an equivalent circuit in Figure 7.11, where G should be seen as the external conductance attached to the lossless LC resonator. This circuit actually resembles the I/O resonator circuit of Figure 7.2a, so that the external quality factor to be extracted is consistent with that defined when we are forming the general coupling matrix. The reflection coefficient or S_{11} at the excitation port of resonator is

$$S_{11} = \frac{G - Y_{in}}{G + Y_{in}} = \frac{1 - Y_{in}/G}{1 + Y_{in}/G} \qquad (7.73)$$

where Y_{in} is the input admittance of the resonator

$$Y_{in} = j\omega C + \frac{1}{j\omega L} = j\omega_0 C \left(\frac{\omega}{\omega_0} - \frac{\omega_0}{\omega} \right) \qquad (7.74)$$

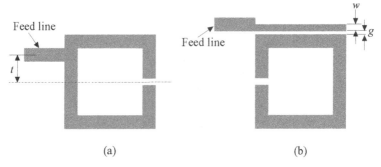

FIGURE 7.10 Typical I/O coupling structures for coupled resonator filters. (**a**) Tapped-line coupling. (**b**) Coupled-line coupling.

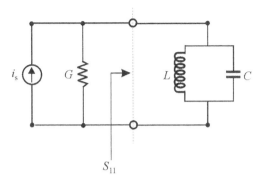

FIGURE 7.11 Equivalent circuit of the I/O resonator with singly loading.

Note that $\omega_0 = 1/\sqrt{LC}$ is the resonant frequency. In the vicinity of resonance, say, $\omega = \omega_0 + \Delta\omega$, Eq. (7.74) may be simplified as

$$Y_{in} = j\omega_0 C \cdot \frac{2\Delta\omega}{\omega_0} \qquad (7.75)$$

where the approximation $(\omega^2 - \omega_0^2)/\omega \approx 2\Delta\omega$ has been used. By substituting Eq. (7.75) into Eq. (7.73) and noting $Q_e = \omega_0 C/G$, we obtain

$$S_{11} = \frac{1 - jQ_e \cdot (2\Delta\omega/\omega_0)}{1 + jQ_e \cdot (2\Delta\omega/\omega_0)} \qquad (7.76)$$

Since we have assumed that the resonator is lossless, the magnitude of S_{11} in Eq. (7.76) is always equal to 1. This is because that in the vicinity of resonance, the parallel resonator of Figure 7.11 behavior likes an open circuit. However, the phase response of S_{11} changes against frequency. A plot of the phase of S_{11} as a function of $\Delta\omega/\omega_0$ is given in Figure 7.12. When the phase is $\pm 90°$ the corresponding value of $\Delta\omega$ is found to be

$$2Q_e \frac{\Delta\omega_\mp}{\omega_o} = \mp 1$$

Hence, the absolute bandwidth between the $\pm 90°$ points is

$$\Delta\omega_{\pm 90°} = \Delta\omega_+ - \Delta\omega_- = \frac{\omega_0}{Q_e}$$

The external quality factor can then be extracted from this relation

$$Q_e = \frac{\omega_0}{\Delta\omega_{\pm 90°}} \qquad (7.77)$$

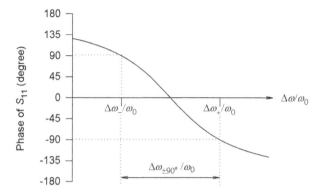

FIGURE 7.12 Phase response of S_{11} for the circuit in Figure 7.11.

It should be commented that the reference plane of S_{11} in the EM simulation may not exactly match that of the equivalent circuit in Figure 7.11, which leads to an extra phase shift such that the phase of the simulated S_{11} does not equal zero at resonance. In this case, the $\Delta\omega_{\mp}$ should be determined from the frequency at which the phase shifts $\pm 90°$ with respect to the absolute phase at ω_0.

Alternatively, the Q_e may be extracted from the group delay of S_{11} at resonance. Let

$$\phi = \tan^{-1}\left(\frac{2Q_e \Delta\omega}{\omega_0}\right)$$

We can rewrite Eq. (7.76) as

$$S_{11} = e^{-j2\phi}$$

The group delay of S_{11} is then given by

$$\tau_{S_{11}}(\omega) = -\frac{\partial(-2\phi)}{\partial \omega} = \frac{4Q_e}{\omega_0} \cdot \frac{1}{1 + (2Q_e \Delta\omega/\omega_0)^2} \qquad (7.78)$$

Recall that $\omega = \omega_0 + \Delta\omega$. At resonance $\Delta\omega = 0$, the group delay in Eq. (7.78) reaches the maximum value

$$\tau_{S_{11}}(\omega_0) = \frac{4Q_e}{\omega_0}$$

Hence, we have

$$Q_e = \frac{\omega_0 \cdot \tau_{S_{11}}(\omega_0)}{4} \qquad (7.79)$$

FORMULATION FOR EXTRACTING EXTERNAL QUALITY FACTOR Q_e 219

Similarly, if the reference plane of simulated S_{11} does not coincide with that of the equivalent circuit in Figure 7.11, an extra group delay may be added, unless the corresponding extra phase shift is frequency independent. Nonetheless, the resonant frequency ω_0 should be determinable from the simulated frequency response of group delay.

7.4.2 Doubly Loaded Resonator

Although the Q_e is defined for a singly loaded resonator, if the resonator is symmetrical, one could add another symmetrical load or port to form a two-port network, as shown in Figure 7.13, where T-T' represents the symmetrical plane and the single LC resonator has been separated into two symmetrical parts. When the symmetrical plane T-T' is short-circuited, we have

$$Y_{ino} = \infty$$
$$S_{11o} = \frac{G - Y_{ino}}{G + Y_{ino}} = -1$$

where Y_{ino} and S_{11o} are the odd-mode input admittance and reflection coefficient at port 1, respectively. On the other hand, replacing the T-T' plane with an open circuit yields the corresponding parameters for the even mode:

$$Y_{ine} = j\omega_0 C \Delta\omega/\omega_0$$
$$S_{11e} = \frac{G - Y_{ine}}{G + Y_{ine}} = \frac{1 - jQ_e\Delta\omega/\omega_0}{1 + jQ_e\Delta\omega/\omega_0}$$

where $\omega_0 = 1/\sqrt{LC}$ and the approximation $(\omega^2 - \omega_0^2)/\omega \approx 2\Delta\omega$ with $\omega = \omega_0 + \Delta\omega$ has been made. Referring to Chapter 2, we can arrive at

$$S_{21} = \frac{1}{2}(S_{11e} - S_{11o}) = \frac{1}{1 + jQ_e\Delta\omega/\omega_0}$$

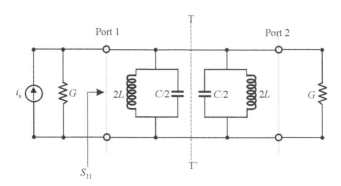

FIGURE 7.13 Equivalent circuit of the I/O resonator with doubly loading.

COUPLED-RESONATOR CIRCUITS

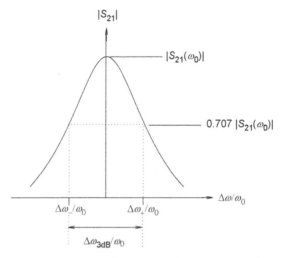

FIGURE 7.14 Resonant amplitude response of S_{21} for the circuit in Figure 7.13.

whose magnitude is given by

$$|S_{21}| = \frac{1}{\sqrt{1 + (Q_e \Delta\omega/\omega_0)^2}} \quad (7.80)$$

Shown in Figure 7.14 is a plot of $|S_{21}|$ against $\Delta\omega/\omega_0$. At resonance, $\Delta\omega = 0$ and, thus, $|S_{21}|$ reaches its maximum value, namely, $|S_{21}(\omega_0)| = 1$. When the frequency shifts such that

$$Q_e \frac{\Delta\omega_\pm}{\omega_0} = \pm 1 \quad (7.81)$$

the value of $|S_{21}|$ has fallen to 0.707 (or −3 dB) of its maximum value according to Eq. (7.80). Define a bandwidth based on Eq. (7.81)

$$\Delta\omega_{3dB} = \Delta\omega_+ - \Delta\omega_- = \frac{\omega_0}{(Q_e/2)} \quad (7.82)$$

where $\Delta\omega_{3dB}$ is the bandwidth for which the attenuation for S_{21} is up 3 dB from that at resonance, as indicated in Figure 7.14. Define a doubly loaded external quality factor Q'_e as

$$Q'_e = \frac{Q_e}{2} = \frac{\omega_0}{\Delta\omega_{3dB}} \quad (7.83)$$

Using Eq. (7.83) to extract the Q'_e first and then the singly loaded external quality factor Q_e is simply the twice of Q'_e.

It should be mentioned that even though the formulations made in this section are based on the parallel resonator, there is no loss of generality, because the same formulas as seen in Eqs. (7.77), (7.79), and (7.83) could be found for the series resonator as well.

7.5 NUMERICAL EXAMPLES

To demonstrate the applications of the above-derived formulas for extracting coupling coefficients and external quality factors, some instructive numerical examples are included in this section. For our purpose, the typical types of coupled microstrip resonators as those shown in Figure 7.15 are employed, without loss of generality. Each of the open-loop resonators is essentially a folded half-wavelength resonator. These coupled structures result from different orientations of a pair of open-loop resonators, which are separated by a spacing s. It is obvious that any coupling in those structures is that of the proximity coupling, which is, basically, through fringe fields. The nature and the extent of the fringe fields determine the nature and the strength of the coupling. It can be shown that at resonance of the fundamental mode, each of the open-loop resonators has the maximum electric field density at the side with an open gap and the maximum magnetic field density at the opposite side. Because the fringe field exhibits an exponentially decaying character outside the region, the electric fringe field is stronger near the side having the maximum electric field distribution, whereas the magnetic fringe field is stronger near the side having

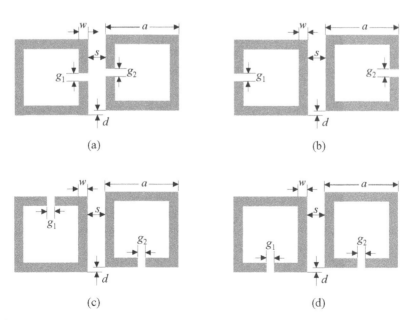

FIGURE 7.15 Typical coupling structures of coupled resonators with (**a**) electric coupling (**b**) magnetic coupling, and (**c–d**) mixed coupling.

the maximum magnetic field distribution. It follows that the electric coupling can be obtained if the open sides of two coupled resonators are proximately placed, as shown in Figure 7.15a. The magnetic coupling can be obtained if the sides with the maximum magnetic field of two coupled resonators are proximately placed, as seen in Figure 7.15b. For the coupling structures in Figures 7.15c and d, the electric and magnetic fringe fields at the coupled sides may have comparative distributions, so that both electric and magnetic couplings occur. In this case, the coupling may be referred to as a mixed coupling. However, it will be demonstrated later that these two coupling structures exhibit distinctive coupling characteristics.

Although the simulated results given below were obtained using a specific commercial full-wave EM simulator [14], any other full-wave EM simulator, or, in fact, experimental measurement, should do the same.

7.5.1 Extracting k (Synchronous Tuning)

The microstrip, square open-loop resonators here have dimensions of $a = 7.0$ mm and $w = 1.0$ mm on a substrate with a relative dielectric constant of 10.8 and thickness of 1.27 mm. It is also assumed that $g_1 = g_2$ for the synchronous tuning and $d = 0$ for a zero offset.

Figure 7.16 shows typical simulated resonant frequency responses of the coupled resonator structures in Figures 7.15a and b, respectively, with $s = 2.0$ mm, where S_{21} denotes the S parameter between the two ports that are very weakly coupled to the coupled resonator structure. We put these two examples together for the comparison because one is for the electric coupling, as shown in Figure 7.16a, and the other the magnetic coupling in shown in Figure 7.16b. In both cases, the two resonant peaks that correspond to the characteristic frequencies f_{p1} and f_{p2}, defined above, are clearly identified from the magnitude responses. From Figure 7.16a, we can find $f_{p1} = 2513.3$ MHz and $f_{p2} = 2540.7$ MHz. Because of synchronous tuning, the coupling coefficient can be extracted using Eq. (7.72) and is $k = 0.01084$. From Figure 7.16b it can be found that $f_{p1} = 2484.2$ MHz and $f_{p2} = 2567.9$ MHz, so that $k = 0.03313$. Hence, with the same coupling spacing s, the magnetic coupling is stronger than the electric coupling. Normally, the stronger the coupling, the wider the separation of the two resonant peaks and the deeper the trough in the middle, as can be seen from the magnitude response. By comparing the phase responses in Figures 7.16a and b, we can observe that both are out of phase. This is evidence that the two coupling coefficients extracted have opposite signs. It might be worth mentioning that to compare the phase responses properly, the port locations with respect to the coupled resonators must be the same in either case. Since the coupled resonator structures considered are symmetrical, we could use another approach, as suggested above, to compare the signs of the two coupling coefficients by looking for the resonant frequencies with electric and magnetic walls inserted, respectively. We performed such simulations and found that $f_e = 2513.3$ MHz and $f_m = 2540.7$ MHz for the electrically coupled resonators, whereas $f_e = 2567.9$ MHz and $f_m = 2484.2$ MHz for the magnetically coupled resonators. Numerically, it shows that $f_e < f_m$ for the electric coupling, but $f_e > f_m$ for the magnetic coupling. The opposite frequency

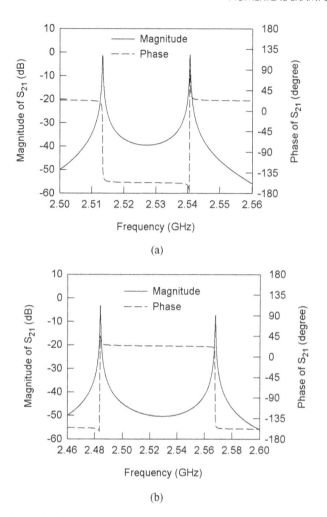

FIGURE 7.16 Typical resonant responses of coupled resonator structures. (**a**) For the structure in Figure 7.15a. (**b**) For the structure in Figure 7.15b.

shifts again indicate that the two resultant coupling coefficients should have the different signs.

Another two pairs of coupled microstrip resonators are considered next; they are the mixed coupling structures shown in Figures 7.15c and d. Coupling coefficients are extracted from the simulated frequency responses, similar to the above, and the results are shown in Figure 7.17, where the extracted coupling coefficient is plotted as a function of the coupling spacing s. It is interesting to note that the two mixed couplings behavior very differently. The coupling coefficient for the mixed coupling structure in Figure 7.15c decreases monolithically with the increase of s, as shown in Figure 7.17a. However, the coupling coefficient for the mixed coupling structure in Figure 7.15d

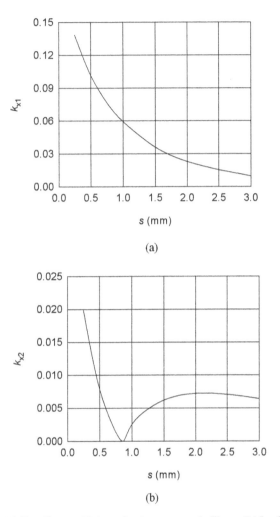

FIGURE 7.17 (a) Coupling coefficients for the structure in Figure 7.15c. (b) Coupling coefficients for the structure in Figure 7.15d.

does not vary monolithically against the s, as illustrated in Figure 7.17b. Furthermore, for the same coupling spacing s, the coupling coefficient in Figure 7.17a is always larger than that in Figure 7.17b. These observations strongly suggest that both the electric and magnetic couplings cancel each other in the coupling structure of Figure 7.15d. To confirm this, Figure 7.18 depicts the simulated resonant responses of this coupled resonator structure with the coupling spacing $s = 0.5$ mm and $s = 2.0$ mm, respectively. These two particular spacing are chosen because we know that the electric coupling decays faster than the magnetic coupling against the spacing [5]. Therefore, as referred to in Figure 7.17b, it would seem that the electric coupling is dominant for the small coupling spacing, say $s < 0.875$ mm, whereas the magnetic coupling becomes dominant when the spacing is larger. As can be seen, the phase response in

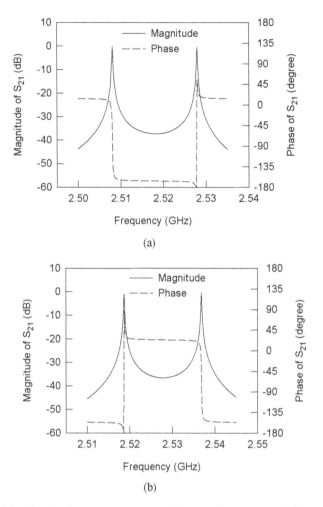

FIGURE 7.18 Simulated resonant responses of the coupling structure in Figure 7.15d for (a) $s = 0.5$ mm, and (b) $s = 2.0$ mm.

Figure 7.18a is, indeed, out of phase with that in Figure 7.18b, showing the opposite signs of the two couplings. This can also be shown by simulating f_e and f_m, since the coupled structure is symmetrical. The simulated results are $f_e = 2508.0$ MHz and $f_m = 2526.9$ MHz and, thus, $f_e < f_m$ when $s = 0.5$ mm; whereas $f_e = 2536.9$ MHz and $f_m = 2518.8$ MHz, so that $f_e > f_m$ when $s = 2.0$ mm. The opposite frequency shifts of f_e or f_m at these two coupling spacings are again the evidence that the resultant coupling coefficients have the opposite signs.

7.5.2 Extracting k (Asynchronous Tuning)

To demonstrate extracting coupling coefficient of asynchronously tuned, coupled microstrip resonators, let us consider the coupled microstrip open-loop resonators in

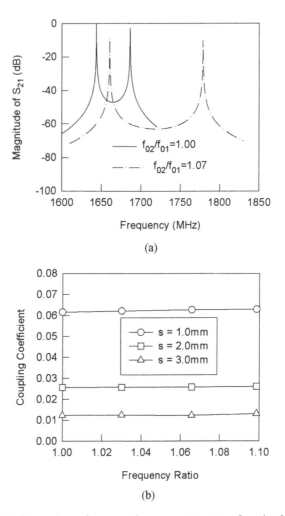

FIGURE 7.19 (a) Comparison of resonant frequency responses of a pair of coupled resonators in Figure 7.15c having been synchronously and asynchronously tuned, respectively. (b) Extracted coupling coefficients with asynchronous tuning.

Figure 7.15c and allow the open gaps indicated by g_1 and g_2 to be different in order to have different self-resonant frequencies f_{01} and f_{02}, respectively. Figure 7.19a shows two typical simulated frequency responses, where the full line is for a synchronously tuned case when $f_{01} = f_{02}$; the dotted line is for an asynchronously tuned case when $f_{01} \neq f_{02}$. In each case, the two resonant frequency peaks that correspond to the two characteristics frequencies of the coupled resonators, that is, f_{p1} and f_{p2}, are clearly identifiable. More numerical results are listed in Table 7.1, and the coupling coefficients extracted using Eq. (7.71) are plotted in Figure 7.19b where the axis of frequency ratio represents a ratio of f_{02} to f_{01}. It seems that the asynchronous tuning in the given range has a very small effect on the coupling. This implies an advantage

TABLE 7.1 Numerical Results of the Coupled Microstrip Open-Loop Resonators of Figure 7.15(c) for g_1 Fixed by 0.4 mm

s(mm)	g_2(mm)	f_{01}(MHz)	f_{02}(MHz)	f_{p1}(MHz)	f_{p2}(MHz)
1	0.4	1664.7	1664.7	1613.2	1715.7
1	1.0	1664.7	1714.7	1631.2	1747.5
1	2.0	1664.7	1774.05	1642.0	1795.5
1	3.0	1664.7	1828.9	1647.5	1845.0
2	0.4	1664.7	1664.7	1643.4	1686.1
2	1.0	1664.7	1714.7	1656.6	1722.9
2	2.0	1664.7	1774.05	1660.4	1778.4
2	3.0	1664.7	1828.9	1661.7	1832.1
3	0.4	1664.7	1664.7	1654.5	1674.9
3	1.0	1664.7	1714.7	1662.6	1716.8
3	2.0	1664.7	1774.05	1663.7	1775.05
3	3.0	1664.7	1828.9	1664.0	1829.8

of using this type of coupled resonator structure to design asynchronously tuned microstrip filters, because the coupling is almost independent of the asynchronous tuning. This makes both the design and tuning of the filter easier. It should be mentioned that if one tries to use Eq. (7.72) to extract the coupling coefficients, a different and incorrect conclusion would be drawn.

7.5.3 Extracting Q_e

As an example, let us consider the I/O coupling structure of Figure 7.10b. Figure 7.20 shows the typical simulated phase and group-delay responses of S_{11}. The resonant frequency can be determined from the peak of the group-delay response and shown

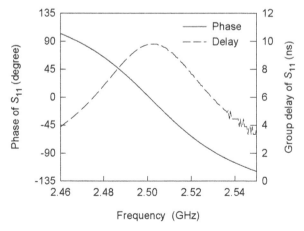

FIGURE 7.20 Typical simulated phase and group-delay responses of S_{11} for a singly loaded resonator.

to be $f_0 = 2502$ MHz. At this resonant frequency, it can be found that the phase is not equal to zero, but $\phi_0 = -4.214°$. This is because of the shift of the reference plane, as explained in Section 7.4. Find the frequencies at which the phase shifts $\pm 90°$ with respect to the ϕ_0 from the phase response. The results of this are $f_- = 2470$ MHz and $f_+ = 2534$ MHz and, thus, $\Delta f_{\pm 90°} = f_+ - f_- = 64$ MHz. The external quality factor is then ready to be extracted using Eq. (7.77). This gives $Q_e = f_0/\Delta f_{\pm 90°} = 2502/64 = 39$.

7.6 GENERAL COUPLING MATRIX INCLUDING SOURCE AND LOAD

Earlier discussion in Section 7.1 only focuses on filter topologies, where the source and the load are coupled to only one resonator each and not to each other. It is well known that such a topology can produce, at most, $n-2$ finite transmission zeros out of n resonators, where n is the degree of filter. The addition of a direct signal path between the source and the load allows the generation of n finite frequency transmission zeros instead of $n-2$. Furthermore, allowing the source or load to be coupled to more than one resonator brings about more diverse topologies for filter designs. Filters of this type have recently attracted more attention [15–19]. Figure 7.21 illustrates a general topology of n-coupled resonators with multiple source/load couplings, where each black node represents a resonator; S and L denote the source and load nodes, respectively. The coupling between two nodes is denoted by m_{ij}, for i or $j = S, 1, 2, \ldots n, L$, and $i \neq j$.

Figure 7.22 shows the so-called $n+2$ or "extended" general coupling matrix for nth-degree coupled resonator filters. The $n+2$ coupling matrix has an extra pair of rows at the top and bottom and an extra pair of columns at the left and right surrounding the "core" coupling matrix. The $n+2$ coupling matrix is an $(n+2) \times (n+2)$ reciprocal matrix (that is, $m_{ij} = m_{ji}$, for $i \neq j$). Nonzero diagonal entries are allowed for an asynchronously tuned filter. The elements of the "core" coupling matrix have the same definitions as discussed in Section 7.1. The extra elements, in the $n+2$ coupling matrix, carry not only the input and output couplings from the

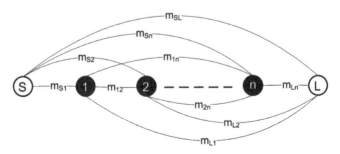

FIGURE 7.21 General topology of n-coupled resonators with multiple source/load couplings.

	S	1	2	...	n-1	n	L
S	0	m_{S1}	m_{S2}	...	$m_{S,n-1}$	m_{Sn}	m_{SL}
1	m_{1S}	m_{11}	m_{12}	...	$m_{1,n-1}$	m_{1n}	m_{1L}
2	m_{2S}	m_{21}	m_{22}	...	$m_{2,n-1}$	m_{2n}	m_{2L}
⋮	⋮	⋮	⋮	⋱	⋮	⋮	⋮
n-1	$m_{n-1,S}$	$m_{n-1,1}$	$m_{n-1,2}$	...	$m_{n-1,n-1}$	$m_{n-1,n}$	$m_{n-1,L}$
n	m_{nS}	m_{n1}	m_{n2}	...	$m_{n,n-1}$	m_{nn}	m_{nL}
L	m_{LS}	m_{L1}	m_{L2}	...	$m_{L,n-1}$	m_{Ln}	0

FIGURE 7.22 $n+2$ general coupling matrix $[m]$.

source and load terminations to resonator nodes in the core matrix, but also a direct coupling between the source and the load (that is, $m_{SL} = m_{LS}$).

The m_{Si} and m_{Li}, which represent the input and output couplings from the source and load to resonator i, can be converted to external quality factors by

$$Q_{e,Si} = \frac{1}{m_{Si}^2 \cdot FBW}$$
$$Q_{e,Li} = \frac{1}{m_{Li}^2 \cdot FBW} \qquad (7.84)$$

where FBW is the fractional bandwidth of bandpass filter.

For a given filter topology, the $n+2$ coupling matrix may be obtained using synthesis methods described [15–18]. Once the coupling matrix $[m]$ is determined, the filter frequency responses can be computed in terms of scattering parameters:

$$S_{21} = -2j \, [A]^{-1}_{n+2,1}$$
$$S_{11} = 1 + 2j \, [A]^{-1}_{11} \qquad (7.85)$$

where $[A]^{-1}_{ij}$ denotes the ith row and jth column element of $[A]^{-1}$. The matrix $[A]$ is given by

$$[A] = [m] + \Omega \, [U] - j \, [q] \qquad (7.86)$$

in which $[U]$ is similar to the $(n+2) \times (n+2)$ identity matrix, except that $[U]_{11} = [U]_{n+2,n+2} = 0$, $[q]$ is the $(n+2) \times (n+2)$ matrix with all entries zeros, except for $[q]_{11} = [q]_{n+2,n+2} = 1$, and Ω is the frequency variable of lowpass prototype. The lowpass prototype response can be transformed to a bandpass response having a fractional bandwidth FBW at a center frequency ω_0 using the well-known frequency transformation:

$$\Omega = \frac{1}{FBW}\left(\frac{\omega}{\omega_0} - \frac{\omega_0}{\omega}\right) \qquad (7.87)$$

To extract the direct source-load coupling m_{SL}, a useful formulation can be derived as follows. Consider an I/O structure that only involves the source-load coupling. From Figure 7.22, the general coupling matrix for the structure under consideration would have a form

$$[m] = \begin{bmatrix} 0 & m_{SL} \\ m_{SL} & 0 \end{bmatrix}$$

which corresponds to a zero-order filter case for $n = 0$. From Eq. (7.86), we have

$$[A] = \begin{bmatrix} 0 & m_{SL} \\ m_{SL} & 0 \end{bmatrix} + \Omega \begin{bmatrix} 0 & 0 \\ 0 & 0 \end{bmatrix} - j \begin{bmatrix} 1 & 0 \\ 0 & 1 \end{bmatrix} = \begin{bmatrix} -j & m_{SL} \\ m_{SL} & -j \end{bmatrix}$$

An analytical solution for $[A]^{-1}$ can be found to be

$$[A]^{-1} = \frac{\begin{bmatrix} j & m_{SL} \\ m_{SL} & j \end{bmatrix}}{1 + m_{SL}^2}$$

Using Eq. (7.85), we obtain

$$S_{21} = -2j\frac{m_{SL}}{1 + m_{SL}^2}$$

or

$$|S_{21}| = \pm\frac{2m_{SL}}{1 + m_{SL}^2}$$

The negative sign is for $m_{SL} < 0$. Solving the above equation for m_{SL} yields

$$m_{SL} = \pm\frac{1 - \sqrt{1 - |S_{21}|^2}}{|S_{21}|} \qquad (7.88)$$

which is a valid solution by taking into account that $m_{SL} = 0$ for $|S_{21}| = 0$ and $0 \le |S_{21}| \le 1$. The choice of a sign is rather relative, which would depend on the signs of other elements in the $n + 2$ general coupling matrix. In any case, it can easily be determined from the desired frequency response.

REFERENCES

[1] A. E. Atia and A. E. Williams, Narrow-bandpass waveguide filters, *IEEE Trans.* **MTT-20**, 1972, 258–265.

[2] L. Accatino, G. Bertin, and M. Mongiardo, A four-pole dual mode elliptic filter realized in circular cavity without screws, *IEEE Trans.* **MTT-44**, 1996, 2680–2687.

[3] C. Wang, H.-W. Yao, K. A. Zaki, and R. R. Mansour, Mixed modes cylindrical planar dielectric resonator filters with rectangular enclosure, *IEEE Trans.* **MTT-43**, 1995, 2817–2823.

[4] H.-W. Yao, C. Wang, and K. A. Zaki, Quarter wavelength ceramic combline filters, *IEEE Trans.* **MTT-44**, 1996, 2673–2679.

[5] J.-S. Hong and M. J. Lancaster, Couplings of microstrip square open-loop resonators for cross-coupled planar microwave filters, *IEEE Trans.* **MTT-44**, 1996, 2099–2109.

[6] J.-S. Hong and M. J. Lancaster, Theory and experiment of novel microstrip slow-wave open-loop resonator filters, *IEEE Trans.* **MTT-45**, 1997, 2358–2665.

[7] J.-S. Hong and M. J. Lancaster, Cross-coupled microstrip hairpin-resonator filters, *IEEE Trans.* **MTT-46**, 1998, 118–122.

[8] J.-S. Hong, M. J. Lancaster, D. Jedamzik, and R. B. Greed, On the development of superconducting microstrip filters for mobile communications applications, *IEEE Trans.* **MTT-47**, 1999, 1656–1663.

[9] P. Blondy, A. R. Brown, D. Cros, and G. M. Rebeiz, Low loss micromachined filters formillimeter-wave telecommunication systems, *IEEE MTT-S Dig.*, 1998, 1181–1184.

[10] A. E. Atia, A. E. William, and R. W. Newcomb, Narrow-band multi-coupled cavity synthesis, *IEEE Trans.* **CAS-21**, 1974, 649–655.

[11] R. J. Cameron, General coupling matrix synthesis methods for Chebyshev filtering functions, *IEEE Trans.* **MTT-47**, 1999, 433–442.

[12] J.-S. Hong, Couplings of asynchronously tuned coupled microwave resonators, *IEE Proc. Microwaves, Antennas Propag* **147**, 2000, 354–358.

[13] C. G. Montgomery, R. H. Dicke, and E. M. Purcell, *Principle of Microwave Circuits*, McGraw-Hill, New York, 1948, Chapter 4.

[14] *EM User's Manual*, Sonnet Software Inc., Liverpool, New York, 2009.

[15] J. R. Montejo-Garai, Synthesis of N-even order symmetric filters with N transmission zeros by means of source-load cross coupling, *Electron. Lett.* **36** (3), 2000, 232–233.

[16] S. Amari, Direct synthesis of folded symmetric resonator filters with source–load coupling, *IEEE Microwave Wireless Comp. Lett.* **11**, 2001, 264–266.

[17] R. J. Cameron, Advanced coupling matrix synthesis techniques for microwave filters, *IEEE Trans. Microwave Theory Tech.* **51**(1), 2003, 1–10.

[18] A. Garia-Lamperez, S. Llorente-Romano, M. Salazar-Palma, M. J. Padilla-Cruz, and I. H. Carpintero, Synthesis of cross-coupled lossy resonator filters with multiple input/output couplings by gradient optimization, *IEEE AP-S Int. Symp.* **2**, 2003, 52–55.

[19] C.-K. Liao and C. -Y.Chang, Design of microstrip quadruplet filters with source-load coupling, *IEEE Trans. Microwave Theory Tech.* **53**(7), 2005, 2302–2308.

CHAPTER EIGHT

CAD for Low-Cost and High-Volume Production

There have been extraordinary recent advances in computer-aided design (CAD) of RF/microwave circuits, particularly in full-wave electromagnetic (EM) simulations. They have been implemented both in commercial and specific in-house software and are being applied to microwave filters simulation, modeling, design, and validation [1,2]. The developments in this area are certainly stimulated by ever-increasing computer power.

Another driving force for the developments is the requirement of CAD for low-cost and high-volume production [3,4]. In general, besides the investment for tooling, the cost of filter production is mainly affected by materials and labor. Microstrip filters using conventional printed circuit boards are of low cost. Using better materials such as superconductors can give better performance of filters, but is normally more expensive. This may then be evaluated by a cost-effective factor in terms of the performance. Labor costs include those for design, fabrication, testing, and tuning. Here the weights for design and tuning can be reduced greatly by using CAD. For instance, in addition to controlling fabrication processes, to tune or not not-to-tune is also more of a question of design accuracy; tuning can be very expensive and time-consuming for mass production. CAD can provide more accurate design with less design iterations, leading to first-pass or tuneless filters. This not only reduces the labor intensiveness and also the cost, but also decreases the time from design to production. The latter can be crucial for winning a market in which there is stiff competition. Furthermore, if the materials used are expensive, the first-pass design or less iteration afforded by CAD will reduce the extra cost of materials and other factors necessary for developing a satisfactory prototype.

Generally speaking, any design involves using computers may be termed as CAD. This may include computer simulation and/or computer optimization. The intention

---

Microstrip Filters for RF/Microwave Applications by Jia-Sheng Hong
Copyright © 2011 John Wiley & Sons, Inc.

TABLE 8.1   Some Commercially Available CAD Tools

Company	Typical product (all trademarks)	Type
Sonnet Software (www.sonnetsoftware.com)	Sonnet Suites Sonnet Lite	3D planar EM software Free 3D Planar EM simulator
Applied Wave Research (AWR) (web.awrcorp.com)	Microwave Office	Integrated package including linear and nonlinear circuit simulators, optimizers, and EM analysis tools
Agilent Technologies (www.agilent.com)	Advanced Design System (ADS)	Integrated package including 3D EM simulators
Ansoft (www.ansoft.com)	HFSS	3D EM simulation
Zeland Software (www.zeland.com)	IE3D	Planar and 3D EM simulation, optimization, and synthesis package
CST (www.cst.com)	CST Microwave Studio	3D EM simulation
QWED (www.qwed.com.pl)	QuickWave-3D	3D EM simulation

of this chapter is to discuss some basic concepts, methods, and issues regarding filter design by CAD. Typical examples of the applications will be described. As a matter of fact, many more CAD examples, in particular, those based on full-wave EM simulation, can be found for many filter designs described in other chapters of this book.

8.1   COMPUTER-AIDED DESIGN (CAD) TOOLS

CAD tools can be developed in house for particular applications. They can be as simple as a few equations written using any common math software, such as Mathcad [5]. For example, the formulations for network connections provided in Chapter 2 can be programmed in this way for analyzing numerous filter networks. There is also now a large range of commercially available RF/microwave CAD tools, which are more sophisticated and powerful, and might include a linear circuit simulator, analytical modes in a vendor library, a 2D or 3D EM solver, and optimizers. Some vendors with their key products for RF/microwave filter CAD are listed in Table 8.1.

8.2   COMPUTER-AIDED ANALYSIS (CAA)

8.2.1   Circuit Analysis

Since most filters are comprised of linear elements or components, linear simulations based on the network or circuit analyses, described in Chapter 2, are simple and fast for computer-aided analysis (CAA). Linear simulations analyze frequency responses of microwave filters or elements based on their analytical circuit models. Analytical models are fast. However, they are normally only valid in certain ranges of frequency and physical parameters.

To demonstrate how a linear simulator usually analyzes a filter, let us consider the stepped-impedance microstrip lowpass filter shown in Figure 8.1a, where W_0 denotes the terminal line width, W_1 and l_1 are the width and length of the inductive-line element, and W_2 and l_2 are the width and length of the capacitive-line element.

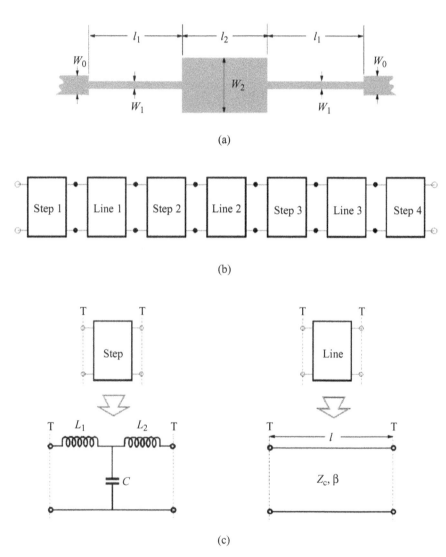

FIGURE 8.1 (a) Stepped-impedance microstrip lowpass filter. (b) Its network representation with cascaded subnetworks for network analysis. (c) Equivalent circuits for the subnetworks.

TABLE 8.2 Circuit Parameters of the Filter in Figure 8.1

Subnetwork		Circuit Parameters					
No.	Name	Z_c (Ω)	β (rad/mm)	l (mm)	L_1 (nH)	L_2 (nH)	C (pF)
1	Step 1				0.085	0.151	0.056
2	Line 1	93	$0.05340 f$	9.81			
3	Step 2				0.493	0.142	0.087
4	Line 2	24	$0.05961 f$	7.11			
5	Step 3				0.142	0.493	0.087
6	Line 3	93	$0.05340 f$	9.81			
7	Step 4				0.151	0.085	0.056

For the linear simulation, the microstrip filter structure is subdivided into cascaded elements and represented by a cascaded network, as illustrated in Figure 8.1b. We note that in addition to the three line elements, four-step discontinuities along the filter structure have been taken into account. Each of the subnetworks is described by the corresponding equivalent circuit shown in Figure 8.1c. The analytical models or closed-form expressions, such as those given in Chapter 4, are used to compute the circuit parameters, that is, L_1, L_2, and C for the microstrip step discontinuities, the characteristic impedance Z_c, and the propagation constant β for the microstrip line elements. The *ABCD* parameters of each subnetwork can be determined by the formulations given in Figure 2.2 of the Chapter 2. The *ABCD* matrix of the composite network of Figure 8.1b is then computed by multiplying the *ABCD* matrices of the cascaded subnetworks which can be converted into the *S* matrix according to the network analysis discussed in Chapter 2. In this way, the frequency responses of the filter are analyzed.

For a numerical demonstration, recall the filter design given in Figure 5.2a of Chapter 5. We have all the physical dimensions for analyzing the filter, as follows: $W_0 = 1.1$ mm, $W_1 = 0.2$ mm, $l_1 = 9.81$ mm, $W_2 = 4.0$ mm, and $l_2 = 7.11$ mm on a 1.27-mm thick substrate with a relative dielectric constant $\varepsilon_r = 10.8$. Using the closed-form expressions given in Chapter 4, we can find the circuit parameters of the subnetworks as referred to in Figure 8.1, which are listed in Table 8.2, where f is the frequency in GHz.

The *ABCD* matrix for each of the line subnetworks (lossless) is

$$\begin{bmatrix} \cos \beta l & j Z_c \sin \beta l \\ j \sin \beta l / Z_c & \cos \beta l \end{bmatrix} \quad (8.1)$$

For each of the step subnetworks, the *ABCD* matrix is given by

$$\begin{bmatrix} 1 - \omega^2 C L_1 & (j\omega L_1 + j\omega L_2) - j\omega^3 C L_1 L_2 \\ j\omega C & 1 - \omega^2 C L_2 \end{bmatrix} \quad (8.2)$$

The *ABCD* matrix of the whole filter network is computed by

$$\begin{bmatrix} A & B \\ C & D \end{bmatrix} = \prod_{i=1} \begin{bmatrix} A & B \\ C & D \end{bmatrix}_i \quad (8.3)$$

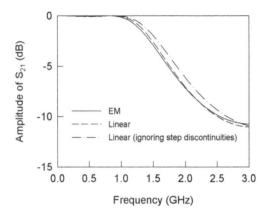

FIGURE 8.2 Computer simulated frequency responses of a microstrip lowpass filter.

where i denotes the number of the subnetworks as consecutively listed in Table 8.2 and the *ABCD* matrices on the right-hand side for the subnetworks are given by either Eq. (8.1) for the line subnetworks or Eq. (8.2) for the step subnetworks. The transmission coefficient of the filter is computed by

$$S_{21} = \frac{2}{A + B/Z_0 + CZ_0 + D} \tag{8.4}$$

where the terminal impedance $Z_0 = 50\ \Omega$. Figure 8.2 shows the linear simulations of the filter as compared to the EM simulation obtained previously in Figure 5.2b. Note that the broken line represents the linear simulation that takes all the discontinuities into account, whereas the dotted line is for the linear simulation ignoring all the discontinuities. As can be seen, the former agrees better with the EM simulation.

Another useful example is shown in Figure 8.3a. This is a three-pole microstrip bandpass filter using parallel-coupled, half-wavelength resonators, as discussed in the Chapter 5. For simplicity, we assume here that all the coupled lines have the same width W. The filter is subdivided into cascaded subnetworks, as depicted in Figure 8.3b, for linear simulation. The computation of the *ABCD* matrices for the step subnetworks are similar to that discussed above. The *ABCD* parameters for each of the coupled-line subnetworks may be computed by Zysman and Johnson [6].

$$\begin{aligned}
A &= D = \frac{Z_{0e} \cot\theta_e + Z_{0o} \cot\theta_o}{Z_{0e} \csc\theta_e - Z_{0o} \csc\theta_o} \\
B &= \frac{j}{2} \frac{Z_{0e}^2 + Z_{0o}^2 - 2Z_{0e}Z_{0o}(\cot\theta_e \cot\theta_o + \csc\theta_e \csc\theta_o)}{Z_{0e} \csc\theta_e - Z_{0o} \csc\theta_o} \\
C &= \frac{2j}{Z_{0e} \csc\theta_e - Z_{0o} \csc\theta_o}
\end{aligned} \tag{8.5}$$

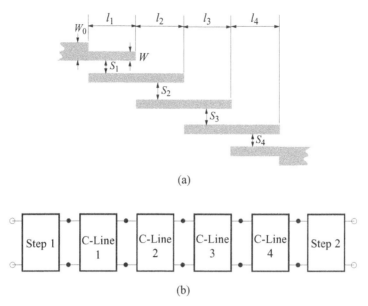

FIGURE 8.3 (a) Microstrip bandpass filter. (b) Its network representation with cascaded subnetworks for network analysis.

where Z_{0e} and Z_{0o} are the even-mode and odd-mode characteristic impedances and θ_e and θ_o are the electrical lengths of the two modes, as discussed in Chapter 4. Numerically, consider a microstrip filter of the form in Figure 8.3a having the dimensions: $W_0 = 1.85$ mm, $W = 1.0$ mm, $s_1 = s_4 = 0.2$ mm, $l_1 = l_4 = 23.7$ mm, $s_2 = s_3 = 0.86$ mm, and $l_2 = l_3 = 23.7$ mm on a GML1000 dielectric substrate with a relative dielectric constant $\varepsilon_r = 3.2$ and a thickness $h = 0.762$ mm. It is important to note that the effect because of the open end of the lines must be taken into account when θ_e and θ_o are computed [7]. This can be done by increasing the line length such that $l \to l + \Delta l$, where Δl may be approximated by the single-line open end, described in Chapter 4, or more accurately by the even- and odd-mode open-end analysis as described in Kirschning and Jansen [8]. Figure 8.4 plots the frequency responses of the filter as analyzed.

It should be mentioned that in addition to the errors in analytical models, particularly when the various elements that make up a microstrip filter are packed tightly together, there are several extra potential sources of errors in the analysis. Circuit simulators assume that discontinues are isolated elements fed by single-mode microstrip lines. However, there can be electromagnetic coupling between various elements in the network because of induced voltages and currents. It takes time and distance to reestablish the normal microstrip current distribution after it passes through a discontinuity. If another discontinuity is encountered before the normal current distribution is reestablished, the "initial conditions" for the second discontinuities are now different from the isolated case because of the interaction of higher modes whose effects

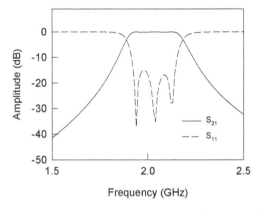

FIGURE 8.4 Computer-simulated frequency responses of a microstrip bandpass filter.

are no longer negligible. All these potential interactions suggest caution whenever we subdivide a filter structure for either circuit analysis or EM simulation.

8.2.2 Electromagnetic Simulation

Electromagnetic (EM) simulation solves the Maxwell's equations with the boundary conditions imposed upon the RF/microwave structure to be modeled. Most commercially available EM simulators use numerical methods to obtain the solution. These numerical techniques include the method of moments (MoM) [9,10], the finite-element method (FEM) [11], the finite-difference time-domain method (FDTD) [12], and the integral equation (boundary element) method (IE/BEM) [13,14]. Each of these methods has its own advantages and disadvantages and is suitable for a class of problems [15–18]. It is not our intention to present these methods here; the interested reader may refer to the references for details. However, we will concentrate on the appropriate utilization of the EM simulations.

EM-simulation tools can accurately model a wide range of RF/microwave structures and can be more efficiently used if the user is aware of sources of error. One principle error, which is common to most all the numerical methods, is because of the finite cell or mesh sizes. These EM simulators divide a RF/microwave filter structure into subsections or cells with 2D or 3D meshing, and then solve Maxwell's equations upon these cells. Larger cells yield faster simulations, but at the expense of larger errors. Errors are diminished by using smaller cells, but at the cost of longer simulation times. It is important to learn if the errors in the filter simulation are because of mesh-size errors. This can be done by repeating the EM simulation using different mesh sizes and comparing the results, which is known as a convergence analysis [19,20].

For demonstration, consider a microstrip pseudointerdigital bandpass filter [19] shown in Figure 8.5. The filter is designed to have 500-MHz bandwidth at a center frequency of 2.0 GHz and is composed of three identical pairs of pseudointerdigital resonators. All pseudointerdigital lines have the same width of 0.5 mm. The coupling

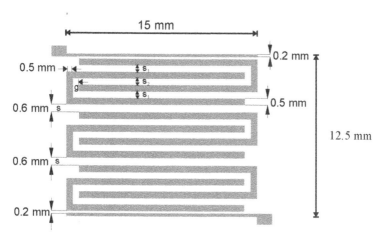

FIGURE 8.5 Layout of a microstrip pseudointerdigital bandpass filter for EM simulation. The filter is on a 1.27-mm thick substrate with a relative dielectric constant of 10.8.

spacing $s_1 = s_2 = 0.5$ mm for each pair of the pseudointerdigital resonators. The coupling spacing between contiguous pairs of the pseudointerdigital resonators is denoted by s, in this case, $s = 0.6$ mm. Two feeding lines, which are matched to the 50-Ω input/output ports, are 15-mm long and 0.2-mm wide. The feeding lines are coupled to the pseudointerdigital structure through 0.2-mm separations. The whole size of the filter is 15 × 12.5 mm on a RT/Duriod substrate having a thickness of 1.27 mm and a relative dielectric constant of 10.8. This size is about $\lambda_g/4 \times \lambda_g/4$, where λ_g is the guided wavelength at the midband frequency on the substrate. For this type of compact filter, the cross coupling of all resonators would be expected. Therefore, it is necessary to use EM simulation to achieve more accurate analysis.

This filter was simulated using a 3D-planar EM simulator [21], but other analogous products could also have been utilized. Similar to most EM simulators, one of the main characteristics of the EM simulator used is the simulation grid or mesh, which can be defined by the user and is imposed on the analyzed structure during numerical EM analysis. Like any other numerical technique based on full-wave EM simulators, there is convergence issue for the EM simulator used. That is, the accuracy of the simulated results depends on the fineness of the grid. Generally speaking, the finer the grid (smaller the cell size), the more accurate the simulation results, but the longer the simulation time and the larger the computer memory required. Therefore, it is very important to consider how small a grid or cell size is needed for obtaining accurate solutions from the EM simulator. To determine a suitable cell size, Figure 8.6 shows the simulated filter frequency responses, that is, the transmission loss and the return loss for different cell sizes. As can be seen, when the cell size is 0.5 × 0.25 mm, the simulation results (full lines) are far from the convergence and give an incorrect prediction. However, as the cell size becomes smaller, the simulation results approach the convergent ones and show no significant changes when the cell size is further reduced below the cell size of 0.25 × 0.1 mm, since the curves for the cell size of 0.5 × 0.1 mm almost overlap those for the cell size of 0.25 × 0.1 mm. This

240 CAD FOR LOW-COST AND HIGH-VOLUME PRODUCTION

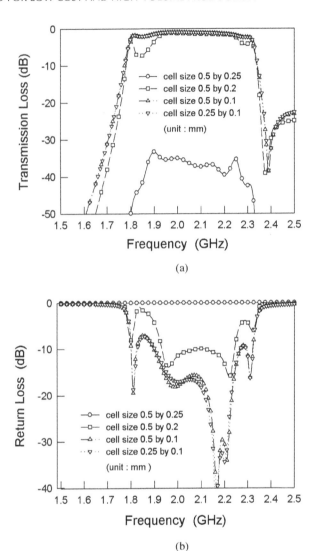

FIGURE 8.6 Convergence analysis for EM simulations of the filter in Figure 8.5.

cell size, in terms of λ_g, is about $0.0045 \times 0.0018\ \lambda_g$. The computational time and the required computer memory for the different cell sizes are the other story. Using a SPARC-2 computer, a computing time of 29 s/frequency and 1 Mb/385 subsections are needed when the cell size is 0.5×0.25 mm. In any case, the smaller cell size results in a larger number of the subsections. Using the same computer, the computing times are 47, 238, and 675 s/frequency and the required computer memories are 1 Mb/482 subsections, 4 Mb/920 subsections, and 7 Mb/1298 subsections for the cell sizes of 0.5×0.2 mm, 0.5×0.1 mm, and 0.25×0.1 mm, respectively. As can be seen, both the computational time and computer memory increase much faster as the cell size becomes smaller. To make the EM simulation not only accurate but also efficient, using a cell size of 0.5 × 0.1 mm should, in this case, be adequate. It should be noticed that how small a cell size, which is measured in physical units (say mm) by the EM simulator, should be specified for convergence is also dependent on operation frequency. In general, the lower the frequency, the larger is the cell size that would be adequate for the convergence. For this reason it would not be wise to specify a very wide operation frequency range (say 1–10 GHz) for simulation, because it would require a very fine grid or small cell in order to obtain a convergent simulation at the highest frequency, and such a fine grid would be more than adequate for the convergence at the lower frequency band, so that a large unnecessary computation time would result.

To verify the accuracy of the electromagnetic analysis, the simulated results using a cell size of 0.5 × 0.1 mm are plotted in Figure 8.7 together with the measured

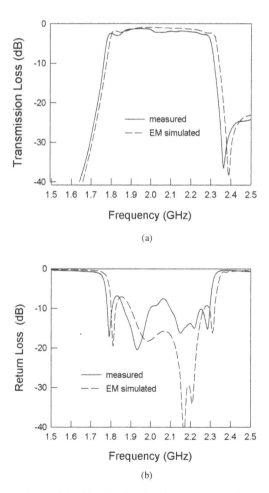

FIGURE 8.7 Comparison of the EM-simulated and -measured performances of the filter in Figure 8.5. The simulation uses a cell size of 0.5 × 0.1 mm.

results for comparison. Good agreement, except for some frequency shift between the measured and the simulated results, can be observed. The frequency shift between the measured and simulated responses is most likely because of the tolerances in the fabrication and substrate material and/or to the assumption of zero metal strip thickness by the EM simulator used [19].

In many practical computer-aided designs, to speed up a filter design, EM simulation is used to accurately model individual components that are implemented in a filter. The initial design is then entirely based on these circuit models and the simulation of the whole filter structure may be performed as a final check [22–31]. In fact, we have applied this approach for many filter designs described in Chapters 5 and 6. We will demonstrate more in the rest of this book. This CAD technique works well in many cases, but caution should be taken when we break the filter structure into several parts for the EM simulation. This is because, as mentioned earlier, the interface conditions at a joint of any two separately simulated parts can be different from that when they are simulated together in the larger structure. Also, when we use this technique, we assume that the separated parts are isolated elements, but in the real filter structure they may be coupled to one another; these unwanted couplings may have significant effects on the entire filter performance, especially in microstrip filters [27].

8.3 FILTER SYNTHESIS BY OPTIMIZATION

8.3.1 General Description

In many practical cases, it is desirable to define the topology of the filter to conform to certain mechanical and packaging constraints. Realization of asymmetric structures as well as asymmetric frequency responses with the minimum number of elements is also desirable. In these cases, it is often not possible to use conventional synthesis techniques, such as those in [32–36] to achieve the desired designs because of the failure of convergence [38]. Therefore, more powerful filter synthesis procedures based on computer optimization may be preferred [37–40].

In general, a computer-aided analysis (CAA) model for the prescribed filter topology is required in order to synthesize the filter using optimization. Assume that the scattering parameters generated by a CAA model for a two-port filter topology are $S_{21}^{CAA}(f, \underline{\Phi})$ and $S_{11}^{CAA}(f, \underline{\Phi})$, where f is the frequency and $\underline{\Phi}$ represents all design variables of the prescribed filter topology, which are to be synthesized by optimization. To do so, an error or objective function may be defined for the least-square case as

$$EF(f, \underline{\Phi}) = \left\{ \sum_{i=1}^{I} \left| S_{21}^{CAA}(f_i, \underline{\Phi}) - S_{21}(f_i) \right|^2 + \sum_{j=1}^{J} \left| S_{11}^{CAA}(f_j, \underline{\Phi}) - S_{11}(f_j) \right|^2 \right\}$$

(8.6)

where $S_{21}(f_i)$ and $S_{11}(f_j)$ are the desired filter frequency responses at sample frequencies f_i and f_j. The optimization-based filter synthesis is then to minimize the error function of Eq. (8.6) by searching for a set of optimal design variables defined in $\underline{\Phi}$ using an optimization algorithm.

For multiple coupled-resonator filters, the expressions of S parameters described in Chapter 7 can be used for synthesizing a prescribed filter topology by optimization. In this case, the general coupling matrix $[m]$ will serve as a topology matrix. Two numerical examples of synthesis by optimization based on the coupling matrix $[m]$ will be described below. Of course, filter synthesis by optimization can be performed directly with many commercial CAD tools. An example of this will follow.

8.3.2 Synthesis of a Quasielliptic-Function Filter by Optimization

As a first example, a four-pole quasielliptic-function filter to be synthesized has a pair of finite frequency attenuation poles at $p = \pm j2.0$ and a return loss better than -26 dB over the passband. Furthermore, both the topology and the frequency response are required to be symmetrical. By referring to Chapter 7, we may express the S parameters of the desired filter topology as

$$S_{21}^{CAA}(\Omega, \underline{\Phi}) = \frac{2}{q_e}[A]_{4,1}^{-1}$$

$$S_{11}^{CAA}(\Omega, \underline{\Phi}) = 1 - \frac{2}{q_e}[A]_{1,1}^{-1} \quad (8.7)$$

$$[A] = j\begin{bmatrix} \Omega - j/q_e & 0 & 0 & 0 \\ 0 & \Omega & 0 & 0 \\ 0 & 0 & \Omega & 0 \\ 0 & 0 & 0 & \Omega - j/q_e \end{bmatrix} - j\begin{bmatrix} 0 & m_{12} & 0 & m_{14} \\ m_{12} & 0 & m_{23} & 0 \\ 0 & m_{23} & 0 & m_{12} \\ m_{14} & 0 & m_{12} & 0 \end{bmatrix}$$

where Ω is the normalized lowpass frequency variable and q_e is the scaled external quality factor that is the same at the filter input and output for the symmetry. As can be seen, the coupling matrix prescribes the desired filter topology. For instance, it has nonzero entry of m_{14} for the desired cross coupling and forces $m_{34} = m_{12}$ for the symmetry. The design parameters to be synthesized by optimization are defined by

$$\underline{\Phi} = \begin{bmatrix} m_{14} & m_{12} & m_{23} & q_e \end{bmatrix} \quad (8.8a)$$

The initial values of $\underline{\Phi}$, which can be estimated from a Chebyshev filter with the same order, are

$$\underline{\Phi}^{initial} = \begin{bmatrix} 0 & 1.07 & 0.79 & 0.7246 \end{bmatrix} \quad (8.8b)$$

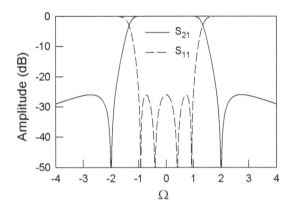

FIGURE 8.8 Frequency response of a four-pole quasielliptic function filter synthesized by optimization.

The error function for this synthesis example is formulated according to Eq. (8.6) at some characteristic frequencies. The optimum design parameters obtained by optimization are found to be

$$\underline{\Phi}^* = \begin{bmatrix} -0.26404 & 0.99305 & 0.86888 & 0.75147 \end{bmatrix} \quad (8.8c)$$

Substituting the optimum design parameters from Eqs. (8.8c) into (8.7), the frequency responses of $S_{21}^{CAA}(\Omega, \underline{\Phi}^*)$ and $S_{11}^{CAA}(\Omega, \underline{\Phi}^*)$ can be computed; the results are plotted in Figure 8.8.

8.3.3 Synthesis of an Asynchronously Tuned Filter by Optimization

The second example of filter synthesis by optimization is a three-pole cross-coupled resonator filter having asymmetrical frequency selectivity. The desired filter response will exhibit a single finite frequency attenuation pole at $p = j3.0$ and a return loss better than -26 dB over the passband. Similarly, from Chapter 7, the S parameters of the desired filter topology may be expressed as

$$S_{21}^{CAA}(\Omega, \underline{\Phi}) = \frac{2}{q_e}[A]_{3,1}^{-1}$$

$$S_{11}^{CAA}(\Omega, \underline{\Phi}) = 1 - \frac{2}{q_e}[A]_{1,1}^{-1} \quad (8.9)$$

$$[A] = j\begin{bmatrix} \Omega - j/q_e & 0 & 0 \\ 0 & \Omega & 0 \\ 0 & 0 & \Omega - j/q_e \end{bmatrix} - j\begin{bmatrix} m_{11} & m_{12} & m_{13} \\ m_{12} & m_{22} & m_{12} \\ m_{13} & m_{12} & m_{11} \end{bmatrix}$$

Although the desired frequency response of the filter is asymmetrical, a symmetrical filter topology is rather desirable and has been imposed in Eq. (8.9) by letting $q_{e1} = q_{e3} = q_e$, $m_{33} = m_{11}$, and $m_{23} = m_{12}$. In this case, the coupling matrix will

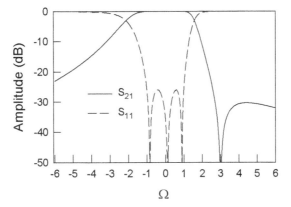

FIGURE 8.9 Frequency response of a three-pole cross-coupled filter synthesized by optimization.

have nonzero entry of m_{13} for the desired cross coupling. It will also have nonzero diagonal elements accounting for asynchronous tuning of the filter. For optimization, the design parameters of this filter are defined by

$$\underline{\Phi} = \begin{bmatrix} m_{13} & m_{11} & m_{22} & m_{12} & q_e \end{bmatrix} \quad (8.10a)$$

The initial values of $\underline{\Phi}$, estimated from a three-pole Chebyshev filter, are given by

$$\underline{\Phi}^{initial} = \begin{bmatrix} 0 & 0 & 0 & 1.28 & 0.6291 \end{bmatrix} \quad (8.10b)$$

Again, the error function for this synthesis example is formed according to Eq. (8.6) at some characteristic frequencies. The results of optimization-based synthesis are

$$\underline{\Phi}^* = \begin{bmatrix} -0.54575 & -0.15222 & 0.46453 & 1.17633 & 0.64000 \end{bmatrix} \quad (8.10c)$$

Figure 8.9 shows the frequency responses of the synthesized filter, which are computed by substituting the optimum design parameters from Eqs. (8.10c) into (8.9).

8.3.4 Synthesis of a UMTS Filter by Optimization

This CAD example is to demonstrate the application of commercial software for filter synthesis. The filter is designed for the Universal Mobile Telecommunication System (UMTS) base-station applications. The specifications for the filter are as follows

Passband frequencies 1950.4–1954.6 MHz
Passband return loss < -20 dB
Rejection ≥ 40 dB for $f \leq 1949.5$ and $f \geq 1955.5$ MHz
Rejection ≥ 50 dB for $f \leq 1947.5$ and $f \geq 1957.5$ MHz
Rejection ≥ 65 dB for $f \leq 1945.6$ and $f \geq 1959.4$ MHz

This is a highly selective bandpass filter. Therefore, a six-pole quasielliptic function filter with a single pair of attenuation poles at finite frequencies as described in Section 9.1 is chosen as an initial design. The attenuation poles at the normalized lowpass frequencies are also chosen as $\Omega = \pm 1.5$ to meet the selectivity. The initial values for the lowpass prototype filter, which can be obtained from Table 9.2 for $\Omega_a = 1.5$ are

$$g_1 = 1.00795, \quad g_2 = 1.43430, \quad g_3 = 2.03664, \quad J_2 = -0.18962, \quad J_3 = 1.39876$$

The corresponding design parameters for the bandpass filter are calculated by using Eq. (9.9) for a fractional bandwidth $FBW = 0.00215$. This results in

$$\begin{aligned} M_{12} &= M_{56} = 0.00179 & M_{34} &= 0.00148 \\ M_{23} &= M_{45} = 0.00126 & M_{25} &= -0.00028 \\ Q_{ei} &= Q_{eo} = 468.57648 \end{aligned} \quad (8.11)$$

The commercial software used for this example is a linear simulator from Microwave Office, as introduced in Section 8.1, although other similar software tools can be used. A schematic circuit for the filter synthesis by optimization is shown in Figure 8.10. The lumped *RLC* elements represent the six synchronously tuned resonators and the quarter-wavelength transmission lines, which have electrical length $EL = \pm 90°$ at the midband frequency f_0, are used to represent the couplings. There are two cross couplings, one between resonators 2 and 5, and the other between resonators 1 and 6. The cross coupling between resonators 1 and 6, which is

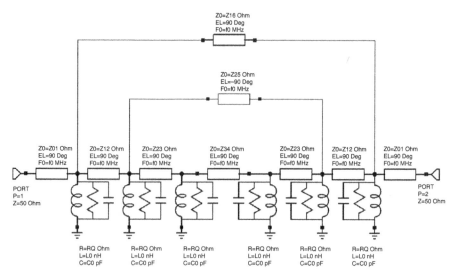

FIGURE 8.10 Microwave Office schematic circuit for computer-aided synthesis of a six-pole UMTS filter.

not available in the initial filter design, is introduced to improve the filter rejection. The circuit parameters in Figure 8.10 can be related to the bandpass filter design parameters by the following equations

Lumped RLC elements of resonators:

$$C0 = \frac{Q_e}{\omega_0 Z} \times 10^{12} \text{ (pF)} \qquad L0 = \frac{Z}{\omega_0 Q_e} \times 10^9 \text{ (nH)}$$

$$RQ = Z \frac{Q_u}{Q_e} \text{ (ohm)}$$

Characteristic impedance of quater-wavelength transmission lines:

$$\begin{aligned}
Z01 &= Z \text{ (ohm)} & Z12 &= \frac{Z}{Q_e M_{12}} \text{ (ohm)} \\
Z23 &= \frac{Z}{Q_e M_{23}} \text{ (ohm)} & Z34 &= \frac{Z}{Q_e M_{34}} \text{ (ohm)} \\
Z25 &= \frac{Z}{Q_e |M_{25}|} \text{ (ohm)} & Z16 &= \frac{Z}{Q_e M_{16}} \text{ (ohm)}
\end{aligned} \qquad (8.12)$$

where $Z = 50$ (Ω) is the terminal impedance at the I/O ports, $\omega_0 = 2\pi f_0$ (rad/s) is the angular frequency at the midband frequency of filter, and Q_u is the unloaded quality factor of resonators. For this example, we assume $Q_u = 10^5$ for all resonators. It should be mentioned that the characteristic impedance for quarter-wavelength transmission lines must be positive, and, hence, for the negative coupling M_{25} the corresponding electrical length is set to $-90°$, as can be seen from Figure 8.10.

The filter response with the initial design parameters of Eq. (8.11) are given in Figure 8.11a against the optimization goals set by slash-line strips according to the filter specifications. This response with the two attenuation poles near the passband edges tends to meet the selectivity, however, it does not meet the 50- and 65-dB rejection requirements at all. The design parameters, including M_{16}, are then optimized to meet the all specifications. The optimum design parameters obtained by optimization are

$$\begin{aligned}
M_{12} &= M_{56} = 0.00178641 & M_{34} &= 0.00149392 \\
M_{23} &= M_{45} = 0.00124961 & M_{25} &= -0.000311472 \\
Q_{ei} &= Q_{eo} = 468.325 & M_{16} &= 1.23902 \times 10^{-5}
\end{aligned}$$

The resultant filter response, shown in Figure 8.11b, meets the all specifications. With the nonzero entry of M_{16}, another pair of attenuation poles are actually placed at the finite frequencies as well. This can clearly be seen from the wide-band frequency response of the synthesized filter, as plotted in Figure 8.12.

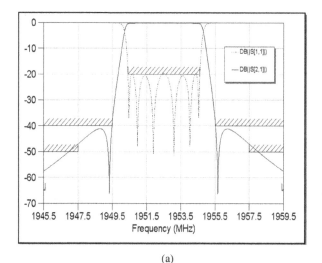

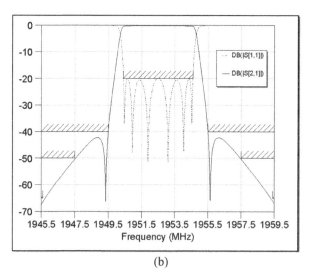

FIGURE 8.11 Frequency response of the UMTS filter against the design goals. (**a**) Before optimization; (**b**) after optimization.

8.4 CAD EXAMPLES

8.4.1 Example One (Chebyshev Filter)

A microstrip bandpass filter is designed to meet the specifications:

Center frequency	4.5 GHz
Passband bandwidth	1 GHz
Passband return loss	<-15 dB
Rejection bandwidth (35 dB)	>2 GHz

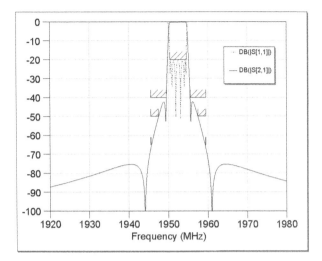

FIGURE 8.12 Wide-band frequency response of the UMTS filter after optimization.

A six-pole symmetrical Chebyshev filter is synthesized using the design equations given in Section 9.1. The design parameters for the bandpass filter are obtained as

$$Q_e = 4.44531$$
$$M_{12} = M_{56} = 0.18867$$
$$M_{23} = M_{45} = 0.13671$$
$$M_{34} = 0.13050$$

The microstrip filter is also required to fit into a circuit size of 39.1 × 21.5 mm on commercial copper-clapped RT/Duroid substrates with a relative dielectric constant of 10.5 ± 0.25 and a thickness of 0.635 ± 0.0254 mm. The narrowest line width and the narrowest spacing between lines are restricted to 0.2 mm. A configuration of edge-coupled, half-wavelength resonator filter is chosen for the implementation and the optimization design is performed using the commercial CAD tool of Microwave Office. Figure 8.13 is the schematic circuit of the filter for optimization. The first and the last resonators are bent by a length of L5 in order to fit the filter into the specified size. The input and output (I/O) are realized by two 50-Ω tapped lines. The effect of microstrip open end is taken into account as indicated. The 50-Ω line width is fixed by 0.6 mm and all resonators have the same line width of 0.26 mm. The other dimensions are optimizable and the initial values are determined as follows. According to the microstrip design equations given in Chapter 4, the half-wavelength is about 12.4 mm; hence, initially L1 = L2 = L3 = 6.2 mm, as well as L4 + L5 = 6.2 mm. The tapped location L4 is estimated to be 2.1 mm using the formulation

$$L4 = \frac{\lambda_g}{2\pi} \sin^{-1}\left(\sqrt{\frac{\pi}{2} \frac{Z_t}{Z_r} \frac{1}{Q_e}}\right) \quad (8.13)$$

250 CAD FOR LOW-COST AND HIGH-VOLUME PRODUCTION

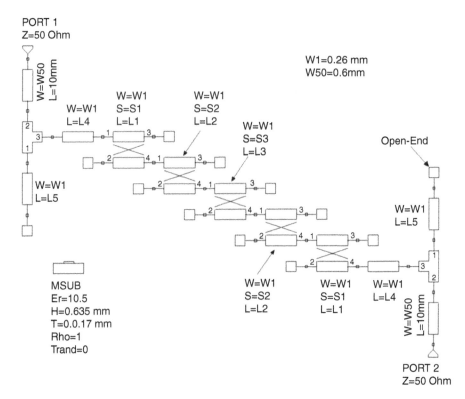

FIGURE 8.13 Schematic circuit for the edge-coupled microstrip bandpass filter for CAD.

where λ_g and Z_r are the microstrip guided wavelength and the microstrip characteristic impedance of the I/O resonators, respectively, Z_t represents the characteristic impedance of the I/O tapped microstrip lines, and Q_e is the external Q given above. For L4 + L5 = 6.2 mm, we have L5 = 4.1 mm. The coupling spacing may be evaluated from

$$\frac{Z_{0e} - Z_{0o}}{Z_{0e} + Z_{0o}} = \frac{\pi}{2} M \qquad (8.14)$$

where Z_{0e} and Z_{0o} are the even- and odd-mode impedances of coupled microstrip lines; M represents the coupling coefficient given above. Applying the design equations for coupled microstrip lines given in Chapter 4, it can be obtained that S1 = 0.254 mm, S2 = 0.413 mm, and S3 = 0.432 mm.

After the optimization design, the optimal physical parameters are listed in Table 8.3 against the initial ones for comparison.

The filter layout resulting from the optimization design is shown in Figure 8.14, where the 50-Ω tapped lines have been bent to the desirable locations of the I/O ports. The analyzed and measured filter responses are plotted in Figure 8.15. It

TABLE 8.3 Physical Parameters Before and After the Optimization

Physical parameter	Initial (mm)	Optimal (mm)
L1	6.2	6.418
L2	6.2	5.936
L3	6.2	6.236
L4	2.1	1.582
L5	4.1	3.364
S1	0.254	0.283
S2	0.413	0.404
S3	0.432	0.460

should be emphasized that because of the approximation of the circuit model and the tolerances of the substrates and fabrication, the analyzed responses normally do not match the measured ones. The main discrepancies may lie in the center frequency, bandwidth, and return loss. To ensure that the fabricated filter will be able to meet the specifications, a simple, yet effective, approach is to design a filter with slightly different specifications to compensate for the discrepancies. For this particular example, the filter responses shown in Figure 8.15a have been optimized with the slightly higher center frequency, wider bandwidth, and smaller return loss. This makes the fabricated filter satisfy the required specifications as indicated in Figure 8.15b.

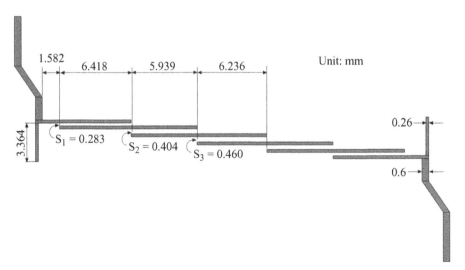

FIGURE 8.14 Layout of the designed edge-coupled microstrip filter on a 0.635-mm thick substrate with a relative dielectric constant of 10.5.

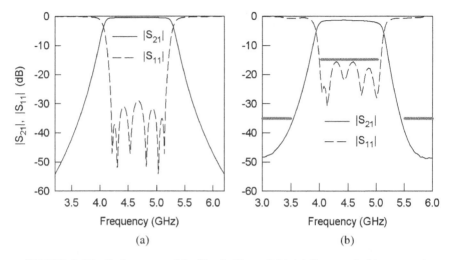

FIGURE 8.15 Performance of the filter in Figure 8.14. (**a**) Computed; (**b**) measured.

8.4.2 Example Two (Cross-Coupled Filter)

When a microstrip filter design involves full-wave EM simulations, it is computationally superior to decompose the filter into different parts that are simulated individually by the EM simulator to extract the desired design parameters according to a prescribed general coupling matrix, as discussed in Chapter 7. They are then combined to obtain the response of the overall filter. This CAD approach, which is particularly effective for narrowband filter designs, is demonstrated with a four-pole cross-coupled filter shown in Figure 8.16a, which consists of four folded quarter-wavelength resonators resulting in a compact filter topology on a dielectric substrate with a thickness denoted by h. Each of the four resonators, as numbered to indicate their sequence in the main coupling path, has a short circuit (via-hole grounding) at one end and an open circuit at the other end. Resonators 1 and 4 are the input and output (I/O) resonators, respectively, and there exists a cross coupling between them.

The four-pole microstrip cross-coupled filter is designed based on the following prescribed general coupling matrix:

$$[m] = \begin{bmatrix} 0 & 0.88317 & 0 & -0.12355 \\ 0.88317 & 0 & 0.74907 & 0 \\ 0 & 0.74907 & 0 & 0.88317 \\ -0.12355 & 0 & 0.88317 & 0 \end{bmatrix} \quad (8.15)$$

with the scaled external quality factor $q_{e1} = q_{e4} = 0.94908$. The filter may be seen as a canonical or single section of CQ filter (refer to Chapter 9). For the filter design, assume a fractional bandwidth $FBW = 0.06$ at a center frequency $f_0 = 1.26$ GHz

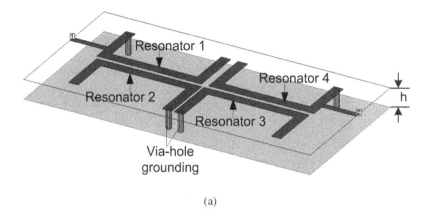

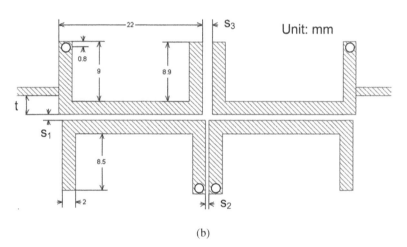

FIGURE 8.16 A four-pole microstrip cross-coupled resonator filter. (**a**) 3D structure. (**b**) Layout on a dielectric substrate with a relative dielectric constant of 3 and a thickness of 0.5 mm.

and a dielectric substrate with a relative dielectric constant of 3 and a thickness $h = 0.5$ mm. Thus, the desired design parameters can be found from Eqs. (7.8) and (7.9):

$$\begin{aligned} M_{12} &= M_{34} = 0.05299 \\ M_{23} &= 0.04494 \\ M_{14} &= -0.00741 \\ Q_{e1} &= Q_{e4} = 15.818 \end{aligned} \qquad (8.16)$$

Full-wave EM simulations are carried out to extract the desired external quality factors and coupling coefficients, using the methods described in Chapter 7. For

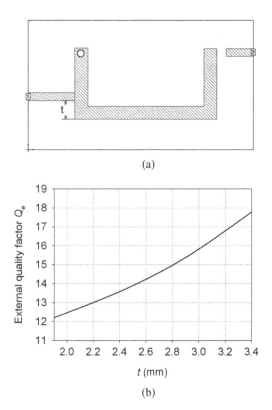

FIGURE 8.17 (a) Arrangement for extracting external quality factor. (b) Design curve for Q_e.

this CAD example, all EM simulations are performed using commercially available software, *Sonnet Suites*, as listed in Table 8.1.

Shown in Figure 8.17a is an arrangement to extract the external quality factor of the I/O resonator. The resonator, which is assumed to be lossless in the simulation, is excited at port 1 through a 50-Ω tapped line at a location indicated by t. Port 2 is very weakly coupled to the resonator in order to find a 3-dB bandwidth of the magnitude response of S_{21} for extracting the single-loaded external quality factor Q_e. A design curve for Q_e against t can be obtained, as shown in Figure 8.17b. In this case, as t increases, the tapped line is moved toward the short circuit (via-hole grounding) of the resonator and, hence, the coupling from the source is weaker so that Q_e increases.

The coupling between resonators 1 and 2 of the filter configuration in Figure 8.16 can be extracted from EM simulation using an arrangement of Figure 8.18a. The two resonators have the same size and the same resonant frequency. The coupling is mainly controlled by the spacing s_1 between them. An offset d is often required because of the necessity of implementing other couplings for the filter design, as we will see later on. For the EM simulation, the coupled resonators are very weakly

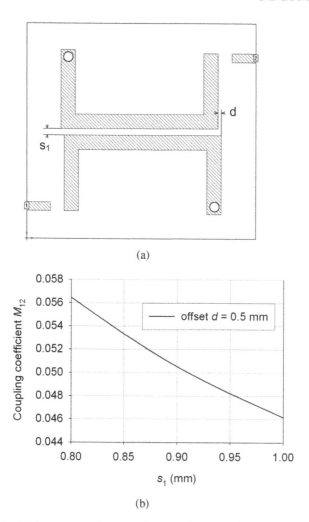

FIGURE 8.18 (a) Arrangement for extracting coupling coefficient M_{12}. (b) Design curve for M_{12}.

excited by the two ports as arranged. Two resonant peaks, which result from the mode split because of the coupling between the two resonators (see Chapter 7), can clearly be observed from the EM-simulated frequency responses. The coupling coefficient can then be extracted using Eq. (7.72). Figure 8.18b shows the design curve for M_{12} for $d = 0.5$ mm. It is obvious that the coupling decreases as the spacing between the coupled resonators increases. It can also be shown that a small offset d has little effect on the coupling.

Figure 8.19a depicts an arrangement for the EM simulation to determine the coupling between resonators 2 and 3 of the filter configuration in Figure 8.16. For the given orientations of the two coupled resonators, the coupling between them at

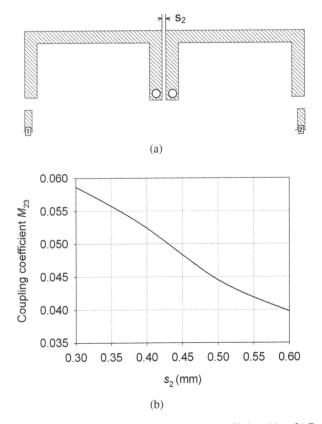

FIGURE 8.19 (a) Arrangement for extracting coupling coefficient M_{23}. (b) Design curve for M_{23}.

fundamental resonance is dominated by the magnetic field and, therefore, can be referred to as magnetic coupling, as discussed in Chapter 7. Using the same approach described above for extracting the inter-resonator coupling, a design curve for M_{23} can be produced and is plotted in Figure 8.19b.

To extract the cross coupling between resonators 1 and 4 of the filter configuration in Figure 8.16, we use an arrangement of Figure 8.20a for EM simulation. The orientations of these two coupled resonators are opposite to that of Figure 8.19a. In this case, the coupling between them is dominated by the electric field and, hence, is referred to as electric coupling. This implementation for the cross coupling is necessary because M_{14} and M_{23} have to be of opposite signs in order to realize a pair of transmission zeros at finite frequencies. Figure 8.20b shows the design curve for M_{14} for the filter design.

From the design curves obtained above, all the physical dimensions associated with the desired design parameters, namely, the external quality factors and coupling coefficients, given in Eq. (8.16), are readily determined, which are $t = 2.9$ mm,

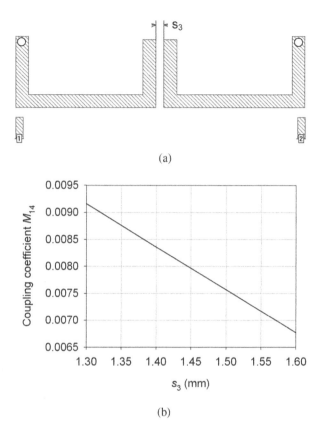

FIGURE 8.20 (a) Arrangement for extracting coupling coefficient M_{14}. (b) Design curve for M_{14}.

$s_1 = 0.9$ mm, $s_2 = 0.5$ mm, and $s_3 = 1.5$ mm, as referring to the layout in Figure 8.16b. Note that the dimensions are rounded off to have a resolution of 0.1 mm and all the vias have a diameter of 1.0 mm. Since the designed filter needs to be synchronously tuned, the folded arm toward the open end of the I/O resonator is slightly longer to compensate for the frequency shift because of the tapped line I/O arrangement. Figure 8.21 shows the EM-simulated performance of the overall filter that is designed by this CAD procedure. The designed filter exhibits a desired frequency response with the two finite transmission zeros as expected.

The above CAD example takes into account only the couplings within the desired signal paths. In reality, however, stray or unwanted cross couplings generate interactions between different parts of the filter outside the desired signal path. Moreover, simulation deviations could exist between an isolated section and the section embedded in the complete filter. Therefore, depending on a particular topology of filter, the inaccuracy inherited in the decomposition approach can be too large to neglect when it is applied to some demanding filter designs. To improve the design accuracy

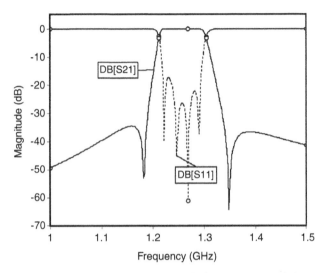

FIGURE 8.21 EM-simulated performance of the designed cross-coupled resonator filter in Figure 8.16 with $t = 2.9$ mm, $s_1 = 0.9$ mm, $s_2 = 0.5$ mm, and $s_3 = 1.5$ mm.

of the decomposition approach, an effective engineering design process is to modify the design against some more accurate model, which may result from more accurate EM simulations of the entire filter or directly from measurements of the filter. Such a design process is normally an iterative process in order to achieve the desirable filter responses in the real world.

REFERENCES

[1] Special Issue on RF and Microwave Filters, Modeling and Design, *Int. J. RF Microwave Computer-Aided Eng.* **17**, 2007.

[2] D. G. Swanson, Jr. and W. J. R. Hoefer, *Microwave Circuit Modeling Using Electromagnetic Field Simulation*, Artech House Publ., Norwood, MA, 2003.

[3] D. G. Swanson, Jr., First pass CAD of microstrip filters cuts development time. Microwave & RF'95, London; 1995, 8–12.

[4] Filters for the masses, *IEEE MTT-S Workshop (WSFL)*. 1999.

[5] *Mathcad User's Guide*. MathSoft Inc.; Cambridge, MA, 1992.

[6] G. I. Zysman and A. K. Johnson, Coupled transmission line networks in an inhomogeneous dielectric medium. *IEEE Trans.* **MTT-17**, 1969, 753–759.

[7] J.-S. Hong, H. P. Feldle, and W. Wiesbeck, Computer-aided design of microwave bandpass filters. In: *Proceedings of the 8th Colloquium on Microwave Communication*, Budapest; 1986, 75–76.

[8] M. Kirschning and R. H. Jansen, Accurate wide-range design equations for the frequency-dependent characteristic of parallel coupled microstrip lines. *IEEE Trans.* **MTT-32**, 1984, 83–90.

[9] R. F. Harringdon, *Field Commutation by Moment Methods*. Macmillian, New York, 1968.

[10] J. C. Rautio and R. F. Harrington, An electromagnetic time-harmonic analysis of arbitrary microstrip circuits. *IEEE Trans.* **MTT-35**, 1987, 726–730.

[11] M. Koshiba, K. Hayata, and M. Suzuki, Finite-element formulation in terms of the electric-field vector for electromagnetic waveguide problems, *IEEE Trans.* **MTT-33**, 1985, 900–905.

[12] K. S. Kunz and R. J. Leubleers, *The Finite Difference Time Domain Method for Electromagnetics*, CRC Press, Boca Raton, FL, 1993.

[13] J. S. Bagby, D. P. Nyquist, and B. C. Drachman, Integral formulation for analysis of integrated dielectric waveguides, *IEEE Trans.* **MTT-33**, 1985, 906–915.

[14] M. Koshiba and M. Suzuki, Application of the boundary-element method to waveguide discontinuities, *IEEE Trans.* **MTT-34**, 1986, 301–307.

[15] Numerical methods. *IEEE Trans.* **MTT-33**, 1985 [SI].

[16] T. Itoh, *Numerical Techniques for Microwave and Millimeter Wave Passive Structures*, John Wiley & Sons, New York, 1989.

[17] D. G. Swanson, Jr., Simulating EM fields, *IEEE Spectrum*, **28**, 1991, 34–37.

[18] D. G. Swanson, Jr. (Guest Editor), *Engineering applications of electromagnetic field solvers*, Int. J. Microwave Millimeter-Wave Computer-Aided Eng. **5**(5), 1995 [SI].

[19] J.-S. Hong and M. J. Lancaster, Investigation of microstrip pseudo-interdigital bandpass filters using a full-wave electromagnetic simulator. *Int. J. Microwave Millimeter-Wave Computer-Aided Eng.* **7**(3), 1997, 231–240.

[20] J. C. Rautio and G. Mattaei, Tracking error sources in HTS filter simulations, *Microwave RF*, **37**, 1998, 119–130.

[21] *EM User's Manual.* Sonnet Software Inc. Liverpool, New York, 1993.

[22] R. Levy, Filter and component synthesis using circuit element models derived from EM simulation. *IEEE MTT-S Workshop (WMA)*, 1997.

[23] G. L. Mattaei and R. J. Forse, A note concerning the use of field solvers for the design of microstrip shunt capacitances in lowpass structures. *Int. J. Microwave Millimeter-Wave Computer-Aided Eng.* **5**(5), 1995, 352–358.

[24] G. L. Mattaei, Techniques for obtaining equivalent circuits for discontinuities in planar microwave circuits. *IEEE MTT-S Workshop (WSC)*, 2000.

[25] J.-S. Hong and M. J. Lancaster, Couplings of microstrip square open-loop resonators for cross-coupled planar microwave filters, *IEEE Trans.* **MTT-44**, 1996, 2099–2109.

[26] J.-S. Hong and M. J. Lancaster, Theory and experiment of novel microstrip slow-wave open-loop resonators filters, *IEEE Trans.* **MTT-45**, 1997, 2358–2365.

[27] J.-S. Hong, M. J. Lancaster, D. Jedamzik, R. B. Greed, and J.-C. Mage, On the performance of HTS microstrip quasi-elliptic function filters for mobile communications application, *IEEE Trans.* **MTT-48**, 2000, 1240–1246.

[28] N. Thomson, J.-S. Hong, R. B. Greed, and D. C. Voyce, Practical approach for designing miniature interdigital filters. In *35th European Microwave Conference Proceedings,* Paris, 2005, 1251–1254.

[29] N. Thomson, J.-S. Hong, R. B. Greed, and D. C. Voyce, Design of miniature wideband interdigital filters. *Int. J. RF Microwave Computer-Aided Eng.* **17**(1), 2007, 90–95.

[30] S. Al-otaibi and J.-S. Hong, Novel substrate integrated folded waveguide filter. *Microwave Optical Technol. Lett.* **50**(4), 2008, 1111–1114.

[31] S. Al-otaibi and J.-S. Hong, Electromagnetic design of folded-waveguide resonator filter with single finite-frequency transmission zero. *Int. J. RF Microwave Computer-Aided Eng.* **18**(1), 2008, 1–7.

[32] B. R. Smith and G. C. Temes, An iterative approximation procedure for automatic filter synthesis. *IEEE Trans. Circuit Theory* **CT-12**, 1965, 107–112.

[33] R. J. Cameron, General prototype network synthesis methods for microwave filters. *ESA J.* **6**, 1982, 193–206.

[34] D. Chambers and J. D. Rhodes, A low pass prototype network allowing the placing of integrated poles at real frequencies. *IEEE Trans.* **MTT-31**, 1983, 40–45.

[35] H. C. Bell, Jr., Canonical asymmetric coupled-resonator filters. *IEEE Trans.* **MTT-30**, 1982, 1335–1340.

[36] A. E. Atia and A. E. Williams, Narrow-bandpass waveguide filters. *IEEE Trans.* **MTT-20**, 1972, 258–265.

[37] H. L. Thal, Jr., Design od microwave filters with arbitrary responses. *Int. J. Microwave Millimeter-Wave Computer-Aided Eng.* **7**(3), 1997, 208–221.

[38] W. A. Atia, K. A. Zaki, and A. E. Atia, Synthesis of general topology multiple coupled resonator filters by optimization, *IEEE MTT-S Dig.* 1998, 821–824.

[39] J.-S. Hong, Computer-aided synthesis of mixed cascaded quadruplet and trisection (CQT) filters, In *Proceedings of 31st European Microwave conference*, London, **3**, 2001, 5–8.

[40] J.-S. Hong, Decomposition approach for computer-aided synthesis of high-order cross-coupled resonator filters. *Int. J. RF Microwave Computer-Aided Eng.* **17**(1), 2007, 13–19.

CHAPTER NINE

Advanced RF/Microwave Filters

There have been increasing demands for advanced RF/microwave filters other than conventional Chebyshev filters in order to meet stringent requirements from RF/microwave systems, particularly from wireless communications systems. In this chapter, we will discuss the designs of some advanced filters. These include selective filters with a single pair of transmission zeros, cascaded quadruplet (CQ) filters, trisection and cascaded trisection (CT) filters, cross-coupled filters using transmission-line inserted inverters, linear-phase filters for group-delay equalization, extracted-pole filters, canonical filters, and multiband filters.

9.1 SELECTIVE FILTERS WITH A SINGLE PAIR OF TRANSMISSION ZEROS

9.1.1 Filter Characteristics

The filter having only one pair of transmission zeros (or attenuation poles) at finite frequencies gives much improved skirt selectivity, making it a viable intermediate case between the Chebyshev and elliptic-function filters, yet with little practical difficulty of physical realization [1–4]. The transfer function of this type of filter is

$$|S_{21}(\Omega)|^2 = \frac{1}{1 + \varepsilon^2 F_n^2(\Omega)}$$

$$\varepsilon = \frac{1}{\sqrt{10^{-\frac{L_R}{10}} - 1}}$$

$$F_n(\Omega) = \cosh\left\{(n-2)\cosh^{-1}(\Omega) + \cosh^{-1}\left(\frac{\Omega_a \Omega - 1}{\Omega_a - \Omega}\right) + \cosh^{-1}\left(\frac{\Omega_a \Omega + 1}{\Omega_a + \Omega}\right)\right\}$$

(9.1)

Microstrip Filters for RF/Microwave Applications by Jia-Sheng Hong
Copyright © 2011 John Wiley & Sons, Inc.

where Ω is the frequency variable that is normalized to the passband cut-off frequency of lowpass prototype filter, ε is a ripple constant related to a given return loss $L_R = 20 \log |S_{11}|$ in dB, and n is the degree of the filter. It is obvious that $\Omega = \pm\Omega_a$ ($\Omega_a > 1$) are the frequency locations of a pair of attenuation poles. Note that if $\Omega_a \to \infty$, the filtering function $F_n(\Omega)$ degenerates to the familiar Chebyshev function. The transmission frequency response of the bandpass filter may be determined using the frequency mapping, as discussed in Chapter 3, that is,

$$\Omega = \frac{1}{FBW} \cdot \left(\frac{\omega}{\omega_0} - \frac{\omega_0}{\omega} \right)$$

in which ω is the frequency variable of bandpass filter, ω_0 is the midband frequency, and FBW is the fractional bandwidth. The locations of two finite frequency attenuation poles of the bandpass filter are given by

$$\omega_{a1} = \omega_0 \frac{-\Omega_a FBW + \sqrt{(\Omega_a FBW)^2 + 4}}{2}$$
$$\omega_{a2} = \omega_0 \frac{\Omega_a FBW + \sqrt{(\Omega_a FBW)^2 + 4}}{2}$$
(9.2)

Figure 9.1 shows some typical frequency responses of this type of filters for $n = 6$ and $L_R = -20$ dB, as compared to that of the Chebyshev filter. As can be seen, the improvement in selectivity over the Chebyshev filter is evident. The closer the attenuation poles to the cut-off frequency ($\Omega = 1$), the sharper the filter skirt, and the higher the selectivity.

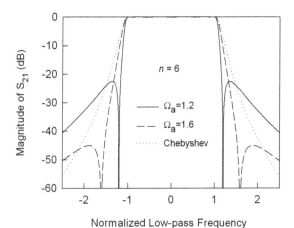

FIGURE 9.1 Comparison of frequency responses of the Chebyshev filter and the design filter with a single pair of attenuation poles at finite frequencies ($n = 6$).

SELECTIVE FILTERS WITH A SINGLE PAIR OF TRANSMISSION ZEROS

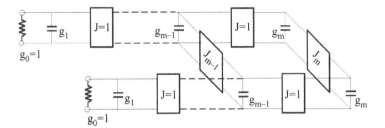

FIGURE 9.2 Lowpass prototype filter for the filter synthesis.

9.1.2 Filter Synthesis

The transmission zeros of this type of filter may be realized by cross coupling a pair of nonadjacent resonators of the standard Chebyshev filter. Levy [2] has developed an approximate synthesis method based on a lowpass prototype filter shown in Figure 9.2, where the rectangular boxes represent ideal admittance inverters with characteristic admittance J. The approximate synthesis starts with the element values for Chebyshev filters

$$g_1 = \frac{2 \sin \frac{\pi}{2n}}{\gamma}$$

$$g_i g_{i-1} = \frac{4 \sin \frac{(2i-1)\pi}{2n} \sin \frac{(2i-3)\pi}{2n}}{\gamma^2 + \sin^2 \frac{(i-1)\pi}{n}}$$

$$(i = 1, 2, \ldots, m), \quad m = n/2 \tag{9.3}$$

$$\gamma = \sinh\left(\frac{1}{n}\sinh^{-1}\frac{1}{\varepsilon}\right)$$

$$S = \left(\sqrt{1+\varepsilon^2} + \varepsilon\right)^2 \quad \text{(the passband VSWR)}$$

$$J_m = 1/\sqrt{S}$$

$$J_{m-1} = 0$$

In order to introduce transmission zeros at $\Omega = \pm\Omega_a$, the required value of J_{m-1} is given by

$$J_{m-1} = \frac{-J'_m}{(\Omega_a g_m)^2 - J'^2_m} \tag{9.4}$$

Introduction of J_{m-1} mismatches the filter and to maintain the required return loss at midband it is necessary to slightly change the value of J_m according to the formula

$$J'_m = \frac{J_m}{1 + J_m J_{m-1}} \tag{9.5}$$

TABLE 9.1 Element Values of Four-Pole Prototype ($L_R = -20$ dB)

Ω_a	g_1	g_2	J_1	J_2
1.80	0.95974	1.42192	−0.21083	1.11769
1.85	0.95826	1.40972	−0.19685	1.10048
1.90	0.95691	1.39927	−0.18429	1.08548
1.95	0.95565	1.39025	−0.17297	1.07232
2.00	0.95449	1.38235	−0.16271	1.06062
2.05	0.95341	1.37543	−0.15337	1.05022
2.10	0.95242	1.36934	−0.14487	1.04094
2.15	0.95148	1.36391	−0.13707	1.03256
2.20	0.95063	1.35908	−0.12992	1.02499
2.25	0.94982	1.35473	−0.12333	1.0181
2.30	0.94908	1.35084	−0.11726	1.01187
2.35	0.94837	1.3473	−0.11163	1.00613
2.40	0.94772	1.34408	−0.10642	1.00086

where J'_m is interpreted as the updated J_m. Equations (9.5) and (9.4) are solved iteratively with the initial values of J_m and J_{m-1} given in Eq. (9.3). No other elements of the original Chebyshev filter are changed.

The above method is simple, yet quite useful, in many cases, for design of selective filters. However, it suffers from inaccuracy, and can even fail for very highly selective filters that require moving the attenuation poles closer to the cut-off frequencies of the passband. This necessitates the use of a more accurate synthesis procedure. Alternatively, one may use a set of more accurate design data tabulated in Tables 9.1–9.3, where the values of the attenuation pole frequency Ω_a cover a wide range of practical designs for highly selective microstrip bandpass filters [4]. For less selective filters, which require a larger Ω_a, the element values can be obtained using the above approximate synthesis procedure.

TABLE 9.2 Element Values of Six-Pole Prototype ($L_R = -20$ dB)

Ω_a	g_1	g_2	g_3	J_2	J_3
1.20	1.01925	1.45186	2.47027	−0.39224	1.95202
1.25	1.01642	1.44777	2.30923	−0.33665	1.76097
1.30	1.01407	1.44419	2.21	−0.29379	1.63737
1.35	1.01213	1.44117	2.14383	−0.25976	1.55094
1.40	1.01051	1.43853	2.09713	−0.23203	1.487
1.45	1.00913	1.43627	2.0627	−0.20901	1.43775
1.50	1.00795	1.4343	2.03664	−0.18962	1.39876
1.55	1.00695	1.43262	2.01631	−0.17308	1.36714
1.60	1.00606	1.43112	2.00021	−0.15883	1.34103

SELECTIVE FILTERS WITH A SINGLE PAIR OF TRANSMISSION ZEROS   265

TABLE 9.3   Element Values of Eight-Pole Prototype ($L_R = -20$ dB)

Ω_a	g_1	g_2	g_3	g_4	J_3	J_4
1.20	1.02947	1.46854	1.99638	1.96641	−0.40786	1.4333
1.25	1.02797	1.46619	1.99276	1.88177	−0.35062	1.32469
1.30	1.02682	1.46441	1.98979	1.82834	−0.30655	1.25165
1.35	1.02589	1.46295	1.98742	1.79208	−0.27151	1.19902
1.40	1.02514	1.46179	1.98551	1.76631	−0.24301	1.15939
1.45	1.02452	1.46079	1.98385	1.74721	−0.21927	1.12829
1.50	1.024	1.45995	1.98246	1.73285	−0.19928	1.10347
1.55	1.02355	1.45925	1.98122	1.72149	−0.18209	1.08293
1.60	1.02317	1.45862	1.98021	1.71262	−0.16734	1.06597

For computer synthesis, the following explicit formulas are obtained by curve fitting for $L_R = -20$ dB:

$$g_1(\Omega_a) = 1.22147 - 0.35543 \cdot \Omega_a + 0.18337 \cdot \Omega_a^2 - 0.0447 \cdot \Omega_a^3 + 0.00425 \cdot \Omega_a^4$$

$$g_2(\Omega_a) = 7.22106 - 9.48678 \cdot \Omega_a + 5.89032 \cdot \Omega_a^2 - 1.65776 \cdot \Omega_a^3 + 0.17723 \cdot \Omega_a^4$$

$$J_1(\Omega_a) = -4.30192 + 6.26745 \cdot \Omega_a - 3.67345 \cdot \Omega_a^2 + 0.9936 \cdot \Omega_a^3 - 0.10317 \cdot \Omega_a^4$$

$$J_2(\Omega_a) = 8.17573 - 11.36315 \cdot \Omega_a + 6.96223 \cdot \Omega_a^2 - 1.94244 \cdot \Omega_a^3 + 0.20636 \cdot \Omega_a^4$$

$$(n = 4 \text{ and } 1.8 \leq \Omega_a \leq 2.4)$$

(9.6)

$$g_1(\Omega_a) = 1.70396 - 1.59517 \cdot \Omega_a + 1.40956 \cdot \Omega_a^2 - 0.56773 \cdot \Omega_a^3 + 0.08718 \cdot \Omega_a^4$$

$$g_2(\Omega_a) = 1.97927 - 1.04115 \cdot \Omega_a + 0.75297 \cdot \Omega_a^2 - 0.245447 \cdot \Omega_a^3 + 0.02984 \cdot \Omega_a^4$$

$$g_3(\Omega_a) = 151.54097 - 398.03108 \cdot \Omega_a + 399.30192 \cdot \Omega_a^2 - 178.6625 \cdot \Omega_a^3 + 30.04429 \cdot \Omega_a^4$$

$$J_2(\Omega_a) = -24.36846 + 60.76753 \cdot \Omega_a - 58.32061 \cdot \Omega_a^2 + 25.23321 \cdot \Omega_a^3 - 4.131 \cdot \Omega_a^4$$

$$J_3(\Omega_a) = 160.91445 - 422.57327 \cdot \Omega_a + 422.48031 \cdot \Omega_a^2 - 188.6014 \cdot \Omega_a^3 + 31.66294 \cdot \Omega_a^4$$

$$(n = 6 \text{ and } 1.2 \leq \Omega_a \leq 1.6)$$

(9.7)

$$g_1(\Omega_a) = 1.64578 - 1.55281 \cdot \Omega_a + 1.48177 \cdot \Omega_a^2 - 0.63788 \cdot \Omega_a^3 + 0.10396 \cdot \Omega_a^4$$

$$g_2(\Omega_a) = 2.50544 - 2.64258 \cdot \Omega_a + 2.55107 \cdot \Omega_a^2 - 1.11014 \cdot \Omega_a^3 + 0.18275 \cdot \Omega_a^4$$

$$g_3(\Omega_a) = 3.30522 - 3.25128 \cdot \Omega_a + 3.06494 \cdot \Omega_a^2 - 1.30769 \cdot \Omega_a^3 + 0.21166 \cdot \Omega_a^4$$

$$g_4(\Omega_a) = 75.20324 - 194.70214 \cdot \Omega_a + 194.55809 \cdot \Omega_a^2 - 86.76247 \cdot \Omega_a^3 + 14.54825 \cdot \Omega_a^4$$

$$J_3(\Omega_a) = -25.42195 + 63.50163 \cdot \Omega_a - 61.03883 \cdot \Omega_a^2 + 26.44369 \cdot \Omega_a^3 - 4.3338 \cdot \Omega_a^4$$

$$J_4(\Omega_a) = 82.26109 - 213.43564 \cdot \Omega_a + 212.16473 \cdot \Omega_a^2 - 94.28338 \cdot \Omega_a^3 + 15.76923 \cdot \Omega_a^4$$

$$(n = 8 \text{ and } 1.2 \leq \Omega_a \leq 1.6)$$

(9.8)

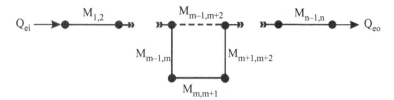

FIGURE 9.3 General coupling structure of the bandpass filter with a single pair of finite frequency zeros.

The design parameters of the bandpass filter, i.e., the coupling coefficients and external quality factors, as referred to in the general coupling structure of Figure 9.3, can be determined by the formulas

$$Q_{ei} = Q_{eo} = \frac{g_1}{FBW}$$

$$M_{i,i+1} = M_{n-i,n-i+1} = \frac{FBW}{\sqrt{g_i g_{i+1}}} \quad \text{for } i = 1 \text{ to } m-1$$

$$M_{m,m+1} = \frac{FBW \cdot J_m}{g_m}$$

$$M_{m-1,m+2} = \frac{FBW \cdot J_{m-1}}{g_{m-1}}$$

(9.9)

9.1.3 Filter Analysis

Having obtained the design parameters of bandpass filter, we may use the general formulation for cross-coupled resonator filters given in Chapter 7 to analyze the filter frequency response. Alternatively, the frequency response can be calculated by

$$S_{21}(\Omega) = \frac{Y_o(\Omega) - Y_e(\Omega)}{(1 + Y_e(\Omega)) \cdot (1 + Y_o(\Omega))}$$

$$S_{11}(\Omega) = \frac{1 - Y_e(\Omega) \cdot Y_o(\Omega)}{(1 + Y_e(\Omega)) \cdot (1 + Y_o(\Omega))}$$

(9.10)

where Y_e and Y_o are the even- and odd-mode input admittance of the filter in Figure 9.2. It can be shown that when the filter is open/short-circuited along its symmetrical plane; the admittance at the two cross-admittance inverters are $\mp J_{m-1}$ and $\mp J_m$. Therefore, the Y_e and Y_o can easily be expressed in terms of the elements in a ladder structure such as

$$Y_e(\Omega) = j(\Omega g_1 - J_1) + \frac{1}{j(\Omega g_2 - J_2)}$$

$$Y_o(\Omega) = j(\Omega g_1 + J_1) + \frac{1}{j(\Omega g_2 + J_2)}$$

for $n = 4$ (9.11a)

$$Y_e(\Omega) = j\Omega g_1 + \cfrac{1}{j(\Omega g_2 - J_2) + \cfrac{1}{j(\Omega g_3 - J_3)}}$$
$$Y_o(\Omega) = j\Omega g_1 + \cfrac{1}{j(\Omega g_2 + J_2) + \cfrac{1}{j(\Omega g_3 + J_3)}} \quad \text{for } n = 6 \qquad (9.11b)$$

$$Y_e(\Omega) = j\Omega g_1 + \cfrac{1}{j\Omega g_2 + \cdots + \cfrac{1}{j(\Omega g_{m-1} - J_{m-1}) + \cfrac{1}{j(\Omega g_m - J_m)}}}$$
$$Y_o(\Omega) = j\Omega g_1 + \cfrac{1}{j\Omega g_2 + \cdots + \cfrac{1}{j(\Omega g_{m-1} + J_{m-1}) + \cfrac{1}{j(\Omega g_m + J_m)}}} \quad \text{for } \begin{array}{l} n = 8, 10, \ldots \\ (m = n/2) \end{array}$$
$$(9.11c)$$

The frequency locations of a pair of attenuation poles can be determined by imposing the condition of $|S_{21}(\Omega)| = 0$ upon Eq. (9.10). This requires $|Y_o(\Omega) - Y_e(\Omega)| = 0$ or $Y_o(\Omega) = Y_e(\Omega)$ for $\Omega = \pm\Omega_a$. From Eq. (9.11) we have

$$j(\Omega_a g_{m-1} + J_{m-1}) + \frac{1}{j(\Omega_a g_m + J_m)} = j(\Omega_a g_{m-1} - J_{m-1}) + \frac{1}{j(\Omega_a g_m - J_m)} \qquad (9.12)$$

This leads to

$$\Omega_a = \frac{1}{g_m}\sqrt{J_m^2 - \frac{J_m}{J_{m-1}}} \qquad (9.13)$$

As an example, from Table 9.2, where $m = 3$, we have $g_3 = 2.47027$, $J_2 = -0.39224$, and $J_3 = 1.95202$ for $\Omega_a = 1.20$. Substituting these element values into Eq. (9.13) yields $\Omega_a = 1.19998$, which is an excellent match. It is more interesting to note from Eq. (9.13) that even if J_m and J_{m-1} exchange signs, the locations of attenuation poles are not changed. Therefore, and more importantly, the signs for the coupling coefficients $M_{m,m+1}$ and $M_{m-1,m+2}$ in Eq. (9.9) are rather relative; it does not matter which one is positive or negative as long as their signs are opposite. This makes the filter implementation easier.

9.1.4 Microstrip Filter Realization

Figure 9.4 shows some filter configurations comprised of microstrip open-loop resonators to implement these types of filtering characteristics in a microstrip. Here the numbers indicate the sequence of direct coupling. Although only the filters up to eight poles have been illustrated, building up of higher-order filters is feasible. There

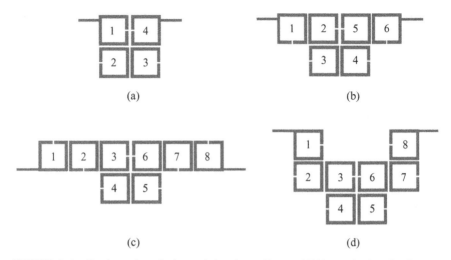

FIGURE 9.4 Configuration of microstrip bandpass filters exhibiting a single pair of attenuation poles at finite frequencies.

are other different filter configurations and resonator shapes that may be used for this implementation.

As an example, an eight-pole microstrip filter is designed to meet the following specifications

Center frequency:	985 MHz
Fractional bandwidth FBW:	10.359%
40-dB Rejection bandwidth:	125.5 MHz
Passband return loss:	-20 dB

The pair of attenuation poles are placed at $\Omega = \pm 1.2645$ in order to meet the rejection specification. Note that the number of poles and Ω_a could be obtained by directly optimizing the transfer function of Eq. (9.1). The element values of the lowpass prototype can be obtained by substituting $\Omega_a = 1.2645$ into Eq. (9.8) and found to be $g_1 = 1.02761, g_2 = 1.46561, g_3 = 1.99184, g_4 = 1.86441, J_3 = -0.33681$, and $J_4 = 1.3013$. Theoretical response of the filter may then be calculated using Eq. (9.10). From Eq. (9.9), the design parameters of this bandpass filter are found

$$M_{1,2} = M_{7,8} = 0.08441 \qquad M_{2,3} = M_{6,7} = 0.06063$$
$$M_{3,4} = M_{5,6} = 0.05375 \qquad M_{4,5} = 0.0723$$
$$M_{3,6} = -0.01752 \qquad Q_{ei} = Q_{eo} = 9.92027$$

The filter is realized using the configuration of Figure 9.4c on a substrate with a relative dielectric constant of 10.8 and a thickness of 1.27 mm. To determine the physical dimensions of the filter, the full-wave EM simulations are carried out to extract the coupling coefficients and external quality factors using the approach described in Chapter 7. The simulated results are plotted in Figure 9.5, where the size

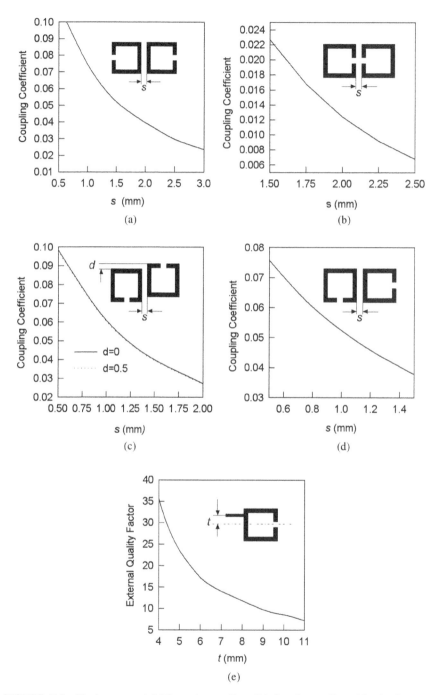

FIGURE 9.5 Design curve. (**a**) Magnetic coupling; (**b**) electric coupling; (**c**) mixed coupling I; (**d**) mixed coupling II; (**e**) external quality factor. (All resonators have a line width of 1.5 mm and a size of 16 × 16 mm on the 1.27-mm thick substrate with a relative dielectric constant of 10.8.)

of each square microstrip open-loop resonator is 16 × 16 mm with a line width of 1.5 mm on the substrate. The coupling spacing s for the required $M_{4,5}$ and $M_{3,6}$ can be determined from Figure 9.5a for the magnetic coupling and 9.5b for the electric coupling, respectively. We have shown in Chapter 7 that both couplings result in the opposite signs of coupling coefficients, which is what we need for implementation of this type of filter. The other filter dimensions, such as the coupling spacing for $M_{1,2}$ and $M_{3,4}$ can be found from Figure 9.5c, while the coupling spacing for $M_{2,3}$ is obtained from 9.5d. The tapped-line position for the required Q_e is determined from Figure 9.5e. It should be mentioned that the design curves in Figure 9.5 may be used for the other filter designs as well. Figure 9.6a is a photograph of the fabricated filter using copper microstrip. The size of the filter is $0.87 \times 0.29\ \lambda_{g0}$. The measured performance is shown in Figure 9.6b. The midband insertion loss is about 2.1 dB, which is attributed to the conductor loss of copper. The two attenuation poles near the cut-off frequencies of the passband are observable, which improves the selectivity. High rejection at the stopband is also achieved.

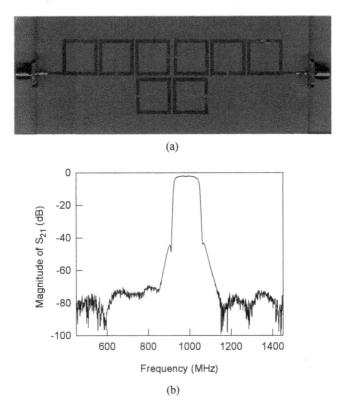

FIGURE 9.6 (a) Photograph of the fabricated eight-pole microstrip bandpass filter designed to have a single pair of attenuation poles at finite frequencies. The size of the filter is about 120 × 50 mm on a 1.27-mm thick substrate with a relative dielectric constant of 10.8. (b) Measured performance of the filter.

9.2 CASCADED QUADRUPLET (CQ) FILTERS

When high selectivity and/or other requirements cannot be met by the filters with a single pair of transmission zeros, as described in the above section, a solution is to introduce more transmission zeros at finite frequencies. In this case, the cascaded quadruplet or CQ filter may be desirable. A CQ bandpass filter consists of cascaded sections of four resonators, each with one cross coupling. The cross coupling can be arranged in such a way that a pair of attenuation poles are introduced at the finite frequencies to improve the selectivity, or it can be arranged to result in group delay self-equalization. Figure 9.7 illustrates typical coupling structures of CQ filters, where each node represents a resonator, the full lines indicate the main path couplings, and the broken line denotes the cross couplings. M_{ij} is the coupling coefficient between the resonators i and j and Q_{e1} and Q_{en} are the external quality factors in association with the input and output couplings, respectively. For higher-degree filters, more resonators can be added in quadruplets at the end. As compared with other types of filters that involve more than one pair of transmission zeros, the significant advantage of CQ filters lies in their simpler tunability, because the effect of each cross coupling is independent [5,6].

9.2.1 Microstrip CQ Filters

As examples of realizing the coupling structure of Figure 9.7a in a microstrip, two microstrip CQ filters are shown in Figure 9.8, where the numbers indicate the sequences of the direct couplings. The filters are comprised of microstrip open-loop

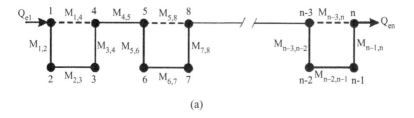

(a)

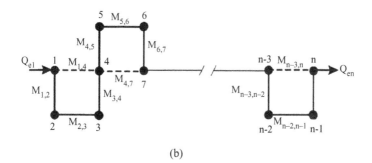

(b)

FIGURE 9.7 Typical coupling structures of CQ filters.

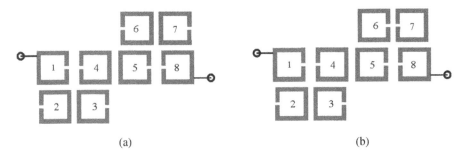

FIGURE 9.8 Configurations of two eight-pole microstrip CQ filters.

resonators; each has a perimeter about a half-wavelength. Note that the shape of resonators need not be square. It may be rectangular, circular, or a meander open loop, so it can be adapted for different substrate sizes. The inter-resonator couplings are realized through the fringe fields of the microstrip open-loop resonators. The CQ filter of Figure 9.8a will have two pairs of attenuation poles at finite frequencies, because both the couplings for M_{23} and M_{14} and the couplings for M_{67} and M_{58} have opposite signs, resulting in a highly selective frequency response. The CQ filter of Figure 9.8b will have only one pair of attenuation poles at finite frequencies, because of the opposite sign of M_{23} and M_{14}, but will exhibit group-delay self-equalization as well because M_{67} and M_{58} have the same sign. This type of filtering characteristic is attractive for high-speed digital transmission systems such as Software Defined Radio (SDR) for minimization of linear distortion while prescribed channel selectivity is being maintained. Although only the eight-pole microstrip CQ filters are illustrated, the building up of filters with more poles and other configurations is feasible.

9.2.2 Design Example

For the demonstration, a highly selective eight-pole microstrip CQ filter with the configuration of Figure 9.8a has been designed, fabricated, and tested. The target specification of the filter was:

Passband frequencies	820–880 MHz
Passband return loss	−20 dB
50-dB Rejection bandwidth	77.5 MHz
65-dB Rejection bandwidth	100 MHz

Therefore, the fractional bandwidth is $FBW = 0.07063$. For a 60-MHz passband bandwidth, the required 50 and 65 dB rejection bandwidths set the selectivity of the filter. To meet this selectivity, the filter was design to have two pairs of attenuation poles near the passband edges, which correspond to $p = \pm j1.3202$ and $p = \pm 1.7942$ on the imaginary axis of the normalized complex lowpass frequency plane. The general coupling matrix and the scaled quality factors of the filter are synthesized by optimization, as described in Chapter 8, and found to be

$$[m] = \begin{bmatrix} 0 & 0.80799 & 0 & -0.10066 & 0 & 0 & 0 & 0 \\ 0.80799 & 0 & 0.6514 & 0 & 0 & 0 & 0 & 0 \\ 0 & 0.6514 & 0 & 0.52837 & 0 & 0 & 0 & 0 \\ -0.10066 & 0 & 0.52837 & 0 & 0.53064 & 0 & 0 & 0 \\ 0 & 0 & 0 & 0.53064 & 0 & 0.49184 & 0 & -0.2346 \\ 0 & 0 & 0 & 0 & 0.49184 & 0 & 0.73748 & 0 \\ 0 & 0 & 0 & 0 & 0 & 0.73748 & 0 & 0.77967 \\ 0 & 0 & 0 & 0 & -0.2346 & 0 & 0.77967 & 0 \end{bmatrix}$$

$q_{e1} = q_{1n} = 1.02828$

(9.14)

According to the discussions in the Chapter 7, the design parameters for the filter with $FBW = 0.07063$ are

$$\begin{aligned} M_{12} &= 0.05707 & M_{56} &= 0.03474 \\ M_{23} &= 0.04601 & M_{67} &= 0.05209 \\ M_{34} &= 0.03732 & M_{78} &= 0.05507 \\ M_{14} &= -0.00711 & M_{58} &= -0.01657 \\ M_{45} &= 0.03748 & Q_{ei} = Q_{eo} &= 14.5582 \end{aligned}$$

(9.15)

The filter response can be calculated using the general formulation for the cross-coupled resonator filters, given in the Chapter 7, and is depicted in Figure 9.9 together with an eight-pole Chebyshev filter response (dotted line) for comparison. The designed filter meets all the specification parameters. The Chebyshev filter has the same return loss level, but its 50-dB and 65-dB rejection bandwidths are 100 and 120 MHz, respectively, which obviously does not meet the rejection requirements.

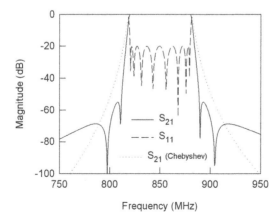

FIGURE 9.9 Comparison of the frequency responses of an eight-pole CQ filter with two pairs of attenuation poles at finite frequencies and an eight-pole Chebyshev filter.

Having determined the design parameters, the next step is to find the physical dimensions for the microstrip CQ filter. For reducing conductor loss and increasing power handling capability, a wider microstrip would be preferable. Hence, the microstrip line width of open-loop resonators used for the filter implementation is 3.0 mm. Full-wave electromagnetic (EM) simulations are performed to extract the coupling coefficients and external quality factors using the formulas described in Chapter 7. This enables us to determine the physical dimensions of the filter. Figure 9.10a shows the filter layout with the dimensions, where all the microstrip open-loop resonators have a size of 20 × 20 mm.

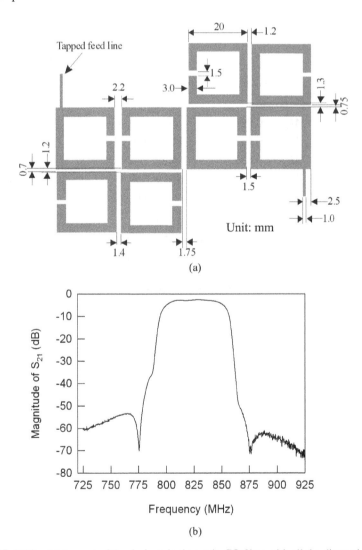

FIGURE 9.10 (a) Layout of the designed microstrip CQ filter with all the dimensions on the 1.27-mm thick substrate with a relative dielectric constant of 10.8.
(b) Measured performance of the microstrip CQ filter.

The designed filter was fabricated using copper microstrip on a RT/Duroid substrate with a relative dielectric constant of 10.8 and a thickness of 1.27 mm. The filter was measured using an HP network analyzer. The measured performance is shown in Figure 9.10b. The midband insertion loss is about −2.7dB, which is mainly attributed to the conductor loss of copper. The two pairs of attenuation poles near the cut-off frequencies of the passband are observable, which improve the selectivity. The measured center frequency was 825 MHz, which was about 25 MHz (2.94%) lower than the designed one. This discrepancy can easily be eliminated by slightly adjusting the open gap of resonators, which hardly affects the couplings.

9.3 TRISECTION AND CASCADED TRISECTION (CT) FILTERS

9.3.1 Characteristics of CT Filters

Figure 9.11 shows two typical coupling structures of cascaded trisection or CT filters, where each node represents a resonator. The full line between nodes indicates main or direct coupling and the broken line indicates the cross coupling. Each CT section is comprised of three directly coupled resonators with a cross coupling. It is this cross coupling that will produce a single attenuation pole at finite frequency. With an assumption that the direct coupling coefficients are positive, the attenuation pole is on the low side of the passband if the cross coupling is also positive, whereas

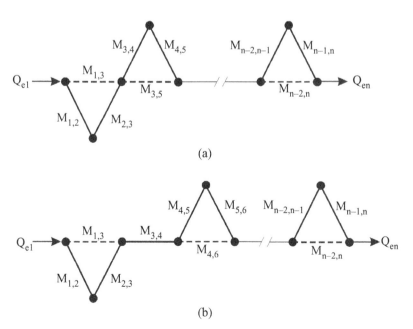

FIGURE 9.11 Typical coupling structures of the cascaded trisection or CT filters.

the attenuation pole will be on the high side of the passband for the negative cross coupling. The transfer function of a CT filter may be expressed as

$$|S_{21}| = \frac{1}{\sqrt{1 + \varepsilon^2 F_n^2(\Omega)}}$$

$$F_n = \cosh\left(\sum_{i=1}^{n} \cosh^{-1}\left(\frac{\Omega - 1/\Omega_{ai}}{1 - \Omega/\Omega_{ai}}\right)\right)$$

(9.16)

where ε is the ripple constant, Ω is the frequency variable of the lowpass prototype filter, Ω_{ai} is the ith attenuation pole, and n is the degree of the filter. Note that the number of the finite frequency attenuation poles is less than n. Therefore, the remainder poles should be placed at infinity of Ω.

The main advantage of a CT filter is its capability of producing asymmetrical frequency response, which is desirable for some applications requiring only a higher selectivity on one side of the passband, but less or none on the other side of the passband [8–15]. In such case, a symmetrical frequency response filter results in a larger number of resonators with a higher insertion loss in the passband, a larger size, and higher cost.

As a demonstration, Figure 9.12 shows a comparison of different types of bandpass filter responses to meet simple specifications of a rejection larger than 20 dB for the normalized frequencies ≥ 1.03, and a return loss ≤ -20 dB over a fractional bandwidth $FBW = 0.035$. As can be seen in Figure 9.12a the four-pole Chebyshev filter does not meet the rejection requirement, but the five-pole Chebyshev filter does. The four-pole elliptic-function response filter with a pair of attenuation poles at finite frequencies meets the specifications. However, the most notable thing is that the three-pole filter with a single CT section having an asymmetrical frequency response not only meets the specifications, but also results in the smallest passband insertion loss as compared to the other filters. The later is clearly illustrated in Figure 9.12b.

9.3.2 Trisection Filters

A three-pole trisection filter is not only the simplest CT filter, but also the basic unit for construction of higher-degree CT filters. Therefore, it is important to understand how it works. For the narrow-band case, an equivalent circuit of Figure 9.13a may represent a trisection filter. The couplings between adjacent resonators are indicated by the coupling coefficients M_{12} and M_{23}, while the cross coupling is denoted by M_{13}. Q_{e1} and Q_{e3} are the external quality factors denoting the input and output couplings, respectively. Note that because the resonators are not necessary synchronously tuned for this type of filter, $1/\sqrt{L_i C_i} = \omega_{0i} = 2\pi f_{0i}$ is the resonant angular frequency of resonator i for $i = 1, 2$, and 3. Although we want the frequency response of trisection filters to be asymmetrical, the physical configuration of filter can be kept symmetrical. Therefore, for simplicity, we can let $M_{12} = M_{23}$, $Q_{e1} = Q_{e3}$, and $\omega_{01} = \omega_{03}$.

The above coupled resonator circuit may be transferred to a lowpass prototype filter shown in Figure 9.13b. Each of the rectangular boxes represents a frequency-invariant

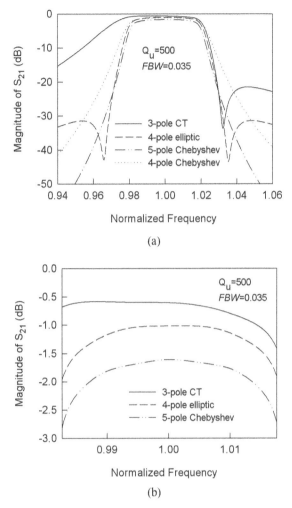

FIGURE 9.12 Comparison of different types of bandpass filter responses to meet simple specification: rejection larger than 20 dB for normalized frequency ≥ 1.03 and return loss loss ≤ -20 dB over fractional bandwidth of 0.035. (**a**) Transmission response. (**b**) Details of passband response.

immittance inverter, with J the characteristic admittance of the inverter. In our case, $J_{12} = J_{23} = 1$ for the inverters along the main path of the filter. The bypass inverter with a characteristic admittance J_{13} accounts for the cross coupling. The capacitance and the frequency-invariant susceptance of the lowpass prototype filter are denoted by g_i and B_i ($i = 1,2,3$), respectively, and g_0 and g_4 are the resistive terminations.

Assume a symmetrical two-port circuit of Figure 9.13b, thus, $g_0 = g_4$, $g_1 = g_3$, and $B_1 = B_3$. Let $g_0 = g_4 = 1.0$ also be the normalized terminations. The scattering parameters of the symmetrical circuit may be expressed in terms of the even- and

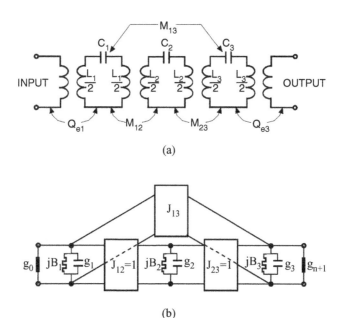

FIGURE 9.13 (a) Equivalent circuit of a trisection bandpass filter. (b) Associated lowpass prototype filter.

odd-mode parameters of one-port circuit formed by inserting open- or short-circuited plane along its symmetrical plane. This results in

$$S_{21} = S_{12} = \frac{S_{11e} - S_{11o}}{2}$$
$$S_{11} = S_{22} = \frac{S_{11e} + S_{11o}}{2} \quad (9.17)$$

with

$$S_{11e} = \frac{1 - \left(g_1 p + jB_1 - jJ_{13} + \dfrac{2J_{12}}{g_2 p + jB_2}\right)}{1 + \left(g_1 p + jB_1 - jJ_{13} + \dfrac{2J_{12}}{g_2 p + jB_2}\right)}$$

$$S_{11o} = \frac{1 - (g_1 p + jB_1 + jJ_{13})}{1 + (g_1 p + jB_1 + jJ_{13})} \quad (9.18)$$

where S_{11e} and S_{11o} are the even- and odd-mode scattering parameters; $p = j\Omega$, with Ω the frequency variable of the lowpass prototype filter. With Eqs. (9.17) and (9.18), the unknown element values of a symmetrical lowpass prototype may be determined by a synthesis method or through an optimization process.

TRISECTION AND CASCADED TRISECTION (CT) FILTERS

At the frequency Ω_a where the finite frequency attenuation pole is located, let $|S_{21}| = 0$ or $S_{11e} = S_{11o}$. Imposing this condition upon Eq. (9.18) and solving for Ω_a, we obtain

$$\Omega_a = -\frac{1}{g_2}\left(\frac{J_{12}}{J_{13}} + B_2\right) \quad (9.19)$$

This attenuation pole has to be outside the passband, namely $|\Omega_a| > 1$.

Because of the cascaded structure of CT filters with more than one trisection, we can expect the similar formulation for each attenuation pole produced by every trisection. Let i, j, and k be the sequence of direct coupling of each trisection; the associated attenuation pole may be expressed as

$$\Omega_a = -\frac{1}{g_j}\left(\frac{J_{ij}}{J_{ik}} + B_j\right) \quad (9.20)$$

Applying the lowpass to bandpass frequency mapping, we can find the corresponding attenuation pole at the bandpass frequency

$$f_a = f_0 \frac{FBW \cdot \Omega_a + \sqrt{(FBW \cdot \Omega_a)^2 + 4}}{2} \quad (9.21)$$

Here f_0 is the midband frequency and FBW is the fractional bandwidth of bandpass filters.

To transfer the lowpass elements to the bandpass ones, let us first transfer a nodal capacitance and its associated frequency-invariant susceptance of the lowpass prototype filter of Figure 9.13b into a shunt resonator of the bandpass filter of Figure 9.13a. Using the lowpass to bandpass frequency transformation, we have

$$\frac{1}{FBW} \cdot \left(\frac{j\omega}{\omega_0} + \frac{\omega_0}{j\omega}\right) \cdot g_i + jB_i = j\omega C_i + \frac{1}{j\omega L_i} \quad (9.22)$$

where $\omega_0 = 2\pi f_0$ is the midband angular frequency. Derivation of Eq. (9.22) with respect to $j\omega$ yields

$$\frac{1}{FBW} \cdot \left(\frac{1}{\omega_0} - \frac{\omega_0}{(j\omega)^2}\right) \cdot g_i = C_i - \frac{1}{(j\omega)^2 L_i} \quad (9.23)$$

Letting $\omega = \omega_0$ in Eqs. (9.22) and (9.23) gives

$$B_i = \omega_0 C_i - \frac{1}{\omega_0 L_i}$$
$$\frac{1}{FBW} \cdot \left(\frac{2}{\omega_0}\right) \cdot g_i = C_i + \frac{1}{\omega_0^2 L_i} \quad (9.24)$$

280 ADVANCED RF/MICROWAVE FILTERS

We can solve for C_i and L_i

$$C_i = \frac{1}{\omega_0}\left(\frac{g_i}{FBW} + \frac{B_i}{2}\right)$$
$$L_i = \frac{1}{\omega_0}\left(\frac{g_i}{FBW} - \frac{B_i}{2}\right)^{-1}$$
(9.25)

and thus

$$\omega_{0i} = \frac{1}{\sqrt{L_i C_i}} = \omega_0 \cdot \sqrt{1 - \frac{B_i}{g_i/FBW + B_i/2}}$$
(9.26)

From Eq. (9.26), we can clearly see that the effect of frequency-invariant susceptance is the offset of resonant frequency of a shunt resonator from the midband frequency of a bandpass filter. Therefore, as we mentioned before, the bandpass filter of this type is, in general, asynchronously tuned.

In order to derive the expressions for the external quality factors and coupling coefficients, we define a susceptance slope parameter of each shunt resonator in Figure 9.13a, as discussed in Chapter 3:

$$b_i = \frac{\omega_{0i}}{2} \cdot \frac{d}{d\omega}\left(\omega C_i - \frac{1}{\omega L_i}\right) \omega = \omega_{0i}$$
(9.27)

Substituting Eqs. (9.25) into (9.27) yields

$$b_i = \omega_{0i} C_i = \frac{\omega_{0i}}{\omega_0} \cdot \left(\frac{g_i}{FBW} + \frac{B_i}{2}\right)$$
(9.28)

By using the definitions similar to those described previously in Chapter 7 for the external quality factor and the coupling coefficient, we have

$$Q_{e1} = \frac{b_1}{g_0} = \frac{\omega_{01}}{\omega_0 g_0} \cdot \left(\frac{g_1}{FBW} + \frac{B_1}{2}\right)$$

$$Q_{en} = \frac{b_n}{g_{n+1}} = \frac{\omega_{0n}}{\omega_0 g_{n+1}} \cdot \left(\frac{g_n}{FBW} + \frac{B_n}{2}\right)$$

$$M_{ij}\Big|_{i \neq j} = \frac{J_{ij}}{\sqrt{b_i b_j}}$$

$$= \frac{\omega_0}{\sqrt{\omega_{0i}\omega_{0j}}} \frac{FBW \cdot J_{ij}}{\sqrt{(g_i + FBW \cdot B_i/2) \cdot (g_j + FBW \cdot B_j/2)}} \quad (i, j = 1, 2, \cdots n)$$
(9.29)

where n is the degree of the filter or the number of the resonators.

Note that the design equations of Eqs. (9.26) and (9.29) are general, since they are applicable for general coupled resonator filters when the equivalent circuits in Figure 9.13 are extended to higher-order filters.

9.3.3 Microstrip Trisection Filters

Microstrip trisection filters with different resonator shapes, such as open-loop [14] and triangularpatch resonators [15], can produce asymmetric frequency responses with an attenuation pole of finite frequency on either side of the passband.

9.3.3.1 Trisection Filter Design: Example One For our demonstration, the filter is designed to meet the following specifications:

Midband or center frequency	905 MHz
Bandwidth of pass band	40 MHz
Return loss in the pass band	< -20 dB
Rejection	> 20 dB for frequencies ≥ 950 MHz

Thus, the fractional bandwidth is 4.42%. A three-pole bandpass filter with an attenuation pole of finite frequency on the high side of the passband can meet the specifications. The element values of the lowpass prototype filter are found to be

$$g_1 = g_3 = 0.695 \quad B_1 = B_3 = 0.185$$
$$g_2 = 1.245 \quad B_2 = -0.615$$
$$J_{12} = J_{23} = 1.0 \quad J_{13} = -0.457$$

From Eqs. (9.26) and (9.29) we obtain

$$f_{01} = f_{03} = 899.471 \text{ MHz}$$
$$f_{02} = 914.713 \text{ MHz}$$
$$Q_{e1} = Q_{e3} = 15.7203$$
$$M_{12} = M_{23} = 0.04753$$
$$M_{13} = -0.02907$$

We can see that the resonant frequency of resonators 1 and 3 is lower than the midband frequency, whereas the resonant frequency of resonator 2 is higher than the midband frequency. The frequency offsets amount to -0.61% and 1.07%, respectively. For $f_0 = 905$ MHz and $FBW = 0.0442$ (see Chapter 7), the generalized coupling matrix and the scaled external quality factors are

$$[m] = \begin{bmatrix} -0.27644 & 1.07534 & -0.65769 \\ 1.07534 & 0.48564 & 1.07534 \\ -0.65769 & 1.07534 & -0.27644 \end{bmatrix} \quad (9.30)$$

$$q_{e1} = q_{e3} = 0.69484$$

The filter frequency response can be computed using the general formulation Eq. (7.30) for the cross-coupled resonator filters. At this stage, it is interesting to point out that if we reverse the sign of the generalized coupling matrix in Eq. (9.30), we can obtain an image frequency response of the filter with the finite frequency attenuation pole moved to the low side of the passband. This means that the design parameters of Eq. (9.30) have dual usage, and one may take the advantage of this to design the filter with the image frequency response.

Having obtained the required design parameters for the bandpass filter, the physical dimensions of the microstrip trisection filter can be determined using full-wave EM simulations to extract the desired coupling coefficients and external quality factors, as described in Chapter 7. Figure 9.14a shows the layout of the designed microstrip filter with the dimensions on a substrate having a relative dielectric constant of 10.8 and a thickness of 1.27 mm. The size of the filter is about $0.19 \times 0.27\ \lambda_{g0}$, where λ_{g0} is the guided wavelength of a 50 Ω line on the substrate at the midband frequency. This size is evidently very compact. Figure 9.14b shows the measured results of the filter. As can be seen, an attenuation pole of finite frequency on the upper side of the passband leads to a higher selectivity on this side of the passband. The measured midband insertion loss is about -1.15 dB, which is mainly because of the conductor loss of copper microstrip.

9.3.3.2 Trisection Filter Design: Example Two
The filter is designed to meet the following specifications:

Midband or center frequency	910 MHz
Bandwidth of pass band	40 MHz
Return loss in the pass band	< -20 dB
Rejection	> 35 dB for frequencies ≤ 843 MHz

A three-pole bandpass filter with an attenuation pole of finite frequency on the low side of the passband can meet these specifications. The element values of the lowpass prototype filter for this design example are

$$g_1 = g_3 = 0.645 \quad B_1 = B_3 = -0.205$$
$$g_2 = 0.942 \quad B_2 = 0.191$$
$$J_{12} = J_{23} = 1.0 \quad J_{13} = 0.281$$

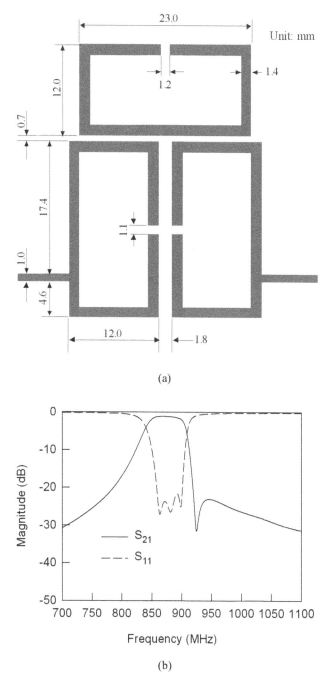

FIGURE 9.14 (a) Layout of the microstrip trisection filter designed to have a higher selectivity on high side of the passband on the 1.27-mm thick substrate with a relative dielectric constant of 10.8. (b) Measured performance of the filter.

Note that, in this case, the frequency-invariant susceptances and the characteristic admittance of cross-coupling inverter have the opposite signs, as referred to those of the above filter design. Similarly, design parameters for the bandpass filter can be found from Eqs. (9.26) and (9.29)

$$f_{01} = f_{03} = 916.159 \text{ MHz}$$
$$f_{02} = 905.734 \text{ MHz}$$
$$Q_{e1} = Q_{e3} = 14.6698$$
$$M_{12} = M_{23} = 0.05641$$
$$M_{13} = 0.01915$$

In contrast to the previous design example, the resonant frequency of resonators 1 and 3 is higher than the midband frequency, whereas the resonant frequency of resonator 2 is lower than the midband frequency. The frequency offsets are 0.68% and -0.47%, respectively. Moreover, the cross-coupling coefficient is positive. For $f_0 = 910$ MHz and $FBW = 0.044$, the generalized coupling matrix and the scaled external quality factors are

$$[m] = \begin{bmatrix} 0.30764 & 1.28205 & 0.43523 \\ 1.28205 & -0.21309 & 1.28205 \\ 0.43523 & 1.28205 & 0.30764 \end{bmatrix} \quad (9.31)$$

$$q_{e1} = q_{e3} = 0.64547$$

Similarly, the frequency response can be computed using Eq. (7.30), and reversing the sign of the generalized coupling matrix in Eq. (9.31) results in an image frequency response of the filter, with the finite frequency attenuation pole having been moved to the high side of the passband.

Figure 9.15a is the layout of the designed filter with all dimensions on a substrate having a relative dielectric constant of 10.8 and a thickness of 1.27 mm. The size of the filter amounts to $0.41 \times 0.17 \, \lambda_{g0}$. The measured results of the filter are plotted in Figure 9.15b. The attenuation pole of finite frequency does occur on the low side of the passband so that the selectivity on this side is higher than that on the upper side. The measured midband insertion loss is about -1.28 dB. Again, the insertion loss is mainly attributed to the conductor loss.

9.3.4 Microstrip CT Filters

It would be obvious that the two microstrip trisection filters described above could be used for constructing the microstrip CT filters with more than one trisection. Of course, by combination of the basic trisections that have the opposite frequency characteristics, a CT filter can also have the finite-frequency attenuation poles on the both sides of the passband. For instance, Figure 9.16 shows two possible configurations of microstrip CT filters. Figure 9.16a is a six-pole CT filter, which has the coupling

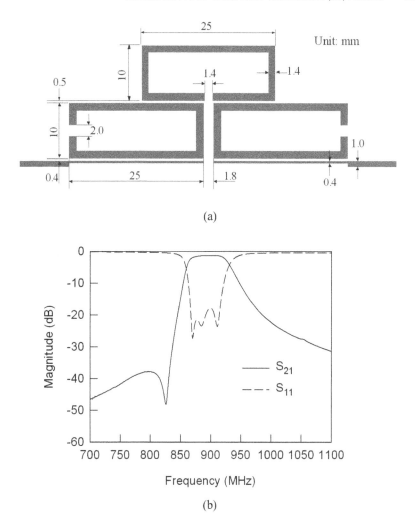

FIGURE 9.15 (a) Layout of the microstrip trisection filter designed to have a higher selectivity on low side of the passband on the 1.27-mm thick substrate with a relative dielectric constant of 10.8. (b) Measured performance of the filter.

structure of Figure 9.11b. It can have two finite-frequency attenuation poles on the high side of the passband. While Figure 9.16b is a five-pole CT filter, it has the coupling structure of Figure 9.11a. This filter structure is able to produce two finite-frequency attenuation poles, one on the low side of the passband and the other on the high side of the passband. A five-pole microstrip CT bandpass filter of this type has been demonstrated [16]. The filter is designed to have two asymmetrical poles, $\Omega_{a1} = -2.0$ and $\Omega_{a2} = 1.8$, placed on opposite sides of the passband for a center frequency $f_0 = 3.0$ GHz and 3.33% fractional bandwidth, and to have a coupling structure of Figure 9.11a with two trisections.

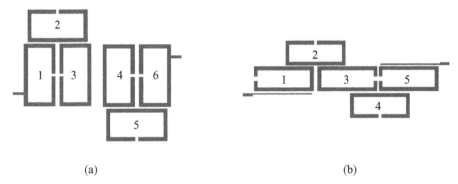

FIGURE 9.16 Configurations of microstrip CT filters.

The element values of the lowpass prototype filter for this design example are $g_0 = g_6 = 1.0$, and

$$g_1 = 0.9834 \quad B_1 = 0.0028 \quad J_{12} = J_{23} = 1.0$$
$$g_2 = 1.586 \quad B_2 = 0.6881 \quad J_{13} = 0.4026$$
$$g_3 = 1.882 \quad B_3 = 0.0194 \quad J_{34} = J_{45} = 1.0$$
$$g_4 = 1.6581 \quad B_4 = -0.7965 \quad J_{35} = -0.4594$$
$$g_5 = 0.9834 \quad B_5 = 0.0028$$

Using Eq. (9.20) to verify the above design yields

$$\Omega_{a1} = -\frac{1}{g_2}\left(\frac{J_{12}}{J_{13}} + B_2\right) = -\frac{1}{1.586}\left(\frac{1.0}{0.4026} + 0.6881\right) = -1.99997$$

$$\Omega_{a2} = -\frac{1}{g_4}\left(\frac{J_{34}}{J_{35}} + B_4\right) = -\frac{1}{1.6581}\left(\frac{1.0}{-0.4594} - 0.7965\right) = 1.80001$$

which match almost exactly to the prescribed locations of finite-frequency attenuation poles.

Applying Eqs. (9.26) and (9.29) to transfer the known lowpass elements to the bandpass design parameters for $f_0 = 3.0$ GHz and $FBW = 0.0333$ gives

$$f_{01} = 2.999858 \text{ GHz} \quad M_{12} = 0.02666$$
$$f_{02} = 2.978406 \text{ GHz} \quad M_{23} = 0.01927$$
$$f_{03} = 2.999485 \text{ GHz} \quad M_{34} = 0.01889$$
$$f_{04} = 3.024183 \text{ GHz} \quad M_{45} = 0.02613$$
$$f_{05} = 2.999858 \text{ GHz} \quad M_{13} = 0.00985$$
$$Q_{e1} = Q_{e5} = 29.53153 \quad M_{35} = -0.01125$$

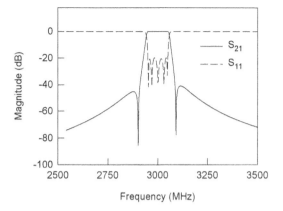

FIGURE 9.17 Theoretical response of a five-pole CT filter with finite frequency attenuation poles on both sides of passband.

The generalized coupling matrix and the scaled external quality factors are

$$[m] = \begin{bmatrix} -0.00285 & 0.80074 & 0.29594 & 0 & 0 \\ 0.80074 & -0.4323 & 0.57882 & 0 & 0 \\ 0.29594 & 0.57882 & -0.01031 & 0.56718 & -0.33769 \\ 0 & 0 & 0.56718 & 0.48415 & 0.78463 \\ 0 & 0 & -0.33769 & 0.78463 & -0.00285 \end{bmatrix} \quad (9.32)$$

$$q_{e1} = q_{e5} = 0.9834$$

With Eq. (9.32) the filter frequency response can be computed using the general formulation Eq. (7.30); the results are plotted in Figure 9.17.

9.4 ADVANCED FILTERS WITH TRANSMISSION-LINE INSERTED INVERTERS

9.4.1 Characteristics of Transmission-Line Inserted Inverters

An ideal immittance inverter has a constant 90° phase shift. To obtain other phase characteristics, we may insert a transmission line on the symmetrical plane of an ideal immittance inverter, as shown in Figure 9.18. The rectangular boxes at the

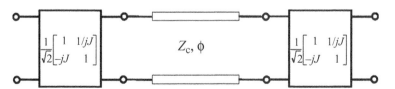

FIGURE 9.18 Transmission line-inserted immittance inverter.

input and output represent the two symmetrical halves of the ideal inverter with a characteristic admittance of J, and the matrix inside each box is the $ABCD$ matrix of the half-inverter. Z_c and ϕ are the characteristic impedance and electrical length of the transmission line. The $ABCD$ matrix of the modified inverter is given by

$$\begin{bmatrix} A & B \\ C & D \end{bmatrix} = \frac{1}{2} \begin{bmatrix} 1 & \frac{1}{jJ} \\ -jJ & 1 \end{bmatrix} \begin{bmatrix} \cos\phi & jZ_c \sin\phi \\ \frac{j\sin\phi}{Z_c} & \cos\phi \end{bmatrix} \begin{bmatrix} 1 & \frac{1}{jJ} \\ -jJ & 1 \end{bmatrix} \quad (9.33)$$

Letting Z_0 be the source and load impedance, the transmission S parameter is

$$S_{21} = \frac{2}{A + B/Z_0 + Z_0 C + D} \quad (9.34)$$

with

$$A = D = \left(JZ_c + \frac{1}{JZ_c}\right)\sin\phi$$

$$B = \frac{2\cos\phi}{jJ} + \left(jZ_c + \frac{1}{jZ_c J^2}\right)\sin\phi$$

$$C = -2jJ\cos\phi + \left(\frac{j}{Z_c} - jZ_c J^2\right)\sin\phi$$

Assuming a constant-phase velocity, the electrical length of the transmission line may be expressed as

$$\phi = \phi_0 \frac{f}{f_0}$$

where f_0 is a reference frequency, or, in our case, the midband frequency of filter, and ϕ_0 is the electrical length at f_0. Figure 9.19 plots the computed frequency responses for a normalized admittance of $JZ_0 = 0.005$ and normalized impedance of $Z_c/Z_0 = 1.0$. These two parameters are chosen to mimic a more realistic scenario in which the inverter should be weakly coupled to external resonators. The frequency axis is normalized to f_0. Since we are more concerned with the phase characteristics, let us look first at the phase responses in Figure 9.19a and b. As can be seen for $\phi_0 < 180°$, the inverter has a phase characteristic of about constant 90° phase shift, which is almost identical to that of the ideal inverter. When $\phi_0 = 180°$, because the transmission line resonates at its fundamental mode (in other words, it behaves like a half-wavelength resonator), there is a 180° phase change at f_0. Afterward and before $\phi_0 = 360°$, the inverter has an almost constant phase, but which is 180° out of that of the ideal inverter. When $\phi_0 = 360°$, there is another 180° phase change because of the resonance of the transmission line. As can be seen, the modified inverter has a changeable phase characteristic. In practice, this type of inverter is

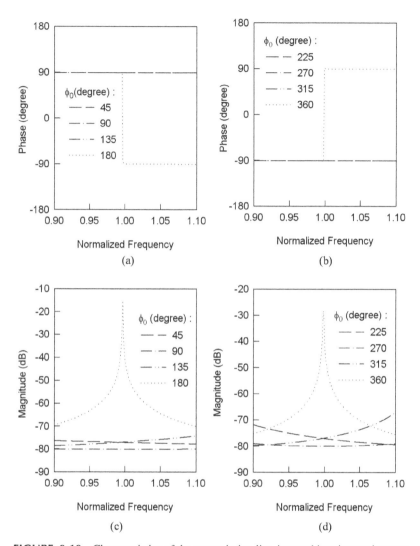

FIGURE 9.19 Characteristics of the transmission line-inserted immittance inverter.

quite useful for construction of filters with advanced filtering characteristics. This will be demonstrated in the next section. The magnitude responses in Figure 9.19c and d are also interesting. In general, the coupling strength of this type of inverter depends on J, Z_c, and ϕ. When $\phi_0 = 180°$ and $360°$, the coupling is strongly dependent on the frequency and reaches its maximum when the transmission line resonates.

9.4.2 Filtering Characteristics with Transmission-Line Inserted Inverters

For the demonstration, let us consider an equivalent circuit of four-pole cross-coupled resonator bandpass filter of Figure 9.20, which may be extended to higher-order

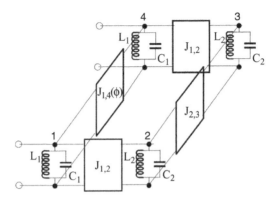

FIGURE 9.20 Equivalent circuit of four-pole cross-coupled resonator filter implemented with a transmission line-inserted immittance inverter.

filters. The filter shown in Figure 9.20 has a symmetrical configuration, but resonators on each side of the folded resonator array may be asynchronously tuned. The admittance inverter represents the coupling between resonators, where J denotes the characteristic admittance. The numbers indicate the sequences of direct couplings. There is one cross coupling between resonators 1 and 4, denoted by $J_{1,4}(\phi)$, which is the transmission line-inserted inverter introduced above.

If $J_{1,4}$ has an opposite sign to $J_{2,3}$ the filter frequency response shows two finite-frequency transmission zeros located at low and high stopband, respectively, resulting in higher selectivity on both sides of the passband. On the other hand, if $J_{1,4}$ and $J_{2,3}$ have the same sign, the filter exhibits linear-phase characteristics, which leads to a self-equalization of group delay. However, with a phase-dependent inverter, other advanced filtering characteristics can be achieved as well. The two-port S parameters of the filter in Figure 9.20 are

$$S_{21} = S_{12} = \frac{Y_e - Y_o}{(1/Z_0 + Y_e)(1/Z_0 + Y_o)}$$

$$S_{11} = S_{22} = \frac{1 - Y_e \cdot Y_o}{(1/Z_0 + Y_e)(1/Z_0 + Y_o)}$$

(9.35)

with

$$Y_e = \left(j\omega C_1 + \frac{1}{j\omega L_1} - jJ_e(\phi)\right) + \left(\frac{J_{1,2}^2}{j\omega C_2 + \frac{1}{j\omega L_2} - jJ_{2,3}}\right)$$

$$Y_o = \left(j\omega C_1 + \frac{1}{j\omega L_1} + jJ_o(\phi)\right) + \left(\frac{J_{1,2}^2}{j\omega C_2 + \frac{1}{j\omega L_2} + jJ_{2,3}}\right)$$

where $\omega = 2\pi f$. $J_e(\phi)$ and $J_o(\phi)$ are the even- and odd-mode characteristic admittance of the transmission line-inserted inverter, which are given by

$$J_e(\phi) = J_{1,4} \frac{Z_c J_{1,4} - \tan\left(\frac{\phi}{2}\frac{f}{f_0}\right)}{Z_c J_{1,4} + \tan\left(\frac{\phi}{2}\frac{f}{f_0}\right)}$$

$$J_o(\phi) = J_{1,4} \frac{1 + Z_c J_{1,4} \tan\left(\frac{\phi}{2}\frac{f}{f_0}\right)}{1 - Z_c J_{1,4} \tan\left(\frac{\phi}{2}\frac{f}{f_0}\right)}$$

(9.36)

With Eq. (9.35) we can use an optimization procedure to obtain the desired filter responses. Figure 9.21 plots the frequency responses of four different kinds of filtering characteristics achievable with the filter circuit of Figure 9.20. The optimized circuit parameters, with fixed $Z_0 = 50\ \Omega$ and $Z_c = 100\ \Omega$, are given as follows.

For the response in Figure 9.21a ($f_0 = 965$ MHz):

$\phi_0 = 167.6°$

$C_1 = 55.76843$ pF, $L_1 = 0.48678$ nH

$C_2 = 56.05285$ pF, $L_2 = 0.48936$ nH

$J_{1,2} = 0.014099$, $J_{2,3} = 0.00895$, $J_{1,4} = 0.00075$ $1/\Omega$

For the response in Figure 9.21b ($f_0 = 950$ MHz):

$\phi_0 = 356.515°$

$C_1 = 56.96285$ pF, $L_1 = 0.48939$ nH

$C_2 = 57.14843$ pF, $L_2 = 0.48685$ nH

$J_{1,2} = 0.01358$, $J_{2,3} = 0.00886$, $J_{1,4} = 0.000974$ $1/\Omega$

For the response in Figure 9.21c ($f_0 = 950$ MHz):

$\phi_0 = 166.395°$

$C_1 = 64.79581$ pF, $L_1 = 0.43128$ nH

$C_2 = 64.56043$ pF, $L_2 = 0.43034$ nH

$J_{1,2} = 0.01215$, $J_{2,3} = 0.01115$, $J_{1,4} = 0.00133$ $1/\Omega$

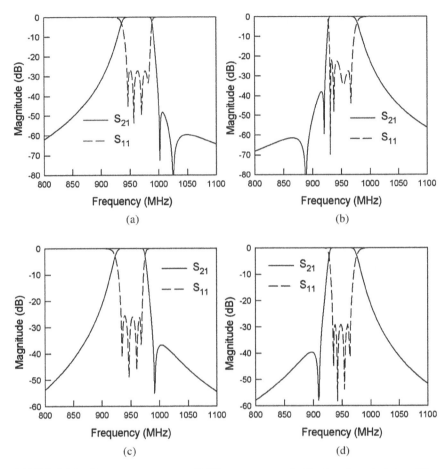

FIGURE 9.21 Frequency responses of different asymmetric filtering characteristics of the filter circuit in Figure 9.20.

For the response in Figure 9.21d ($f_0 = 950$ MHz):

$$\phi_0 = 337.17°$$
$$C_1 = 64.98783 \text{ pF}, \quad L_1 = 0.42973 \text{ nH}$$
$$C_2 = 65.47986 \text{ pF}, \quad L_2 = 0.43151 \text{ nH}$$
$$J_{1,2} = 0.01225, \quad J_{2,3} = 0.0106, \quad J_{1,4} = 0.0017 \text{ 1/}\Omega$$

All the four filters show asymmetrical frequency responses with one or two finite-frequency attenuation poles on the either side of the passband. This is because the transmission line- inserted inverters resonate at around the midband frequency of filter, which also makes the filter responses more sensitive to the electrical length of the inserted transmission line.

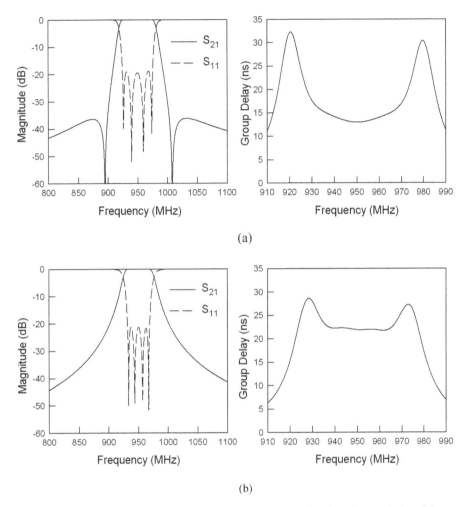

FIGURE 9.22 Frequency responses of different symmetric filtering characteristics of the filter circuit in Figure 9.20.

Figure 9.22 shows the frequency responses of the other two kinds of filtering characteristics, including their group delay, attainable with the transmission line-inserted inverters. Both the filters have the same midband frequency $f_0 = 950$ MHz. The one filter exhibits an elliptic-function response, as shown in Figure 9.22a, with the optimized circuit parameters:

$$\phi_0 = 230°$$
$$C_1 = 60.46767 \text{ pF}, \quad L_1 = 0.46088 \text{ nH}$$
$$C_2 = 60.93767 \text{ pF}, \quad L_2 = 0.46088 \text{ nH}$$
$$J_{1,2} = 0.01695, \quad J_{2,3} = 0.01448, \quad J_{1,4} = 0.00371 \text{ 1}/\Omega$$

The other filter, which has a linear-phase characteristic as depicted in Figure 9.22b, has been optimized to have the following circuit parameters:

$$\phi_0 = 40°$$
$$C_1 = 72.4059 \text{ pF}, \quad L_1 = 0.38476 \text{ nH}$$
$$C_2 = 72.8859 \text{ pF}, \quad L_2 = 0.38476 \text{ nH}$$
$$J_{1,2} = 0.01655, \quad J_{2,3} = 0.01095, \quad J_{1,4} = 0.00438 \ 1/\Omega$$

As can be seen, the main advantage of using the transmission line-inserted inverter lies in its flexibility to achieve different kinds of filtering characteristics with the same coupling structure. This feature should be useful when filter designers encounter difficulties in arranging the cross coupling to achieve desired filtering characteristics, especially when planar filter structures are considered.

9.4.3 General Transmission-Line Filter

A general transmission-line filter configuration, similar to that shown in Figure 9.23a, has been used to implement the all above filtering characteristics based on changing the crossing-line length between resonators 1 and 4 [17]. Note that the resonators with different shapes than the straight half-wavelength transmission line may be used and the crossing line may be meandered if necessary. For transmission-line filter design, the above circuit parameters may be transformed to the external quality factor and coupling coefficients by

$$Q_e = \omega_{01} C_1 Z_0$$
$$M_{1,2} = \frac{J_{1,2}}{\sqrt{b_1 b_2}} = \frac{J_{1,2}}{\sqrt{\omega_{01} C_1 \omega_{02} C_2}}$$
$$M_{2,3} = \frac{J_{2,3}}{\sqrt{b_2 b_3}} = \frac{J_{2,3}}{\omega_{02} C_2} \tag{9.37}$$
$$M_{1,4}(\phi = 0) = \frac{J_{1,4}}{\sqrt{b_1 b_4}} = \frac{J_{1,4}}{\omega_{01} C_1}$$

where b_i represents the susceptance of ith resonator resonating at $\omega_{0i} = 1/\sqrt{L_i C_i}$.

A microstrip filter of this type on a substrate with a relative dielectric constant of 10.8 and a thickness of 1.27 mm has been designed using ADS; its physical dimensions are given in Figure 9.23a, and its simulated frequency response is plotted in Figure 9.23b. Note that all the microstrip lines have the same width of 1.1 mm, except for the crossing line, which has a width of 0.5 mm. A short microstrip line at the bottom is added to enhance the coupling between two adjacent resonators. It may be pointed out that in the ADS simulation, the model of three coupled lines should be used whenever it is appropriate. More examples, including experimental

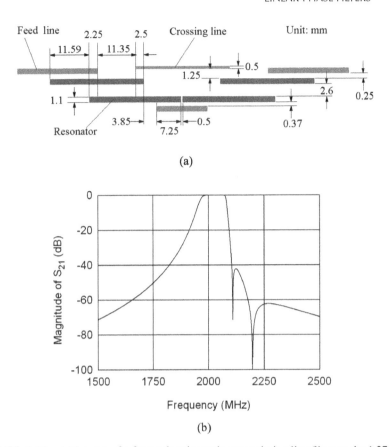

FIGURE 9.23 (a) Layout of a four-pole microstrip transmission line filter on the 1.27-mm thick substrate with a relative dielectric constant of 10.8. (b) Simulated performance of the filter.

demonstrations of this type of microstrip filter with different filtering characteristics, can be found [17].

9.5 LINEAR-PHASE FILTERS

In many RF/microwave communications systems, flat group delay of a bandpass filter is demanded, in addition to its selectivity. In order to achieve a flat group-delay characteristic, two approaches are commonly used. The first one is to use an external delay equalizer cascaded with a bandpass filter that is designed only to meet the amplitude requirement. In this case, the external equalizer would ideally be an all-pass network with an opposite delay characteristic to that of the cascaded bandpass filter. The second approach is to design a bandpass filter with an imposed linear-phase requirement in addition to the amplitude requirement. This type of filter is referred to as a self-equalized or linear-phase filter. The two-port network of a linear-phase filter is a nonminimum phase network, which is opposite to a minimum-phase, two-port network that physically has only one signal path from input to output, or mathematically has no right-half plane-transmission zeros on the complex frequency plane. Hence, physically, a linear-phase filter network must have more than one path of transmission. Mathematically, its transfer function has to contain distinct right-half plane zeros. This can be achieved by introducing cross coupling between nonadjacent resonators to produce more than one signal path and by controlling the sign of the cross coupling to place the transmission zero on the right-half plane.

9.5.1 Prototype of Linear-Phase Filter

A lowpass prototype network for the design of narrow-band RF/microwave linear-phase bandpass filter, introduced by Rhodes [18], is shown in Figure 9.24. The two-port network has a symmetrical structure with an even-degree of $n = 2m$. The terminal impedance is normalized to 1 Ω and denoted by g_0. The frequency-dependent elements are shunt capacitors, while all coupling elements are frequency-invariant immittance inverters, represented by rectangular boxes. Without loss of generality, the characteristic admittance of main-line immittance inverters is normalized to a unity value, indicated by $J = 1$. The immittance inverter denoted with a characteristic admittance J_m is for the direct coupling of two symmetrical parts of the network, whereas the immittance inverters with characteristic admittance J_1 to J_{m-1} are for the cross couplings between nonadjacent elements or resonators. Therefore, an n-pole filter can have a maximum number $n/2 - 1$ of crosscouplings. The scattering-transfer function of the network is

$$S_{21}(p) = \frac{Y_o(p) - Y_e(p)}{[1 + Y_e(p)] \cdot [1 + Y_o(p)]} \quad (9.38)$$

where $p = \sigma + j\Omega$ is the complex frequency variable of the lowpass prototype filter and $Y_e(p)$ and $Y_o(p)$ are the input admittance of the even- and odd-mode form of the

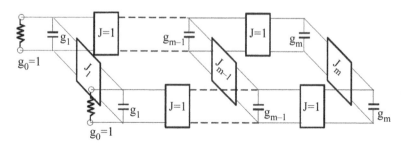

FIGURE 9.24 A lowpass prototype for design of RF/microwave linear-phase filters.

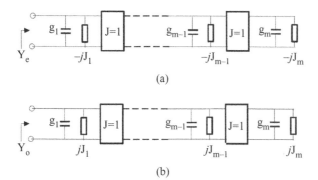

FIGURE 9.25 Even- and odd-mode networks of the lowpass prototype in Figure 9.24. (**a**) Even-mode; (**b**) odd-mode.

lowpass prototype network. This is shown in Figure 9.25 and given by

$$Y_e(p) = pg_1 - jJ_1 + \cfrac{1}{pg_2 - jJ_2 + \cfrac{1}{\cdots + \cfrac{1}{pg_{m-1} - jJ_{m-1} + \cfrac{1}{pg_m - jJ_m}}}}$$

$$Y_o(p) = pg_1 + jJ_1 + \cfrac{1}{pg_2 + jJ_2 + \cfrac{1}{\cdots + \cfrac{1}{pg_{m-1} + jJ_{m-1} + \cfrac{1}{pg_m + jJ_m}}}}$$

The transfer function may be expanded into a rational function whose numerator and denominator are polynomials in the complex frequency variable p. It can be shown that for the symmetrical even-degree, lowpass prototype network of Figure 9.24, the transfer function numerator must be an even polynomial in the complex frequency variable. This condition constrains locations of finite-frequency zeros on the complex frequency plane and results in two basic location patterns for the linear-phase filters [19]. The first one is shown in Figure 9.26a, where finite real or σ axis zeros must occur in a pair with symmetry about the imaginary axis at $p = \pm\sigma_a$. The second location pattern for transmission zeros is shown in Figure 9.26b, where finite complex plane zeros must occur with quadrantal symmetry at $p = \pm\sigma_a \pm j\Omega_a$.

The simplest form of the linear-phase filters may consists of only a single cross coupling that results in a single pair of finite real zeros along the σ axis. This type of the filter could give almost perfectly a flat group delay over the central 50% of the passband, which would meet the requirement for many communication links. To obtain the element values of the lowpass prototype, the Levy's approximate synthesis approach [2], described in Section 9.1, may be used. However, in this case, the

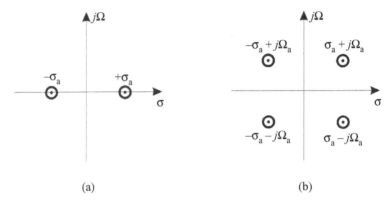

FIGURE 9.26 Location patterns of finite frequency zeros for linear-phase filters on the complex frequency plane. (a) Real or σ-axis zeros occurring in pair with symmetry about the imaginary axis. (b) Complex zeros occurring with quadrantal symmetry.

cross-coupling J_{m-1} will have the same sign as J_m for a single pair of transmission zeros at $\sigma = \pm\sigma_a$, where σ is the real axis of the complex frequency plane. Thus, the formulation for J_{m-1} in Eq. (9.4) should be expressed as

$$J_{m-1} = \frac{-J'_m}{(j\sigma_a g_m)^2 - J'^2_m} \tag{9.39}$$

Similarly, this approximate design approach will slightly deteriorate the return loss at the passband edges.

To achieve a good approximation to a linear-phase response over a larger part of the passband, more than one cross coupling is needed. In that case, the complex transmission zeros with quadrantal symmetry may be desirable to optimize the phase response. In general, the more the cross couplings, which produce the finite-frequency zeros with locations as shown in Figure 9.26, the better the group-delay equalization may be achieved. However, many requirements, especially the selectivity, can allow deterioration in the phase performance at the edges of the passband. In many high-capacity communication systems, phase linearity requirements over the central region of the passband are very severe and it is essential to have a good linear-phase response in this region. Normally, group-delay ripples must be of the order of a fraction of a percentage. For this purpose, the optimized element values for a single pair of real finite zeros, as well as four complex zeros with quadrantal symmetry, are tabulated in Tables 9.4–9.7. These prototype filters are designed for equal ripple with –20 and –30 dB return loss at the passband.

Although the tables of element values show that all the coupling elements are positive, they may be negative polarity as well as long as they have the same sign; this provides flexibility in the filter designs. Using Eq. (9.38) and the tabulated element values, the group delay and the amplitude responses can be computed and are plotted in Figures 9.27–9.30, together with those of conventional Chebyshev

TABLE 9.4 Element Values of the Lowpass Prototype for Linear-Phase Filters ($nz = 2$, $|S_{11}| = -20$ dB)

	$n = 4$	$n = 6$	$n = 8$	$n = 10$
g_1	0.90467	0.98469	1.01360	1.02637
J_1	0.17667	0.00000	0.00000	0.00000
g_2	1.25868	1.39274	1.44312	1.46361
J_2	0.77990	0.16358	0.00000	0.00000
g_3		1.91438	1.95262	1.98809
J_3		0.93623	0.21760	0.00000
g_4			1.70352	1.68037
J_4			0.75577	0.22613
g_5				2.21111
J_5				0.88444

TABLE 9.5 Element Values of the Lowpass Prototype for Linear-Phase Filters ($nz = 2$, $|S_{11}| = -30$ dB)

	$n = 4$	$n = 6$	$n = 8$	$n = 10$
g_1	0.59653	0.68080	0.71345	0.72931
J_1	0.15670	0.00000	0.00000	0.00000
g_2	1.07941	1.27862	1.35378	1.38764
J_2	0.84116	0.17316	0.00000	0.00000
g_3		1.59325	1.67790	1.73157
J_3		0.87564	0.22254	0.00000
g_4			1.71645	1.71157
J_4			0.79702	0.23753
g_5				1.96956
J_5				0.82892

TABLE 9.6 Element Values of the Lowpass Prototype for Linear-Phase Filters ($nz = 4$, $|S_{11}| = -20$ dB)

	$n = 6$	$n = 8$	$n = 10$
g_1	0.96575	1.00495	1.02205
J_1	0.07341	0.00000	0.00000
g_2	1.35981	1.42838	1.45674
J_2	0.28063	0.06035	0.00000
g_3	2.05257	1.93246	1.97644
J_3	0.80032	0.37282	0.08414
g_4		1.90473	1.67627
J_4		0.64210	0.32978
g_5			2.52393
J_5			0.76405

TABLE 9.7 Element Values of the Lowpass Prototype for Linear-Phase Filters ($nz = 4$, $|S_{11}| = -30$ dB)

	$n = 6$	$n = 8$	$n = 10$
g_1	0.66506	0.70449	0.72381
J_1	0.06225	0.00000	0.00000
g_2	1.24018	1.33337	1.37599
J_2	0.28515	0.06820	0.00000
g_3	1.66324	1.64945	1.71338
J_3	0.75977	0.34517	0.08830
g_4		1.88446	1.69889
J_4		0.68707	0.35892
g_5			2.26798
J_5			0.70501

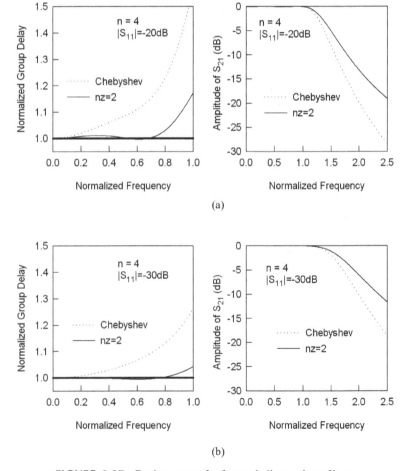

FIGURE 9.27 Design curves for four-pole linear-phase filters.

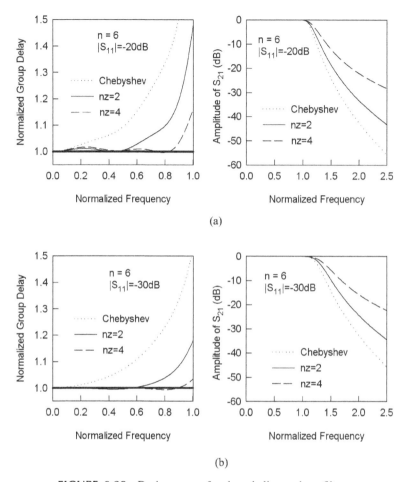

FIGURE 9.28 Design curves for six-pole linear-phase filters.

filters for comparison. The group delay is normalized to that at the center frequency ($\Omega = 0$) and the horizontal line for the unity-normalized group delay over entire passband indicates the ideal linear-phase response, which serves as a reference. In accordance with the tables, n is the degree of filter, nz is the number of finite zeros, and $|S_{11}|$ indicates the return loss at the passband. Because of symmetry in frequency response, only the responses for $\Omega \geq 0$ are plotted. As can be seen, for a given degree of the filter, increasing the number of the finite zeros enhances the group-delay equalization over a larger part of the passband. Moreover, the smaller passband return loss reduces the ripples of group delay, resulting in an even flatter group-delay response. However, in either case, selectivity is reduced. Hence, there will be a trade-off between the linear phase and selectivity in the filter designs. These plots may be used by the filter designers to meet a given specification; the element values are then found from the corresponding table.

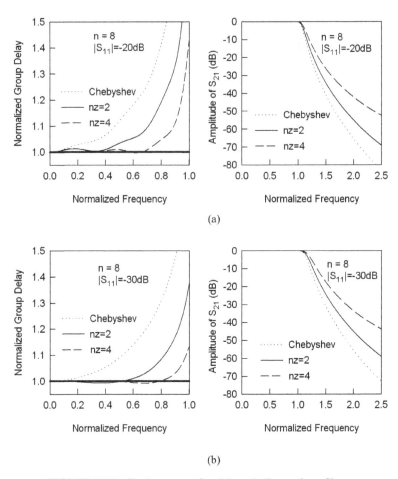

FIGURE 9.29 Design curves for eight-pole linear-phase filters.

9.5.2 Microstrip Linear-Phase Bandpass Filters

From the circuit elements of the lowpass prototype filter, the design parameters, namely, the external quality factor and coupling coefficients for the linear-phase bandpass filters, can be determined by

$$Q_e = \frac{g_1}{FBW}$$

$$M_{i,i+1} = M_{n-i,n-i+1} = \frac{FBW}{\sqrt{g_i g_{i+1}}} \quad \text{for } i = 1, 2, \ldots, \frac{n}{2} - 1 \qquad (9.40)$$

$$M_{i,n+1-i} = \frac{FBW \cdot J_i}{g_i} \quad \text{for } i = 1, 2, \ldots, \frac{n}{2}$$

where n is the degree of the filter.

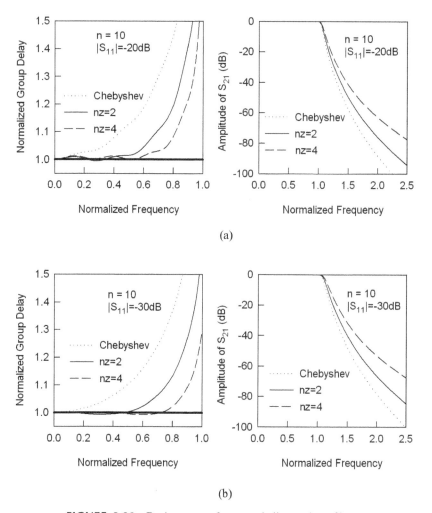

FIGURE 9.30 Design curves for ten-pole linear-phase filters.

For example, a microstrip filter is required to meet the following specification:

Passband frequencies	920–975 MHz
Passband return loss	−20 dB
Group delay variation	< 0.5% over the central 50% of the passband
Rejection at $f_0 \pm 68$ MHz	> 15 dB

The center frequency f_0 can be calculated by

$$f_0 = \sqrt{920 \times 975} = 947.10084 \text{ MHz}$$

and the fractional bandwidth is

$$FBW = \frac{975 - 920}{f_0} = 0.05807$$

At $f_0 \pm 68$ MHz, the normalized lowpass frequencies are determined by

$$\Omega = \frac{1}{FBW}\left(\frac{f_0 \pm 68}{f_0} - \frac{f_0}{f_0 \pm 68}\right)$$

which results in $\Omega_1 = 2.39$ for $f_0 + 68$ MHz and $\Omega_2 = -2.57$ for $f_0 - 68$ MHz. By referring to the design curves of Figure 9.27, we find that the required specification can be met with a four-pole linear phase filter whose lowpass element values, which are available from Table 9.4 for $n = 4$, are

$$g_1 = 0.90467 \quad J_1 = 0.17667$$
$$g_2 = 1.25868 \quad J_2 = 0.77990$$

Using Eq. (9.40), the values of design parameters for the bandpass filter are calculated

$$Q_e = \frac{0.90467}{0.05807} = 15.57896$$

$$M_{1,2} = M_{3,4} = \frac{0.05807}{\sqrt{0.90467 \times 1.25868}} = 0.05442$$

$$M_{1,4} = \frac{0.05807 \times 0.17667}{0.90467} = 0.01134$$

$$M_{2,3} = \frac{0.05807 \times 0.77990}{1.25868} = 0.035981$$

Shown in Figure 9.31a is the designed four-pole microstrip linear phase filter on a substrate with a relative dielectric constant of 10.8 and a thickness of 1.27 mm. The filter, which resembles the configuration of Figure 9.23a, is vertically symmetrical, and all lines have the same width of 1.1 mm, except for a narrower crossing line, as indicated. The simulated and measured performances of the filter are given in Figure 9.31b, where the simulated one is obtained using ADS linear simulator. The results show group-delay equalization about the central 50% of the passband as designed.

9.6 EXTRACTED POLE FILTERS

It has been demonstrated previously that the cross-coupled double array network, such as that shown in Figures 9.2 and 9.24, is a quite natural lowpass prototype for

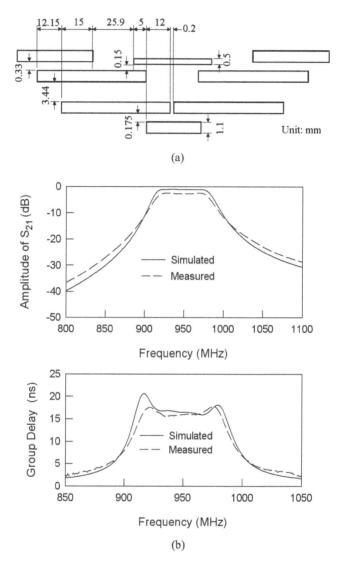

FIGURE 9.31 (a) Layout of a microstrip four-pole linear-phase bandpass filter on the 1.27-mm thick substrate with a relative dielectric constant of 10.8. (b) Simulated and measured performances of the linear phase filter.

introducing the finite-frequency transmission zeros in the complex-frequency variable p with the cross couplings between nonadjacent elements. When the finite-frequency transmission zeros occur on the imaginary axis of the complex p plane, which may be referred to as the finite j-axis zeros, then each zero at either side of the passband is not independently tunable by a single element in the network. Since the location of this type of transmission zero is very sensitive, from a practical viewpoint, it would be desirable to extract them, in an independent manner, from either above or below the passband. For this purpose, an extracted pole synthesis procedure, based on Rhodes and Cameron [20] is described in the next section.

9.6.1 Extracted Pole Synthesis Procedure

Since most transfer functions may be realized by lowpass prototype filters with a symmetrical cross-coupled double array network, this will enable the network to be synthesized with complex conjugate symmetry. Such a restriction is implied in the following synthesis procedurewhich is developed in terms of the cascaded or *ABCD* matrix. At any stage in the synthesis process, the *ABCD* matrix of the remaining network will be of the form

$$\begin{bmatrix} A & B/Z_0 \\ CZ_0 & D \end{bmatrix} = \frac{1}{e}\begin{bmatrix} a & b \\ c & d \end{bmatrix} \quad (9.41)$$

where Z_0 denotes the two-port terminal impedance. Hence, a, b, c, d, and e are normalized parameters, which are, in general, the functions of the complex frequency variable p.

At the beginning of the synthesis, the scattering transfer function of the lowpass prototype filter may be given by

$$S_{21} = \frac{2e}{a+b+c+d}$$

Let e have the same finite roots as S_{21}, so that e will contain as factors all the finite- frequency transmission zeros of the network. Besides, since the initial network considered is symmetrical, then all finite j-axis zeros (attenuation poles) are of complex conjugate symmetry, such as $e = 0$ at $p = \pm j\Omega_k$.

There are two kinds of extraction cycles required in the synthesis procedure. The first is aimed at extracting a conjugate pair of finite j-axis zeros from the remaining cross-coupled network in a cascaded manner. This kind of cycle may be repeated until either all or desired finite j-axis zeros have been extracted. After that, the second kind of extraction cycle, which will synthesize the remaining network as a cross-coupled network with complex conjugate symmetry is performed and repeated until the network has been completely synthesized.

In order to extract a pair of j-axis zeros from the network, we need to perform the equivalent transformations as shown in Figures 9.32 and 9.33. Referring to these two figures, it might be worth mentioning that the idea behind the extraction of an element from the network is simply to add another element with the opposite sign at the location where the element will be extracted. In Figure 9.32b, the remaining network denoted by N', from which a phase shifter of angle $-\phi$ has been extracted

EXTRACTED POLE FILTERS   307

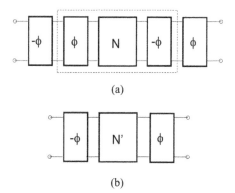

FIGURE 9.32 Equivalent networks for extracting phase shifters.

from the input and a phase shifter of angle ϕ has been extracted from the output, has a new normalized *ABCD* matrix of the form

$$\frac{1}{e}\begin{bmatrix} a' & b' \\ c' & d' \end{bmatrix} = \frac{1}{e}\begin{bmatrix} \cos\phi & j\sin\phi \\ j\sin\phi & \cos\phi \end{bmatrix} \cdot \begin{bmatrix} a & b \\ c & d \end{bmatrix} \cdot \begin{bmatrix} \cos\phi & -j\sin\phi \\ -j\sin\phi & \cos\phi \end{bmatrix}$$

$$= \frac{1}{e}\begin{bmatrix} a\cos^2\phi + d\sin^2\phi + j\sin\phi\cdot\cos\phi\cdot(c-b) & b\cos^2\phi + c\sin^2\phi + j\sin\phi\cdot\cos\phi\cdot(d-a) \\ c\cos^2\phi + b\sin^2\phi + j\sin\phi\cdot\cos\phi\cdot(a-d) & d\cos^2\phi + a\sin^2\phi + j\sin\phi\cdot\cos\phi\cdot(b-c) \end{bmatrix}$$

(9.42)

In Figure 9.33b, a shunt admittance of Y_1 and Y_2 have been further extracted from the input and output of the network N', respectively. The normalized *ABCD* matrix of the remaining network denoted by N'' is

$$\frac{1}{e''}\begin{bmatrix} a'' & b'' \\ c'' & d'' \end{bmatrix} = \frac{1}{e}\begin{bmatrix} 1 & 0 \\ -Y_1 & 1 \end{bmatrix} \cdot \begin{bmatrix} a' & b' \\ c' & d' \end{bmatrix} \cdot \begin{bmatrix} 1 & 0 \\ -Y_2 & 1 \end{bmatrix}$$

$$= \frac{1}{e}\begin{bmatrix} a' - b'Y_2 & b' \\ c' - a'Y_1 - d'Y_2 + b'Y_1Y_2 & d' - b'Y_1 \end{bmatrix}$$

(9.43)

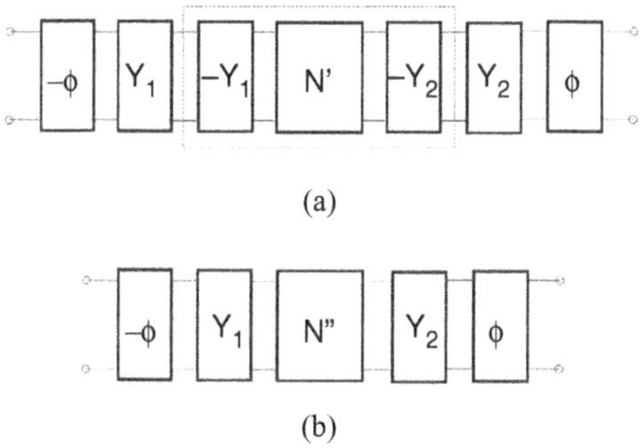

FIGURE 9.33 Equivalent networks for extracting shunt admittances.

The pair of transmission zeros of $p = \pm j\Omega_k$ have been extracted by the shunt admittances

$$Y_1 = \frac{b_k}{p - j\Omega_k}$$
$$Y_2 = \frac{b_k}{p + j\Omega_k} \tag{9.44}$$

where b_k is a constant, which will be determined later. Therefore, e'', which will not contain the transmission zeros of $p = \pm j\Omega_k$, is given by

$$e'' = \frac{e}{(p + j\Omega_k)(p - j\Omega_k)} = \frac{e}{(p^2 + \Omega_k^2)} \tag{9.45}$$

Furthermore, from Eq. (9.43) we have

$$\frac{b''}{e''} = \frac{b'}{e} = \frac{b'}{e''(p^2 + \Omega_k^2)}$$

Because b''/e'' is supposed to be analytical at $p = \pm j\Omega_k$, b' must possess a factor $(p^2 + \Omega_k^2)$. Hence, from Eq. (9.42), find a ϕ equal to ϕ_1 such that

$$b\cos^2\phi_1 + c\sin^2\phi_1 + j\sin\phi_1 \cdot \cos\phi_1 \cdot (d - a)|_{p=\pm j\Omega_k} = 0$$

Dividing the both sides by $\cos^2\phi_1$ yields

$$b + c\tan^2\phi_1 + j(d - a)\tan\phi_1|_{p=\pm j\Omega_k} = 0$$
$$\tan\phi_1 = \frac{-j(d - a) \pm \sqrt{-(d - a)^2 - 4bc}}{2c}\bigg|_{p=\pm j\Omega_k} \tag{9.46}$$

Since $ad - bc = e^2$ for the reciprocity and $e = 0$ for $p = \pm j\Omega_k$ from Eq. (9.46), we can obtain one useful solution

$$\tan\phi_1 = \frac{b}{ja}\bigg|_{p=-j\Omega_k} \quad \text{or} \quad \phi_1 = \tan^{-1}\left(\frac{b}{ja}\bigg|_{p=-j\Omega_k}\right) \tag{9.47}$$

This solution results in d' having a factor of $(p + j\Omega_k)$. From Eqs. (9.43) to (9.45), we have

$$d''(p^2 + \Omega_k^2) = d' - b'\frac{b_k}{p - j\Omega_k}$$

EXTRACTED POLE FILTERS   309

Because d'' is supposed to be analytical at $p = j\Omega_k$ so that $\lim_{p \to j\Omega_k} d''(p^2 + \Omega_k^2) = 0$, the constant b_k is actually a residue given by

$$b_k = \lim_{p \to j\Omega_k} (p - j\Omega_k) \frac{d'}{b'} \tag{9.48}$$

A complete cycle of the extraction of a pair of finite j-axis zeros has now been completed with a determined $ABCD$ matrix of Eq. (9.43) for the remaining network. This basic cycle may now be repeated until either all or the desired pairs of finite j-axis zeros have been extracted in this manner.

Afterward, the remaining network is to be synthesized as a cross-coupled double array with complex conjugate symmetry. Assume that there is at least a pair of transmission zeros at $p = \pm j\infty$, which will be extracted in the second kind of extraction cycle. However, before further processing, it is necessary to extract a pair of unity impedance phase shifters to allow this pair of zeros at infinity to be extracted next. Imagining that the pair of zeros at infinity could be obtained at $p = \pm j\Omega$ for $\Omega \to \infty$, the procedure of the extraction is much the same as seen in Figure 9.32. The desired phase shift is now given by

$$\phi_2 = \tan^{-1}\left(\frac{b}{ja}\bigg|_{p=-j\infty}\right) \tag{9.49}$$

Note that a and b in Eq. (9.49) are the parameters of the remaining network after completing all desired cycles of the first kind of extraction cycle. The updated $ABCD$ matrix will now be in the form given in Eq. (9.42) with $\phi = \phi_2$, where the degree of b' is now two lower than the degree of c'.

The second kind of extraction cycle starts with the extraction of a pair of admittances

$$\begin{aligned} Y_1 &= pC_1 + jB_1 \\ Y_2 &= PC_1 - jB_1 \end{aligned} \tag{9.50}$$

where C_1 is the capacitance and B_1 is the frequency invariant susceptance. The extraction process, which resembles that illustrated in Figure 9.33, yields

$$C_1 = \frac{d' + a'}{2b'p}\bigg|_{p=j\infty} \quad \text{and} \quad B_1 = \frac{d' - a'}{2jb'}\bigg|_{p=j\infty} \tag{9.51}$$

Thus, the admittances of Eq. (9.50) are determined and the $ABCD$ matrix of the remaining network can be updated from Eq. (9.43) with $e'' = e$, in this case.

To continue the extraction cycle, an immittance inverter of characteristic admittance J_{12} must be extracted in parallel with the network, as shown in Figure 9.34a,

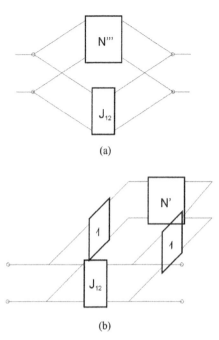

FIGURE 9.34 (a) An immittance inverter has been extracted from the remaining network. (b) A pair of unity immittance inverters has been extracted from the remaining network.

such that the remaining two-port network possesses another pair of transmission zeros at infinity, if there is any. This results in

$$J_{12} = \frac{e''}{jb''}\Big|_{p=j\infty} \tag{9.52}$$

and the remaining two-port $ABCD$ matrix is

$$\frac{1}{e'''}\begin{bmatrix} a''' & b''' \\ c''' & d''' \end{bmatrix} = \frac{1}{e'' - jb''J_{12}}\begin{bmatrix} a'' & b'' \\ c'' + 2je''J_{12} + b''J_{12}^2 & d'' \end{bmatrix} \tag{9.53}$$

To complete an extraction cycle of the second kind, a pair of unity immittance inverters is extracted further, as illustrated in Figure 9.34b, and the $ABCD$ matrix of the remaining network is updated as

$$\frac{1}{e'}\begin{bmatrix} a' & b' \\ c' & d' \end{bmatrix} = \frac{-1}{e'''}\begin{bmatrix} d''' & c''' \\ b''' & a''' \end{bmatrix} \tag{9.54}$$

Figure 9.35 shows the filter network after performing one extraction cycle of the first kind and one extraction cycle of the second kind, where each finite-frequency

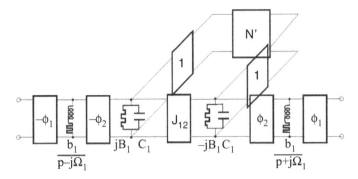

FIGURE 9.35 Extracted-pole filter network after one extraction cycle of the first kind and one extraction cycle of the second kind.

attenuation pole is extracted as an inductor in series with a frequency-invariant reactive component. The second kind of extraction cycle is then repeated until the filter network is completely synthesized.

9.6.2 Synthesis Example

An example of a four-pole, lowpass quasielliptic function prototype having a single pair of attenuation poles at $p = \pm j2.0$ will be demonstrated. The element values of the cross-coupled filter, which can be obtained from Table 9.1, are

$$g_1 = 0.95449 \quad J_1 = -0.16271$$
$$g_2 = 1.38235 \quad J_2 = 1.06062$$

The initial *ABCD* matrix of the filter may be determined by

$$\frac{1}{e}\begin{bmatrix} a & b \\ c & d \end{bmatrix} = \frac{1}{Y_o - Y_e}\begin{bmatrix} Y_e + Y_o & 2 \\ 2Y_e Y_o & Y_e + Y_o \end{bmatrix} \quad (9.55)$$

where Y_e and Y_o are the even- and odd-mode input admittances that can be calculated by Eq. (9.11), which results in

$$e = j(-0.62184p^2 - 2.48731)$$
$$a = d = 3.64785p^3 + 4.91214p$$
$$b = 3.82178p^2 + 2.24983$$
$$c = 3.48184p^4 + 7.42865p^2 + 2.74986$$

Using Eq. (9.47) with $\Omega_1 = 2.0$ yields

$$\phi_1 = 33.95908°$$

From Eq. (9.42) with $\phi = \phi_1$ the ABCD matrix for the remaining network is

$e = j(-0.62184p^2 - 2.48731)$
$a' = 3.64785p^3 + 4.91214p + j(1.613206p^4 + 1.671135p^2 + 0.2316739)$
$b' = 1.0864385p^4 + 4.947232p^2 + 2.405854$
$c' = 2.395401p^4 + 6.303198p^2 + 2.593836$
$d' = 3.64785p^3 + 4.91214p - j(1.613206p^4 + 1.671135p^2 + 0.2316739)$

The residue b_1 is then found from Eq. (9.48)

$$b_1 = 2.5849$$

and, therefore,

$$Y_1 = \frac{2.5849}{p - j2.0} \quad \text{and} \quad Y_2 = \frac{2.5849}{p + j2.0}$$

The ABCD matrix of the remaining network, after extracting the transmission zeros at $p = \pm j2.0$ is calculated using Eqs. (9.43) and (9.45). This leads to

$$\begin{aligned}
e'' &= -j0.62182 \\
a'' &= 0.839233p + j(1.61301p^2 + 0.83522) \\
b'' &= 1.08645p^2 + 0.60147 \\
c'' &= 2.39517p^2 + 1.802791 \\
d'' &= 0.839233p - j(1.61301p^2 + 0.83522)
\end{aligned} \quad (9.56)$$

Since there is only one pair of finite j-axis zeros in this example, the synthesis continues to extract a pair of transmission zeros at infinity, and the ABCD matrix is updated by

$$\frac{1}{e}\begin{bmatrix} a & b \\ c & d \end{bmatrix} = \frac{1}{e''}\begin{bmatrix} a'' & b'' \\ c'' & d'' \end{bmatrix}$$

Another pair of phase shifters is then extracted with the phase angle given by Eq. (9.49). By inspection of $a = a''$ and $b = b''$ from Eq. (9.56), we can immediately say

$$\phi_2 = \tan^{-1}\left(\frac{b}{ja}\right)\Big|_{p=j\infty} = \tan^{-1}\left(\frac{1.08645}{-1.61301}\right) = -33.9624°$$

The *ABCD* matrix of the remaining network is again determined by Eq. (9.42) with $\phi = \phi_2$, which gives

$$e' = -j0.62182$$
$$a' = 0.839252p - j0.2426229$$
$$b' = 0.2023413$$
$$c' = 3.481545p^2 + 2.20192$$
$$d' = 0.839252p + j0.2426229$$

From Eq. (9.51), we have

$$C_1 = 4.14777 \quad \text{and} \quad B_1 = 1.19892$$

With Y_1 and Y_2 defined in Eq. (9.50), the *ABCD* matrix of the network, after extracting the pair of transmission zeros at infinity is found from Eq. (9.43)

$$\frac{1}{e''}\begin{bmatrix} a'' & b'' \\ c'' & d'' \end{bmatrix} = \frac{1}{-j0.62182}\begin{bmatrix} 0 & 0.202341 \\ 1.910934 & 0 \end{bmatrix}$$

An immittance inverter is extracted next, and using Eq. (9.52), we have

$$J_{12} = \frac{-0.62182}{0.202341} = -3.07313$$

9.6.3 Microstrip-Extracted Pole Bandpass Filters

The design of a four-pole microstrip extracted pole bandpass filter will be demonstrated next. All necessary design equations and information can be used for the design of higher-order extracted pole bandpass filters as well.

The lowpass prototype of four-pole extracted pole filter is illustrated in Figure 9.36a. Note that the two-phase shifters with $\mp\phi_1$ at the input and output have been omitted because they do not affect the overall response of the filter. The lowpass prototype is then transformed to a lumped-element bandpass filter, shown in Figure 9.36b. As can be seen, the extracted poles in the lowpass prototype have been converted into a pair of series-resonant shunt branches and the shunt admittances into two shunt resonators. By applying the frequency transformation and equating slope

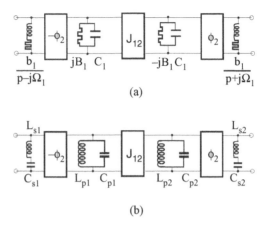

FIGURE 9.36 (a) Lowpass prototype of four-pole extracted pole filter. (b) Transformed lumped-element bandpass filter with extracted attenuation poles.

reactances as well as slope susceptances, we can obtain

$$C_{s1} = \frac{b_1}{\omega_0}\left(\frac{1}{FBW} + \frac{\Omega_1}{2}\right)^{-1} \quad L_{s1} = \frac{1}{\omega_0 b_1}\left(\frac{1}{FBW} - \frac{\Omega_1}{2}\right)$$

$$C_{s2} = \frac{b_1}{\omega_0}\left(\frac{1}{FBW} - \frac{\Omega_1}{2}\right)^{-1} \quad L_{s2} = \frac{1}{\omega_0 b_1}\left(\frac{1}{FBW} + \frac{\Omega_1}{2}\right)$$

$$C_{p1} = \frac{1}{\omega_0}\left(\frac{C_1}{FBW} + \frac{B_1}{2}\right) \quad L_{p1} = \frac{1}{\omega_0}\left(\frac{C_1}{FBW} - \frac{B_1}{2}\right)^{-1}$$

$$C_{p2} = \frac{1}{\omega_0}\left(\frac{C_1}{FBW} - \frac{B_1}{2}\right) \quad L_{p2} = \frac{1}{\omega_0}\left(\frac{C_1}{FBW} + \frac{B_1}{2}\right)^{-1}$$

(9.57)

where ω_0 is the midband angular frequency of bandpass filter and FBW is the fractional bandwidth.

For design distributed-element bandpass filters, the following design parameters would be preferable

$$\omega_{si} = 2\pi f_{si} = \frac{1}{\sqrt{L_{si}C_{si}}} \quad \omega_{pi} = 2\pi f_{pi} = \frac{1}{\sqrt{L_{pi}C_{pi}}}$$

$$Q_{esi} = \omega_{si}L_{si} \quad Q_{epi} = \omega_{pi}C_{pi}$$

for $i = 1$ and 2,

$$M_{12} = \frac{J_{12}}{\sqrt{\omega_{p1}C_{p1} \cdot \omega_{p2}C_{p2}}}$$

(9.58)

where ω_{si} and Q_{esi} are the angular resonant frequency and the external quality factor of the ith series resonators, ω_{pi} and Q_{epi} are the angular resonant frequency and the external quality factor of the ith shunt resonators, and M_{12} denotes the coupling coefficient of the two shunt resonators.

For our purpose, the microstrip bandpass filter is designed to have a fractional bandwidth $FBW = 0.05$ at midband frequency $f_0 = 2000$ MHz; the synthesized results in Section 9.6.2 will be used. Recall the results obtained in Section 9.6.2

$$\Omega_1 = 2.0 \qquad b_1 = 2.5849$$
$$C_1 = 4.14777 \qquad B_1 = 1.19892$$
$$\phi_2 = -33.9624° \qquad J_{12} = -3.07313$$

From Eqs. (9.57) and (9.58) we have

$$C_{s1} = 9.79523 \times 10^{-12} \text{ F} \qquad L_{s1} = 5.84925 \times 10^{-10} \text{ H}$$
$$C_{s2} = 1.08263 \times 10^{-11} \text{ F} \qquad L_{s2} = 6.46496 \times 10^{-10} \text{ H}$$
$$C_{p1} = 6.64908 \times 10^{-9} \text{ F} \qquad L_{p1} = 9.66263 \times 10^{-13} \text{ H}$$
$$C_{p2} = 6.55368 \times 10^{-9} \text{ F} \qquad L_{p2} = 9.52398 \times 10^{-13} \text{ H}$$

and

$$f_{s1} = 2102.63 \text{ MHz} \qquad f_{s2} = 1902.38 \text{ MHz}$$
$$f_{p1} = 1985.60 \text{ MHz} \qquad f_{p2} = 2014.51 \text{ MHz}$$
$$Q_{es1} = 7.72757 \qquad Q_{es2} = 7.72757$$
$$Q_{ep1} = 82.95323 \qquad Q_{ep2} = 82.95323$$
$$M_{12} = -0.03705$$

It should be noticed that all the resonators have different resonant frequencies and, therefore, the filter is asynchronously tuned. The external quality factors evaluated at the individual resonant frequencies show the same value for the pair of series resonators and the same value for the two shunt resonators.

For microstrip realization of the extracted pole filter, the series-resonant shunt branch may be transformed into another one-port equivalent network, as shown in Figure 9.37a, which may then be realized with the microstrip structure in Figure 9.37b, where λ_g is the guided wavelength of microstrip at the desired resonant frequency. The equivalence can be achieved by equating the normalized input impedance denoted with z_{in}. The quarter-wavelength microstrip line denoted by l_1 functions as a unity immittance inverter and, hence, its characteristic impedance should be the same as the normalization impedance, which is commonly 50 Ω for microstrip circuits. The half-wavelength microstrip resonator indicated by l_3 is coupled externally through a coupled line portion with length l_2 and spacing s, which are so determined as to achieve a desired external quality factor. The characteristic impedance of the microstrip resonator, including the coupled line portion is not necessarily 50 Ω, and,

316   ADVANCED RF/MICROWAVE FILTERS

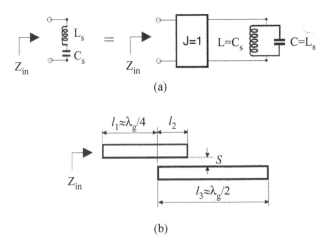

FIGURE 9.37 (a) Equivalent circuits for the series-resonant shunt branch. (b) Microstrip realization of the series-resonant shunt branch.

of course, other forms of microstrip resonator and coupled structures, such as those introduced in Figure 7.10, can be used.

Shown in Figure 9.38a is the designed four-pole microstrip extracted pole filter on a substrate with a relative dielectric constant of 10.8 and a thickness of 1.27 mm. Note that the width for all wide microstrip lines is the same and equal to 1.1 mm and all narrow microstrip lines have the same width of 0.5 mm. The required external quality factors and the inter-resonator coupling are implemented using coupled microstrip line structures, and the phase shifters are realized simply with 50-Ω microstrip lines. It should be mentioned that the phase lengths may be partly absorbed by adjacent circuits and for a convenient physical realization they may be modified by an arbitrary number of half-wavelengths, namely, $\phi \to \phi \pm k\pi$. Figure 9.38b plots the simulated frequency response of the designed microstrip extracted pole bandpass filter, which was obtained using Microwave Office, a commercial CAD tool introduced in the Chapter 8. As can be seen, the two extracted attenuation poles near the passband edges have been successfully realized.

9.7 CANONICAL FILTERS

9.7.1 General Coupling Structure

It has been known that a possible configuration for obtaining the most general bandpass-filtering characteristic from a set of n-multiple-coupled synchronously tuned resonators is the canonical form [21,22]. Figure 9.39 shows the general coupling structure of canonical filters, where each node represents a resonator, the full lines indicate the cascaded (or direct) couplings, and the broken lines denote the cross couplings. In this coupling structure, the resonators are numbered 1 to n, with the

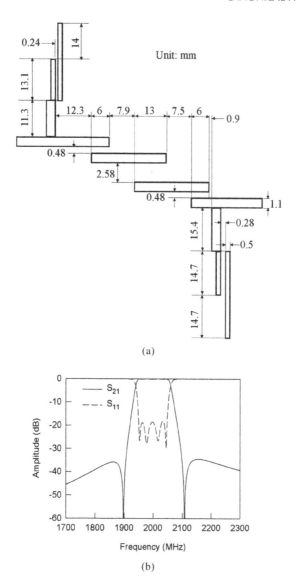

FIGURE 9.38 (a) Layout of a designed four-pole microstrip extracted pole bandpass filter on a 1.27-mm thick substrate with a relative dielectric constant of 10.8. (b) Simulated performance of the microstrip extracted pole bandpass filter.

input and output ports located in resonators 1 and n, respectively. For the canonical realization, the cascaded couplings of the same sign are normally required for consecutively numbered resonators, that is, 1 to 2, 2 to 3, ..., $n-1$ to n (as in the Chebyshev filter). In addition, the cross coupling of arbitrary signs must be provided between resonators 1 and n, 2 and $n-1$, ..., etc. As in the canonical form, the more

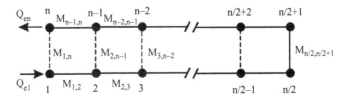

FIGURE 9.39 General coupling structure of canonical filters.

general-frequency responses, such as elliptic-function response and selective linear-phase (group-delay self-equalization) response, which can be obtained from multiple couplings, allow a given filter specification to be met by fewer resonators. This, in turn, leads to minimum weight and volume. However, there is a penalty—unlike a CQ filter as discussed previously— the effect of each cross coupling in a canonical filter is not independent. This could make the filter tuning more difficult.

Referring to Figure 9.39, M_{ij} is the coupling coefficient between the resonators i and j, and Q_{e1} and Q_{en} are the external quality factors of the input and output resonators, respectively.

Since, in most practical cases, the canonical filters are symmetrical, they will have a symmetrical set of couplings, that is, $M_{12} = M_{n-1,n}$, $M_{23} = M_{n-2,n-1}$ and so on, and $Q_{e1} = Q_{en}$. These bandpass design parameters can be obtained by the synthesis methods described in [22–24]. They can also be synthesized by optimization based on the general coupling-matrix formulation, as discussed in Chapter 8.

The structure of general-coupling matrix of a canonical filter may be demonstrated as following for $n = 6$:

$$[m] = \begin{bmatrix} 0 & m_{12} & 0 & 0 & 0 & m_{16} \\ m_{12} & 0 & m_{23} & 0 & m_{25} & 0 \\ 0 & m_{23} & 0 & m_{34} & 0 & 0 \\ 0 & 0 & m_{34} & 0 & m_{45} & 0 \\ 0 & m_{25} & 0 & m_{45} & 0 & m_{56} \\ m_{16} & 0 & 0 & 0 & m_{56} & 0 \end{bmatrix} \quad (9.59)$$

where m_{ij} denotes the normalized coupling coefficient (see Chapter 7). $[m]$ is reciprocal (i.e., $m_{ij} = m_{ji}$), and its all diagonal entries m_{ii} are zero because the canonical filter is synchronously tuned. There are also only $n/2 - 1$ cross couplings. For a give bandpass filter with a fractional bandwidth FBW, the design parameters are

$$\begin{aligned} M_{ij} &= m_{ij} FBW \\ Q_{e1} &= q_{e1}/FBW \\ Q_{en} &= q_{en}/FBW \end{aligned} \quad (9.60)$$

where q_{e1} and q_{en} are the scaled external quality factors as defined in Chapter 7.

9.7.2 Elliptic-Function/Selective Linear-Phase Canonical Filters

As mentioned above, the elliptic-function response or selective linear-phase response can be obtained in the canonical form. For demonstration, consider two typical sets of the normalized coupling coefficients and the scaled external quality factors for these two types of filters.

For the elliptic-function response ($n = 6$):

$$\begin{aligned} m_{12} = m_{56} &= 0.82133 & m_{34} &= 0.78128 \\ m_{23} = m_{45} &= 0.53934 & m_{25} &= -0.28366 \\ q_{e1} = q_{e6} &= 1.01787 & m_{16} &= 0.05776 \end{aligned} \quad (9.61)$$

For the selective linear-phase response ($n = 6$):

$$\begin{aligned} m_{12} = m_{56} &= 0.83994 & m_{34} &= 0.58640 \\ m_{23} = m_{45} &= 0.60765 & m_{25} &= 0.01416 \\ q_{e1} = q_{e6} &= 0.9980 & m_{16} &= -0.03966 \end{aligned} \quad (9.62)$$

These two sets of parameters may be used for designing six-pole ($n = 6$) canonical bandpass filters exhibiting either the elliptic-function response or the selective linear-phase response, with the passband return loss ≤ -20 dB. The elliptic-function response will have an equal-ripple stopband with the minimum attenuation of 40 dB. The selective linear-phase filter will have a fairly flat group delay over 50% of the passband.

For example, canonical bandpass filters of these two types are designed to have a passband from 975 to 1025 MHz. Thus, the center frequency and the fractional bandwidth are

$$f_0 = \sqrt{975 \times 1025} = 999.6875 \text{ MHz}$$

$$FBW = \frac{1025 - 975}{f_0} = 0.05$$

Substituting Eqs. (9.61) and (9.62) into (9.60), we can find the bandpass design parameters for the two filters. For the canonical filter with the elliptic-function response, the design parameters are

$$\begin{aligned} M_{12} = M_{56} &= 0.04108 & M_{34} &= 0.03908 \\ M_{23} = M_{45} &= 0.02698 & M_{25} &= -0.01419 \\ Q_{e1} = Q_{e6} &= 20.35104 & M_{16} &= 0.00289 \end{aligned}$$

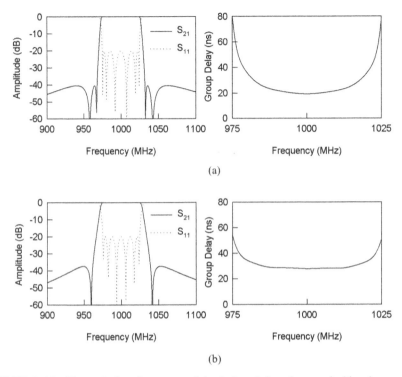

FIGURE 9.40 Theoretical performances of the designed six-pole canonical bandpass filters having a passband from 975 to 1025 MHz. (**a**) The elliptic-function response; (**b**) selective linear-phase response.

and the theoretical performance of this filter is plotted in Figure 9.40a. The canonical filter with the selective linear-phase response has the design parameters

$$M_{12} = M_{56} = 0.04201 \quad M_{34} = 0.02933$$
$$M_{23} = M_{45} = 0.03039 \quad M_{25} = 0.00071$$
$$Q_{e1} = Q_{e6} = 19.95376 \quad M_{16} = -0.00198$$

and its theoretical performance is shown in Figure 9.40b.

Following the approaches described in Chapter 7 for extracting coupling coefficients and external quality factors, the filters can be realized in different forms of microwave structures. For microstrip realization, the configurations proposed in Hong and Lancaster [25] and Jokela [26] may be used.

9.8 MULTIBAND FILTERS

Recent development in wireless communication systems has presented new challenges to operate in multiple separated frequency bands in order to access different services with multimode terminal. In order to accommodate this multiband RF signal reception and transmission into a single RF transceiver, a dual-band or multiband RF front-end circuit is required. To this end, dual-band and multiband filters become an essential component. There are numerous ways to construct such a filter [27–30]; several typical examples are described below.

9.8.1 Filters Using Mixed Resonators

Figure 9.41 shows the coupling structure for a dual-band filter using mixed resonators. The input and output nodes are denoted by S and L, respectively. The channel A is formed by four resonator nodes, denoted by 1^A to 4^A, which resonate at the center frequency of the first desired passband. Similarly, channel B consists of another four resonator nodes, that is, 1^B to 4^B, which resonate at the center frequency of the second passband. Therefore, each group of four resonators forms a single quadruplet filter, where a cross coupling introduced will result in a quasielliptic-function response, as mentioned in earlier sections. For a practical realization of this filter type, dual-band resonators are used at the input and output. The first two resonating modes of each dual-band resonator are deployed for the resonator nodes 1^A and 1^B, or 4^A and 4^B. The other resonator nodes are realized using single-mode resonators. Thus, the direct couplings along each main signal channel can be independently controlled.

As an example, a microstrip dual-band filter of this type is designed to have the first passband centered at $f_{01} = 2.45$ GHz and the second passband centered at $f_{02} = 5.17$ GHz. The two passbands have the same fractional bandwidth $FBW = 0.05$. The same four-pole lowpass prototype with a pair of attenuation poles at $\pm\Omega_a$ for $\Omega_a = 2.4$ is also assumed for each quadruplet section. From Table 9.1, the element

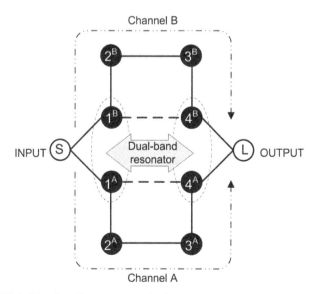

FIGURE 9.41 Coupling structure of a dual-band filter using mixed resonators.

values of the lowpass prototype are found to be $g_1 = 0.94772$, $g_2 = 1.34408$, $J_1 = -0.10642$, and $J_2 = 1.00086$. Using Eq. (9.9), the corresponding design parameters, namely, the coupling coefficients and external quality factors, for $FBW = 0.05$ are given by

$$\begin{aligned} M_{1,2} = M_{3,4} = 0.044 \quad & M_{2,3} = 0.037 \\ M_{1,4} = -5.615 \times 10^{-3} \quad & Q_{ei} = Q_{eo} = 18.954 \end{aligned} \quad (9.63)$$

The microstrip dual-band filter is realized using the configuration of Figure 9.42a on a substrate with a relative dielectric constant of 3.0 and a thickness of 0.508 mm. Resonators 2^A and 3^A are in a form of the so-called slow-wave, open-loop resonator (see Chapter 10) so that they not only resonate at f_{01} with a small size, but also shift their first spurious resonant frequency far beyond f_{02}. Resonators 2^B and 3^B are the hairpin resonators resonating at f_{02}. The dual-band resonators, as indicated in Figure 9.42a, may be seen as a folded stepped-impedance resonator for $W_1 \neq W_2$. By changing the ratio of W_1/W_2 and their associated line lengths, the dual-band resonators can resonate at the two desired frequencies, that is, f_{01} and f_{02}.

To determine the other physical dimensions of the filter, the full-wave EM simulations are carried out to extract the desired coupling coefficients and external quality factors using the approach described in Chapter 7. For example, the coupling spacings sa_{12} and sb_{12} are determined so as to obtain the desired $M_{1,2}$ given in Eq. (9.63). Similarly, the coupling spacings sa_{23} and sb_{23} are found such that the desired $M_{2,3}$ given in Eq. (9.63) can be achieved. The coupling spacing s_{14} is determined to meet the desired cross coupling of $M_{1,4}$ given in Eq. (9.63). The dual-band filter is excited with a special tapped-line arrangement at the I/O ports, as shown in Figure 9.42a, whose position and dimensions need to be properly determined to obtain the external quality factor Q_{ei} or Q_{eo}, specified in Eq. (9.63), for both desired passbands. The final dimensions of the filter are denoted in Figure 9.42a. The EM-simulated performance of the filter is shown in Figure 9.42b. As can be seen, there is a single pair of transmission zeros near each passband because of the designed cross coupling. The other additional transmission zeros observed are attributed to a skew-symmetrical feeding structure as shown in Figure 9.42a, which somehow improve the stopband response.

9.8.2 Filters Using Dual-Band Resonators

Dual-band filters can also be designed using dual-band resonators based on a coupling structure shown in Figure 9.43. Each dual-band resonator operates at two desired frequencies, which are usually the first two resonant modes for a distributed resonator.

There are numerous forms to construct a dual-band resonator; what would need to be considered include size, inter-resonator coupling, as well as I/O coupling for designing a dual-band filter. For example, shown in Figure 9.44a is a compact microstrip dual-band resonator, which consists of an open loop with a central loading element. The first two resonant modes, which resonate at f_{01} and f_{02}, respectively,

MULTIBAND FILTERS 323

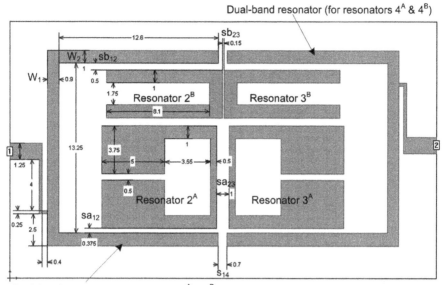

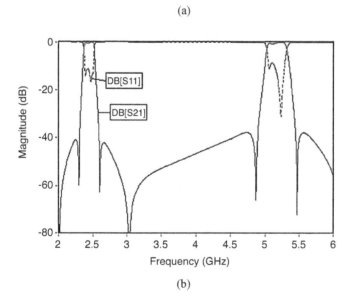

FIGURE 9.42 (a) Layout of the designed microstrip dual-band filter on a 0.508-mm thick substrate with a relative dielectric constant of 3.0. All the dimensions are in millimeters. (b) Simulated performance of the filter.

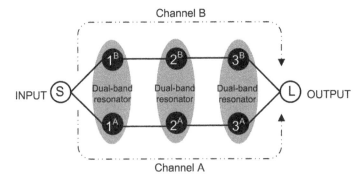

FIGURE 9.43 Coupling structure of dual-band filter based on dual-band resonators.

are to be utilized for the design of dual-band filter. Without the loading element in the center, the open loop is similar to the slow-wave, open-loop resonator described in Chapter 10, and, hence, $f_{02} > 2f_{01}$ can be obtained. With the central-loading element, f_{02} can be shifted down to a desired second band frequency toward f_{01} without changing f_{01}, as demonstrated in Figure 9.44b.

The inter-resonator coupling depends on the orientation of two coupled resonators. Figure 9.45a displays a particular arrangement where the two dual-band resonators are coupled through a spacing s. Since each dual-band resonator operates at two frequency bands, as referred to in Figure 9.43, the simulated coupling coefficients for the two desired bands against the spacing s are plotted in Figure 9.45b. The coupling coefficients are extracted from the simulation using the method described in Chapter 7. For the simulation, the resonators have the same dimensions as that given in Figure 9.44, except for W_{2b} and L_{2b} being fixed at 6.1 and 3.2 mm, respectively. As can be seen, the coupling coefficient for the first band is always larger than that for the second band over the given couple spacing. This implies that when a dual-band filter is designed, the first band will have a fractional bandwidth larger than that of the second band. In general, unlike coupled single-mode resonators, there is somewhat a restriction on the independent control of the couplings for coupled dual-band resonators. To change the coupling characteristics, the dual-band resonator shape or layout usually needs to be changed.

The other important design parameters are the external quality factors for dual-band filter. Figure 9.46a shows an arrangement to excite the input or output dual-band resonator. This arrangement consists of a narrow tapped line with a length of l_t and a width of w_t. The one end of the tapped line is tapped onto the resonator at a distance, denoted by t, which is away from the symmetrical plane of the resonator. The other end of the tapped line is connected to the 50-Ω feed line. The tapped line is also coupled to the resonator through a coupling gap of g_t, as indicated. Using the method introduced in Chapter 7, the external quality factors can be extracted from EM simulation, and the results are plotted in Figure 9.46b, where Q_{e1} is the external quality factor for the first band and Q_{e2} is for the second band. Again, in the simulation, the dielectric substrate and the dimensions used for the resonator are all

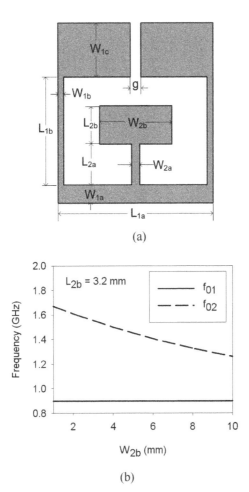

FIGURE 9.44 (a) A compact microstrip dual-band resonator. (b) Its first two resonant frequencies against the central loading element. The microstrip resonator is on a dielectric substrate with a relative dielectric constant of 10.8 and a thickness of 1.27 mm. The other dimensions are $W_{1a} = 1.5$ mm, $W_{1b} = 0.5$ mm, $W_{1c} = 4.5$ mm, $L_{1a} = 13.1$ mm, $L_{1b} = 9$ mm, $g = 0.9$ mm, $W_{2a} = 0.7$ mm, and $L_{2a} = 3.4$ mm.

the same as those given in Figure 9.44, except for $W_{2b} = 6.1$ mm and $L_{2b} = 3.2$ mm. By inspecting Figure 9.46b, two important observations can be found in association with the two physical design parameters t and l_t. First, as t is reduced from 5.15 to 4.15 mm, Q_{e1} increases, whereas Q_{e2} decreases. This suggests that reducing t reduces the I/O coupling for the first band, but increases the I/O coupling for the second. Second, as l_t is increased, Q_{e1} decreases, whereas Q_{e2} increases. This implies that for a longer l_t, the I/O coupling is stronger for the first band, but weaker for the second band. Thus, a combined design of t and l_t allows a more flexible realization of individual external I/O couplings desired for a dual-band filter.

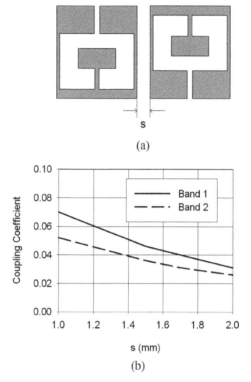

FIGURE 9.45 (a) An arrangement for coupled dual-band resonators. (b) Simulated coupling coefficients.

To demonstrate a design of dual-band filter, based the above discussed dual-band resonator, a three-pole lowpass Chebyshev prototype with $g_0 = g_4 = 1.0$, $g_1 = g_3 = 0.8516$, and $g_2 = 1.1032$ for a passband ripple of 0.04321 dB (refer to Table 3.2) are used for the design. The first band is to have a fractional bandwidth $FBW = 0.04$ centered at 0.9 GHz. The second desired band has a center frequency at 1.4 GHz with $FBW = 0.03$. Applying the design equations of (5.24), we obtain the bandpass design parameters, that is, the coupling coefficients M_{ij} and the external quality factors $Q_{ei/o}$ for each band, which are given as follows.

For the first band:

$$M_{1,2} = M_{2,3} = 0.041 \quad Q_{ei} = Q_{eo} = 21.29$$

For the second band:

$$M_{1,2} = M_{2,3} = 0.031 \quad Q_{ei} = Q_{eo} = 28.387$$

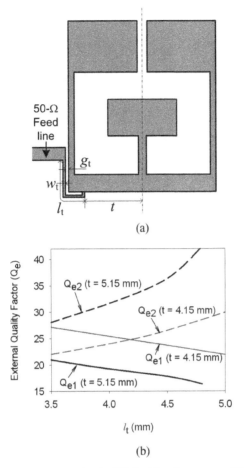

FIGURE 9.46 (a) An arrangement for the I/O dual-band resonator. (b) Simulated external quality factors for $w_t = g_t = 0.2$ mm.

These desired design parameters can be found from a set of design curves, such as those presented in Figures 9.45b and 9.46b, so that the filter dimensions are determined. The designed microstrip filter has a symmetrical structure with respect to the input and output ports, as shown in Figure 9.47a. To compensate for a frequency shift effect because of the I/O feeding arrangement, the central loading element for the I/O resonators is slightly different from that for the resonator in the middle. The designed filter is simulated with a full-wave EM simulator and the simulated performance is presented in Figure 9.47b. Each band shows a desired three-pole Chebyshev filtering characteristic with three transmission poles and good return loss response in the passband. It can be shown that the dual-band filter is spurious-free over a wide frequency range up to 2.83 GHz, which is more than twice the center frequency of the second band.

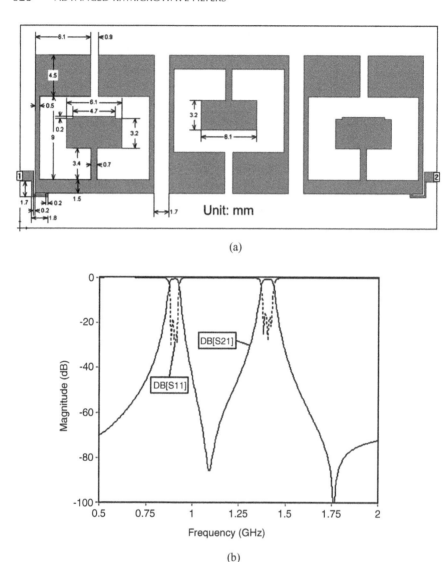

FIGURE 9.47 (a) Layout of a microstrip dual-band filter on a dielectric substrate with a relative dielectric constant of 10.8 and a thickness of 1.27 mm. The dimensions are in millimeters. (b) Simulated performance of the filter.

9.8.3 Filters Using Cross-Coupled Resonators

Dual-band or multiband filters can also be realized using cross-coupled resonators. In this approach, a wide passband is split into two or more subbands using transmission zero(s) [29,30]. In general, this type of multiband filter would be more suitable for a small separation between adjacent bands. For our demonstration, Figure 9.48a

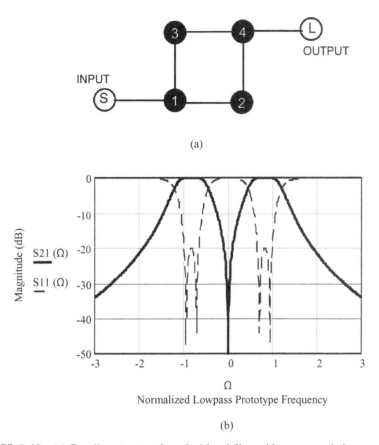

FIGURE 9.48 (a) Coupling structure for a dual-band filter with cross-coupled resonators. (b) Theoretical frequency response.

illustrates a coupling structure for such a dual-band filter, which has four cross-coupled resonators as numbered. S and L denote the source and load nodes, respectively. Because of the cross coupling, there exist two signal paths from the input to the output. The first path goes through resonators 1-2-4, and the second path follows resonators 1-3-4. Referring to Chapter 7, the $n+2$ general coupling matrix for this coupling structure is given in Eq. (9.64). For $m_{s1} = m_{L4} = 0.70$, $m_{12} = m_{34} = 0.86$, and $m_{13} = m_{24} = 0.272$, the scattering parameters against frequencies can be computed using the formulations described in Chapter 7 and the results are plotted in Figure 9.48b, where the frequency axis uses a frequency variable for the normalized lowpass prototype. It can clearly be seen that a transmission zero is introduced in the middle band (at $\Omega = 0$) to divide the ripple passband from $\Omega = -1$ to $\Omega = 1$ into two subbands. Figure 9.49a depicts the layout for a microstrip realization using four half-wavelength resonators, which has a rotation symmetrical structure. The EM-simulated performance of the dual-band microstrip filter is shown in Figure 9.49b,

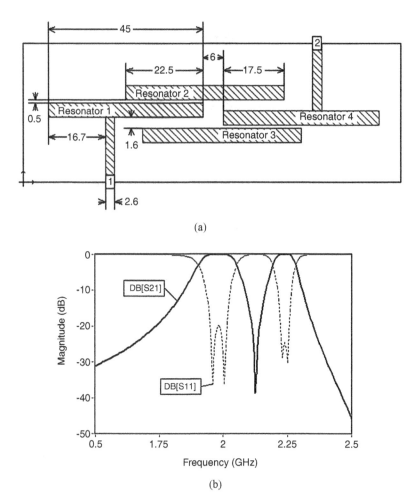

FIGURE 9.49 (a) A microstrip realization of dual-band filter with cross-coupled resonators. The dimensions are in millimeters on a 1.016-mm-thick dielectric substrate with a relative dielectric constant of 3.0. (b) EM-simulated performance.

where the frequency response is somehow asymmetrical because of some unwanted cross couplings.

$$[m] = \begin{bmatrix} 0 & m_{s1} & 0 & 0 & 0 & 0 \\ m_{s1} & 0 & m_{12} & m_{13} & 0 & 0 \\ 0 & m_{12} & 0 & 0 & m_{24} & 0 \\ 0 & m_{13} & 0 & 0 & m_{34} & 0 \\ 0 & 0 & m_{24} & m_{34} & 0 & m_{L4} \\ 0 & 0 & 0 & 0 & m_{L4} & 0 \end{bmatrix} \quad (9.64)$$

Another example of dual-band filter using cross-coupled resonators is shown in Figure 9.50a, where six open-loop microstrip resonators, as numbered, are cross

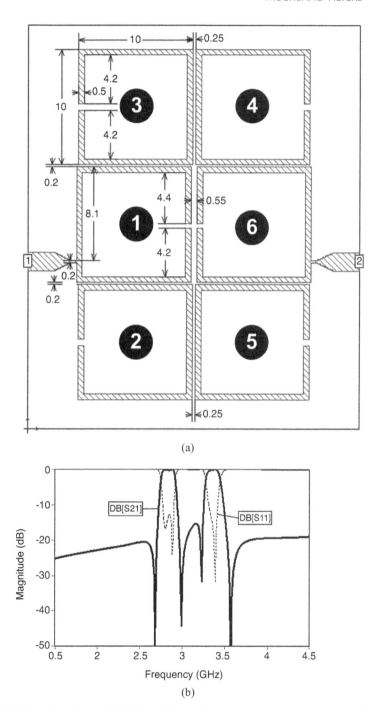

FIGURE 9.50 (a) Microstrip dual-band filter using cross-coupled resonators on a dielectric substrate with a relative dielectric constant of 2.2 and a thickness of 0.508 mm. The dimensions are in millimeters. (b) Simulated performance of the filter.

coupled. The input is located at resonator 1 and the output is from resonator 6. Again, multiple signal paths exist from the input to the output. In this case, the first signal path follows resonators 1-3-4-6 and the second one follows resonators 1-2-5-6. In addition, there is a cross coupling between resonators 1 and 6. As a result of combination of these signal paths, multiple transmission zeros can be generated and some are deployed to form dual bands, as demonstrated in Figure 9.50b.

REFERENCES

[1] R. M. Kurzok, General four-resonator filters at microwave frequencies, *IEEE Trans.* **MTT-14**, 1996, 295–296.

[2] R. Levy, Filters with single transmission zeros at real and imaginary frequencies, *IEEE Trans.* **MTT-24**, 1976, 172–181.

[3] J.-S. Hong, M. J. Lancaster, R. B. Greed, and D. Jedamzik, Highly selective microstrip bandpass filters for HTS and other applications. In 28th *European Microwave conference*, October 1998, Amsterdam, The Netherlands; 1998.

[4] J.-S. Hong and M. J. Lancaster, Design of highly selective microstrip bandpass filters with a single pair of attenuation poles at finite frequencies *IEEE Trans.* **MTT-48**, 2000, 1098–1107.

[5] R. J. Cameron and J. D. Rhodes, Asymmetric realization for dual-mode bandpass filters, *IEEE Trans.* **MTT-29**, 1981, 51–58.

[6] R. Levy, Direct synthesis of cascaded quadruplet (CQ) filters, *IEEE Trans.* **MTT-43**, 1995, 2940–2944.

[7] *EM User's Manual*. Sonnet Software, Inc. Liverpool, New York, 1996.

[8] R. M. Kurzrok, General three-resonator filters in waveguide, *IEEE Trans.* **MTT-14**, 1966, 46–47.

[9] L. F. Franti and G. M. Paganuzzi, Odd-degree pseudoelliptical phase-equalized filter with asymmetric bandpass behavior. In *Proceedings of European Microwave conference*, Amsterdam, Sept.; 1981. p111–116.

[10] R. J. Cameron, Dual-mode realization for asymmetric filter characteristics. *ESA J.* **6**, 1982, 339–356.

[11] R. Hershtig, R. Levy, and K. Zaki, Synthesis and design of cascaded trisection (CT) dielectric resonator filters. In *Proceedings of European Microwave conference*, Jerusalem, Sept.; 1997. p784–791.

[12] R. R. Mansour, F. Rammo, and V. Dokas, Design of hybrid-coupled multiplexers and diplexers using asymmetrical superconducting filters, *IEEE MTT-S Dig.* 1993, 1281–1284.

[13] A. R. Brown and G. M. Rebeiz, A high-performance integrated K-band diplexer, *IEEE MTT-S Dig.* 1999, 1231–1234.

[14] J.-S. Hong and M. J. Lancaster, Microstrip cross-coupled trisection bandpass filters with asymmetric frequency characteristics. *IEE Proc Microwave Antennas Propag.* **146**(1): 1999, 84–90.

[15] J.-S. Hong and M. J. Lancaster, Microstrip triangular patch resonator filters. *IEEE MTT-S Dig.* 2000, 331–334.

[16] C.-C. Yang and C.-Y. Chang, Microstrip cascade trisection filter. *IEEE MGWL* 9, 1999, 271–273.

[17] J.-S. Hong and M. J. Lancaster, Transmission line filters with advanced filtering characteristics, *IEEE MTT-S, Dig.* 2000, 319–322.

[18] J. D. Rhodes, A lowpass prototype network for microwave linear phase filters, *IEEE Trans.* **MTT-18**, 1970, 290–301.

[19] R. J. Wenzel, Solving the approximation problem for narrow-band bandpass filters with equal-ripple passband responses and arbitrary phase responses. *IEEE MTT-S Dig.* 1975, 50.

[20] J. D. Rhodes and R. J. Cameron, General extracted pole synthesis technique with applications to low-loss TE_{011} mode filters, *IEEE Trans.* **MTT-28**, 1980, 1018–1028.

[21] A. E. Williams and A. E. Atia, Dual-mode canonical waveguide filters, *IEEE Trans.* **MTT-25**, 1977, 1021–1026.

[22] G. Pfitzenmaier, Synthesis and realization of narrow-band canonical microwave bandpass filters exhibiting linear phase and transmission zeros, *IEEE Trans.* **MTT-30**, 1982, 1300–1310.

[23] A. E. Atia and A. E. Williams, Narrow bandpass waveguide filters, *IEEE Trans.* **MTT-20**, 1972, 258–265.

[24] A. E. Atia, A. E. Williams, and R. W. Newcomb, Narrow-band multiple coupled cavity synthesis, *IEEE Trans.* **CAS-21**, 1974, 649–655.

[25] J.-S. Hong and M. J. Lancaster, Canonical microstrip filter using square open-loop resonators, *Electron. Lett.* **31**(23), 1995, 2020–2022.

[26] K. T. Jokela, Narrow-band stripline or microstrip filters with transmission zeros at real and imaginary frequencies, *IEEE Trans.* **MTT-28**, 1980, 542–547.

[27] C.-F. Chen, T.-Y. Huang, and R.-B. Wu, Design of dual- and triple-passband filters using alternately cascaded multiband resonators, *IEEE Trans.* **MTT-54**, 2006, 3550–3558.

[28] J.-S. Hong and W. Tang, Dual-band filter based on non-degenerate dual-mode slow-wave open-loop resonators, *IEEE MTT-S Dig.* 2009, 861–864.

[29] S. Al-otaibi, Z.-C. Hao, J.-S. Hong, Multilayer folded-waveguide dual-band filter, *IEEE MTT-S Dig.* 2008, 993–996.

[30] M. Mokhtaari, J. Bornemann, K. Rambabu, and S. Amari, Coupling-matrix design of dual and triple passband filters, *IEEE Trans.* **MTT-54**, 2006, 3940–3946.

CHAPTER TEN

Compact Filters and Filter Miniaturization

Microstrip filters are already small in size compared with other filters, such as waveguide filters. Nevertheless, for some applications where the size reduction is of primary importance, smaller microstrip filters are desirable, even though reducing the size of a filter generally leads to an increase in dissipation losses in a given material and, consequently, reduced performance. Miniaturization of microstrip filters may be achieved by using high dielectric constant substrates or lumped elements. However, very often, for specified substrates, a change in the geometry of filters is required and, therefore, numerous new filter configurations become possible [1]. This chapter is intended to describe novel concepts, methodologies, and designs for compact filters and filter miniaturization. The new types of filters discussed include: compact open-loop and hairpin resonator filters, slow-wave resonator filters, miniaturized dual-mode filters, lumped-element filters, filters using high dielectric constant substrates, and multilayer filters, including those based on low-temperature co-fired ceramic (LTCC) and liquid crystal polymer (LCP) packaging materials.

10.1 MINIATURE OPEN-LOOP AND HAIRPIN RESONATOR FILTERS

In the last chapter we introduced a class of microstrip open-loop resonator filters. To miniaturize this type of filter, one can use so-called meander open-loop resonators [2]. For demonstrate a compact microstrip filter of this type, with a fractional bandwidth of 2% at a midband frequency of 1.47 GHz, has been designed on a RT/Duroid substrate, which has a thickness of 1.27 mm and a relative dielectric constant of 10.8. Figure 10.1 illustrates the layout and the EM-simulated performance of the filter.

Microstrip Filters for RF/Microwave Applications by Jia-Sheng Hong
Copyright © 2011 John Wiley & Sons, Inc.

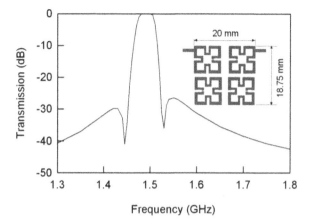

FIGURE 10.1 Layout and simulated performance of a miniature microstrip four-pole elliptic-function filter on a substrate with a relative dielectric constant of 10.8 and a thickness of 1.27 mm.

This filter structure is for realizing an elliptic-function response, constructed from four microstrip meander open-loop resonators (although more resonators may be implemented). Each meander open-loop resonators has a size smaller than $\lambda_{g0}/8 \times \lambda_{g0}/8$, where λ_{g0} is the guided wavelength at the midband frequency. Therefore, to fabricate the filter in Figure 10.1, the required circuit size only amounts to $\lambda_{g0}/4 \times \lambda_{g0}/4$. In this case, the whole size of the filter is 20.0×18.75 mm, which is just about $\lambda_{g0}/4 \times \lambda_{g0}/4$ on the substrate, as expected. This size is quite compact for distributed parameter filters. The filter transmission response exhibits two attenuation poles at finite frequencies, which is a typical characteristic of the elliptic-function filters.

A small size and high-performance eight-pole high temperature superconducting (HTS) filter of this type has also been developed for mobile communication applications [3]. The filter is designed to have a quasielliptic function response with a passband from 1710 to 1785 MHz, which covers the whole receive-band of digital communications system DCS1800. To reduce the cost, it is designed on a 0.33-mm thick r-plane sapphire substrate using an effective isotropic dielectric constant of 10.0556 [4]. Figure 10.2 is the layout of the filter, which consists of eight meander open-loop resonators in order to fit the entire filter onto a specified substrate size of 39×22.5 mm. Although each HTS microstrip meander open-loop resonator has a size only amounting to 7.4×5.4 mm, its unloaded quality factor is over 5×10^4 at a temperature of 60K. The orientations of resonators not only allow meeting the required coupling structure for the filter, but also allow each resonator to experience the same permittivity tensor. This means that the frequency shift of each resonator because of the anisotropic permittivity of sapphire substrate is the same, which is very important for the synchronously tuned narrow-band filter. The HTS microstrip filter is fabricated using 330-nm thick YBCO thin film, which has a critical temperature $T_c = 87.7K$. The fabricated HTS filter is assembled into a test housing for measurement, as shown in Figure 10.3a. Figure 10.3b plots experimental results of the

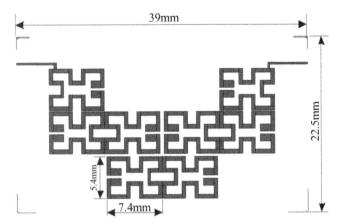

FIGURE 10.2 Layout of eight-pole HTS quasi-elliptic function filter using miniature microstrip open-loop resonators on a 0.33-mm thick sapphire substrate.

superconducting filter, measured at a temperature of 60K and without any tuning. The filter shows the characteristics of the quasielliptical response with two diminishing transmission zeros near the passband edges, resulting in a sharper filter shirt to improve the filter selectivity. The filter also exhibits very low insertion loss in the passband because of the high unloaded quality factor of the resonators.

In a similar fashion, a conventional hairpin resonator in Figure 10.4a may be miniaturized by loading a lumped-element capacitor between the both ends of the resonator, as indicated in Figure 10.4b, or alternatively with a pair of coupled lines folded inside the resonator as shown in Figure 10.4c [5]. It has been demonstrated that the size of a three-pole miniaturized hairpin resonator filter is reduced to one-half that of the conventional one, and the miniature filters of this type have found the application to receiver front-end MIC's [5]. In addition, using interwound hairpin resonators, the so-called pseudointerdigital filter with a compact size has been developed [6–9].

10.2 SLOW-WAVE RESONATOR FILTERS

In general, the size of a microwave filter is proportional to the guided wavelength at which it operates. Since the guided wavelength is proportional to the phase velocity, reducing the phase velocity or obtaining slow-wave propagation can then lead to the size reduction. This follows the development of ladder microstrip line resonators and filters [10–14]. On the other hand, in order to reduce interference by keeping out-of-band signals from reaching a sensitive receiver. A wider upper stopband, including $2f_0$, where f_0 is the midband frequency of a bandpass filter, may also be required. However, many planar bandpass filters that are comprised of half-wavelength resonators inherently have a spurious passband at $2f_0$. A cascaded lowpass filter or

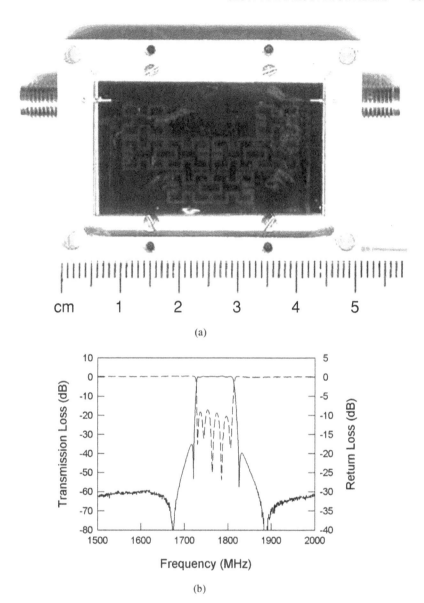

FIGURE 10.3 (a) Photograph of the fabricated HTS filter in test housing. (b) Measured performance of the filter at a temperature of 60K.

bandstop filter may be used to suppress the spurious passband at a cost of extra insertion loss and size. Although quarter-wavelength resonator filters have the first spurious passband at $3f_0$, they require short-circuit (grounding) connections with via-holes, which is not quite compatible with planar fabrication techniques. Lumped-element filters ideally do not have any spurious passband at all, but suffer from higher

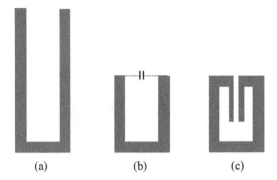

FIGURE 10.4 Structural variations to miniaturize hairpin resonator. (**a**) Conventional hairpin resonator. (**b**) Miniaturized hairpin resonator with loaded lumped capacitor. (**c**) Miniaturized hairpin resonator with folded coupled lines.

loss and poorer power handling capability. Bandpass filters using stepped-impedance [15], or slow-wave resonators, such as end-coupled slow-wave [16] and slow-wave open-loop resonators [17,18], are able to control spurious response with a compact filter size, because of the effects of a slow wave. A general and comprehensive circuit theory for these types of slow-wave resonator is treated next before introducing the filters.

10.2.1 Capacitively Loaded Transmission-Line Resonator

For our purpose, first, let us consider a capacitively loaded lossless transmission-line resonator of Figure 10.5, where C_L is the loaded capacitance and Z_a, β_a, and d are the characteristic impedance, the propagation constant, and the length of the unloaded line, respectively. Thus the electrical length is $\theta_a = \beta_a d$. The circuit response of Figure 10.5 may be described by

$$\begin{bmatrix} V_1 \\ I_1 \end{bmatrix} = \begin{bmatrix} A & B \\ C & D \end{bmatrix} \cdot \begin{bmatrix} V_2 \\ -I_2 \end{bmatrix} \qquad (10.1)$$

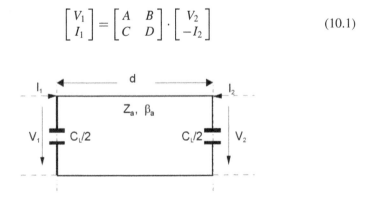

FIGURE 10.5 Capacitively loaded transmission line resonator.

with

$$A = D = \cos\theta_a - \frac{1}{2}\omega C_L Z_a \sin\theta_a \quad (10.2a)$$

$$B = jZ_a \sin\theta_a \quad (10.2b)$$

$$C = j\left(\omega C_L \cos\theta_a + \frac{1}{Z_a}\sin\theta_a - \frac{1}{4}\omega^2 C_L^2 Z_a \sin\theta_a\right) \quad (10.2c)$$

where $\omega = 2\pi f$ is the angular frequency; A, B, C, and D are the network parameters of the transmission matrix, which also satisfy the reciprocal condition $AD - BC = 1$.

Assume that a standing wave has been excited subject to the boundary conditions $I_1 = I_2 = 0$. For no vanished V_1 and V_2, it is required that

$$\frac{C}{A} = \left.\frac{I_1}{V_1}\right|_{I_2=0} = \left.\frac{I_2}{V_2}\right|_{I_1=0} = 0 \quad (10.3)$$

Because

$$A = \left.\frac{V_1}{V_2}\right|_{I_2=0} = \begin{cases} -1 & \text{for the fundamental resonance} \\ 1 & \text{for the first spurious resonance} \end{cases} \quad (10.4)$$

we have from Eq. (10.2a) that

$$\cos\theta_{a0} - \frac{1}{2}\omega_0 C_L Z_a \sin\theta_{a0} = -1 \quad (10.5a)$$

$$\cos\theta_{a1} - \frac{1}{2}\omega_1 C_L Z_a \sin\theta_{a1} = 1 \quad (10.5b)$$

where the subscripts 0 and 1 indicate the parameters associated with the fundamental and the first spurious resonance, respectively. Substituting Eqs. (10.5a) and (10.5b) into Eq. (10.2c), and letting $C = 0$ according to Eq. (10.3), yields

$$\frac{\omega_0 C_L}{2}(1 - \cos\theta_{a0}) = \frac{1}{Z_a}\sin\theta_{a0} \quad (10.6a)$$

$$\frac{\omega_1 C_L}{2}(1 + \cos\theta_{a1}) = -\frac{1}{Z_a}\sin\theta_{a1} \quad (10.6b)$$

These two eigenequations can further be expressed as

$$\theta_{a0} = 2\tan^{-1}\left(\frac{1}{\pi f_0 Z_a C_L}\right) \quad (10.7a)$$

$$\theta_{a1} = 2\pi - 2\tan^{-1}(\pi f_1 Z_a C_L) \quad (10.7b)$$

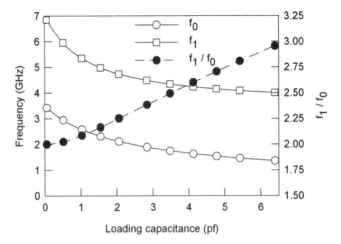

FIGURE 10.6 According to a circuit model, figure shows fundamental and first spurious resonant frequencies of a capacitively loaded transmission-line resonator, as well as their ratio against loading capacitance.

from which the fundamental resonant frequency f_0 and the first spurious resonant frequency f_1 can be determined. Now it can clearly be seen from Eqs. (10.7a) and (10.7b) that $\theta_{a0} = \pi$ and $\theta_{a1} = 2\pi$, when $C_L = 0$. This is the case for the unloaded half-wavelength resonator. For $C_L \neq 0$, it can be shown that the resonant frequencies are shifted down as the loading capacitance is increased, indicating the slow-wave effect. For a demonstration, Figure 10.6 plots the calculated resonant frequencies according to Eqs, (10.7a) and (10.7b), as well as their ratio for different capacitance loading when $Z_a = 52\ \Omega$, $d = 16$ mm and the associated phase velocity $v_{pa} = 1.1162 \times 10^8$ m/s. As can be seen when the loading capacitance is increased, in addition to the decrease of both resonant frequencies, the ratio of the first spurious resonant frequency to the fundamental one is increased. To understand the physical mechanism that underlines this phenomenon, which is important for our applications, we may consider the circuit of Figure 10.5 as a unit cell of a periodically loaded transmission line. This is plausible, as we may mathematically expand a function defined in a bounded region into a periodic function. Let β be the propagation constant of the capacitively loaded lossless periodic transmission line. Applying Floquet's theorem [21], i.e.,

$$V_2 = e^{-j\beta d} V_1$$
$$-I_2 = e^{-j\beta d} I_1 \quad (10.8)$$

to Eq. (10.1) results in

$$\begin{bmatrix} A - e^{j\beta d} & B \\ C & D - e^{j\beta d} \end{bmatrix} \cdot \begin{bmatrix} V_2 \\ -I_2 \end{bmatrix} = \begin{bmatrix} 0 \\ 0 \end{bmatrix} \quad (10.9)$$

A nontrivial solution for V_2, I_2 exists only if the determinant vanishes. Hence,

$$(A - e^{j\beta d})(D - e^{j\beta d}) - BC = 0 \tag{10.10}$$

Since $A=D$ for the symmetry and $AD\text{-}BC = 1$ for the reciprocity, the dispersion equation of (10.10) becomes

$$\cos(\beta d) = \cos\theta_a - \frac{1}{2}\omega C_L Z_a \sin\theta_a \tag{10.11}$$

according to Eqs. (10.2a–10.2c).

Because the dispersion equation governs the wave propagation characteristics of the loaded line, we can substitute Eqs. (10.7a) and (10.7b) into Eq. (10.11) for those particular frequencies. It turns out that $\cos(\beta_0 d) = -1$ for the fundamental resonant frequency and $\cos(\beta_1 d) = 1$ for the first spurious resonant frequency. As $\beta_0 = \omega_0/v_{p0}$ and $\beta_1 = \omega_1/v_{p1}$, where v_{p0} and v_{p1} are the phase velocities of the loaded line at the fundamental and the first spurious resonant frequencies, respectively, we obtain

$$\frac{f_1}{f_0} = 2\frac{v_{p1}}{v_{p0}} \tag{10.12}$$

If there were no dispersion, the phase velocity would be a constant. This is only true for the unloaded line. However, for the periodically loaded line, the phase velocity is frequency-dependent. It would seem that, in our case, the increase in ratio of the first spurious resonant frequency to the fundamental one when the capacitive loading is increased would be attributed to the increase of the dispersion. By plotting dispersion curves according to Eq. (10.11), it can be clearly shown that the dispersion effect indeed accounts for the increase in ratio of the first spurious resonant frequency to the fundamental one [18]. Therefore, this property can be used to design a bandpass filter with a wider upper stopband. It is obvious that, based on the circuit model of Figure 10.5, different resonator configurations may be realized [15–20]. Microstrip filters developed with two different types of slow-wave resonators are described in following sections.

10.2.2 End-Coupled Slow-Wave Resonators Filters

Figure 10.7a illustrates a symmetrical microstrip slow-wave resonator, which is composed of a microstrip line with both ends loaded with a pair of folded open stubs. Assume that the open stubs are shorter than a quarter-wavelength at the frequency considered and the loading is capacitive. The equivalent circuit as shown in Figure 10.5 can then represent the resonator.

To demonstrate the characteristics of this type of slow-wave resonator, a single resonator was first designed and fabricated on a RT/Duroid substrate having a thickness $h = 1.27$ mm and a relative dielectric constant of 10.8. The resonator has dimensions, referring to Figure 10.7a, of $a = b = 12.0$ mm, $w_1 = w_2 = 3.0$ mm, and

342 COMPACT FILTERS AND FILTER MINIATURIZATION

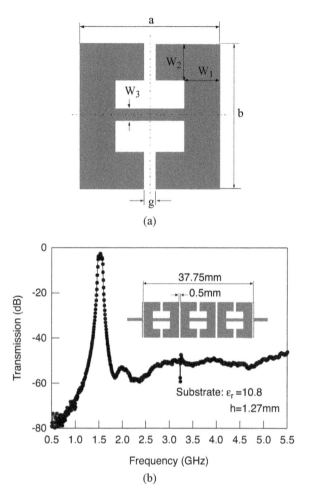

FIGURE 10.7 (a) A microstrip slow-wave resonator. (b) Layout and measured frequency response of end-coupled microstrip slow-wave resonator bandpass filter.

$w_3 = g = 1.0$ mm. The measured frequency response shows the fundamental resonance occurs at $f_0 = 1.54$ GHz and no spurious resonance is observed for frequency even up to $3.5 f_0$. A three-pole bandpass filter that consists of three end-coupled above resonators was then designed and fabricated. The layout and the measured performance of the filter are shown in Figure 10.7b. The size of the filter is 37.75×12 mm. The longitudinal dimension is even smaller than half-wavelength of a 50-Ω line on the same substrate. The filter has a fractional bandwidth of 5% at a midband frequency 1.53 GHz and a wider upper stopband up to 5.5 GHz, which is about 3.5 times of the midband frequency. It is also interesting to note that there is a very sharp notch, like an attenuation pole, loaded at about $2f_0$ in the responses, shown in Figure 10.7b.

10.2.3 Slow-Wave, Open-Loop Resonator Filters

10.2.3.1 Slow-Wave, Open-Loop Resonator A so-called microstrip slow-wave, open-loop resonator, which is composed of a microstrip line with both ends loaded with folded open stubs, is illustrated in Figure 10.8a. The folded arms of open stubs are not only for increasing the loading capacitance to ground, as referred to in Figure 10.5, but also for the purpose of producing interstage or cross couplings. Shown in Figure 10.8b are the fundamental and first spurious resonant frequencies as well as their ratio against the length of folded open stub, obtained using a full-wave EM simulator [22]. Note that in this case the length of folded open stub is defined as $L = L_1$ for $L \leq 5.5$ mm and $L = 5.5 + L_2$ for $L > 5.5$ mm, as referred to in Figure 10.8a. One might notice that the results obtained by the full-wave EM simulation bear close similarity to those obtained by circuit theory, as shown in Figure 10.6. This

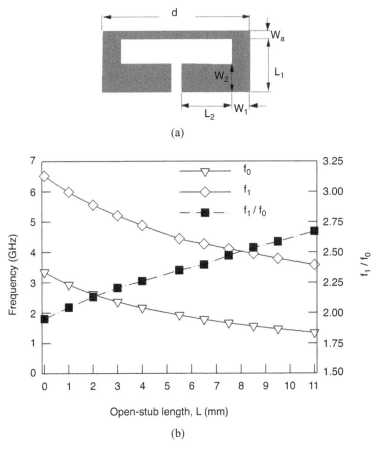

FIGURE 10.8 (a) A microstrip slow-wave, open-loop resonator. (b) Full-wave EM simulated fundamental and first spurious resonant frequencies of a microstrip slow-wave, open-loop resonator, as well as their ratio against the loading open stub.

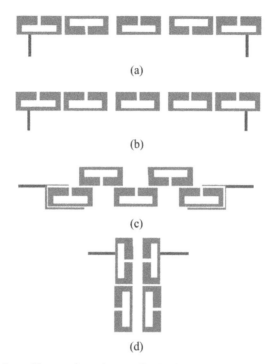

FIGURE 10.9 Some filter configurations realized using microstrip slow-wave, open-loop resonators.

is what would be expected because, in this case, the unloaded microstrip line, which has a length of $d = 16$ mm and a width of $w_a = 1.0$ mm on a substrate with a relative dielectric constant of 10.8 and a thickness of 1.27 mm, exhibits about the same parameters of Z_a and v_{pa} as those assumed in Figure 10.6, and the open stub approximates the lumped capacitor. At this stage, it may be worthwhile to point out that to approximate the lumped capacitor, it is essential that the open stub should have a wider line or lower characteristic impedance. In this case, by referring to Figure 10.8a, we have $w_1 = 2.0$ mm and $w_2 = 3.0$ mm for the folded open-stub. It should be mentioned that the slow-wave, open-loop resonator differs from the miniaturized hairpin resonator primarily in that they are developed from rather different concepts and purposes. The latter is developed from conventional hairpin resonator by increasing capacitance between both ends to reduce the size of the conventional hairpin resonator, as discussed in the last section. The main advantage of microstrip slow-wave open-loop resonator of Figure 10.8a over the previous one is that various filter structures (see Figure 10.9) would be more easily constructed, including cross-coupled resonator filters, which exhibit elliptic- or quasielliptic-function response.

10.2.3.2 Five-Pole Direct Coupled Filter For our demonstration, we will focus on two examples of narrowband, microstrip, slow-wave open-loop resonator filters.

The first one is five-pole direct coupled filter with overlapped coupled slow-wave, open-loop resonators, as shown in Figure 10.9c. This filter was developed to meet the following specifications for an instrumentation application:

Center Frequency	1335 MHz
3-dB Bandwidth	30 MHz
Passband Loss	3 dB Max
Minimum stopband rejection	dc to 1253 MHz, 60 dB
	1457 − 2650 MHz, 60 dB
	2650 −3100 MHz, 30 dB
60-dB Bandwidth	200 MHz Max

As can be seen, a wide upper stopband including $2f_0$ is required and at least 30-dB rejection at $2f_0$ is needed.

The bandpass filter was designed to have a Chebyshev response and the design parameters, such as the coupling coefficients and the external quality factor Q_e, could be synthesized from a standard Chebyshev lowpass prototype filter. Considering the effect of conductor loss, that is, the narrower the bandwidth, the higher the insertion loss, which is even higher at the passband edges because the group delay is usually longer at the passband edges. The filter was then designed with a slightly wider bandwidth, trying to meet the 3-dB bandwidth of 30 MHz, as specified. The resultant design parameters are

$$M_{12} = M_{45} = 0.0339$$
$$M_{23} = M_{34} = 0.0235$$
$$Q_e = 22.4382$$

The next step of the filter design was to characterize the couplings between adjacent microstrip slow-wave, open-loop resonators as well as the external quality factor of the input or output microstrip slow-wave, open-loop resonator. The techniques described in Chapter 7 were used to extract these design parameters with the aid of full-wave EM simulations. Figure 10.10a depicts the extracted coupling coefficient against different overlapped length d for a fixed coupling gap s, where the size of the resonator is 16 × 6.5 mm on a substrate with a relative dielectric constant of 10.8 and a thickness of 1.27 mm. One can see that the coupling increases almost linearly with the overlapped length. It can also be shown that for a fixed d, reducing or increasing coupling gap s increases or decreases the coupling. From the filter configuration of Figure 10.9c, one might expect the cross coupling between nonadjacent resonators. It has been found that the cross coupling between nonadjacent resonators is quite small when the separation between them is larger than 2 mm, as shown by Figure 10.10b. This, however, suggests that the filter structure in Figure 10.9b would be more suitable for very narrow band realization that requires very weak coupling between resonators. The filter was then fabricated on an RT/Duroid substrate. Figure 10.11a shows a photograph of the fabricated filter. The size of this five-pole filter is about $0.70 \times 0.15\ \lambda_{go}$, where λ_{go} is the guided wavelength of a 50-Ω line on the

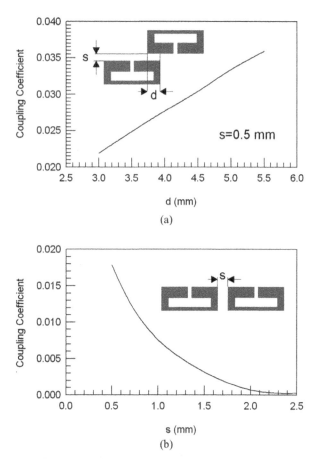

FIGURE 10.10 Modeled coupling coefficients of (a) overlapped coupled and (b) end-coupled slow-wave, open-loop resonators.

substrate at the midband frequency. Figure 10.11b shows experimental results, which represent the first design iteration. The filter had a midband loss less than 3 dB and exhibited the excellent stopband rejection. It can be seen that more than 50-dB rejection at $2f_o$ has been achieved.

10.2.3.3 Four-Pole Cross-Coupled Filter The second trial microstrip slow-wave, open-loop resonator filter is that of four-pole cross-coupled filter, as illustrated in Figure 10.9d. The design parameters are listed below

$$Q_e = 26.975$$
$$M_{12} = M_{34} = 0.0297$$
$$M_{23} = 0.0241$$
$$M_{14} = -0.003$$

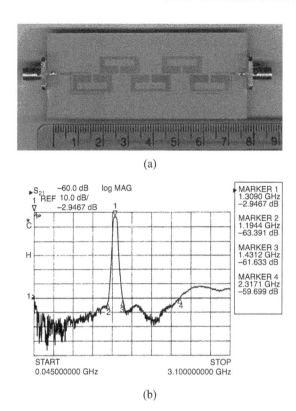

FIGURE 10.11 (a) Photograph of the fabricated five-pole bandpass filter using microstrip slow-wave, open-loop resonators. (b) Measured performance of the filter.

Similarly, the coupling coefficients of three basic coupling structures encountered in this filter were modeled using the techniques described in Chapter 7. The results are depicted in Figure 10.12. Notice that the mixed and magnetic couplings are used to realize $M_{12} = M_{34}$ and M_{23}, respectively, whereas the electric coupling is used to achieve the cross-coupling M_{14}. The tapped line input or output was used in this case and the associated external Q could be characterized by the method mentioned before. The filter was designed and fabricated on a RT/Duroid 6010 substrate with a relative dielectric constant of 10.8 and a thickness of 1.27 mm. Figure 10.13a shows a photograph of the fabricated four-pole cross-coupled filter. In this case, the size of the filter amounts only to 0.18×0.36 λ_{go}. The measured filter performance is illustrated in Figure 10.13b. The measured 3-dB bandwidth is about 4% at 1.3 GHz. The minimum passband loss was approximately 2.7 dB. The filter exhibited a wide upper stopband with a rejection better than 40 dB up to about 3.4 GHz. The two transmission zeros, which are the typical elliptic-function response, can also clearly be observed. However, the locations of transmission zeros are asymmetrical. It has been shown that this mainly resulted from a frequency-dependent cross coupling in this filter example [18].

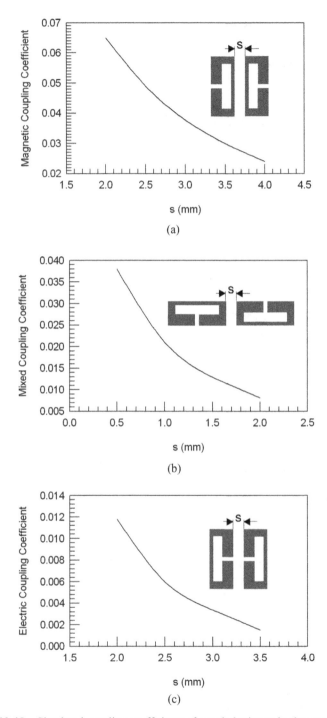

FIGURE 10.12 Simulated coupling coefficients of coupled microstrip slow-wave, open-loop resonators. (**a**) Magnetic coupling; (**b**) mixed coupling; (**c**) electric coupling.

(a)

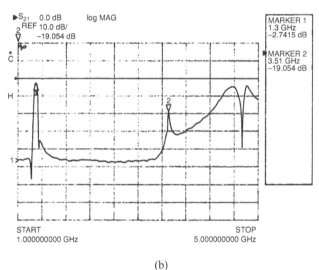

(b)

FIGURE 10.13 (a) Photograph of the fabricated four-pole bandpass filter using microstrip slow-wave open-loop resonators. (b) Measured performance of the filter.

10.3 MINIATURE DUAL-MODE RESONATOR FILTERS

Dual-mode resonators have been widely used to realize many RF/microwave filters [23–37]. A main feature and advantage of this type of resonators lies in the fact that each of dual-mode resonators can used as a doubly tuned resonant circuit, and therefore the number of resonators required for a n-degree filter is reduced by one-half, resulting a compact filter configuration.

10.3.1 Microstrip Dual-Mode Resonators

For our discussion, let us consider a microstrip square patch resonator represented by a Wheeler's cavity model [38], as illustrated by Figure 10.14a, where the top and bottom of the cavity are the perfect electric walls and the remainder sides are the perfect magnetic walls. The EM fields inside the cavity can be expanded in terms of TM_{mn0}^z modes:

$$E_z = \sum_{m=0}^{\infty} \sum_{n=0}^{\infty} A_{mn} \cos\left(\frac{m\pi}{a}x\right) \cos\left(\frac{n\pi}{a}y\right)$$

$$H_x = \left(\frac{j\omega\varepsilon_{eff}}{k_c^2}\right)\left(\frac{\partial E_z}{\partial y}\right), H_y = -\left(\frac{j\omega\varepsilon_{eff}}{k_c^2}\right)\left(\frac{\partial E_z}{\partial x}\right) \quad (10.13)$$

$$k_c^2 = \left(\frac{m\pi}{a}\right)^2 + \left(\frac{n\pi}{a}\right)^2$$

where A_{mn} represents the mode amplitude, ω is the angular frequency, a and ε_{eff} are the effective width and permittivity [38]. The resonant frequency of the cavity is given by

$$f_{mn0} = \frac{1}{2\pi\sqrt{\mu\varepsilon_{eff}}}\sqrt{\left(\frac{m\pi}{a}\right)^2 + \left(\frac{n\pi}{a}\right)^2} \quad (10.14)$$

Note that there is an infinite number of resonant frequencies corresponding to different field distributions or modes. The modes that have the same resonant

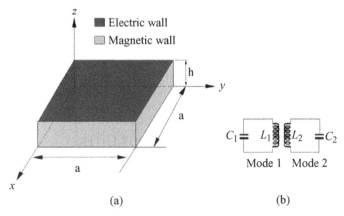

FIGURE 10.14 (a) Cavity model of a dual-mode microstrip resonator. (b) Equivalent circuit of the dual-mode resonator.

frequency are called the degenerate modes. Therefore, the two fundamental modes, that is, TM_{100}^z and TM_{010}^z modes, are a pair of the degenerate modes because

$$f_{100} = f_{010} = \frac{1}{2a\sqrt{\mu\varepsilon_{eff}}} \qquad (10.15)$$

Also note from Eq. (10.13) that the field distributions of these two modes are orthogonal to each other. In order to couple them, some perturbation to the symmetry of the cavity is needed and the two coupled degenerate modes function as two coupled resonators, as depicted in Figure 10.14b.

A microstrip dual-mode resonator is not necessarily square in shape, but usually has a two-dimensional (2D) symmetry. Figure 10.15 shows some typical microstrip dual-mode resonators, where D above each resonator indicates its symmetrical dimension and λ_{g0} is the guided-wavelength at its fundamental resonant frequency in the associated resonator. Note that a small perturbation has been applied to each dual-mode resonator at a location that is assumed at 45° offset from its two orthogonal modes. For instance, a small notch or a small cut is used to disturb the disk- and square-patch resonators, while a small patch is added to the ring, square-loop, and meander-loop resonators, respectively. It should be mentioned that for coupling of the orthogonal modes, the perturbations could also take forms other than

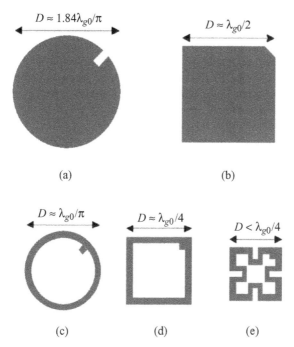

FIGURE 10.15 Some microstrip dual-mode resonators. (**a**) Circular disk; (**b**) square patch; (**c**) circular ring; (**d**) square loop; (**e**) meander loop.

TABLE 10.1 Experimental Microstrip Dual-Mode Resonators on Copper-Clapped RT/Duroid Substrate with a Relative Dielectric Constant of 10.8 and a Thickness of 1.27 mm

Resonator Type	Circuit Size $D \times D$ mm^2	Resonant Frequency (GHz)	Q (uncovered)	Q (covered)
Disk	33.0 × 33.0	1.568	84	246
Square patch	28.5 × 28.5	1.554	100	266
Ring	23.5 × 23.5	1.575	167	208
Square loop	20.5 × 20.5	1.558	161	214
Meander loop	16.0 × 16.0	1.588	186	219

those demonstrated in Figure 10.15. For examples, a small elliptical deformation of a circular patch or disk may be used for coupling the two degenerate modes and, similarly, a square patch may be distorted slightly into a rectangular shape for the coupling.

For comparison, a set of microstrip dual-mode resonators in Figure 10.15 were designed and fabricated on copper-clapped RT/Duroid substrate with a relative dielectric constant of 10.8 and a thickness of 1.27 mm. The line width for the ring and the square loop is 2.0 mm. The meander loop has a line width of 2.0 mm for its four corner arms and a line width of 1.5 mm for the inward meandered lines. Table 10.1 lists some important parameters and measured results. As can be seen, these resonators resonate at about the same fundamental frequency, but occupy different circuit sizes as measured by $D \times D$. The meander-loop resonator has the smallest size with the size reduction of 53%, 68%, and 76% against the ring, the square patch, and the disk, respectively. The quality factor, Q, of each resonator is given by two values measured with and without a copper cover, and the difference between the two would indicate the effect of radiation. In general, the smaller the microstrip resonator, the smaller the radiation loss, but the higher the conductor loss.

10.3.2 Miniaturized Dual-Mode Resonator Filters

Since each dual-mode resonator is equivalent to a doubly tuned resonant circuit, knowing the coupling coefficient between a pair of degenerate modes is essential for the filter design. The coupling coefficient can be extracted from the mode frequency split using the formulation described in Chapter 7 and the information for the mode frequency split may be obtained by EM simulation.

The simplest dual-mode filter is the two-pole bandpass filter using a single dual-mode resonator. To show this, a two-port bandpass filter composed of a dual-mode microstrip meander-loop resonator was designed and fabricated on a RT/Duroid substrate having a thickness of 1.27 mm and a relative dielectric constant of 10.8 [32]. Figure 10.16 shows the layout of the filter and its measured performance. As indicated, a small square patch of size $d \times d$ is attached to an inner corner of the loop for coupling

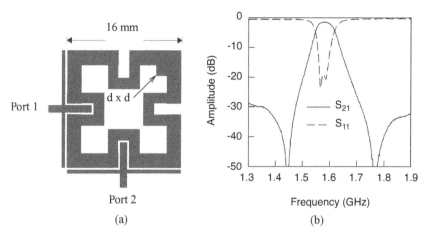

FIGURE 10.16 (a) Layout of a two-pole dual-mode microstrip filter on the 1.27-mm thick substrate with a relative dielectric constant of 10.8. (b) Measured performance of the filter.

a pair of degenerate modes. When $d = 0$, no perturbation is added and only single mode is excited by the either port. The simulated field pattern shows that the excited resonant mode is corresponding to the TM_{100}^z mode in a square-patch resonator when port 1 is excited. If the excitation port is changed to port 2, the field pattern is rotated by 90° for the associated degenerate mode, which corresponds to the TM_{010}^z mode in a square-patch resonator. When $d \neq 0$, no matter what the excitation port is, both the degenerate modes are excited and coupled to each other, which causes the resonance frequency splitting. The degree of coupling modes depends on the size of d, which, in return, controls the mode splitting. For this filter, $d = 2$ mm. The meander loop has a size of 16×16 mm, and is formed using two different line widths of 2 and 1.5 mm. The input/output is introduced by the cross branch having arm widths of 1 and 0.5 mm, respectively. All coupling gaps are 0.25 mm. The filter has a 2.5% fractional bandwidth at 1.58 GHz. The minimum insertion loss is 1.6 dB. This is mainly because of the conductor loss.

Figure 10.17a shows an example of miniaturized four-pole, microstrip dual-mode filter. The filter was designed to fit into a circuit size of 20×10 mm on a substrate with a relative dielectric constant $\varepsilon_r = 9.8$ and a thickness $h = 0.5$ mm. The filter is comprised of two meander-loop, dual-mode resonators, each of which has a size of 7×7 mm. The extracted coupling coefficients for the degenerate modes are plotted in Figure 10.17b, where the horizontal axis is the size of perturbation patch d, and is normalized by the substrate thickness of h. It is interesting to note that (1) the coupling coefficient is almost linearly proportional to the size of the normalized perturbation patch; (2) the coupling coefficient is almost independent of the relative dielectric constant of the substrate, so that the coupling is naturally the magnetic coupling. The filter was designed using direct EM simulation. The simulated frequency response of

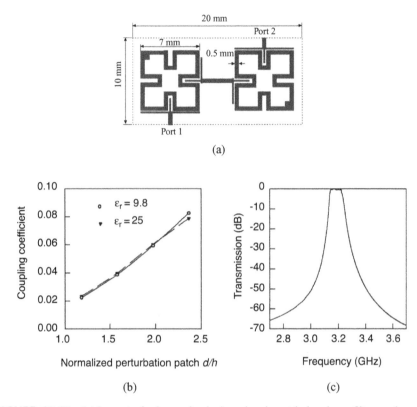

FIGURE 10.17 (a) Layout of a four-pole, dual-mode microstrip bandpass filter on the 0.5-mm thick substrate with a relative dielectric constant of 9.8. (b) Coupling coefficients of the degenerate modes. (c) Simulated frequency response of the filter.

the filter is shown in Figure 10.17c, exhibiting a fractional bandwidth of 2.2% at a center frequency of 3.185 GHz.

It should be pointed out that similar to the cross-coupled single-mode resonator filters, discussed in Chapter 9, by introducing cross coupling between nonadjacent modes of the dual-mode filters, more advanced filtering characteristics, such as elliptic- or quasielliptic- function response can be achieved [27,28,33].

To further miniaturize microstrip dual-mode filters, especially for applications at RF or lower microwave frequencies, a new type of microstrip fractal dual-mode resonator has been proposed [35]. Figure 10.18a shows the layout of a two-pole microstrip bandpass filter comprised of a fractal dual-mode resonator, so-called because its shape matures from a basic repetitive pattern. The filter was fabricated on a RT/Duroid substrate with $\varepsilon_r = 10.8$ and a thickness of 1.27 mm. The measured performance is plotted in Figure 10.18b, showing a midband frequency of 820 MHz. The performance of this filter is similar to what would be expected for the other types of microstrip dual-mode filters. The size of the filter is significantly reduced, as

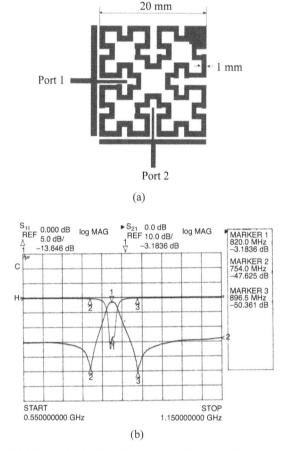

FIGURE 10.18 (a) Miniaturized 820-MHz bandpass filter with the fractal microstrip dual-mode resonator on the 1.27-mm thick substrate with a relative dielectric constant of 10.8. (b) Measured performance of the filter.

compared with a ring resonator on the same substrate and having the same resonant frequency, which gives a size reduction amounting to 80% [35].

10.3.3 Dual-Mode Triangular-Patch Resonator Filters

Two-dimensional patch dual-mode resonators, such as circular disks and square patches, as introduced previously, appear more attractive for filter applications where low loss and high-power handling are of primary concern. Furthermore, at millimeter waves, the size may not be the issue and the use of patch resonators can also ease the fabrication. In addition to the circular- and square-patch resonators, another useful patch resonator is the triangular-patch resonator. Single-mode operations of triangular-patch resonator have been reported [39–41]. This section will focus on dual-mode operations of triangular-patch resonators and their applications for designing dual-mode filters [42,43]. It will be shown that that the dual-mode triangular-patch resonator filter operates differently from the dual-mode square- or circular-patch resonator filter, which offers not only alternative designs, but also results in a compact size and simple coupling topology in a cascaded form.

10.3.3.1 Theoretical Formulation Figure 10.19 shows the geometry of an equilateral triangular-microstrip patch resonator on a dielectric substrate with a ground plane.

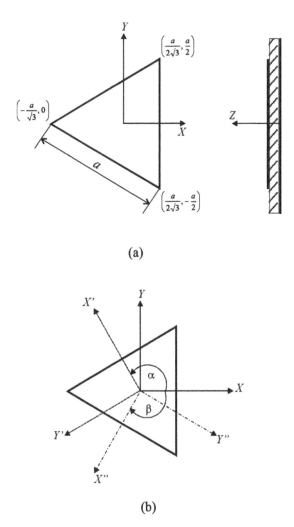

FIGURE 10.19 (a) Equilateral triangular microstrip patch geometry. (b) Rotated coordinate systems.

Similar to dealing with a square-patch microstrip resonator, a Wheeler's cavity mode can be used, where the top and bottom of the cavity are the perfect electric walls and the remaining sides are the perfect magnetic walls. One can then expand the electromagnetic (EM) fields inside the triangular cavity in terms of $TM^z_{m,n,l}$ modes [40],

$$E_z(x,y) = A_{m,n,l} \begin{Bmatrix} \cos\left[\left(\frac{2\pi x}{\sqrt{3}a} + \frac{2\pi}{3}\right)l\right]\cos\left[\frac{2\pi(m-n)y}{3a}\right] \\ + \cos\left[\left(\frac{2\pi x}{\sqrt{3}a} + \frac{2\pi}{3}\right)m\right]\cos\left[\frac{2\pi(n-l)y}{3a}\right] \\ + \cos\left[\left(\frac{2\pi x}{\sqrt{3}a} + \frac{2\pi}{3}\right)n\right]\cos\left[\frac{2\pi(l-m)y}{3a}\right] \end{Bmatrix} \quad (10.16)$$

$$H_x = \frac{j}{\omega\mu_0}\frac{\partial E_z}{\partial y}$$

$$H_y = \frac{-j}{\omega\mu_0}\frac{\partial E_z}{\partial x}$$

where $A_{m,n,l}$ is a constant and a is the length of the triangle side. Note that for the $TM^z_{m,n,l}$ modes $H_z = E_x = E_y = 0$. Unlike the square-patch resonator, the indexes m, n, l in Eq. (10.16) do not represent the number of standing waves along the x-, y-, z-axes of the coordinate system, and the condition of $m + n + l = 0$ must also be imposed to satisfy the wave equation. A fundamental mode, which can be found from Eq. (10.16), is the $TM^z_{1,0,-1}$ mode; its electric field is given by

$$E_z(x,y) = A_{1,0,-1}\left\{2\cos\left(\frac{2\pi x}{\sqrt{3}a} + \frac{2\pi}{3}\right)\cos\left(\frac{2\pi y}{3a}\right) + \cos\left(\frac{4\pi y}{3a}\right)\right\} \quad (10.17)$$

However, our intention is to find out another degenerate mode to pair with Eq. (10.17). It is known that $TM^z_{1,0,0}$ and $TM^z_{0,1,0}$ are a pair of degenerate modes of a square-patch resonator. A question then arises: By similarity, would the $TM^z_{0,1,-1}$ mode be another degenerate mode of an equilateral triangular-patch resonator? Unfortunately, the problem we are facing is not as simple as that and the answer to the question is "no". The reason is as follows. By inspection of Eq. (10.16), we notice that the interchange of the three indexes m,n,l in Eq. (10.16) leaves the field patterns unchanged. Hence, the EM fields for $TM^z_{0,1,-1}$ as well as $TM^z_{-1,0,1}$ and $TM^z_{0,-1,1}$ modes, which all have the same resonant frequency as that of the $TM^z_{1,0,-1}$ mode, are exactly identical to that given in Eq. (10.17). It is envisaged that Eq. (10.16) alone can not predict any degenerate modes, which have the same resonant frequency, but different field patterns to that given by Eq. (10.17). This is because the EM field

358   COMPACT FILTERS AND FILTER MINIATURIZATION

solutions of Eq. (10.16) are, as a matter of fact, not a complete set. To investigate degenerate modes theoretically, we have used the formulation below.

For our purpose, let us consider the dominant mode only and express the vector fields in the (x, y, z) coordinate system

$$\begin{aligned} \underline{E} &= E_z(x, y)\hat{z} \\ \underline{H} &= H_x(x, y)\hat{x} + H_y(x, y)\hat{y} \end{aligned} \qquad (10.18)$$

where $E_z(x, y)$ is given by Eq. (10.17) and all the magnetic field components can be derived from the electric field as indicated in Eq. (10.18). Noticing a rotation symmetry of the equilateral triangular patch resonator, the vector fields can also be expressed in the two rotated coordinate systems, that is, the (x', y', z) and (x'', y'', z) coordinate systems as shown in Figure 10.19b, respectively

$$\begin{aligned} \underline{E'} &= E'_z\left(x', y'\right)\hat{z} \\ \underline{H'} &= H'_x\left(x', y'\right)\hat{x}' + H'_y\left(x', y'\right)\hat{y}' \end{aligned} \qquad (10.19)$$

$$\begin{aligned} \underline{E''} &= E''_z\left(x'', y''\right)\hat{z} \\ \underline{H''} &= H''_x\left(x'', y''\right)\hat{x}'' + H''_y\left(x'', y''\right)\hat{y}'' \end{aligned} \qquad (10.20)$$

where $E'_z\left(x', y'\right)$ and $E''_z\left(x'', y''\right)$ take the same form as Eq. (10.17) in the associated coordinate systems. By far, neither set of the field solutions of Eqs. (10.18), (10.19), and (10.20) alone can predict any degenerate mode operation. However, if there is another degenerate mode other than Eq. (10.17) existing in the (x, y, z) coordinate system, it must result from, according the principle of superposition, a superposition of these fields. In this way, it can be found that

$$\begin{aligned} \underline{E'} &- \underline{E''} \\ \underline{H'} &- \underline{H''} \end{aligned} \qquad (10.21)$$

is, indeed, a field solution for the other fundamental degenerate mode.

In order to present this newly found degenerate mode in the (x, y, z) coordinate system, we need to first project the vector fields of Eq. (10.21) onto the (x, y, z) coordinate systems with the transformations:

$$\begin{aligned} \hat{x}' &= \hat{x}\cos(\alpha) + \hat{y}\sin(\alpha) & \hat{y}' &= -\hat{x}\sin(\alpha) + \hat{y}\cos(\alpha) \\ \hat{x}'' &= \hat{x}\cos(\beta) + \hat{y}\sin(\beta) & \hat{y}'' &= -\hat{x}\sin(\beta) + \hat{y}\cos(\beta) \end{aligned} \qquad (10.22)$$

where $\alpha = 2\pi/3$ and $\beta = -2\pi/3$ are the coordinate rotating angles, which are indicated in Figure 10.19b. Similarly, we also need to express x', y', x'' and y'' in terms of x, y using the similar coordinate transformations. The resultant electric field of the newly found degenerate mode, as a counterpart of Eq. (10.17), is given by

$$E'_z(x, y) - E''_z(x, y) =$$

$$A_{1,0,-1} \begin{Bmatrix} 2\cos\left(\dfrac{2\pi (x\cos(\alpha) + y\sin(\alpha))}{\sqrt{3}a} + \dfrac{2\pi}{3}\right)\cos\left(\dfrac{2\pi (-x\sin(\alpha) + y\cos(\alpha))}{3a}\right) \\ +\cos\left(\dfrac{4\pi (-x\sin(\alpha) + y\cos(\alpha))}{3a}\right) \\ -2\cos\left(\dfrac{2\pi (x\cos(\beta) + y\sin(\beta))}{\sqrt{3}a} + \dfrac{2\pi}{3}\right)\cos\left(\dfrac{2\pi (-x\sin(\beta) + y\cos(\beta))}{3a}\right) \\ -\cos\left(\dfrac{4\pi (-x\sin(\beta) + y\cos(\beta))}{3a}\right) \end{Bmatrix}$$

(10.23)

The resultant magnetic fields can be found accordingly. Thus, Eqs. (10.17) and (10.23) give the basic field solutions of a pair of fundamental degenerate modes in an equilateral triangular-microstrip patch resonator. We should refer to them as mode 1, which is based on Eq. (10.23), and mode 2, which is based on Eq. (10.17) in the following discussions.

10.3.3.2 Field Patterns of Degenerate Modes For applications of a microwave resonator, it is always desirable to know the field pattern of a relevant resonant mode. Using the formulation described in the last section, a computer program was written and used to compute the field patterns of the two degenerate modes of a triangular-patch resonator. Figure 10.20 illustrates the computed electric fields of the modes 1

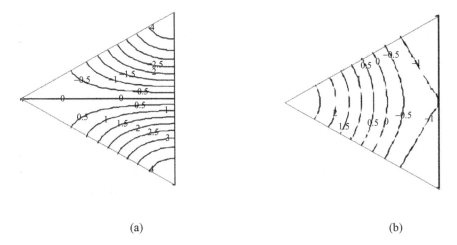

(a)   (b)

FIGURE 10.20 Electric field patterns of the degenerate modes. (**a**) Mode 1 (odd mode); (**b**) mode 2 (even mode).

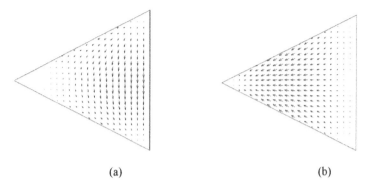

FIGURE 10.21 Current distributions of the degenerate modes. (**a**) Mode 1 (odd mode); (**b**) mode 2 (even mode).

and 2, directly resulting from Eqs. (10.23) and (10.17), respectively. It is interesting to see that the mode 1 exhibits an antisymmetrical field pattern with respect to the horizontal axis, whereas the mode 2 has a symmetrical field pattern. One can also observe that the field pattern of either mode can not be obtained by simply rotating its counterpart's field. This situation is totally different from that of the dual modes of a square- or circular-patch resonator, where the one mode can be obtained by rotating the other mode by 90° in the coordinate system.

Since the magnetic field $\underline{H}$ can be derived from the electric field, we can find the current distribution or density $\underline{J} = \hat{z} \times \underline{H}$ accordingly. The computed current distributions of the two degenerate modes of an equilateral triangular patch resonator are shown in Figure 10.21. Again, we can see that, with respect to the horizontal symmetrical plane, the modes 1 and 2 behave as an odd and an even mode, respectively.

10.3.3.3 Dual-Mode Bandpass Filters To demonstrate the application of the dual-mode microstrip triangular-patch resonator, two-pole bandpass filters with a single microstrip triangular-patch resonator were investigated first. Two filter configurations have been developed, depicted in Figure 10.22.

Note that for an equilateral triangle, $b = \sqrt{3}a/2$. Since we need to split the two degenerate modes of an equilateral triangular-patch resonator for designing a

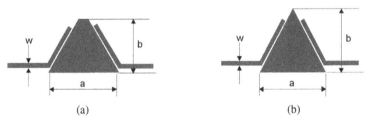

FIGURE 10.22 Two-pole dual-mode microstrip triangular patch resonator filters. (**a**) Configuration 1; (**b**) configuration 2.

bandpass filter, the mode splitting is achieved by introducing a small cut as shown in Figure 10.22a, where $b < \sqrt{3}a/2$, or by deforming the equilateral triangle into an isosceles triangle as the case of Figure 10.22b, where $b \neq \sqrt{3}a/2$. In either case it is found that the resonant frequency of the mode 2 is effectively shifted, whereas the resonant frequency of the mode 1 is almost unchanged. At first, it might be thought that once the modes were split, as the case for a perturbed dual-mode square-patch resonator, there would be some coupling between the two modes. However, after a careful examination of this type of mode splitting according to the theory of asynchronously tuned coupled resonators, that is, if the two split-mode frequencies are equal to the two self-resonant frequencies, respectively, there is no coupling between the two resonators, as discussed in Chapter 7. It was then believed that the two modes were actually hardly coupled to each other for the mode perturbations introduced. This discovery is very important for developing dual-mode filter of this type, as well as for understanding its operation.

The equality between the self-resonant frequencies and split-mode frequencies for the geometrical perturbations used has been investigated using full-wave EM simulations and is demonstrated as follows. By inspecting the field patterns of Figures 10.20 and 10.21, we can see that the mode 1 is actually an odd mode while the mode 2 an even mode. This allows us to simulate the self-resonant frequencies of the two modes by placing an electric or magnetic wall along the symmetrical axis, respectively, which are illustrated in Figures 10.23 and 10.24 for the two mode-split geometries used. For either geometry, when an electric wall is applied, only the odd mode or mode 1 is excited. In this case, the self-resonant frequency response is plotted using the broken line, where the self-resonant frequency can be identified at a resonant peak. On the other hand, if a magnetic wall is applied, only the even mode or mode 2 is excited and the simulated results are plotted using the dotted line. Without applying any electric or magnetic wall, the simulation results show the split-mode frequencies,

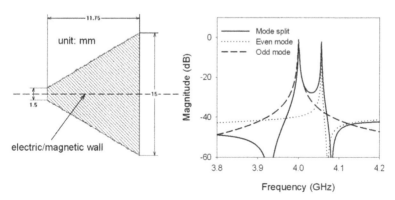

FIGURE 10.23 Simulated split mode frequencies (full-line) and self-resonant frequencies (dotted line for the even mode and broken line for the odd mode) for a perturbed (small cut) dual-mode triangular microstrip resonator on a 1.27-mm thick dielectric substrate with a relative dielectric constant of 10.8.

362 COMPACT FILTERS AND FILTER MINIATURIZATION

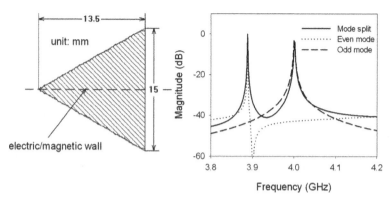

FIGURE 10.24 Simulated split mode frequencies (full-line) and self-resonant frequencies (dotted line for the even mode and broken line for the odd mode) for a perturbed (isosceles) dual-mode triangular microstrip resonator on a 1.27 mm thick dielectric substrate with a relative dielectric constant of 10.8.

which are plotted using the full line. It is evident that the two split-mode frequencies for either geometrical perturbation are equal to the self-resonant frequencies, which confirms the above suggestion that there is no coupling between the two modes.

To this end, a coupling structure for the two-pole dual-mode bandpass filters of Figure 10.22 has been developed, which is shown in Figure 10.25. There are four nodes, labeled with S for the source, L for the load, 1 for the mode 1, and 2 for the mode 2, respectively. The full line indicates a direct coupling between two adjacent nodes and, in this case, the broken line denotes a source-load coupling. A noticeable thing is that the coupling structure does not have a line connected between the nodes 1 and 2 so that there is no coupling between the two resonating modes. This coupling structure, which has a parallel form for the dual modes of interest, is entirely different from that for the dual-mode square- or circular-patch resonator filters in which the main couplings between modes are in a series structure.

Two filter designs having a coupling structure of Figure 10.25 are demonstrated next. The layout of the first design is shown in Figure 10.26a, where the dark arrow from the terminal port indicates a shift of the reference plane in EM simulation [22]. The dual-mode patch resonator is fed by two coupled lines with a coupling gap of 0.25 mm. Figure 10.26b plots theoretical and EM-simulated responses of the filter.

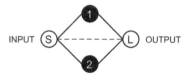

FIGURE 10.25 Coupling structure for two-pole dual-mode microstrip triangular-patch resonator filters.

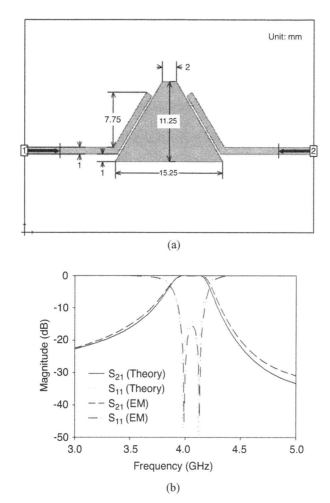

FIGURE 10.26 (a) Layout of a two-pole dual-mode microstrip triangular-patch resonator filter on a 1.27-mm thick dielectric substrate with a relative dielectric constant of 10.8. (b) Theoretical and EM-simulated responses.

The theoretical response is computed, using the formulation introduced in Chapter 7, with the following $n + 2$ coupling matrix

$$[m] = \begin{bmatrix} 0 & 0.7699 & 0.6468 & -0.015 \\ 0.7699 & 1.25 & 0 & -0.7699 \\ 0.6468 & 0 & -1.15 & 0.6468 \\ -0.015 & -0.7699 & 0.6468 & 0 \end{bmatrix} \quad (10.24)$$

The couplings from the source to mode 1 and from mode 1 to the load have the same magnitude, but opposite sign, because of the antisymmetrical field distribution of the mode 1 (odd mode). The filter has a typical Chebyshev filtering characteristic with a ripple passband of 0.2 GHz centered at 4.055 GHz. Using the parameter-extraction technique described in Section 10.3.4, it can be found that the modes 1 and 2 are resonating at 3.93 and 4.17 GHz, and the external quality factors for modes 1 and 2 are 34.2046 and 48.4716, respectively. Therefore, for this design, the odd mode has a resonant frequency lower than that of the even mode.

The layout and performance of the second designed filter are shown in Figure 10.27. As compared to the first design, there is only a small change in the shape

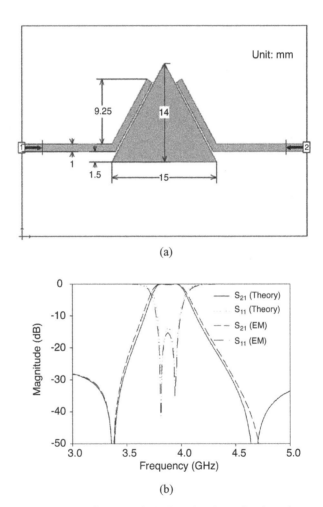

FIGURE 10.27 (a) Layout of a two-pole dual-mode microstrip triangular-patch resonator filter on a 1.27-mm thick dielectric substrate with a relative dielectric constant of 10.8. (b) Theoretical and EM-simulated responses.

of microstrip dual-mode triangular-patch resonator; however, the filtering characteristic is quite different from the previous one. This filter exhibits a quasielliptic function response with two transmission zeros at finite frequencies near the passband, resulting in a better selectivity as compared to the first filter design. The filter has a fractional bandwidth $FBW = 0.049$ and a center frequency $f_0 = 3.88$ GHz. The $n + 2$ coupling matrix for the second filter design is

$$[m] = \begin{bmatrix} 0 & 0.6917 & 0.6572 & -0.025 \\ 0.6917 & -1.1572 & 0 & -0.6917 \\ 0.6572 & 0 & 1.1572 & 0.6572 \\ -0.025 & -0.6917 & 0.6572 & 0 \end{bmatrix} \quad (10.25)$$

As compared with Eq. (10.24), the nonzero diagonal elements have opposite signs. In this case, the modal frequency of the odd mode is higher than that of the even mode. Furthermore, the second filter has a stronger source-load coupling than does the first filter. Again, by employing the parameter extraction technique described in Section 10.3.4, the modal frequencies are found to be 3.99 GHz for the odd mode and 3.77 GHz for the even mode. The external quality factors for mode 1 (odd mode) and mode 2 (even mode) are 42.6577 and 47.2509, respectively.

The unloaded Q of triangular-patch dual mode resonator is expected to be similar to that of dual-mode square-patch resonator. Full-wave simulations are used to extract the unloaded Q for the two modes of a triangular-patch resonator having the dimensions given in Figure 10.23. It turns out that the two modes also have the similar unloaded Q. Considering only the conductor loss by assuming a 10-µm thick copper patch and a 10-µm thick copper ground, both modes have an unloaded Q of around 950. Further taking into account the dielectric loss by assuming a loss tangent of 0.002 for the dielectric substrate used, the simulated unloaded Q for both modes are about 320. For filter applications, the resonators should be assembled in a metal housing to eliminate the radiation loss. Higher-order filters of this type can be developed based on two-pole designs. A four-pole filter with two triangular-patch dual mode resonators has been demonstrated [43].

It might be worth mentioning that there are methods for producing mode splitting other than those described above. For instance, Figure 10.28 shows the simulated characteristics of mode splitting for a dual-mode triangular patch resonator perturbed using a narrow slit along its symmetrical axis. For the simulations, the width of the slit was kept constant as 0.5 mm while the length of the slit, l was changed from 4 to 6 mm. The microstrip substrate has a thickness of 1.27 mm and a relative dielectric constant of 10.8. As can be seen from Figure 10.28, the longer the slit the larger is the mode splitting. Again, it can be shown that the mode self-resonant frequencies coincide with the two split-mode frequencies. This indicates that the two split modes obtained in this way also do not couple with each other. However, in this case, the simulations show that the self-resonant frequency of the even mode is hardly changed against the variation of the slit length, implying that this mode is not perturbed. This can easily be explained in the light of the current distributions shown in Figure 10.21

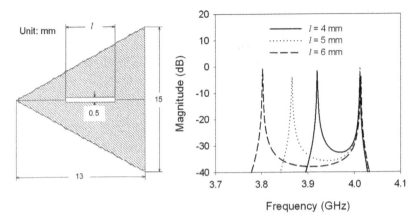

FIGURE 10.28 Characteristics of mode splitting for a dual-mode triangular-patch resonator perturbed with a narrow slit.

and the reason for this is that the slit is cut along the current flowing of the even mode so it in fact does not perturb the modal field distribution. Dual-mode band-reject filters using this type of resonator have been reported [44]. It has also been shown that a microstrip triangular loop can exhibit the similar dual-mode properties for filter applications [45,46].

10.3.4 Dual-Mode Open-Loop Filters

Following the dual-mode triangular patch resonator introduced in the last section, this section will deal with a new type of miniature microstrip dual-mode resonator for filter applications. The new dual-mode resonator is developed from a single-mode (operated) open-loop resonator [47], resulting in a compact size. It has a distinct characteristic, namely, its two operation modes do not couple with each other, which is similar to that of the dual-mode triangular-patch resonator, but differs from that of the other dual-mode resonators discussed earlier. For the applications of this new type of dual-mode resonator, some filter examples are described.

10.3.4.1 Dual-Mode Open-Loop Resonators Figure 10.29 shows the basic schematic of a dual-mode microstrip open-loop resonator (in gray). The dual-mode resonator consists of an open loop with an open gap, g. The open loop has a line width of W_1 and a size of $L_{1a} \times L_{1b}$. A loading element with a form of open stepped-impedance stub, is tapped from inside onto the open loop. The loading element has dimensions of L_{2a} and W_{2a} for the narrow line section and L_{2b} and W_{2b} for the wide-line section. The resonator is symmetrical with respect to the A-A' reference plane. The resonator is coupled to the input and output (I/O) ports with a feed structure having a line width of W and a coupling spacing s. The port terminal impedance is Z_0.

The first two resonating modes, existing in the resonator of Figure 10.29, are referred to as the odd and even modes. Depending on the dimensions of the resonator,

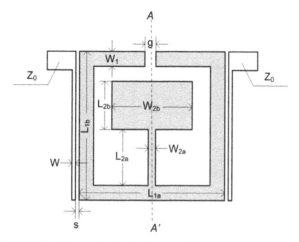

FIGURE 10.29 Schematic of a dual-mode microstrip open-loop resonator.

these two modes can have the same or different modal frequencies and, for the latter, the modal resonant frequency of one mode can be either higher or lower than that of the other one. The characteristics of the dual-mode open-loop resonator are investigated by full-wave electromagnetic (EM) simulation. To excite the resonator, two ports are weakly coupled to the resonator with a large spacing s. For the demonstration, the microstrip structure of Figure 10.29 is simulated on a substrate with a relative dielectric constant of 10.8 and a thickness of 1.27 mm. The dimensions used in the simulation are $s = 2.5$ mm, $W = 0.5$ mm, $g = 0.9$ mm, $W_1 = 1.5$ mm, $L_{1a} = 15.1$ mm, $L_{1b} = 15.0$ mm, $L_{2a} = 5.5$ mm, $W_{2a} = 0.7$ mm, $L_{2b} = 5.5$ mm, and $W_{2b} = 8.1$ mm. The terminal impedance $Z_0 = 50$ Ω. The simulated magnitude response of S_{21} is plotted in Figure 10.30, where two resonant peaks can be observed. In order to show that these two resonant peaks are the one belonging to the odd mode and the other belonging to the even mode, we will apply the symmetrical network analysis, discussed in Chapter 2. From Eq. (2.44), we have

$$S_{11e} = S_{11} + S_{21}$$
$$S_{11o} = S_{11} - S_{21} \quad (10.26)$$

where S_{11e} and S_{11o} are the one-port, even- and odd-modes scattering parameters, defined in Eq. (2.42). Thus, from the EM-simulated two-port S parameters, S_{11e} and S_{11o} can be found. Using Eq. (7.78), the group-delay (GD) responses of S_{11e} and S_{11o} can be computed, which are also plotted in Figure 10.30. As can be seen, the two group-delay peaks coincide with the two magnitude peaks of S_{21}. At resonance, group delay is maximum. Hence, for the simulation example under consideration, the even- is higher than the odd-mode frequency.

It is also interesting to notice from Figure 10.30 that there is a finite-frequency transmission zero when the two modes split. The transmission zero is allocated on the high side of the two modes when the even- is higher than the odd-mode frequency.

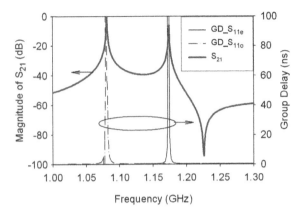

FIGURE 10.30 Magnitude and group-delay responses of a weakly excited dual-mode open-loop resonator.

On the other hand, the transmission zero is on the low side of the two modes when the even- is lower than the odd-mode frequency. This unique property of the dual-mode open-loop resonator can be explored for designing filters with an asymmetrical frequency response, which will be demonstrated later on.

Another distinct characteristic of the dual-mode open-loop resonator is that the two modes are not coupled to each other even after the modes are split. No coupling between the two modes of the dual-mode open-loop resonator is verified following the theory of asynchronously tuned coupled resonators, discussed in Chapter 7. As referred to in Eq. (7.71), the theory states that, if the two split-mode frequencies are equal to the two self-resonant frequencies, respectively, there is no coupling between the two resonators. From the results shown in Figure 10.30, it is evident that the two split-mode frequencies, which are associated with the two resonant peaks shown in the magnitude response of S_{21}, are equal to the two self-resonant frequencies, which are obtainable from the group-delay responses of S_{11e} and S_{11o}. This, therefore, confirms that there is no coupling between the two modes. This characteristic is very important for developing dual-mode filter using this kind of resonator, which is presented next.

10.3.4.2 Two-Pole Dual-Mode Open-Loop Resonator Filters According to the above discussion, a two-pole filter, using a single dual-mode open-loop resonator of Figure 10.29, can be designed with a coupling structure as shown in Figure 10.31,

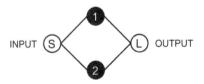

FIGURE 10.31 Coupling structure for a two-pole dual-mode open-loop resonator filter.

where the node S denotes the source or input, the node L denotes the load or output, and the nodes 1 and 2 represent the odd and even modes, respectively.

For this coupling structure, an $n+2$ coupling matrix (see Chapter 7) is given by

$$[m] = \begin{bmatrix} 0 & m_{S1} & m_{S2} & 0 \\ m_{S1} & m_{11} & 0 & m_{L1} \\ m_{S2} & 0 & m_{22} & m_{L2} \\ 0 & m_{L1} & m_{L2} & 0 \end{bmatrix} \quad (10.27)$$

The matrix elements are related to either the external quality factors (Q_e) or the resonant frequencies (f_0) in association with the even and odd modes, respectively. For a specific geometry of Figure 10.29, the modal Q_e and f_0 can be extracted from the group-delay responses for S_{11e} and S_{11o}. Referring to Figure 10.30, the frequency of a mode can be identified at its group-delay maximum. Since S_{11e} and S_{11o} are derived from the two-port scattering parameters, in respect to Eq. (10.26), for a doubly loaded resonator, following the discussion in Section 7.4, we have

$$Q_{e,e/o} = \frac{\omega_{0,e/o} \cdot \tau_{S_{11e/o}}(\omega_{0,e/o})}{2} \quad (10.28)$$

where the subscript e/o denotes either the even mode or the odd mode. Figures 10.32 and 10.33 display the extracted Q_e and f_0 with respect to some geometric parameters of the two-port filtering structure in Figure 10.29.

In all the cases, the Q_e for the even mode is larger than that for the odd mode. This implies that the odd mode has a larger I/O coupling than does the even mode. As can be seen from Figure 10.32a, the Q_e strongly depends on s, which is a primary parameter to control the I/O coupling for both modes. Changing W and W_1 can also alter the Q_e, but less significantly than changing s, especially for the odd mode. For a practical design, s, W, and W_1 can be adjusted to get a desired Q_e for the odd mode. On the other hand, the desired Q_e for the even mode can be obtained by controlling other dimensions such as W_{2a}, W_{2b}, and L_{2a} or L_{2b}, referred to in Figure 10.33, where we can see that the Q_e and the resonant frequency of the odd mode are hardly affected by these dimensions. This is because the tapping point for the loading element inside the open loop (see Figure 10.29), is actually a virtual ground for the odd mode, since the perimeter of the open-loop is about a half-wavelength for the odd mode. As a consequence, the loading element does not affect the odd-mode characteristics, which, however, are used to control the even-mode characteristics in design.

Based on the above discussion, two filter designs are described next. The first one, that is, filter I, is designed to have fractional bandwidth $FBW = 0.046$ at a center frequency of 1015.5 MHz, and to have a finite-frequency transmission zero on the

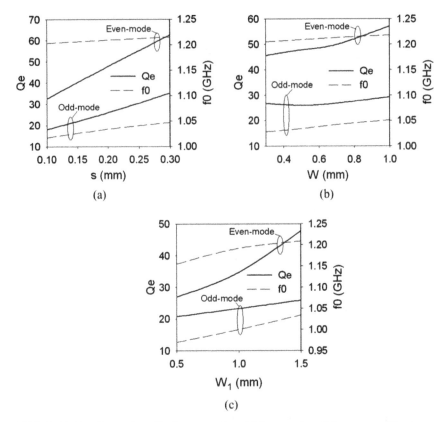

FIGURE 10.32 External quality factors and modal frequencies of the two-port filtering structure in Figure 10.29 on a substrate with a relative dielectric constant of 10.8 and a thickness of 1.27 mm for $Z_0 = 50\ \Omega$, $L_{1a} = 15.1$ mm, $L_{1b} = 15.0$ mm, $g = 0.9$ mm, $L_{2a} = 5.5$ mm, $W_{2a} = 0.7$ mm, $L_{2b} = 5.5$ mm, and $W_{2b} = 8.1$ mm. (a) This is compared with s for $W = 0.5$ mm and $W_1 = 1.5$ mm (b) versus W for $s = 0.2$ mm and $W_1 = 1.5$ mm, and (c) versus W_1 for $s = 0.2$ mm and $W = 0.5$ mm.

low side of the passband. The $n + 2$ coupling matrix for the filter is

$$[m] = \begin{bmatrix} 0 & 0.9024 & 0.5714 & 0 \\ 0.9024 & -1.0918 & 0 & -0.9024 \\ 0.5714 & 0 & 1.3487 & 0.5714 \\ 0 & -0.9024 & 0.5714 & 0 \end{bmatrix} \quad (10.29)$$

Referring to Eq. (10.27), it should be noted that $m_{L1} = -m_{S1}$ for the odd mode and $m_{L2} = m_{S2}$ for the even mode. Using the formulations introduced in

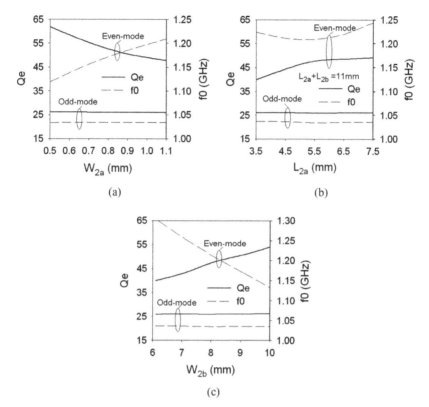

FIGURE 10.33 External quality factors and modal frequencies of the two-port filtering structure in Figure 10.29 on a substrate with a relative dielectric constant of 10.8 and a thickness of 1.27 mm for $Z_0 = 50\,\Omega$, $s = 0.2$ mm, $W = 0.5$ mm, $W_1 = 1.5$ mm, $L_{1a} = 15.1$ mm, $L_{1b} = 15.0$ mm, and $g = 0.9$ mm. (a) This is compared with W_{2a} for $L_{2a} = 5.5$ mm, $L_{2b} = 5.5$ mm and $W_{2b} = 8.1$ mm. (b) versus L_{2a} for $W_{2a} = 1.1$ mm, $W_{2b} = 8.1$ mm, and $L_{2a} + L_{2b} = 11$ mm, and (c) versus W_{2b} for $W_{2a} = 1.1$ mm, $L_{2a} = 5.5$ mm, and $L_{2b} = 5.5$ mm.

Chapter 7, we have

$$Q_{e1} = \frac{1}{m_{S1}^2 \cdot FBW} = \frac{1}{0.9024^2 \times 0.046} = 26.6959$$

$$Q_{e2} = \frac{1}{m_{S2}^2 \cdot FBW} = \frac{1}{0.5714^2 \times 0.046} = 66.5827$$

$$f_{01} = f_0 \cdot \left(1 - \frac{m_{11} \cdot FBW}{2}\right) = 1015.5 \times \left(1 - \frac{-1.0918 \times 0.046}{2}\right) = 1041\,\text{MHz}$$

$$f_{02} = f_0 \cdot \left(1 - \frac{m_{22} \cdot FBW}{2}\right) = 1015.5 \times \left(1 - \frac{1.3487 \times 0.046}{2}\right) = 984\,\text{MHz}$$

where Q_{e1} and f_{01} are the external quality factor and the resonant frequency for the odd mode, while Q_{e2} and f_{02} are the design parameters associated with the even mode. The microstrip filter is to be implemented on a dielectric substrate with a relative dielectric constant of 10.8 and a thickness of 1.27 mm. The desired design parameters, that is, Q_{e1}, Q_{e2}, f_{01}, and f_{02} can be extracted from EM simulations, as discussed above, so as to determine the dimensions of the filter. Figure 10.34a displays the layout of the designed filter. The two small open stubs attached to the loading element inside the open loop are deployed for an additional control of the

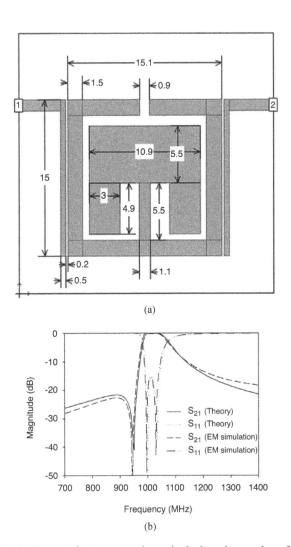

FIGURE 10.34 (a) Layout of a two-port microstrip dual-mode open loop filter (filter I) on a 1.27-mm thick dielectric substrate with a relative dielectric constant of 10.8. (All dimensions are in millimeters.) (b) Theoretical and EM-simulated responses of the filter.

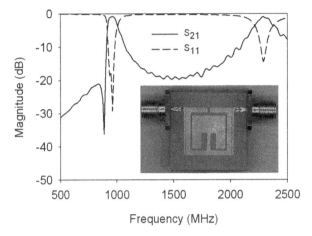

FIGURE 10.35 Measured wideband performance of the fabricated filter I (see the inserted photograph).

even-mode characteristics, which also allow an efficient utilization of the circuit area inside the open loop. Theoretical and EM-simulated responses of the filter are plotted in Figure 10.34b, where the theoretical responses are computed based on the $n+2$ coupling matrix of Eq. (10.29) using the formulation Eq. (7.85).

Figure 10.35 shows the measured performance of the fabricated filter. The filter exhibits a higher selectivity on the low side of the passband because of the finite-frequency transmission zero. The first spurious does not occur at $2f_0$, but at a higher frequency.

The second two-port filter of this type, that is, filter II, is demonstrated to have a finite-frequency transmission zero on the high side of the passband. In this case, the even-mode frequency should be higher than the odd-mode frequency. For comparison, the second filter, or filter II, is obtained by simply removing the two small open stubs from filter I; the resultant layout is shown in Figure 10.36a. Filter II has a fractional bandwidth $FBW = 0.035$ centered at 1063 MHz. Its design parameters can be extracted to be $Q_{e1} = 26.5583$, $f_{01} = 1033$ MHz, $Q_{e2} = 63.003$, and $f_{02} = 1092$ MHz. The $n+2$ coupling matrix for the filter II is

$$[m] = \begin{bmatrix} 0 & 1.0372 & 0.6734 & 0 \\ 1.0372 & 1.6127 & 0 & -1.0372 \\ 0.6734 & 0 & -1.5589 & 0.6734 \\ 0 & -1.0372 & 0.6734 & 0 \end{bmatrix} \quad (10.30)$$

Figure 10.36b shows the theoretical and EM-simulated responses of the filter II. In general, both are in good agreement. Some discrepancy at the upper stopband is likely because of a small parasitic source-load coupling, which is not taken into account in the coupling matrix of Eq. (10.30).

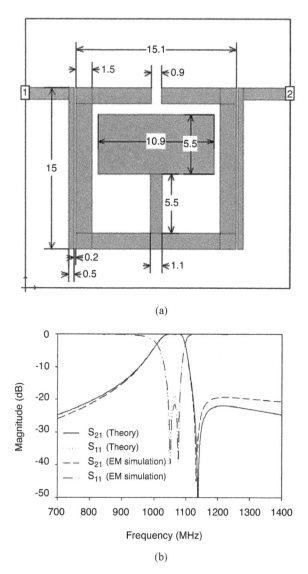

FIGURE 10.36 (a) Layout of a two-port microstrip dual-mode open loop filter (filter II) on a 1.27-mm thick dielectric substrate with a relative dielectric constant of 10.8. (All dimensions are in millimeters.) (b) Theoretical and EM-simulated responses of the filter.

By inspecting the layouts of filters I and II, it is envisaged that there exist paths to facilitate a parasitic source-load coupling, even though very small. To confirm this, assuming $m_{sL} = 0.01$ in the coupling matrixes of Eqs. (10.29) and (10.30), the recomputed theoretical responses are plotted in Figure 10.37 for both filters I and II, respectively. As can be seen, better agreement between the theory and the EM simulation is achieved.

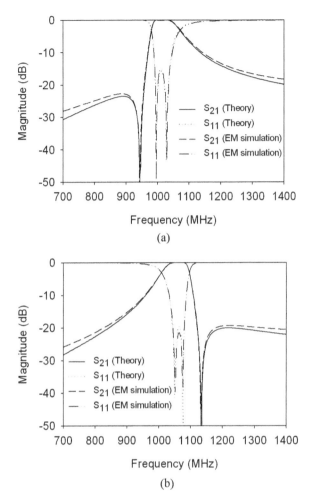

FIGURE 10.37 Improved theoretical responses by taking into account the parasitic source-load coupling for $M_{sL} = 0.01$. (a) Filter I; (b) filter II.

10.3.4.3 Four-Pole Dual-Mode Open-Loop Resonator Filters A simple way to design a high-order filter with more than one dual-mode resonators of this type is to cascade second-order or two-pole blocks through some nonresonating nodes. For example, Figure 10.38 illustrates a four-pole filter of this type and its coupling structure. In this case, the two dual-mode resonators without any direct couplings between them are cascaded through the two nonresonating nodes as indicated. The two dual-mode resonators are also not the same as shown; one will produce a finite-frequency zero on the high side of the passband, whereas the other will produce another finite-frequency zero on the low side of the passband.

Figure 10.39 shows the configuration of the general coupling matrix that includes two nonresonating nodes for this four-pole filter. In general, a nonresonating node

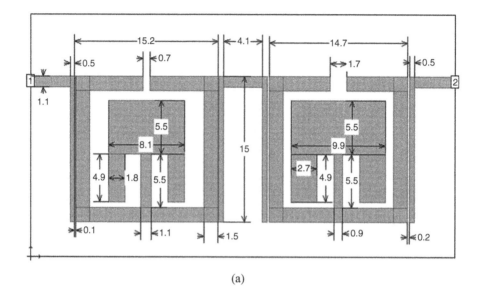

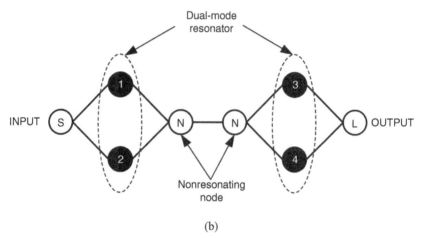

FIGURE 10.38 (a) Layout of a four-pole dual-mode open-loop filter on a 1.27-mm thick dielectric substrate with a relative dielectric constant of 10.8. (All dimensions are in millimeters.) (b) Its coupling structure.

is simply a node that is connected to ground by a frequency-independent reactance [48]. In this case, since the reactances at the two nonresonating nodes are both zero, their diagonal entries in the matrix are zero, similar to that for the source and the load. The element m_{NN} denotes the coupling between the two nonresonating nodes, which represents a connection between the two second-order blocks. Theoretically, m_{NN} can be implemented with a transmission line inverter, whose characteristic

	S	1	2	N	N	3	4	L
S	0	m_{s1}	m_{s2}	0	0	0	0	0
1	m_{s1}	m_{11}	0	m_{1N}	0	0	0	0
2	m_{s2}	0	m_{22}	m_{2N}	0	0	0	0
N	0	m_{1N}	m_{2N}	0	m_{NN}	0	0	0
N	0	0	0	m_{NN}	0	m_{N3}	m_{N4}	0
3	0	0	0	0	m_{N3}	m_{33}	0	m_{3L}
4	0	0	0	0	m_{N4}	0	m_{44}	m_{4L}
L	0	0	0	0	0	m_{3L}	m_{4L}	0

FIGURE 10.39 General coupling matrix including nonresonating nodes.

impedance may or may not be 50 Ω. Depending on the reference planes of respective nonresonating nodes in a practical implementation, the distance between the two second-order blocks can be shorter than a quarter wavelength. In Figure 10.39, there are two shading 4 × 4 submatrixes, each representing the coupling matrix for the respective second-order block. Based on the above discussion on the parameter extraction for the two-pole dual-mode filter, the matrix elements for an individual second-order block can be extracted for the filter given in Figure 10.38a. In other words, for a given general coupling matrix, each second-order block can be designed and implemented separately.

Assume the four-pole filter has a fractional bandwidth $FBW = 0.075$ at the center frequency $f_0 = 1.05$ GHz. The extracted coupling matrix is given by

$$[m] = \begin{bmatrix} 0 & 0.8554 & 0.5805 & 0.01 & 0 & 0 & 0 & 0 \\ 0.8554 & 0.9651 & 0 & -0.8554 & 0 & 0 & 0 & 0 \\ 0.5805 & 0 & -1.0159 & 0.5805 & 0 & 0 & 0 & 0 \\ 0.01 & -0.8554 & 0.5805 & 0 & 1.1 & 0 & 0 & 0 \\ 0 & 0 & 0 & 1.1 & 0 & 0.7408 & 0.46 & 0.01 \\ 0 & 0 & 0 & 0 & 0.7408 & -0.6095 & 0 & -0.7408 \\ 0 & 0 & 0 & 0 & 0.46 & 0 & 1.0667 & 0.46 \\ 0 & 0 & 0 & 0 & 0.01 & -0.7408 & 0.46 & 0 \end{bmatrix}$$

(10.31)

Note that $m_{sN} = 0.01$ and $m_{LN} = 0.01$ are assumed in the above coupling matrix to take into account a small coupling from the source to the first nonresonating node and the second nonresonating node to the load, following the previous discussion on the two-pole filter. The theoretical response can be computed using the formulation, that is, Eq. (7.85), introduced in Chapter 7. In this case,

$$[A] = [m] + [\Omega'] - j[q] \qquad (10.32)$$

where $[m]$ is given by (10.31), $[\Omega']$ and $[q]$ are given as follows:

$$[\Omega'] = \begin{bmatrix} 0 & 0 & 0 & 0 & 0 & 0 & 0 & 0 \\ 0 & \Omega(f) & 0 & 0 & 0 & 0 & 0 & 0 \\ 0 & 0 & \Omega(f) & 0 & 0 & 0 & 0 & 0 \\ 0 & 0 & 0 & 0 & 0 & 0 & 0 & 0 \\ 0 & 0 & 0 & 0 & 0 & 0 & 0 & 0 \\ 0 & 0 & 0 & 0 & 0 & \Omega(f) & 0 & 0 \\ 0 & 0 & 0 & 0 & 0 & 0 & \Omega(f) & 0 \\ 0 & 0 & 0 & 0 & 0 & 0 & 0 & 0 \end{bmatrix} \quad (10.33)$$

$$\Omega(f) = \frac{1}{FBW} \cdot \left(\frac{f}{f_0} - \frac{f_0}{f} \right)$$

$$[q] = \begin{bmatrix} 1 & 0 & 0 & 0 & 0 & 0 & 0 & 0 \\ 0 & 0 & 0 & 0 & 0 & 0 & 0 & 0 \\ 0 & 0 & 0 & 0 & 0 & 0 & 0 & 0 \\ 0 & 0 & 0 & 0 & 0 & 0 & 0 & 0 \\ 0 & 0 & 0 & 0 & 0 & 0 & 0 & 0 \\ 0 & 0 & 0 & 0 & 0 & 0 & 0 & 0 \\ 0 & 0 & 0 & 0 & 0 & 0 & 0 & 0 \\ 0 & 0 & 0 & 0 & 0 & 0 & 0 & 1 \end{bmatrix} \quad (10.34)$$

The computed theoretical response is plotted in Figure 10.40, together with the EM-simulated result, where good agreement can be observed. The designed filter is fabricated and tested. Figure 10.41 shows the experimental results, where the small insert is the photograph of the fabricated filter. The demonstrated filter exhibits a quasielliptic-function response with two finite-frequency transmission zeros, one on each side of the passband.

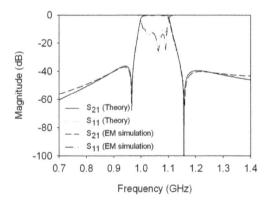

FIGURE 10.40 Theoretical and EM-simulated responses of four-pole dual-mode open-loop filter.

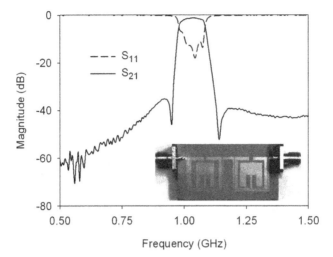

FIGURE 10.41 Experimental results of the fabricated four-pole dual-mode open-loop filter on a 1.27-mm thick dielectric substrate with a relative dielectric constant of 10.8.

10.4 LUMPED-ELEMENT FILTERS

Lumped-element microwave filters exhibit small physical size and broad spurious-free frequency bands. Usually, lumped-element filters are constructed using parallel-plate chip capacitors and air-wound inductors soldered into a small housing. Skilled manual labor is required to build and tune such a filter. It is also often difficult to integrate them into an otherwise all thin-film assembly. To overcome these difficulties, microstrip lumped-element filters, which are fabricated entirely using printed circuit or thin-film technologies, appear to be more desirable [49–54].

Basically, microstrip lumped-element filters can be designed based on lumped-element filter networks, such as those presented in Chapters 3, and constructed using lumped or quasilumped components as described in Chapter 4. A lowpass and a highpass filter of this type have been discussed in Chapter 5 and 6, respectively. Depending on network topologies and component configurations, there are many alternative realizations [49–54].

For instance, the tubular lumped-element bandpass filter topology as shown in Figure 10.42 is popular for realization in coaxial or microstrip forms [49–51]. This

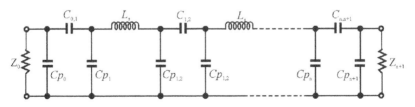

FIGURE 10.42 Circuit topology of tubular lumped-element bandpass filter.

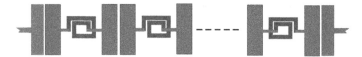

FIGURE 10.43 Microstrip realization of tubular lumped-element bandpass filter.

topology is formed by alternatively cascading the π networks of capacitors and the series inductors. In a simple microstrip form, as illustrated in Figure 10.43, each of the π networks of capacitors is realized using two parallel-coupled microstrip patches, whereas the inductances are realized using loop or spiral inductors. Interdigital or MIM capacitors may be implemented to enhance the series capacitances if the coupled microstrip patches cannot provide adequate series capacitances that would be required. Discussions on these lumped or quasilumped components can be found in Chapter 4.

The tubular filter can be derived from the filter structure in Figure 3.19a, which is comprised of impedance inverters and series resonators. The impedance inverter can use the form of the T network of lumped-element capacitors in Figure 3.21b. The series capacitor of each series resonator is split into two capacitors, one on each side of the inductor. The resulting new T networks of capacitors between the adjacent inductors are then converted to the exact π networks of capacitors, as those shown in Figure 10.42. The formation of the π networks at the input and output needs to be treated in a somewhat different manner. This is because that there is no way of absorbing or realizing the negative capacitance of that would appear in series with the resistor termination. This difficulty can be removed by equating the input admittances of the two one-port networks in Figure 10.44. The series capacitor ($2C_s$) appearing in the network on the left results from the splitting of the series capacitor of the first or last series resonator in Figure 3.19a. Since the equalization is imposed only at the center frequency (ω_0) of the filter, the two networks are approximately comparable. This, however, works satisfactorily for the narrow-band applications.

The derivation of design equations for the tubular filter topology of Figure 10.42 follows the approaches described above. The design procedures and equations for this type of lumped-element bandpass filter are summarized below.

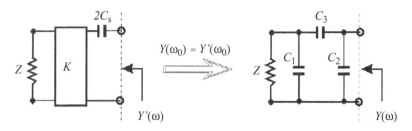

FIGURE 10.44 Approximate transformation of two one-port networks.

Choose the series inductance L_s (assuming the same value for all the inductors). Specify the center frequency ω_0, the fractional bandwidth FBW as defined in Eq. (3.41b), and the termination resistances Z_0 and Z_{n+1}. The formulas given in Figure 3.19a are then calculated as follows:

$$C_s = \frac{1}{\omega_0^2 L_s}$$

$$K_{0,1} = \sqrt{\frac{Z_0 FBW \omega_0 L_s}{\Omega_c g_0 g_1}}$$

$$K_{i,i+1} = \frac{FBW \omega_0 L_s}{\Omega_c} \sqrt{\frac{1}{g_i g_{i+1}}} \quad \text{for } i = 1 \text{ to } n-1 \qquad (10.35)$$

$$K_{n,n+1} = \sqrt{\frac{Z_{n+1} FBW \omega_0 L_s}{\Omega_c g_n g_{n+1}}}$$

where g_i and Ω_c are the element values and the cutoff frequency of a chosen lowpass prototype of the order n.

For determining Cp_0, $C_{0,1}$, and Cp_1, choose a trial capacitance Cx_0. Then:

$$C_{0,1} = \sqrt{\frac{G_0'\left(1 + Z_0^2 \omega_0^2 C x_0^2\right)}{Z_0 \omega_0^2}}$$

$$Cp_0 = Cx_0 - C_{0,1} \qquad (10.36)$$

$$Cp_1 = \frac{1}{\omega_0}\left[B_0' - \frac{\omega_0 C_{0,1}\left(1 + Z_0^2 \omega_0^2 Cp_0 Cx_0\right)}{1 + (Z_0 \omega_0 Cx_0)^2}\right]$$

where

$$G_0' = \frac{Z_0\left(2\omega_0 C_s K_{0,1}\right)^2}{Z_0^2 + \left(2\omega_0 C_s K_{0,1}^2\right)^2} \quad \text{and} \quad B_0' = \frac{Z_0^2\left(2\omega_0 C_s\right)}{Z_0^2 + \left(2\omega_0 C_s K_{0,1}^2\right)^2}$$

Note that the value of Cx_0 must be chosen such that the resultant $C_{0,1}$, Cp_0, and Cp_1 are attainable, that is, not negative. The value of Cx_0 may also be alternated to result in more desirable values for $C_{0,1}$, Cp_0, and Cp_1. For instance, Cx_0 may be alternated to make $Cp_0 = Cp_1$.

For determining $C_{i,i+1}$ and $Cp_{i,i+1}$ between the adjacent inductors:

$$C_{i,i+1} = \left.\frac{N_{i,i+1}}{D_{i,i+1}}\right|_{i=1 \text{ to } n-1}$$

$$Cp_{i,i+1} = \left.\frac{Np_{i,i+1}}{D_{i,i+1}}\right|_{i=1 \text{ to } n-1} \qquad (10.37)$$

where

$$D_{i,i+1} = \frac{4C_s}{1 - 2C_s\omega_0 K_{i,i+1}} + \frac{1}{\omega_0 K_{i,i+1}}$$

$$N_{i,i+1} = \left(\frac{2C_s}{1 - 2C_s\omega_0 K_{i,i+1}}\right)^2$$

$$Np_{i,i+1} = \frac{1}{\omega_0 K_{i,i+1}} \left(\frac{2C_s}{1 - 2C_s\omega_0 K_{i,i+1}}\right)$$

In the case that the filter is not symmetrical, choose a trial capacitance Cx_{n+1} for determining Cp_{n+1}, $C_{n,n+1}$, and Cp_n:

$$C_{n,n+1} = \sqrt{\frac{G'_{n+1}\left(1 + Z_{n+1}^2\omega_0^2 Cx_{n+1}^2\right)}{Z_{n+1}\omega_0^2}}$$

$$Cp_{n+1} = Cx_{n+1} - C_{n,n+1} \tag{10.38}$$

$$Cp_n = \frac{1}{\omega_0}\left[B'_{n+1} - \frac{\omega_0 C_{n,n+1}\left(1 + Z_{n+1}^2\omega_0^2 Cp_{n+1}Cx_{n+1}\right)}{1 + (Z_{n+1}\omega_0 Cx_{n+1})^2}\right]$$

where

$$G'_{n+1} = \frac{Z_{n+1}\left(2\omega_0 C_s K_{n,n+1}\right)^2}{Z_{n+1}^2 + \left(2\omega_0 C_s K_{n,n+1}^2\right)^2} \quad \text{and} \quad B'_{n+1} = \frac{Z_{n+1}^2(2\omega_0 C_s)}{Z_{n+1}^2 + \left(2\omega_0 C_s K_{n,n+1}^2\right)^2}$$

Similarly, the choice of Cx_{n+1} must guarantee that the resultant $C_{n,n+1}$, Cp_{n+1}, and Cp_n are realizable.

In general, these design equations work well for filters with a narrower bandwidth, say $FBW \leq 0.05$, but tend to result in a smaller bandwidth and a lower center frequency when the filter bandwidth is increased. This is demonstrated with the following design examples.

Design Examples

Two four-pole ($n = 4$) tubular lumped-element bandpass filters are designed. The first filter has a passband from 1.95 to 2.05 GHz, and the second filter has a passband from 1.8 to 2.2 GHz. Their fractional bandwidths are 5% and 20%, respectively. A Chebyshev lowpass prototype with a passband ripple of 0.04321 dB (or return loss of -20 dB) is chosen for the designs. The element values of the lowpass prototype, which can be obtained from Table 3.2, are $g_0 = 1.0$, $g_1 = 0.9314$, $g_2 = 1.2920$, $g_3 = 1.5775$, $g_4 = 0.7628$, and $g_5 = 1.2210$ for $\Omega_c = 1$. The tubular bandpass filters are supposed to be terminated with 50-Ω resistors, i.e., $Z_0 = Z_{n+1} = 50\,\Omega$. Choose the series inductance $L_s = 5.0$ nH, and the trial capacitances $Cx_0 = Cx_{n+1} = 3.0$ pF. Apply the design Eqs. (10.35–10.38) to determine the values for all the capacitors.

For the first filter with passband from 1.95 to 2.05 GHz these are:

$Cp_0 = Cp_5 = 1.6037$ pF $Cp_{12} = Cp_{34} = 2.3228$ pF $Cp_{23} = 2.3686$ pF
$Cp_1 = Cp_4 = 1.6165$ pF $C_{12} = C_{34} = 0.2331$ pF $C_{23} = 0.1785$ pF
$C_{01} = C_{45} = 1.3963$ pF

For the second filter with passband from 1.8 to 2.2 GHz, the capacitances are:

$Cp_0 = Cp_5 = 0.4064$ pF $Cp_{12} = Cp_{34} = 1.8724$ pF $Cp_{23} = 1.9964$ pF
$Cp_1 = Cp_4 = 1.3091$ pF $C_{12} = C_{34} = 1.0831$ pF $C_{23} = 0.7825$ pF
$C_{01} = C_{45} = 2.5936$ pF

The resultant tubular filters are symmetrical even though the lowpass prototype is asymmetrical. The determined component values are substituted into the filter circuit in Figure 10.42 for analysis. Figure 10.45 plots the analyzed frequency responses of both filters. The first designed filter exhibits the desired responses, as shown in Figure 10.45a. The second designed filter, which has a wider bandwidth as referring to Figure 10.45b, shows a fractional bandwidth of 19.5% instead of 20% as required. Its center frequency is also shifted down about 1%. Nevertheless, the responses are close to the desired ones.

With the advent of multilayer circuit technologies based on LTCC and LCP materials, miniature filters can be implemented with lumped-element components embedded in a LTCC or LCP package. For example, miniature lumped-element LTCC filters are reported [55] for emerging global-positioning system (GPS) applications. These filters use a true lumped-element topology with three multilayer L–C resonators, occupying only $0.03 \times 0.05 \times 0.004\lambda$ of a low-permittivity LTCC substrate and leading to superior stopband performance. The fully embedded filter displays a

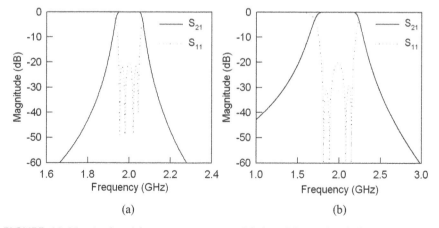

FIGURE 10.45 Analyzed frequency responses of designed four-pole tubular lumped-element bandpass filters (**a**) for 5% bandwidth and (**b**) for 20% bandwidth.

10-dB bandwidth from 1.6 to 1.75 GHz with an insertion loss of 2.8 dB and a return loss of 21.3 dB at the center frequency of 1.64 GHz.

10.5 MINIATURE FILTERS USING HIGH DIELECTRIC-CONSTANT SUBSTRATES

Using high dielectric-constant substrates is another approach for filter miniaturization [56–65]. High dielectric-constant materials, particularly high dielectric constant- and low-loss ceramics, are already in use in the rapidly expanding wireless segment of the electronic industry and their use continues to increase. Constructing microstrip and stripline filters on high dielectric-constant substrates is amenable to inexpensive printed circuit board technology for low-cost mass production. When high dielectric-constant substrates are used to design miniaturized filters, one must pay attention to some design considerations:

- Low-loss filter needed a sufficient line width. This limit is associated with the necessity of using low-value characteristic impedance lines.
- High sensitivity to small variations in physical dimensions.
- Excitation of higher-order modes.
- Difficulty in the realization of high characteristic impedance lines.

In addition, temperature stability associated with high dielectric substrates becomes more of a concern because it is important for reduction of temperature-induced drift in filter-operating frequencies. The temperature stability of a substrate is described by two thermal coefficients. One is the thermal coefficient of dielectric constant, given by $\Delta\varepsilon_r/(\varepsilon_r \Delta T)$, where ε_r is the relative dielectric constant and $\Delta\varepsilon_r$ is the small linear change in ε_r caused by the temperature change ΔT. The other is the thermal coefficient of expansion, given by $\Delta l/(l\Delta T)$, where l represents a physical dimension and Δl is the small linear change in l because of the temperature change ΔT. Similarly, the temperature stability of a filter operating frequency can be defined by $\Delta f/(f\Delta T)$, where Δf is the small linear change in the operating frequency f caused by the temperature change ΔT. Conventionally, these parameters are measured by ppm/°C or 10^{-6}/°C.

End-coupled half-wavelength resonator filters (see Chapter 5) on high dielectric-constant ($\varepsilon_r = 38$) substrates have been investigated [58]. The substrate material is composed of zirconium-tin-titanium oxide [(ZrSn)TiO$_4$] and possesses a relative dielectric constant of 38. The loss tangent of the material as quoted by the manufacturer is 0.0001. This corresponds to a dielectric quality factor of the order of 10,000. The material is also very stable with change in temperature, demonstrating a temperature variation of 6 ppm/°C. Seven-pole filters centered at 6.04 and 8.28 GHz were designed for 140-MHz 3-dB bandwidths. This particular design was chosen because of its eventual application in the design of compact multiplexers. The low-loss nature of the material is adequate for the intended purpose. Conductor losses also limit the Q_u and vary with the choice of transmission line. Stripline was chosen since it offered adequate performance electrically and reasonable mechanical requirements. The impedance level is typically found to be in the range of 10 to 20 Ω. The conductor losses for stripline are acceptable for these impedance levels. The final consideration in the design involves consideration of higher-order modes. The waveguide cutoff frequencies are lowered with the presence of higher dielectric-constant substrates. They are, however, still high enough for useful operation at the frequencies of interest.

Miniature microstrip hairpin-line bandpass filters have been reported using high dielectric constant ($\varepsilon_r = 80-90$) substrates to achieve superior performance, smaller size, and lower cost [60]. Design of hairpin-line filters is similar to that described in Chapter 5. Realization of hairpin-line filters using high-ε_r materials poses another serious problem when 50-Ω input/output tapping lines are designed. Those line widths become unrealizable small. Therefore, the tapping-line impedance is made smaller than 50 Ω so that the lines can be conveniently realized without any etching and tolerance problems. As a result, the input/output reference planes are shifted outward from the conventional 50-Ω tapping points for connection to a 50-Ω system. These factors are taken into consideration when the external Q–factor is evaluated and modeled. A five-pole filter is designed and fabricated on a 2-mm thick substrate with a relative dielectric constant of 80. The substrate material is a solid solution of barium and strontium titanates and has a dielectric Q of 10,000 (or loss tangent $\tan\delta = 10^{-4}$). Experimentally, the filter shows a midband frequency at 905 MHz with 46-MHz bandwidth and −20-dB return loss. The measured midband insertion loss is about 2.4 dB. Another similar five-pole filter is designed and realized on niobium–niodinum titanate substrate [60]. The dielectric constant of the material is 90. The substrate thickness is 1 mm.

A small-size microstrip combline (referring to Chapter 5) bandpass filter has been demonstrated by using a high dielectric-constant ($\varepsilon_r = 92$) substrate [59]. The substrate material is composed of $BaO\text{-}TiO_2\text{-}Nd_2O_3$ with a relative dielectric constant of 92. The filter was designed to have a three-pole Chebyshev response with 0.01-dB ripple and 51-MHz passband centered at 1.2 GHz. Shield lines between coupled resonators were used to achieve a smaller coupling coefficient with smaller spacing. The filter was fabricated with a substrate size of 7.4 × 20 × 3 mm. The measured bandwidth in which the return loss is smaller than −25 dB is about 50 MHz and the insertion loss is 1.6 dB at the center frequency of 1.185 GHz. This type of bandpass filter can be produced at low cost and is useful for portable radio equipment [59].

A miniature four-pole microstrip, stepped-impedance resonator bandpass filter has been developed for mobile communications [63]. The filter is realized on a dielectric substrate with a high relative dielectric constant $\varepsilon_r = 89$ and a thickness of 2 mm. In consideration of conductor loss and spurious response, the half-wavelength stepped-impedance resonators have lower characteristic impedance with a line width of 3 mm in the center part and higher characteristic impedance with a line width of 1.5 mm on the both sides. The unloaded factor of the resonators is about 400 at 1.5 GHz. The filter is designed to have a Chebyshev response with 0.01-dB ripple in a passband of 45 MHz at center frequency $f_0 = 1.575$ GHz. The measured midband insertion loss is 2 dB. The first spurious response of this filter appears at $1.78 f_0$.

386 COMPACT FILTERS AND FILTER MINIATURIZATION

Small integrated, microwave, multichip modules (MCMs) using high dielectric substrates have also been developed for the electronic toll collection (ETC) system [64]. A variety of dielectric substrates whose polycrystal structure and surface are suitable for thin film process can be selected for the specifications of products according to dielectric constants, Q-factors, temperature stability, etc. For miniaturization of a microstrip filter chip, a very high dielectric-constant ($\varepsilon_r = 110$) substrate is used. The microstrip filter is basically a three-pole interdigital bandpass filter, consisting of a few coupled quarter-wavelength microstrip resonators [64]. The filter was designed for a center frequency of 5.8 GHz and fabricated on a 0.3-mm thick substrate. The chip size of this filter is only 1×2 mm. Measured insertion loss of 2.4 dB and return loss of -14 dB have been achieved.

10.6 MULTILAYER FILTERS

There has been increasing interest in multilayer filters to meet the challenges of meeting size, performance, and cost requirements. Multilayer filter technology also provides another dimension in the flexible design and integration of other microwave components, circuits, and subsystems. Multilayer filters may be implemented in various manners. For example, they may be composed of various coupled-line resonators or circuit elements that are located at different layers without any ground plane inserted between the adjacent layers [66–70]. In contrast, another category of the multilayer filters utilizes aperture couplings on common ground between adjacent layers [71–75]. Patterning ground(s) to form slotted or so-called defect ground structures allows additional filtering functionality to be realized [76–106]. Using metallic via-holes to integrate the rectangular waveguide into the microstrip substrate [107,108], leads to the development of substrate integrated waveguide filters [109–120]. Recent advances in packaging materials, such as the low-temperature co-fired ceramic (LTCC) and liquid crystal polymer (LCP), has also stimulated the development of multilayer filters [131–192]. Typical types of multilayer filters will be described in the following subsections.

10.6.1 Aperture-Coupled Resonator Filters

Narrowband multilayer bandpass filters, including aperture-coupled dual-mode microstrip or stripline resonators filters [71], aperture-coupled quarter-wavelength microstrip line filters [72], and aperture-coupled microstrip, open-loop, resonator filters [74,75] have been developed. In what follows, we will discuss the design of the aperture-coupled, microstrip, open-loop resonator filters, although the design methodology is applicable to the other narrow-band multilayer bandpass filters.

Figure 10.46 shows the structure of an aperture-coupled, microstrip, open-loop resonator bandpass filter, which consists of two arrays of microstrip open-loop resonators that are located on the outer sides of two dielectric substrates with a common ground plane in between. Apertures on the ground plane are introduced to couple the

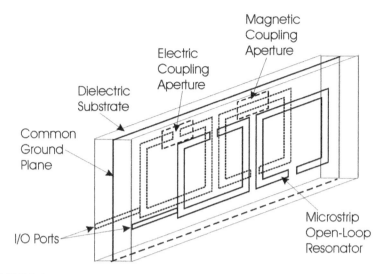

FIGURE 10.46 Structure of aperture-coupled microstrip open-loop resonator filter.

resonators between the two resonator arrays. Depending on the arrangement of the apertures, different filtering characteristics can easily be realized.

To design this class of filters requires knowledge of mutual couplings between coupled microstrip open-loop resonators. The two types of aperture couplings that are normally encountered in the filter design are shown in Figure 10.47, where h is the substrate thickness; a, w, and g are the dimensions of microstrip open-loop resonator, and dx and dy are the dimensions of aperture on the common ground plane. In Figure 10.47a the aperture is centered at a position where the magnetic field is strongest for the fundamental resonant mode of the pair of microstrip open-loop resonators on the both sides. Hence, the resultant coupling is the magnetic coupling and the aperture may be referred to as the magnetic aperture. In Figure 10.47b the aperture is centered at a position where the electric field is strongest and, thus, the resultant coupling is the electric coupling and the aperture may be referred to as the electric aperture.

Full-wave EM simulations have been performed to understand the characteristics of these two types of aperture couplings [75]. Two split resonant-mode frequencies, as discussed in Chapter 7, are easily identified by the two resonant peaks. The larger the aperture size, the wider the separation of the two modes and the stronger the coupling. However, it is noticeable that the high-mode frequency of the magnetic coupling and the low-mode frequency of the electric coupling remain unchanged regardless of the aperture size or coupling strength. This situation is different from that observed in the coupled microstrip open-loop resonators on the single layer, where the two resonant-mode frequencies are always changed against coupling strength. It has been found that the difference is because of the effect of the coupling aperture on the

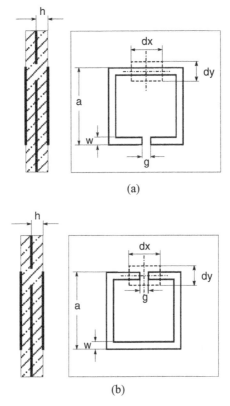

FIGURE 10.47 Two alternative aperture couplings of back-to-back microstrip open-loop resonators. (**a**) Magnetic coupling; (**b**) electric coupling.

resonant frequency of uncoupled resonators. With an aperture on the ground plane, the resonator inductance increases, whereas the resonator capacitance decreases as the aperture size is increased. Therefore, one would expect that the resonant frequency of microstrip open-loop resonators is either decreased against the magnetic aperture or increased against the electric aperture. The EM-simulated resonant frequencies of the decoupled resonators with the presence of a coupling aperture have verified this [75].

In order to extract coupling coefficients of the aperture-coupled resonators, following the formulation described in the Chapter 7, the two equivalent circuits of Figs. 7.4b and 7.5b can be employed with new definitions of self-inductance and self-capacitance [75]. In the magnetic-coupling circuit of Figure 7.5b, the self-inductance is defined by $L = L_0 + L_m$, with L_0 representing the resonator inductance without the coupling aperture. On the other hand, the self-capacitance in the electric coupling circuit of Figure 7.4b is defined by $C = C_0 - C_m$, with C_0 representing the resonator capacitance when the coupling aperture is not present. They are defined so as to account for the aperture effect. Now, if the symmetry plane T - T' in Figure 7.5b is subsequently replaced by electric and magnetic walls, we can obtain the following resonant-mode frequencies

$$f_e = \frac{1}{2\pi\sqrt{L_0 C}}$$
$$f_m = \frac{1}{2\pi\sqrt{(L_0 + 2L_m)C}}$$
(10.39)

As can be seen, the high-mode frequency f_e is independent of coupling and the low-mode frequency decreases as L_m or the coupling is increased. These two resonant frequencies are observable from the full-wave EM simulation and the magnetic coupling coefficient can be extracted by

$$k_m = \frac{L_m}{L} = \frac{f_e^2 - f_m^2}{f_e^2 + f_m^2}$$
(10.40)

Similarly, if we replace the symmetry plane T - T' in Figure 7.4b with an electric and a magnetic wall, respectively, we obtain the two resonant-mode frequencies

$$f_e = \frac{1}{2\pi\sqrt{LC_0}}$$
$$f_m = \frac{1}{2\pi\sqrt{L(C_0 - 2C_m)}}$$
(10.41)

In this case, the high-mode frequency f_m is increased with an increase of C_m or the coupling while the low-mode frequency f_e is kept unchanged as observed in the full-wave simulation. The electric coupling coefficient is then extracted by

$$k_e = \frac{C_m}{C} = \frac{f_m^2 - f_e^2}{f_e^2 + f_m^2}$$
(10.42)

Figure 10.48 shows some numerical results of the coupling coefficients together with the normalized center frequency of the aperture-coupled, microstrip open-loop resonators. The normalized center frequency is defined as $f_0 = (f_m + f_e)/2f_r$, with f_r the resonator frequency for a zero-aperture size. With the same size of the aperture, the magnetic coupling is stronger than the electric coupling. Both the couplings increase as the aperture sizes are increased. However, when the aperture sizes are increased, f_0 with the magnetic aperture decreases whereas f_0 with the electric aperture increases. This must be taken into account in the filter design. To compensate for frequency shifting, it is found that a more practical way is to adjust the open-gap dimension g of the open-loop resonators in Figure 10.47. This is because the slightly changing the open-gap g hardly changes the couplings, yet very efficiently tunes the frequency [75].

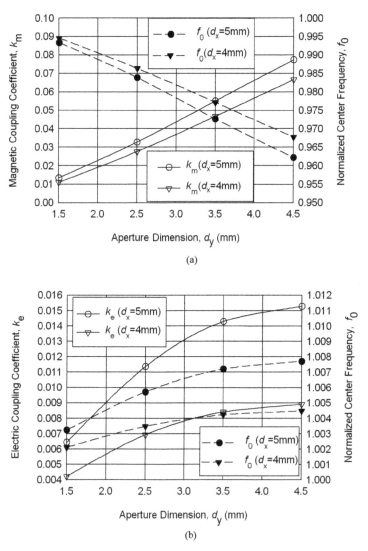

FIGURE 10.48 Simulated coupling coefficient and normalized center frequency of aperture coupling structures in Figure 10.47 with $h = 1.27$, $a = 16$, $w = 1.5$, $g = 1.0$ mm and a substrate relative dielectric constant of 10.8. (a) With magnetic coupling aperture. (b) With electric coupling aperture.

Design Examples

In order to demonstrate the feasibility and capability of this class of microstrip filters, three experimental four-pole bandpass filters having different filtering characteristics have been designed, fabricated, and tested. All the filters have a fractional bandwidth of 4.146% at a center frequency of 965 MHz. The first filter is designed to have a Chebyshev response having the following coupling matrix and external quality factor

$$M_{Chebyshev} = \begin{bmatrix} 0 & 0.0378 & 0 & 0 \\ 0.0378 & 0 & 0.0290 & 0 \\ 0 & 0.0290 & 0 & 0.0378 \\ 0 & 0 & 0.0378 & 0 \end{bmatrix} \quad (10.43)$$

$$Q_e = 22.5045$$

A single magnetic aperture coupling as described above is used to realize $M_{23} = M_{32}$. The second filter is designed to have an elliptic-function response having the coupling matrix and Q_e

$$M_{Elliptic} = \begin{bmatrix} 0 & 0.03609 & 0 & -0.00707 \\ 0.03609 & 0 & 0.03181 & 0 \\ 0 & 0.03181 & 0 & 0.03609 \\ -0.00707 & 0 & 0.03609 & 0 \end{bmatrix} \quad (10.44)$$

$$Q_e = 23.0221$$

In addition to a magnetic aperture coupling for $M_{23} = M_{32}$, in this case, an extra electric aperture coupling is utilized to realize $M_{14} = M_{41}$. The third filter is designed to have a linear phase response and its coupling matrix and external Q are given by

$$M_{Linear\ phase} = \begin{bmatrix} 0 & 0.03776 & 0 & 0.00804 \\ 0.03776 & 0 & 0.02494 & 0 \\ 0 & 0.02494 & 0 & 0.03776 \\ 0.00804 & 0 & 0.03776 & 0 \end{bmatrix} \quad (10.45)$$

$$Q_e = 22.5045$$

Note that the cross coupling $M_{14} = M_{41}$ is positive as compared with the negative one for the above elliptic-function filter. Thus, both $M_{23} = M_{32}$ and $M_{14} = M_{41}$ are realized using the magnetic aperture couplings. For comparison, the theoretical responses of the three experimental filters are plotted together in Figure 10.49, showing distinguishable filtering characteristics that are demanded for different applications.

The filters are fabricated using copper metallization on RT/Duroid substrates with a relative dielectric constant of 10.8 and a thickness of 1.27 mm. Figure 10.50 is a photograph of a four-pole experimental filter, where only two microstrip open-loop resonators on the top layer are observable. The dimensions of the resonators are $a = 16$ mm and $w = 1.5$ mm. It should be mentioned that except for a difference in arranging apertures on a common ground plane, the three filters have a very similar

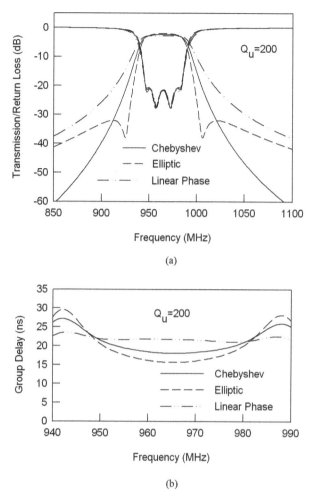

FIGURE 10.49 Theoretical responses of experimental filters with a unloaded resonator quality factor $Q_u = 200$. (**a**) Transmission and reflection loss. (**b**) Group delay.

outlook. The designed Chebyshev filter uses only a single magnetic aperture with a size of $dx = 4.0$ mm and $dy = 2.55$ mm. For the elliptic-function filter, both magnetic and electric apertures are needed, having sizes of 4.5 × 2.55 mm and 4.0 × 2.55 mm, respectively, whereas the linear-phase filter uses two magnetic apertures of 4.0 × 2.4 mm and 4.0 × 1.3 mm.

The measured filter performances, including the group-delay responses, are plotted in Figure 10.51. In general, good filter performance has been achieved from this single iteration of design and fabrication. In Figure 10.51a, the measured minimum passband insertion loss for the Chebyshev filter was about 2.3 dB. This loss is expected mainly because of conductor loss. The measured bandwidth was slightly wider,

FIGURE 10.50 Photograph of a fabricated four-pole back-to-back microstrip open-loop resonator filter.

which would be attributed to stronger couplings. The measured elliptic-function filter in Figure 10.51(b) showed the two desirable transmission zeros. However, it exhibited an asymmetrical frequency response. This is most likely to have been caused by frequency-dependent couplings, especially the cross coupling of M_{14}. The minimum passband insertion loss for this filter was also measured to be 2.3 dB. Figure 10.51c shows the measured performance of the linear-phase filter. The measured minimum passband insertion loss was about 2.6 dB. The loss is slightly higher than that of the previous two filters, which is expected from the calculated performance in Figure 10.49. The measured filter did show a linear group delay in the passband, but it also showed an asymmetrical frequency response. The latter was again attributed to the frequency-dependent couplings. The issue of frequency-dependent couplings has been intensively investigated [75], with some useful suggestions to improve the filter asymmetrical response.

10.6.2 Filters with Defected Ground Structures

In recent years, there has been a growing interest in defected ground structures in planar transmission lines that exhibiting multielectromagnetic band gaps or stopbands in frequency responses. The so-called defected ground structure (DGS) is actually a patterned or slotted structure introduced on the ground plane of microstrip or coplanar waveguide (CPW), which allows interaction among microstrip, CPW, and slotline. Novel types of compact filtering devices can be designed based on periodic or nonperiodic DGSs [76–106].

10.6.2.1 Circuit Model for Defected Ground Structures Some examples of DGSs are illustrated in Figures 10.52 and 10.53. The structures of Figures 10.52a and 10.53a may be referred to as the microstrip and CPW L-shaped DGSs, respectively. The structures of Figures 10.52b and 10.53b can be seen as the metal-loaded dumbbell-shaped DGSs in microstrip and CPW, respectively. These unit cells of

394 COMPACT FILTERS AND FILTER MINIATURIZATION

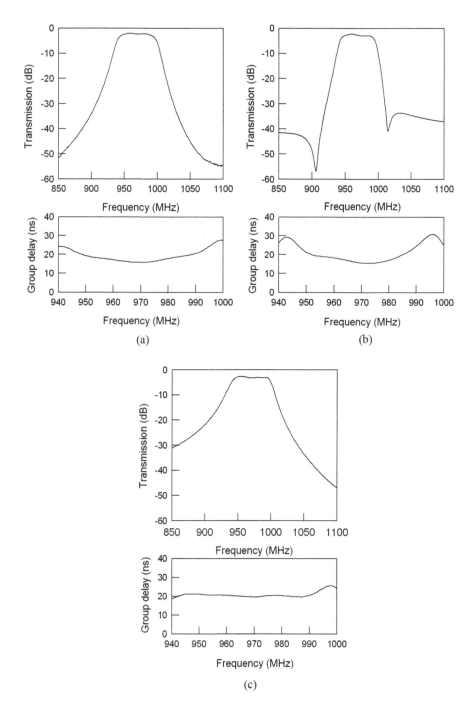

FIGURE 10.51 Measured performances of the experimental four-pole filters. (**a**) Chebyshev response. (**b**) Elliptic-function response. (**c**) Linear-phase response.

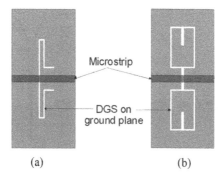

FIGURE 10.52 Typical microstrip defected ground structures. (**a**) L-shaped DGS. (**b**) Metal-loaded dumbbell-shaped DGS.

DGSs possess some interesting characteristics for developing DGS-based filtering devices. For instance, the floating metals inside the dumbbell-shaped DGSs can be used not only to control the separation of the first two resonant modes, but also to facilitate DC bias for electronic tuning.

A direct optimum design of an array of DGSs using full-wave EM simulations can be time consuming, particularly when the number of DGSs in a device is large. In this case, the optimization based on an equivalent circuit of the device is highly desirable. To this end, the key issue is to obtain a simple and accurate circuit model for unit cells of DGSs. A lumped-element circuit model has been proposed to model the first two resonant modes of a very specific slot resonator in the ground plane for microstrip structures [76]. More recently, circuit models based on slotlines have been reported for microstrip DGSs with slotted ground plane [77,78]. Although this type of circuit model using slotline elements can present a periodic frequency response, it requires accurate models for slotline effective dielectric constant and characteristic impedance, since both parameters are highly frequency-dependent. Hence, this type

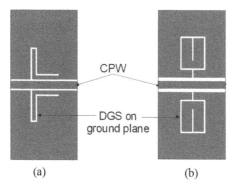

FIGURE 10.53 Typical coplanar waveguide (CPW) defected ground structures. (**a**) L-shaped DGS. (**b**) Metal-loaded dumbbell-shaped DGS.

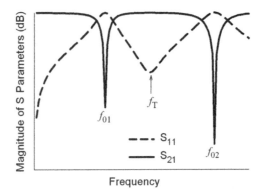

FIGURE 10.54 Typical frequency response of the DGSs that exhibit multi stop bands.

of circuit model would be more suitable for simple transverse slots [78]. In what follows, a more general circuit model [83] that is able to represent varieties of DGSs in either microstrip or CPW is described.

Many DGSs including those depicted in Figure 10.52 and 10.53 exhibit multi stopbands in frequencies. Figure 10.54 shows the typical frequency response up to the second stopband for such DGSs, where f_{01} and f_{02} are the first and second resonant frequencies, respectively, and f_T denotes a transit frequency. For many DGS device designs, a frequency range up to the second stopband is most interesting. Therefore, the circuit model, as shown in Figure 10.55, can be used to take into account the first and second resonant modes. Thus, a unit cell DGS, either in microstrip or CPW, is modeled by the two LC resonators, i.e., L_1 and C_1, L_2, and C_2, in connection with a T-network consisting of C_p, Ls_1, and Ls_2. The T-network is essential here to represent the interaction between the two resonators. Below the transit frequency f_T the first resonator dominates the frequency characteristics, whereas the second resonator is dominant for the frequency above f_T.

In Figure 10.55, Z_0 is the characteristic impedance of the transmission line. To extract all other circuit parameters, the following data are sufficient and can be easily found from EM simulations: f_{01}, f_{02}, f_T, Δf_{3dB_1} (the 3-dB bandwidth at f_{01}), Δf_{3dB_2}

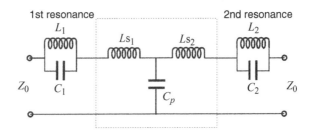

FIGURE 10.55 General circuit model for the DGSs.

(the 3-dB bandwidth at f_{02}), X_{11}, X_{22}, and X_{21}, which are the imaginary parts of three Z parameters at f_T. It can be shown by circuit theory that

$$C_i = \frac{1}{Z_0} \cdot \frac{1}{4\pi \Delta f_{3dB\_i}} \quad \text{for} \quad i = 1, 2 \tag{10.46}$$

$$L_i = \frac{1}{(2\pi f_{0i})^2 C_i}$$

$$C_p = -\frac{1}{2\pi f_T X_{21}}$$

$$Ls_i = \frac{X_{ii} - X_{21}}{2\pi f_T} + \frac{L_i}{(f_T/f_{0i})^2 - 1} \quad \text{for} \quad i = 1, 2 \tag{10.47}$$

The above circuit model has been used to model various DGSs and excellent agreement with EM simulations and/or measurements have been obtained. Some examples are given next.

Modeling Example 1

The first modeling example is for a microstrip L-shaped DGS shown in Figure 10.56a, where all the dimensions are in millimeters. The substrate used has a thickness of 1.27 mm and a relative dielectric constant of 10.8. The microstrip line has a characteristic impedance of 50 Ω at the reference planes. In this case, the circuit parameters

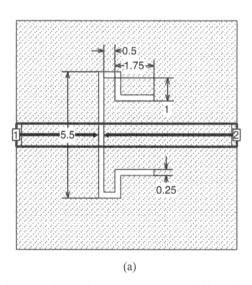

(a)

FIGURE 10.56 (a) Layout of the microstrip DGS modeled (dimensions in millimeters). (b) and (c) Magnitude and phase responses of theory (circuit model) and full-wave EM simulation. (*Continued*)

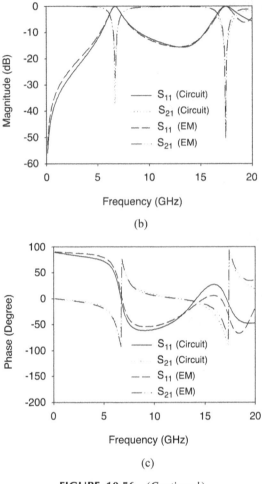

FIGURE 10.56 (*Continued*)

extracted from the EM simulation are $C_1 = 1.349\,\text{pF}$, $L_1 = 0.4184\,\text{nH}$, $C_2 = 1.02\,\text{pF}$, $L_2 = 0.082\,\text{nH}$, $C_p = 0.028\,\text{pF}$, $Ls_1 = -0.749\,\text{nH}$, and $Ls_2 = 0.5745\,\text{nH}$. Note that the negative value for Ls_1 is perfectly allowed for the circuit modeling. This is similar to a lumped-element inverter with negative elements in which the negative elements physically may be absorbed by the adjacent reactance components in the circuit, as discussed in Chapter 3 for immittance inverters. This DGS has two widely separated resonant modes. Nevertheless, excellent agreement between the theory (circuit model) and full-wave EM simulation, for both magnitude and phase responses, can be observed from DC to 20 GHz in Figure 10.56b and c, respectively.

Modeling Example 2

This example demonstrates the application of the above circuit model for modeling a period coplanar waveguide (CPW) GDS, as shown in Figure 10.57a, which is on a 1.27-mm thick dielectric substrate with a relative dielectric constant of 10.8. The unit cell of DGS is that of Figure 10.53a. With the dimensions shown, the circuit parameters extracted from the EM simulation are $C_1 = 2.059$ pF, $L_1 = 0.8746$ nH, $C_2 = 1.457$ pF, $L_2 = 0.2303$ nH, $C_p = 0.0385$ pF, $Ls_1 = -1.009$ nH, and $Ls_2 = 1.093$ nH. The periodicity of the unit cells is 7.75 mm. An equivalent circuit for this three-cell DGS device is simply obtained by cascading the unit cell circuit model in Figure 10.55 and the 7.75-mm long transmission line alternatively. Figure 10.57b show the circuit modeled and EM-simulated S parameters.

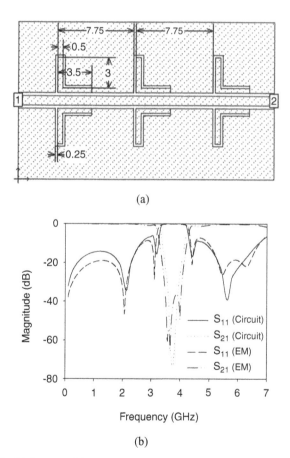

FIGURE 10.57 (a) Layout of the three-cell period DGS in CPW (dimensions in millimeters). (b) Magnitude responses of theory (circuit model) and full-wave EM simulation.

UWB Filter Using Nonuniform Periodical Slotted Ground Structure Defected or slotted ground structures have been deployed in developing compact filters [83–106]. For example, an ultra-wideband (UWB) bandpass filter with a multilayer nonuniform periodical structure on liquid crystal polymer (LCP) substrates has been developed [99,100]. Shown in Figure 10.58a is the basic unit of a proposed bandpass filtering structure, which includes three metal layers. Microstrip input and output (or excitation ports) are at the top layer. Broadside coupled stub lines are designed on the top and middle layers, respectively. A slotted ground structure is realized on the bottom ground layer. The primary operation of this bandpass filter unit may be explained in light of an equivalent circuit of Figure 10.58b. The series capacitor $C2$ represents the coupling between the broadside coupled open-circuited stubs, which

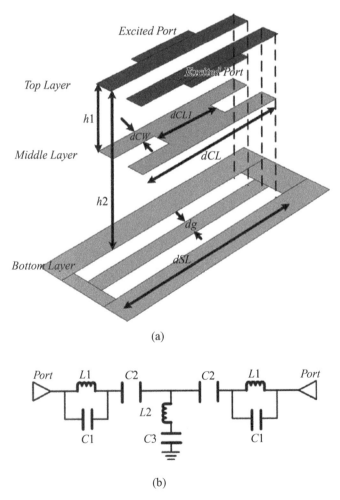

FIGURE 10.58 Bandpass filter unit with slotted ground structure. (**a**) 3-D view; (**b**) equivalent circuit.

provide highpass characteristic; the shunt capacitor $C3$ and inductor $L2$ are used to model the equivalent capacitance and inductance of the metal strip pattern on the middle layer with respect to the ground. The inductance $L1$ and capacitance $C1$ result from the etched slots on the ground. Since $L1$ and $C1$ only model the first or fundamental resonance of the slotted ground structure, in order to more accurately model a wideband frequency response, the equivalent circuit of Figure 10.58b would need to be modified in a way similar to that addressed in Figure 10.55.

The slotted ground structure unit is investigated by using full-wave simulation. In the simulation, liquid crystal polymer substrate, which has a relative dielectric constant $\varepsilon_r = 3.15$ and a loss tangent $\tan\delta = 0.0025$, is used to support the metal layers. The physical dimensions set in the simulation, as referred to in Figure 10.58, are: $h1 = 25$ μm, $h2 = 800$ μm, $dCW = 0.7$ mm, $dg = 0.7$ mm, $dCL = 4.0$ mm, and $dCL1 = 2.0$ mm. Figure 10.59 shows the full-wave simulated scattering parameter of S_{21} with varied slot length dSL. For a given dSL, a transmission zero can be observed at upper stopband, which is generated by the resonance of the slot. As the slot length dSL changes, the upper passband cutoff frequency shifts because of the shift of the transmission zero. This indicates the upper cutoff frequency of the proposed bandpass filter unit is determined by the slot length dSL.

On the other hand, it can be shown that the lower passband cutoff frequency is determined by the series capacitance $C2$ or the broadside coupled stub lines, which can be controlled by the length dCL, the width dCW, and the separation $h1$, referred to in Figure 10.58. The series capacitance $C2$ can be extracted by matching the response of a series capacitor to the full-wave simulated S_{21} of broadside coupled stubs. Longer and wider coupled lines with a smaller separation of $h1$ result in larger series capacitance. For a wider filter passband, a larger series capacitance is required, which can be easily implemented in the broadside coupling structure.

Several bandpass units of Figure 10.58 can be cascaded to design a UWB bandpass filter with improved performance. An example of this type of filter using 10 units

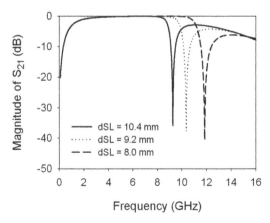

FIGURE 10.59 Full-wave simulated S_{21} with varied slot length dSL for the bandpass filter unit with slotted ground structure.

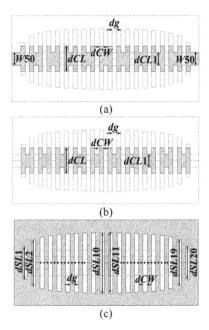

FIGURE 10.60 Layouts for 10-cell UWB bandpass filter with slotted ground structure. (a) Top layer; (b) middle layer; (c) bottom layer.

is given in Figure 10.60. The top layer of Figure 10.60a and the middle layer of Figure 10.60b have a similar array of open-circuited stubs. However, the patterns between the two layers are alternatively placed in order to facilitate a cascade signal path through broadside coupling between each pair of open-circuited stubs, which are overlapped on the two layers. The bottom metal layer of Figure 10.60c serves as the ground, which has a nonuniform slot pattern in order to produce multiple transmission zeros over a wider frequency range to improve the upper stopband performance.

In principle, the filter of Figure 10.60 is a multilayer nonuniform periodical structure, which allows a very compact UWB bandpass filter with a wide upper stopband to be built. For designing the 10-unit UWB filter, the nonuniform slot lengths are chosen as

$$dSL_i = dSL_{2n+1-i} = dSL_n \cdot C^{(\frac{1}{n}-\frac{1}{i})} \quad \text{for} \quad i = 1, 2, \ldots 10 \tag{10.48}$$

where dSL_n is a reference slot length, determined by the upper cutoff frequency of the filter. C is a constant and n is the number of units used in the design. In this case, $n = 10$, and $C = 2.0$ is adopted in the design to achieve a good in-band matching. The UWB filter is designed on multilayer liquid crystal polymer substrate with a relative dielectric constant of 3.15. The thickness between the top and middle layers is 25 μm, while the thickness between the middle and ground layers is 775 μm. In the design, very narrow lines are avoided to relax the fabrication tolerance for a

TABLE 10.2 Geometry Parameters (in mm) for the UWB Filter of Figure 10.60

$W50$	2.0	$dCL1$	2.0
dCW	0.7	$dSL10$	9.6
dg	0.7	$dSL11$	9.6
dCL	4.0	$h2$	0.8
$h1$	0.025		

low-cost fabrication. As a result, the broadside coupling stubs with a width of 0.7 mm are chosen. The stub length, dCL, is then designed to meet the desired capacitance that affects the lower cutoff frequency. The upper cutoff frequency is decided by the longest slots on the ground, which have a length of $dSL10 = dSL11$. The slot width is the same as the stub width. The length $dSL10$ is then designed such that a transmission zeros at the upper cutoff frequency of 10.6 GHz is obtained. Table 10.2 lists the dimensions of the designed filter. The other slot lengths can be found from Eq. (10.48).

Since the 25-μm thickness liquid crystal polymer film has a tolerance in the fabrication, it has been shown that the changing of the thickness between top and second layers only has an effect on the series capacitance, which mainly affects the lower cutoff frequency [100].

The designed UWB bandpass filter has been fabricated by using multilayer liquid crystal polymer lamination technology. Figure 10.61 shows the photograph of fabricated filter, which has a circuit size of 27.3 ×15.6 mm, excluded the 50-Ω microstrip feeding lines.

The fabricated filter is measured using HP8510 vector network analyzer (VNA). Anritsu 3680 universal test fixture is used to connect fabricated prototype with VNA in the measurement, although a filter of this type could be supported like a suspended stripline device in a system box. Thru-reflect-line calibration technique is adopted to remove the effects of the test fixture from the measurement. The measured results are plotted in Figure 10.62. The measured filter has a wide passband of approximate 7 GHz and an insertion loss of 0.66 dB at a center frequency of 6.85 GHz. It also has an excellent selectivity at upper passband edge. The rejection level increases from 10 to 55 dB in a narrow stopband, that is, 10.3 − 10.5 GHz. A wide stopband is

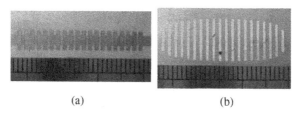

(a)   (b)

FIGURE 10.61 Photographs for fabricated 10-cell UWB bandpass filter with slotted ground structure. (**a**) Top view; (**b**) bottom view.

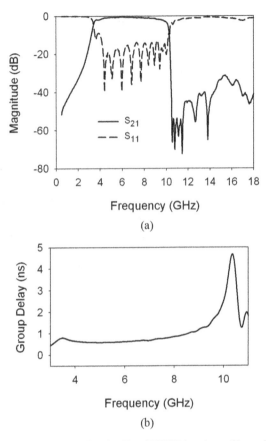

FIGURE 10.62 Measured results for the 10-cell UWB bandpass filter with slotted ground structure. (**a**) S parameters; (**b**) group delay.

obtained from 10.4 to 18.0 GHz, with a rejection level higher than 31 dB. Shown in Figure 10.62b is the measured group delay, which is 0.67 ns at 6.85 GHz with a variation within 0.2 ns from 3.15 to 8.3 GHz. At the edge of upper passband, an increased group delay, caused by the sharp selectivity, is observed.

10.6.3 Substrate-Integrated Waveguide Filters

Air-filled rectangular waveguide can be used to design high-performance filters, but requires complex transitions to integrated planar circuits in addition to a bulky size. A straightforward solution is to integrate the rectangular waveguide into the microstrip substrate, forming the so-called substrate integrated waveguide (SIW) [107]. This will certainly reduce the Q factor of the waveguide because of dielectric filling and volume reduction, but the entire circuit including planar circuit, transition, and waveguide can be constructed using standard print circuit board (PCB) or other multilayer processing techniques, such as LTCC and LCP. A common method to form the SIW sidewalls is to use metallic via-holes or postwalls [107,108]. Practical techniques have been developed to design and implement substrate integrated waveguide filters [109–120]. As compared with a normal microstrip line filter, SIW filers are expected to be of low-loss and high-power handling.

10.6.3.1 Substrate Integrated Cavity Figure 10.63 shows a typical SIW cavity or resonator. The cavity is constructed entirely on a dielectric substrate with a thickness of h. The cavity has a width of W and length of L. Usually, h is smaller than W. The cavity's top and bottom walls are formed by the metallic plates on the both sides of the substrate, whereas its sidewalls are formed by metallic vias or posts. The posts have a diameter of d and the pitch length or the separation between adjacent posts is denoted by p. These are two important parameters that should be properly chosen in order to minimize the radiation loss. The study in Dealandes and Wu [109] reveals that for an electrically small post, that is, $d < 0.2\lambda$, where λ is the wavelength in the dielectric material, the radiation loss is negligible with a ratio d/p of 0.5. The radiation loss tends to decrease as the post gets smaller for a constant ratio d/p, which is conditioned by the fabrication process.

For the dominant TE_{101} mode, the size of the SIW cavity may be determined by the corresponding resonance frequency from [110]

$$f_{TE_{101}} = \frac{c}{2\sqrt{\mu_r \varepsilon_r}} \sqrt{\left(\frac{1}{W_{eff}}\right)^2 + \left(\frac{1}{L_{eff}}\right)^2} \qquad (10.49)$$

where W_{eff} and L_{eff} are the equivalent width and length of the SIW cavity, respectively, and given by

$$W_{eff} = W - \frac{d^2}{0.95p}, \quad L_{eff} = L - \frac{d^2}{0.95p} \qquad (10.50)$$

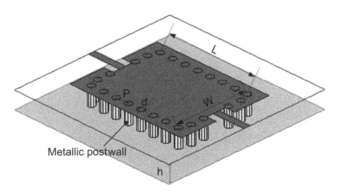

FIGURE 10.63 Configuration of substrate integrated waveguide cavity using metallic post sidewalls.

where W and L are the real width and length of the SIW cavity, c is the velocity of light in free space, and μ_r and ε_r are the relative permeability and relative permittivity of the substrate. In most cases, the dielectric substrate is a nonmagnetic material and, hence, $\mu_r = 1$.

For example, consider a SIW cavity on a 1-mm thick dielectric substrate with a relative dielectric constant $\varepsilon_r = 3$, $W = 22$ mm, $L = 30$ mm, $d = 2$ mm, and $d/p = 0.5$ or $p = 4$ mm. Applying Eqs. (10.49) and (10.50), the resonance frequency for the TE_{101} mode is found to be 5.103 GHz, which agrees with the full-wave EM simulated result. Theoretically, this resonance frequency is independent of the substrate thickness. In reality, the substrate thickness may affect the resonance frequency, but very slightly. As can be shown by EM simulation, in this case, when the substrate thickness h varies from 0.5 to 1.5 mm, the resonance frequency only changes less then 0.2% with respect to the resonance frequency for $h = 1$ mm. The thinner substrate tends to result in a slightly lower frequency. However, the substrate thickness can have more of an impact on the loss of the SIW cavity. Assume a loss tangent $\tan\delta = 0.002$ for the substrate and a conductivity $\sigma = 5.8 \times 10^7$ S/m for all the metals (copper) with 17-μm thickness. The EM simulation shows that the substrate thickness h varies from 0.5 to 1.5 mm and the unloaded quality factor changes from 250 to 360. The thicker substrate leads to a lower loss or higher Q. It can also be shown that the dielectric loss appears to be more than the conductor loss, while the radiation loss is negligible.

10.6.3.2 Substrate-Integrated Waveguide Filters To design a SIW filter, shunt-inductive coupling can easily be realized, resembling that in an air-filled rectangular waveguide filter. The shunt-inductive coupling can be in the form of either a shunt-inductive post or an iris (aperture). The development of these types of SIW filters have been reported [109–116]. For example, Figure 10.64 is a configuration of three-pole SIW bandpass filter using shunt-inductive posts. The SIW filter consists of four coupling posts centered in the guide and two microstrip-to-SIW transitions at the input and output, respectively. The small posts facilitate the input and output couplings, while the large posts are for the inter-resonator couplings.

The design of the type of filter may follow a procedure similar to that for designing an air-filled waveguide filter. It may also be designed based on the coupling-matrix

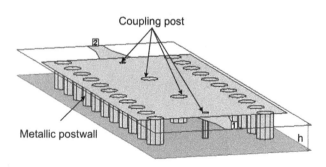

FIGURE 10.64 Configuration of SIW filter using shunt-inductive posts for coupling.

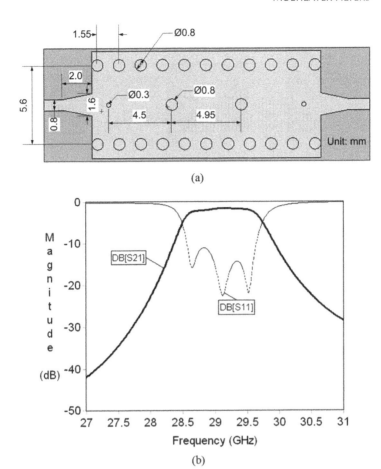

FIGURE 10.65 (a) Layout of a three-pole SIW filter on a dielectric substrate with a relative dielectric constant of 2.2 and a thickness of 0.25 mm. (b) EM simulated filter performance.

method, as discussed in Chapter 7. Figure 10.65a shows the layout of a designed SIW filter that has a configuration of Figure 10.64. The filter is designed on a dielectric substrate with a relative dielectric constant of 2.2 and a thickness of 0.25 mm. Figure 10.65b plots the EM-simulated performance of the filter, obtained using a commercially available simulation tool [22]. The simulation assumes a loss tangent $\tan\delta = 0.001$ for the substrate and a conductivity $\sigma = 5.8 \times 10^7$ S/m for all the metals (copper) with 17-μm thickness.

Generally, an inductive-post filter may have to use a number of posts of different diameters, all centered in the guide. This is necessary for control couplings, for which a smaller diameter results in a larger coupling. Nevertheless, in order to reduce both fabrication complexity and cost, the filter geometry can be replaced by an offset postarrangement with the same post diameter. A filter of this type is demonstrated in Dealandes and Wu [109].

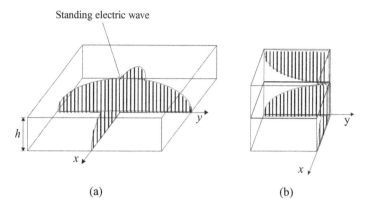

FIGURE 10.66 (a) Conventional TE_{101} waveguide resonator with standing waves. (b) Folded-waveguide quarter-wavelength resonator with standing waves.

Folded-Waveguide Resonator Filters When designed at a low microwave frequency, a SIW filter may require a large circuit area or footprint. To overcome this problem and to best exploit a multilayer circuit technology, a new concept of folded-waveguide resonator, as reported [121–123], can be deployed for miniaturizing SIW filters. To explain this concept, Figure 10.66a illustrates standing waves in a conventional TE_{101}-mode waveguide resonator. If we can fold this conventional TE_{101} waveguide resonator along its x-axis and then remove one-half of the folded waveguide along its y-axis while maintaining the original standing waves, we obtain the so-called folded-waveguide quarter-wavelength resonator of Figure 10.66b, which has only one-quarter of the footprint of the conventional TE_{101} waveguide resonator. This is feasible if the two faces along the axes in Figure 10.66b are perfect magnetic walls or perfect magnetic conductors (PMC).

The above folded-waveguide resonator concept is attractive for constructing compact-cavity resonators and filters with small footprints. The question is then how to implement this concept. Figure 10.67 illustrates a practical realization, where a conceptual folded-waveguide quarter-wavelength resonator has two orthogonal slots

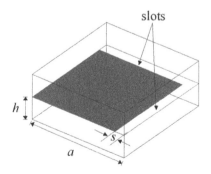

FIGURE 10.67 The folded waveguide quarter-wavelength resonator.

open on the common metal plate. It is that the two slots mimic magnetic walls to allow electromagnetic fields to continue from one half to the other of the folded waveguide cavity so that the desired TE_{101} resonant mode can be maintained. It can be shown by full-wave EM simulation that the standing wave pattern in the folded-waveguide resonator is one-quarter of its counterpart in a conventional TE_{101} waveguide resonator [123].

Substrate integrated waveguide filters using folded-waveguide resonators have been demonstrated [118–120]. For example, Figure 10.68 shows a configuration

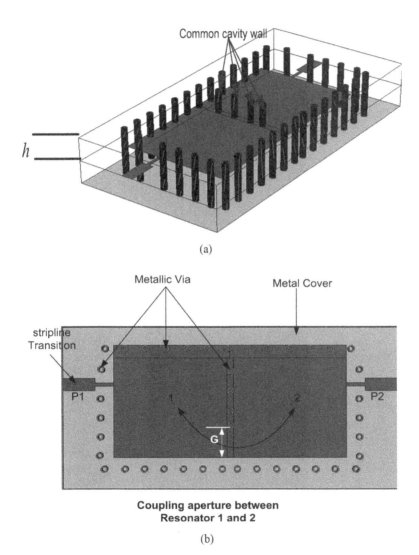

FIGURE 10.68 (a) 3-D view of a two-pole substrate integrated folded-waveguide filter. (b) Plane view.

of two-pole substrate-integrated folded-waveguide resonator filter. It includes two-substrate integrated folded-waveguide resonators, which are separated with a common cavity wall-formed metallic vias. The two resonators 1 and 2 are coupled to each other by coupling aperture (G) as indicated in Figure 10.68b. It can be shown that the coupling of this type is basically a magnetic coupling. To ensure the filter can be conveniently integrated with other microwave circuits, a microstrip-stripline transition is necessary at the input or output, where the stripline section is directly tapped at the input or output resonator.

The filter is designed on a dielectric substrate material with a relative dielectric constant of 3.2 and a thickness of $h = 0.762$ mm for each layer. Shown in Figure 10.69 is the view of three metal layers of the filter. The filter is simulated using commercially available EM-simulation software [22]; the results are plotted in Figure 10.70.

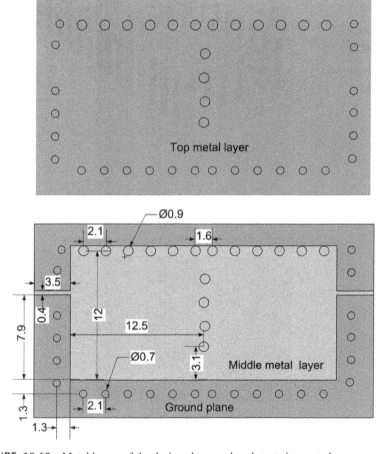

FIGURE 10.69 Metal layers of the designed two-pole substrate integrated folded-waveguide filter on substrate material with a relative dielectric constant of 3.2 and a thickness of 0.762 mm between adjacent metal layers (all dimensions are in millimeters).

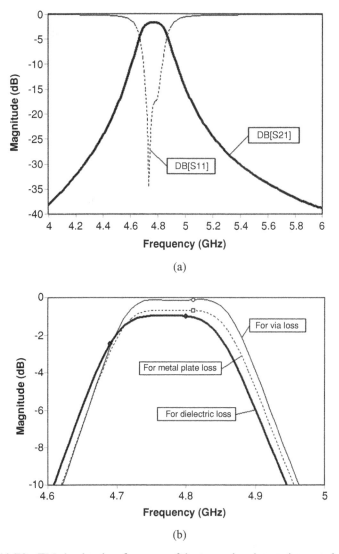

FIGURE 10.70 EM-simulated performance of the two-pole substrate integrated folded-waveguide filter. (**a**) Response with all material losses included. (**b**) Passband insertion loss when individual loss mechanisms are considered.

EM simulations are also carried out to investigate the loss effect on the filter because of individual loss mechanisms, namely, the conductor loss in the metal layers, the dielectric loss in the substrate, and the conductor loss in the vias, respectively. Figure 10.70b shows the simulated results. Each curve shows the passband response when only one loss mechanism is considered. To simulate the conductor loss, 35-μm thick copper with a conductivity of $\sigma = 5.8 \times 10^7$ S/m is assumed. To consider the dielectric loss in the simulation, a dielectric loss tangent of 0.003 has been chosen. From Figure 10.70b, one can see that the dielectric loss is still significant in this miniature filter.

Higher-order filters of this type can be designed and constructed based on various topologies, including those reported in Al-otaibi and Hong [124,125]. Other miniature folded-waveguide filters [1,126–128] can also be implemented with integrated substrates.

10.6.4 LTCC and LCP Filters

Recent development of multilayer filters has been driven by the emerging device technologies based on two advanced materials, namely, the low-temperature co-fired ceramic (LTCC) and liquid crystal polymer (LCP) [129,130]. These technologies enable the creation of monolithic, 3D, cost-effective microwave circuits/packages [99,100,131–192].

10.6.4.1 LTCC Technology and Filters Low-temperature co-fired ceramic (LTCC) technology, which borrows on thick-film materials but builds structures based on co-fired ceramic designs, is developed for firing temperature (typically <1000°C) below the melting points of many good conductor materials, such as gold, silver, and copper. This allows the low–loss metallizations to be made and is a significant advantage over high-temperature co-fired ceramics (HTCC) technology. HTCC materials require firing temperature >1600°C, which limit the metallizations to more refractory, lower conductivity metallization, such as tungsten and manganese. These metallizations are neither wire-bondable nor solderable and, therefore, require subsequent plating.

LTCC is a glass-ceramics composite that provides a unique and versatile approach to highly integrated, high-performance electronic packaging. The technology offers significant benefits in terms of design flexibility, density, and reliability. The multilayer capability of LTCC technology allows multiple circuitry to be handled in a single self-contained, hermetic package. Monolithic LTCC structures incorporating buried components allow increased design flexibility by providing a mechanism for establishing microstrip, stripline, coplanar waveguide, and DC lines within the same medium. The ability to integrate digital, analog, RF, microwave, and buried passive components in this manner reduces assembly complexity and improves overall component and system reliability by reducing part count and interconnections. In addition, the reduced weight of LTCC packages and the low loss characteristics of the dielectric and conductors make LTCC an ideal candidate for high-performance commercial and military electronic systems.

Unlike thick-film process, where successive lamination and firing steps cause bowing and line degradation, the single-step lamination and firing of LTCC produces a circuit pattern at substrate with fine, high-quality line definition. In addition, the elimination of costly repeated firings greatly increases the number of conductive layers achievable. The current state of the LTCC technology allows high-density circuitry (as fine as 50-μm lines and spaces) interconnected with conductive vias (as fine as 50-μm diameter).

The dielectric properties of LTCC materials critical for performance of microwave circuits are dielectric loss, dielectric constant, insulation resistance, and dielectric strength. Low loss is desirable for transmit and receive signals, while low dielectric constant is important for high-speed signal processing. High insulation resistance and dielectric strength are also desirable. These properties are a function of chemical composition, processing, and interactions with conductor materials. For instance, LTCC materials based on Ca-B-Si-O glass have a composite relative dielectric constant of 6 ± 0.3 and low dielectric losses comparable to that of 99% alumina [133]. LTCC materials based on Pb-B-Si-O, Ba-Al-Si-O have the relative dielectric constants $\approx$ 8-10, but higher dielectric losses as compared with Ca-B-Si-O-based glass ceramics. LTCC materials with much higher dielectric constants are also available [135].

For circuit fabrications, LTCC technology uses thick-film dielectric materials that are cast as a tape in place of a screen-printable thick-film dielectric composition. For example, the 951 Green Tape (Green Tape is a trademark of DuPont Corporation) technology has been used to manufacture multilayer circuits with more than 50 layers [144]. The tape is blanked to size and registration holes are then punched. Vias are formed in the dielectric tape by punching or drilling and conductor lines are screen-printed on the tape. When all layers have been punched and printed, the tape layers are registered, laminated, and co-fired. The co-fire process (i.e., the dielectric and conductor are fired in a single step) involves fewer firing steps, enables inspection of punched and printed layers prior to lamination, and has greater layer capability than conventional sequential thick-film technology. LTCC applications are in volume production using gold (Au) and silver (Ag), as well as combined Au and Ag metallurgies.

LTCC technology can be used to obtain all of the circuit design advantages of stripline without the need for any mechanical mechanism to hold the circuit together. These circuits can then be directly wire bonded to ceramic-based hybrids. The use of LTCC technology also eliminates air gaps or other discontinuities present in conventional stripline assemblies, which will enhance the repeatability and performance of these circuits. All of these factors combined can greatly reduce the size, weight, cost, and mechanical complexity of systems that require stripline design advantages. For example, the design and fabrication of a LTCC stripline edge-coupled bandpass filter has been described [132]. The stripline filter was designed to realize a fourth-order Chebyshev prototype response from 3.4 to 4.6 GHz with 0.05-dB ripple. This filter was fabricated using DuPont Type 851A3 Green Tape and each sheet of this LTCC material has a postfired thickness of 0.165 mm and a relative dielectric constant $\varepsilon_r = 7.9$. Circuits can use anywhere from 4 to 20 sheets of this material, which allows a great deal of design flexibility. The substrate thickness was chosen to be the maximum of 20 sheets (3.3-mm thick) so that the required coupled line gaps would be as large as possible for easy fabrication. The stripline and microstrip conductors were patterned with a 400-mesh thick film screen using co-fired ink. The top and bottom ground planes were then connected on all edges, except for the area around the microstrip transitions to the filter input and output, by painting wrap-around grounds with postfired ink. The measured midband insertion loss was 1.5 dB and the filter achieved approximately 50 dB of ultimate rejection. The second harmonic response was also reduced because of the equivalence of even- and odd-mode phase velocities in the stripline structure. Several seven-pole LTCC stripline edge-coupled bandpass filters were also designed with 100% yield to tolerances in materials and manufacturing processes [135]. The measured results indicated that this type of filter has great promise for miniaturization of advanced receivers and signal sources.

Miniature LTCC filters, which are usually second- or third-order, can be made for portable telephone applications [139–143]. A two-pole combline bandpass filter is reported [139]. It is made up of five layers of low-temperature co-fired ceramics. The LTCC material is a Bi-Ca-Nb-O system and its relative permittivity is 58 with the temperature coefficient regarding stripline resonator frequency of $+20$ ppm/°C. This material is co-fired with silver electrode-conductive paste at 950°C. The filters are designed at 950 MHz and 1.9 GHz. The dimensions of fabricated LTCC bandpass filter are $4.5 \times 3.2 \times 2.0$ mm. The measured insertion losses are less than 1.8 dB at 950 MHz and less than 1.2 dB at 1.9 GHz, respectively. The return losses are smaller than -20 dB in the each passband. The transmission characteristics have an attenuation pole near the passband. Therefore, large attenuation of more than 30 dB is obtained at the frequency offsetting 100 MHz under 950 MHz and at the frequency offsetting 300 MHz under 1.9 GHz. Further, the experimental results show that the temperature variation of the pole frequency is within a megahertz from -20°C to $+60$°C.

Another miniature LTCC bandpass filter for 2.4-GHz wireless LAN (WLAN) applications is also described [131]. The filter was designed to have a center frequency of 2.45 GHz with an operating bandwidth of at least 100 MHz. Figure 10.71 depicts the decomposed as well as complete views for the filter physical layout. This is a second-order filter based on two capacitor-loaded transmission-line resonators and three capacitive inverters. In Figure 10.71, the diagram in the top-left corner gives the grounding structure of the filter. It consists of three conductor plates at three different layers connecting through multiple via-holes. The structure in the top-right corner is the two resonators. The meandered lines are sandwiched between the bottom two ground planes to form two striplines (or inductors). Moreover, each line is loaded with a capacitor at one of its ends. Their other ends are grounded through the use of via-holes. Coupling capacitors, shown on the bottom-left, are implemented by three conducting plates sandwiched between the two conductors of the loading capacitors. Finally, the complete structure is shown on the bottom-right. The resulting filter has an overall size of 1.1×2.3 mm with a total of ten dielectric layers with a dielectric constant of 7.8. The measured in-band insertion loss is around 2 dB with reasonably good return loss response.

10.6.4.2 LCP Technology and Filters

Liquid crystal polymer (LCP) is a new and promising organic thermoplastic material. Table 10.3 shows typical characteristics of commercially available LCP materials. ULTRALAM 3850 is a core film with 315°C melting temperature [193], whereas ULTRALAM 3908 is a bonding film with 280°C melting temperature [194]. Generally, LCP has extremely low moisture-absorption

MULTILAYER FILTERS 415

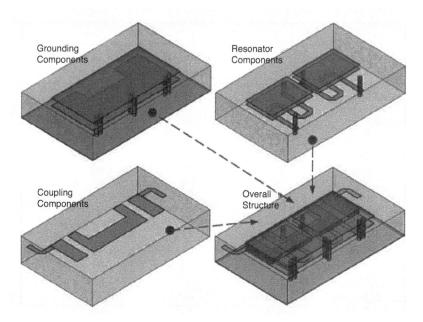

FIGURE 10.71 Physical structure of a miniature two-pole LTCC filter. (Taken from [131], © 2008 IEEE.)

character, which can reduce the baking time and maintain stable electrical and mechanical properties in humid environment. It exhibits good dimensional stability and is suitable for the mechanical fabrication. It has extraordinary barrier properties comparable to that of glass and low coefficient of thermal expansion, which are important to packaging applications. Moreover, the LCP film has excellent electrical characteristics, such as stable low dielectric constant and low dissipation factor, which are desired for the microwave application. It is shown that, from 30 to 110 GHz, the LCP film has a dielectric constant of 3.16 ± 0.05 and a low loss tangent from 0.0028 to 0.0045 [130]. These merits make LCP film an ideal material for millimeter wave applications as well.

The cost of liquid crystal polymer is comparable to that of conventional-print circuit-board material and is cheaper than LTCC. The active and passive devices can be integrated in compact, vertically integrated RF modules by using homogeneous multilayer LCP technology at a low temperature (about 290°C), which would be more challenging for the LTCC technology because of its much higher ($\sim$850°C) processing temperature. The unique combination of properties, such as low cost and excellent electrical and mechanical characteristics makes liquid crystal polymer technology ideally suitable for implementing compact high-density system-in-package applications.

To implement the high-density RF module, one can use two types of liquid crystal polymer films for the fabrication, for example, the bonding film ULTRALAM 3908

TABLE 10.3 Typical Characteristics of Liquid Crystal Polymer (LCP) Films [193,194]

Property		ULTRALAM 3850	ULTRALAM 3908	Unit
Mechanical Properties				
Dimensional	MD	−0.06	<0.1	%
Stability	CMD	−0.03	<0.1	
Tensile strength		200	216	MPa
Tensile modules		2255	2450	MPa
Density		1.4		gm/cm^3
Thermal Properties				
Coefficient of mhermal	X	17	17	
expansion, CET	Y	17	17	ppm/°C
(30°C to 150°C)	Z	150	150	
Melting temperature		315	280	°C
Relative thermal index,	Mechanical	190	190	°C
RTI	Electrical	240	240	
Thermal conductivity		0.5		W/m/°K
Electrical Properties				
Dielectric constant (10 GHz, 23°C)		2.9	2.9	
Dissipation factor (10 GHz, 23°C)		0.0025	0.0025	
Surface resistivity		1.0×10^{10}	1.0×10^{12}	MOhm
Volume resistivity		1.0×10^{12}	2.6×10^{14}	MOhm cm
Dielectric breakdown strength		1378	118	KV/cm
Environment Properties				
Chemical resistance		98.7	98.7	%
Water absorption (23°C, 24-h)		0.04	0.04	%
Coefficient of hygroscopic expansion, CHE (60°C)		4	4	ppm/%RH
Flammability		VTM-0	VTM-0	

with 280°C melting temperature and the core film ULTRALAM 3850 with 315°C melting temperature. The bonding film has a typical thickness of 25 or 50 µm, while the core film has several available thicknesses, such as 100, 50, and 25 µm. By properly choosing the combination of the core and bonding films, desired thickness can be achieved. The fabrication, imaging, and etching process of LCP technology are compatible with most of available wet-process systems. However, since most of the LCP films are very thin, vias have to be carefully drilled using laser with proper speed to avoid the overheating of the sidewall surface, which can lead to smearing of the LCP film. In the lamination, the bonding film is inserted between core films to bond them together. To obtain a good lamination quality, the lamination process requires

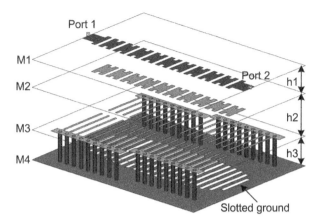

FIGURE 10.72 Multilayer structure of LCP UWB filter with multiple notch bands.

carefully controlled pressure and temperature. The proper temperature makes the bonding film melt and keeps the core film solid. The proper pressure can help the melting bonding film to be seamlessly touched by the core film without any splashing. A recommended lamination process for ULTRALAM 3908 and ULTRALAM 3850 can be found [195].

UWB filters have been constructed [99,100,184–194], using multilayer LCP technology. With the multilayer capability, not only broadside coupling can easily be implemented for designing an ultra-wideband bandpass filter, but also more functionality can be integrated in a compact structure. For example, notch bands can be realized in a UWB bandpass filter to suppress unwanted interferences. Figure 10.72 shows a 3D structure of such a filter. It consists of four metal layers (denoted M1 to M4). The top (M1) and second (M2) layers accommodate two offset arrays of similar patterns, which attain a highpass signal path through broadside coupling between the two layers, which controls the low-cutoff frequency of the UWB passband. The bottom metal layer (M4) serves as the ground, which has a nonuniform slot pattern that controls the high-cutoff frequency as well as the upper stopband of the UWB bandpass filter. An additional metal layer (M3) is used, where one pair of short-circuited stubs or quarter-wavelength resonators is integrated underneath the second layer. The function of the short-circuited quarter-wavelength stub is similar to that of a resonator electromagnetically coupled to the main signal line to form a narrowband bandstop filter.

In order to obtain a desired notch band, the length of short-circuited stubs can initially be chosen as a quarter-wavelength of the center frequency of the desired notch band. The notch bandwidth depends on the coupling of the short-circuited stubs to the main signal path implemented between the top and second layers. This coupling is mainly controlled by h_2, the distance from the third to the second layer. It can be shown that a smaller h_2 results in a wider notch band with higher rejection as a result of stronger coupling between short-circuited stubs and broadside-coupled microstrip lines. Hence, it is a primary design parameter used to control the notch bandwidth. Another design parameter, which is the offset of the

418 COMPACT FILTERS AND FILTER MINIATURIZATION

open end of the short-circuited stub from the center, can have a smaller effect on the coupling or notch bandwidth as well, which can be adjusted in the design for fine tuning [100].

Figure 10.73 shows the planar layouts for a designed UWB filter. The thicknesses between the adjacent metal layers are $h_1 = 25$ μm, $h_2 = 575$ μm, and $h_3 = 200$ μm. The third metal layer of Figure 10.73c has four groups of integrated short-circuited stub resonators. With short-circuited stubs that have the same or varied lengths, a UWB bandpass filter with single- or multiple-notch bands can be designed. For example, for a designed UWB filter with three notch bands, $L_{n,1} = L_{n,2} = 5.64$ mm,

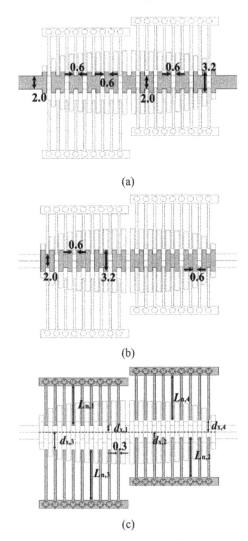

FIGURE 10.73 Planar view of the multilayer LCP UWB filter with multiple-notch bands. (**a**) Top layer; (**b**) second layer; (**c**) third layer. (All dimensions are in millimeters.)

$L_{n,3} = 6.4$ mm, and $L_{n,4} = 7.14$ mm with the offsets $d_{x,1} = d_{x,2} = d_{x,3} = 0.8$ mm and $d_{x,4} = 0.6$ mm.

For the filter fabrication, four types of liquid crystal polymer films are used, including seven layers of 50-μm bonding films with a melting temperature of 290°C, two layers of 25-μm, four layers of 50-μm, and two layers of 100-μm core films. The core films have a melting temperature of 315°C. In the bonding process, bonding films are inserted between core films and a temperature of 290°C is used to bond all films together. Throughout the fabrication process, laser-drilled alignment holes have been used to maintain the alignment between different layers. The alignment holes were machined using a Spectra physics Inazuma ns-pulse duration laser system delivering approximately 1.8 W (average power) of light at 355 nm. The pulse repetition rate was 15 kHz and cutting was done by many (approximately 30) passes of a relatively high-scan speed of 250 mm/s to minimize thermal damage to the polymer, which has a melting point of 290°C (bonding film) or 315°C (core film). A Nutfield XLR8 two-axis optical scan head was used, which enables precise and repeatable location of the beam across a work piece.

Figure 10.74 shows the photograph of fabricated filter, which has a size of 23.4 × 17.34 mm (0.95 by 0.7 λ_g, where λ_g is the guided wavelength at 6.85 GHz), excluding the 50-ohm feeding microstrip lines. The performance of the filter, measured by using the HP 8510 network analyzer, is presented in Figure 10.75. The filter exhibits a good UWB passband with three designed notch bands. The first notch band is measured at 5.4 GHz with a rejection level of 23.99 dB and a 10-dB bandwidth of 1.11% (from 5.37 to 5.43 GHz). The measured second notch band locates at 5.98 GHz with a rejection level of 23.97 dB and a 10-dB bandwidth of 1.34% (from 5.95 to 6.03 GHz). The third notch band is measured at 6.76 GHz with a rejection level of 32.78 dB and a 10-dB bandwidth of 2.95% (from 6.65 to 6.85 GHz). The measured insertion losses of four subpassbands are 2.15 dB at 4.7 GHz, 1.4 dB at 5.65 GHz, 1.5 dB at 6.4 GHz, and 0.6 dB at 8.5 GHz, respectively. The measured group-delay response of the filter is depicted in Figure 10.75b, where the three peaks observed correspond to the three notch bands.

More examples of LCP UWB filters can be found in Chapter 12. Many other types of microwave filters have also been developed using LCP technology [164–183]. The technology is also very promising for development of millimeter-wave front

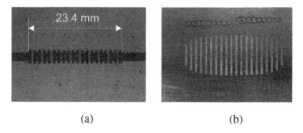

FIGURE 10.74 Photograph for fabricated LCP UWB bandpass filter with triple notch bands. (**a**) Top layer; (**b**) bottom layer.

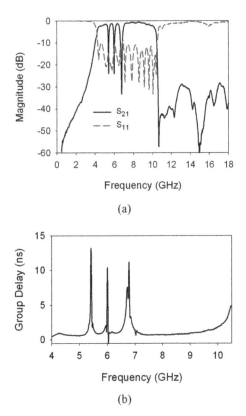

FIGURE 10.75 Measured performance of the triple notch-band UWB filter using multilayer LCP technology. (a) Magnitude; (b) group delay.

ends that including antennas and filters. For example, the development of integrated passive devices for V-band (50 – 75 GHz) transceiver applications on low-lost LCP technology is reported [180]. Filters, matching networks, duplexers, and antenna arrays have been designed, implemented, and characterized. The developed duplexer consists of two channel filters, which are four-pole quasielliptic-function microstrip filters using cross-coupled folded-half-wavelength open-loop resonators (see Chapter 9). The two channel filters have a bandwidth of 2 GHz centered at 60 and 63 GHz, respectively; hence, they are the narrowband filters with a fractional bandwidth around 3%. The measured duplexer shows an average passband insertion loss of 4 dB, while simultaneously exhibiting steep stopband attenuation to provide the necessary isolation between the two closely spaced transmit and receive channels. Slot-loaded patch antenna element and arrays have also been developed providing a low-profile solution, while still meeting the bandwidth specifications. The integrated module is compact, conformal, shows good matching and isolation characteristics, and is a good candidate for low-cost, short-range wireless applications.

REFERENCES

[1] J.-S. Hong, Recent progress in miniature microwave filters (Invited paper), In *2008 IEEE MTT-S International Microwave Workshop Series on Art of Miniaturizing RF and Microwave Passive Components (IMWS)*, 14–15, Dec. 2008, 1–6.

[2] J.-S. Hong, and M. J. Lancaster, Compact microwave elliptic function filter using novel microstrip meander open-loop resonators, *Electron. Lett.* **32**, 1996, 563–564.

[3] J.-S. Hong, M. J. Lancaster, and Y. He, Superconducting quasi-elliptic function filter on r-plane sapphire substrate, In *ICMMT' 2000*, Beijing, September 2000, 167–171.

[4] I. B. Vendik, O. G. Vendik, and S. S. Gevorgian, Effective dielectric permittivity of r-cut sapphire microstrip, In *Proceedings of the 24th European Microwave Conference*, 1994, 395–400.

[5] M. Sagawa, K. Takahashi, and M. Makimoto, Miniaturized hairpin resonator filters and their application to receiver front-end MIC's, *IEEE Trans.* **MTT-37**, 1989, 1991–1997.

[6] J.-S. Hong, and M. J. Lancaster, Development of new microstrip pseudo-interdigital bandpass filters, *IEEE MGWL*, **5**(8), 1995, 261–263.

[7] J.-S. Hong, and M. J. Lancaster, Investigation of microstrip pseudo-interdigital bandpass filters using a full-wave electromagnetic simulator, *Intern J. Microwave Millimeter-Wave Computer-Aided Eng.* **7**(3), 1997, 231–240.

[8] M. H. Weng, H. W. Wu, and C. Y. Hung, A compact pseudointerdigital SIR bandpass filter with improved stopband performance, *Microwave Optical TechnolLett.* **47**(5), 2005, 462–464.

[9] Y.-M. Chen, and S.-F. Chang, A compact stepped-impedance pseudo-interdigital bandpass filter with controllable transmission zero and wide stopband range, In *2009 European Microwave Conference*, 783–786.

[10] J.-S. Hong, and M. J. Lancaster, A novel microwave periodic structure — The ladder microstrip line, *Microwave Optical Technol. Lett.* **9**, 1995, 207–210.

[11] J.-S. Hong, and M. J. Lancaster, Capacitively loaded microstrip loop resonator, *Electron. Lett.* **30**, 1994, 1494–1495.

[12] J.-S. Hong, and M. J. Lancaster, Edge-coupled microstrip loop resonators with capacitive loading, *IEEE WGWL* **5**, 1995, 87–89.

[13] J.-S. Hong, M. J. Lancaster, A. Porch, B. Avenhaus, P. Woodall, and F. Wellhofer, New high temperature superconductive microstrip lines and resonators, *Appl. Superconductivity Inst. Phys. Conf. Ser. No.* **148**, 1195–1198.

[14] J.-S. Hong, and M. J. Lancaster, Novel slow-wave ladder microstrip line filters, In *Proceedings of the 26th European Microwave Conference*, 1996, 431–434.

[15] M. Makimoto, and S. Yamshita, Bandpass filters using parallel coupled stripline stepped impedance resonators, *IEEE Trans.* **MTT-28**, 1980, 1413–1417.

[16] J.-S. Hong, and M. J. Lancaster, End-coupled microstrip slow-wave resonator filter, *Electron. Lett.* **32**(16), 1996, 1494–1496.

[17] J. S. Hong, and M. J. Lancaster, Microstrip slow-wave resonator filters, *IEEE MTT-S Digest*, 1997, 713–716.

[18] J.-S. Hong, and M. J. Lancaster, Theory and experiment of novel microstrip slow-wave open-loop resonator filters, *IEEE Trans.* **MTT-45**, 1997, 2358–2365.

[19] L. Zhu, and K. Wu, Accurate circuit model of interdigital capacitor and its application to design of new quasi-lumped miniaturized filters with suppression of harmonic resonance. *IEEE Trans.* **MTT-48**, 2000, 347–356.

[20] S.-Y. Lee, and C.-M. Tsai, A new network model for miniaturized hairpin resonators and its applications. *IEEE MTT-S Digest*, 2000, 1161–1164.

[21] A. F. Harvey, Periodic and guided structures at microwave frequencies. *IRE Trans.* **MTT-8**, 1960, 30–61.

[22] EM User's Manual. Sonnet Software, Inc., Liverpool, New York, 2009.

[23] A. E. Williams, and A. E. Atia, Dual-mode canonical waveguide filters. *IEEE Trans.* **MTT-25**, 1977, 1021–1026.

[24] S. J. Fiedziuszko, Dual-mode dielectric resonator loaded cavity filters. *IEEE Trans.* **MTT-30**, 1982, 1311–1316.

[25] C. Wang, K. A. Zaki, and A. E. Atia, Dual-mode conductor-loaded cavity filters. *IEEE Trans.* **MTT-45**, 1997, 1240–1246.

[26] I. Wolff, Microstrip bandpass filter using degenerate modes of a microstrip ring resonator. *Electron. Lett.* **8**(12), 1972, 302–303.

[27] J. A. Curtis, and S. J. Fiedziuszko, Miniature dual mode microstrip filters. *IEEE MTT-S Digest*, 1991, 443–446.

[28] J. A. Curtis, and S. J. Fiedziuszko, Multi-layered planar filters based on aperture coupled dual-mode microstrip or stripline resonators, *IEEE MTT-S Digest* 1992, 1203–1206.

[29] R. R. Mansour, Design of superconductive multiplexers using single-mode and dual-mode filters, *IEEE Trans.* **MTT-42**, 1994, 1411–1418.

[30] U. Karacaoglu, I. D. Robertson, and M. Guglielmi, An improved dual-mode microstrip ring resonator filter with simple geometry, In *Proceedings of European Microwave Conference*, 1994, 472–477.

[31] J.-S. Hong, and M. J. Lancaster, "Bandpass characteristics of new dual-mode microstrip square loop resonators, *Electron. Lett.* **31**(11), 1995, 891–892.

[32] J.-S. Hong and M. J. Lancaster, Microstrip bandpass filter using degenerate modes of a novel meander loop resonator, *IEEE Microwave Guided Wave Lett.* **5**(11), 1995, 371–372.

[33] J.-S. Hong, and M. J. Lancaster, Realization of quasielliptic function filter using dual-mode microstrip square loop resonators, *Electron. Lett.* **31**(24), 1995, 2085–2086.

[34] H. Yabuki, M. Sagawa, M. Matsuo, and M. Makimoto, Stripline dual-mode ring resonators and their application to microwave devices, *IEEE Trans.* **MTT-44**, 1996, 723–729.

[35] J.-S. Hong, and M. J. Lancaster, Recent advances in microstrip filters for communications and other applications, *IEE Colloq. Advan. Passive Microwave Components*, 1997, 22.

[36] Z. M. Hejazi, P. S. Excell, and Z. Jiang, Compact dual-mode filters for HTS satellite communication systems, IEEE *Microwave Guided Wave Lett.* **8**(8), 1996, 275–277.

[37] R. R. Mansour, S.Ye, S. Peik, V. Dokas, and B. Fitzpatrick, Quasi dual-mode resonators, *IEEE MTT-S Digest*, 2000, 183–186.

[38] H. A. Wheeler, Transmission line properties of parallel strips separated by a dielectric sheet, *IEEE Trans.* **MTT-13**, 1965, 172–185; Transmission line properties of a strip on a dielectric sheet on a plane, *IEEE Trans.* **MTT-25**, 1977, 631–647.

[39] J.-S. Hong, and M. J. Lancaster, Microstrip triangular patch resonators filters, In *2000 IEEE MTT-S International Microwave Symposium Digest*, 331–334.

[40] M. Cuhaci, and D. S. James, Radiation from triangular and circular resonators in microstrip, *IEEE MTT-S Digest* 1977, 438–441.

[41] J. Helszajn, and D. S. James, Planar triangular resonators with magnetic walls, *IEEE Trans.* **MTT-26**(2), 1978, 95–100,

[42] J.-S. Hong, and S. Li, Dual-mode microstrip triangular patch resonators and filters, In *2003 IEEE MTT-S International Microwave Symposium Digest*, 1901–1904.

[43] J.-S. Hong, and S. Li, Theory and experiment of dual-mode microstrip triangular patch resonators and filters, *IEEE Trans. Microwave Theory Techn.* **MTT-52**, 2004, 1237–1243.

[44] J.-S. Hong, Microstrip dual-mode band reject filter, In *2005 IEEE MTT-S International Microwave Symposium Digest*, 945–948.

[45] C. Lugo, and J. Papapolymerou, Bandpass filter design using a microstrip triangular loop resonator with dual-mode operation, *IEEE Microwave Wireless Components Lett.* **15**(7), 2005, 475–477.

[46] R. Wu, and S. Amari, New triangular microstrip loop resonators for bandpass dual-mode filter applications, *IEEE MTT-S Dig.* 2005, 941–944.

[47] J.-S. Hong, H. Shaman, and Y.-H. Chun, Dual-mode microstrip open-loop resonators and filters, *IEEE Trans. Microwave Theory Techn.* **MTT-55**(8), 2007, 1764–1770.

[48] S. Amari, U. Rosenberg, and J. Bornemann, Singlets, cascaded singlets and the nonresonating node model for advanced modular design of elliptic filters, *IEEE Microwave Wireless Components Lett.* **14**(5), 2004, 237–239.

[49] D. Swanson, Thin-film lumped-element microstrip filters, *IEEE MTT-S Dig.* 1989, 671–674.

[50] D. G. Swanson, Jr., R. Forse, and B. J. L. Nilsson, A 10 GHz thin film lumped element high temperature superconductor filter, *IEEE MTT-S Dig.* 1992, 1191–1193.

[51] Q. Huang, J.-F. Liang, D. Zhang, and G.-C. Liang, Direct synthesis of tubular bandpass filters with frequency-dependent inductors, *IEEE MTT-S Dig.* 1998, 371–374.

[52] A. F. Sheta, K. Hettak, J. ph. Coupez, C. Person, and S. Toutain, A new semi-lumped microwave filter structure, *IEEE MTT-S Dig.* 1995, 383–386.

[53] G. L. Hey-Shipton, Quasi-lumped element bandpass filters using DC isolated shunt inductors, *IEEE MTT-S Dig.* 1996, 1493–1496.

[54] T. Patzelt, B. Aschermann, H. Chaloupka, U. Jagodzinski, and B. Roas, High-temperature superconductive lumped-element microwave all-pass sections, *Electron. Lett.* **29**(17), 1993, 1578–1579.

[55] G. Brzezina, L. Roy, and L. MacEachern, Design enhancement of miniature lumped-element LTCC bandpass filters, *IEEE Trans.* **MTT-57**, 2009, 815–823.

[56] G. D. Vendelin, High-dielectric substrates for microwave hybrid integrated circuitry, *IEEE Trans.* **MTT-15**, 1967; 750–752.

[57] K. C. Wolters, and P. L. Clar, Microstrip transmission lines on high dielectric constant substrates for hybrid microwave integrated circuits, *IEEE MTT-S Dig.* 1967, 129–131.

[58] F. J. Winter, J. J. Taub, and M. Marcelli, High dielectric constant strip line band pass filters, *IEEE Trans.* **MTT-39**, 1991, 2182–2187.

[59] K. Konno, Small-size comb-line microstrip narrow BPF, *IEEE MTT-S Dig.* 1962, 917–920.

[60] P. Pramanik, Compact 900-MHz hairpin-line filters using high dielectric constant microstrip line, *IEEE MTT-S Dig.* 1993, 885–888.

[61] I. C. Hunter, S. R. Chandler, D. Young, and A. Kennerley, Miniature microwave filters for communication systems, *IEEE Trans.* **MTT-43**, 1995, 1751–1757.

[62] A. F. Sheta, K. Hettak, J. P. Coupez, and S. Toutain, A new semi-lumped filter structure, *IEEE MTT-S Dig.* 1995, 383–386.

[63] A. F. Sheta, J. P. Coupez, G. Tanne, and S. Toutain, Miniature microstrip stepped impedance resonaror bandpass filters and diplexers for mobile communications, *IEEE MTT-S Dig.* 1996, 607–610.

[64] M. Murase, Y. Sasaki, A. Sasabata, H. Tanaka, and Y. Ishikawa, Multi-chip transmitter/receiver module using high dielectric substrates for 5.8 GHz ITS applications, *IEEE MTT-S Dig.* 1999, 211–214.

[65] . A. C. Kundu, and K. Endou, "TEM-mode planar dielectric waveguide resonator BPF for W-CDMA," *IEEE MTT-S Dig.* 2000, 191–194.

[66] W. Schwab, and W. Menzel, Compact bandpass filters with improved stop-band characteristics using planar multilayer structures, *IEEE MTT-S Dig.* 1992, 1207–1210.

[67] C. Person, A. Sheta, J. Ph. Coupez, and S.Toutain, Design of high performance band-pass filters by using multi-layer thick-film technology, In *Proceedings of the 24th European Microwave Conference*, 1994, Cannes, France, 466–471.

[68] W. Schwab, F. Boegelsack, and W. Menzel, Multilayer suspended stripline and coplanar line filters, *IEEE Trans.* **MTT-42**, 1994, 1403–1406.

[69] O. Fordham, M.-J. Tsai, and N. G. Alexopoulos, Electromagnetic synthesis of overlap-gap-coupled microstrip filters, *IEEE MTT-S Dig.* 1995, 1199–1202.

[70] C. Cho and K. C. Gupta, Design methodology for multilayer coupled line filters, *IEEE MTT-S Dig.* 1997, 785–788.

[71] J. A. Curtis, and S. J. Fiedzuszko, Multi-layered planar filters based on aperture coupled, dual mode microstrip or stripline resonators, *IEEE MTT-S Dig.* 1992, 1203–1206.

[72] S. J. Yao, R. R. Bonetti, and A. E. Williams, Generalized dual-plane milticoupled line filters, *IEEE Trans.* **MTT-41**, 1993, 2182–2189.

[73] H.-C. Chang, C.-C. Yeh, W.-C. Ku, and K.-C. Tao, A multilayer bandpass filter integrated into RF module board, *IEEE MTT-S Dig.* 1996, 619–622.

[74] J.-S. Hong, and M. J. Lancaster, Back-to-back microstrip open-loop resonator filters with aperture couplings, *IEEE MTT-S Dig.* 1999, 1239–1242.

[75] J.-S. Hong, and M. J. Lancaster, Aperture-coupled microstrip open-loop resonators and their applications to the design of novel microstrip bandpass filters, *IEEE Trans.* **MTT-47**, 1999. 1848–1855.

[76] C. C. Chang, C. Caloz, and T. Itoh, Analysis of compact slot resonator in the ground plane for microstrip structures, In *2001 Asia-Pacific Microwave Conference* 2001, 1100–1103.

[77] C. S. Kim, J. S. Lim, S. Nam, K. Y. Kang, and A. Ahn, Equivalent circuit modeling of spiral defected ground structure for microstrip line, *Electron. Lett.*, **38**(19), 2002, 1109–1110.

[78] C. Caloz, H. Okabe, T. Iwai, and T. Itoh, A simple and accurate model for microstrip structures with slotted ground plane, *IEEE Microwave Wireless Components Lett.* **14**(4), 2004, 133–135.

[79] D. Ahn, J. S. Park, C. S. Kim, Y. Qian, and T. Itoh, A design of the low-pass filter using the novel microstrip defected ground structures, *IEEE Trans. Microwave Theory Tech.* **49**(10), 2001, 86–93.

[80] B. M. Karyamapudi, and J.-S. Hong, Coplanar waveguide periodic structures with resonant elements and their application in microwave filters, *IEEE MTT-S Intern. Microwave Symp. Dig.* 2003, 1619–1622.

[81] Y. Chung, S. S. Jeon, S. Kim, D. Ahn, J. I. Choi, and T. Itoh, Multifunctional microstrip transmission lines integrated with defected ground structure for RF front-end application, *IEEE Trans. Microwave Theory Tech.* **52**(5), 2004, 1425–1432.

[82] B. M. Karyamapudi, and J.-S. Hong, Characterization and applications of a compact CPW defected ground structures, *Microwave Optical Technol. Lett.* **47**(1), 2005, 26–31.

[83] J.-S. Hong, and B. M. Karyamapudi, A general circuit model for defected ground structures in planar Transmission lines, *IEEE Microwave Wireless Components Lett.*, **15**(10), 2005, 706–708.

[84] S.-W. Ting, K.-W. Tam, and R. P. Martins, Miniaturized microstrip lowpass filter with wide stopband using double equilateral U-shaped defected ground structure, *IEEE Microwave Wireless Components Lett.* **16**(5), 2006, 240–242.

[85] J.-S. Lim, C.-S. Kim, D. Ahn, Y.-C. Jeong, and S. Nam, Design of low-pass filters using defected ground structure, *IEEE Trans. Microwave Theory Techn.* **53**(8), 2005, 2539–2545.

[86] N. C. Karmakar, S. M. Roy, and I. Balbin, Quasi-static modeling of defected ground structure, *IEEE Trans. Microwave Theory Techn.* **54**(5), 2006, 2160–2168.

[87] M. K. Mandal, and S. Sanyal, A novel defected ground structure for planar circuits, *IEEE Microwave Wireless Components Lett.* **16**(2), 2006, 93–95.

[88] J.-S. Park, J.-S. Yun, and D. Ahn, A design of the novel coupled-line bandpass filter using defected ground structure with wide stopband performance, *IEEE Trans. Microwave Theory Techn.* **50**(9), 2002, 2037–2043.

[89] A. Abdel-Rahman, A. K. Verma, A. Boutejdar, and A. S. Omar, Compact stub type microstrip bandpass filter using defected ground plane, *IEEE Microwave Wireless Components Lett.* **14**(4), 2004, 136–138.

[90] D.-J. Woo, T.-K. Lee, J.-W. Lee, C.-S. Pyo, and W.-K. Choi, Novel U-slot and V-slot DGSs for bandstop filter with improved Q factor, *IEEE Trans. Microwave Theory Techn.* **54**(6), Part 2, 2006, 2840–2847.

[91] H.-J. Chen, T.-H. Huang, C.-S. Chang, L.-S. Chen, N.-F. Wang, Y.-H. Wang, and M.-P. Houng, A novel cross-shape DGS applied to design ultra-wide stopband low-pass filters, *IEEE Microwave Wireless Components Lett.* **16**(5), 2006, 252–254.

[92] J.-X. Chen, J.-L. Li, K.-C. Wan, and Q. Xue, Compact quasi-elliptic function filter based on defected ground structure, *IEE Proc. Microwaves, Antennas Propagation* **153**(4), 2006, 320–324.

[93] W.-T. Liu, C.-H. Tsai, T.-W. Han, and T.-L. Wu, An embedded common-mode suppression filter for GHz differential signals using periodic defected ground plane, *IEEE Microwave Wireless Components Lett.* **18**(4), 2008, 248–250.

[94] H. B. El-Shaarawy, F. Coccetti, R. Plana, M. El Said, and E. A. Hashish, Compact bandpass ring resonator filter with enhanced wide-band rejection characteristics using defected ground structures, *IEEE Microwave Wireless Components Lett.* **18**(8), 2008, 500–502.

[95] Y. Liu, C. H. Liang, and Y. J. Wang, Ultra-wideband bandpass filter using hybrid quasi-lumped elements and defected ground structure, *Electron. Lett.* **45**(17), 2009, 899–900.

[96] A. Boutejdar, A. Elsherbini, and A. S. Omar, Method for widening the reject-band in low-pass/band-pass filters by employing coupled C-shaped defected ground structure, *IET Microwaves Antennas Propagation* **2**(8), 2008, 759–765.

[97] C. M. Yang, R. Jin, J. Geng, X. Huang, and G. Xiao, Ultra-wideband bandpass filter with hybrid quasi-lumped elements and defected ground structure, *IET Microwaves Antennas Propagation* **1**(3), 2007, 733–736.

[98] G.-M. Yang, R. Jin, C. Vittoria, V. G. Harris, and N. X. Sun, Small ultra-wideband (UWB) bandpass filter with notched band, *IEEE Microwave Wireless Components Lett.* **18**(3), 2008, 176–178.

[99] Z.-C. Hao, and J.-S. Hong, UWB bandpass filter with a multilayer non-uniform periodical structure on LCP substrates, In *2009 IEEE MTT-S International Microwave Symposium Digest*, 7–12 June 2009, 853–856.

[100] Z.-C. Hao, J.-S. Hong, J. P. Parry, and D. P. Hand, Ultra-wideband bandpass filter with multiple notch bands using nonuniform periodical slotted ground structure, *IEEE Trans. Microwave Theory Techn.* **57**(12), Part 2, 2009, 3080–3088.

[101] J. Park, J.-P. Kim, and S. Nam, Design of a novel harmonic-suppressed microstrip low-pass filter, *IEEE Microwave Wireless Components Lett.* **17**(6), 2007, 424–426.

[102] A. Balalem, A. R. Ali, J. Machac, and A. Omar, Quasi-elliptic microstrip low-pass filters using an interdigital dgs slot, *IEEE Microwave Wireless Components Lett.* **17**(8), 2007, 586–588.

[103] S.-J. Wu, C.-H. Tsai, T.-L. Wu, and T. Itoh, A novel wideband common-mode suppression filter for gigahertz differential signals using coupled patterned ground structure, *IEEE Trans. Microwave Theory Techn.* **57**(4), Part 1, 2009, 848–855.

[104] S. Y. Huang, and Y. H. Lee, A compact e-shaped patterned ground structure and its applications to tunable bandstop resonator, *IEEE Trans Microwave Theory Techn.* **57**(3), 2009, 657–666.

[105] Y.-H. Chun, J.-S. Hong, P. Bao, T. J. Jackson, and M. J. Lancaster, BST varactor tuned bandstop filter with slotted ground structure, In *2008 IEEE MTT-S International Microwave Symposium Digest*, 15–20 June 2008, 1115–1118.

[106] Y.-H. Chun, J.-S. Hong, P. Bao, T. J. Jackson, and M. J. Lancaster, Tunable slotted ground structured bandstop filter with BST varactors, *IET Microwaves Antennas Propagation* **3**(5), 2009, 870–876.

[107] D. Deslandes, and K. Wu, Integrated microstrip and rectangular waveguide in planar form, *IEEE Microwave Wireless Component Lett.* **11**(2), 2001, 68–70.

[108] J. R. Bray, and L. Roy, Resonant frequencies of post-wall waveguide cavities, *Proc. Inst. Elect. Eng.* **150**(10), 2003, 365–368.

[109] D. Dealandes, and K. Wu, Single-substrate integration techniques for planar circuits and waveguide filters, *IEEE Trans. Microwave Theory Tech.* **51**(2), 2003, 593–596.

[110] X. Chen, W. Hong, T. Cui, J. Chen, and K. Wu, Substrate integrated waveguide (SIW) linear phase filter, *IEEE Microwave Wireless Component Lett.* **15**(11), 2005, 787–789.

[111] T. B. Cao, P. Lorenz, M. Saglam, W. Kraemer, and R. H. Jansen, Investigation of symmetry influence in substrate integrated waveguide (SIW) band-pass filters using

symmetric inductive posts, In *38th European Microwave Conference, EuMC 2008.* 27–31 Oct. 2008, 492–495.

[112] X. Chen, W. Hong, T. Cui, J. Chen, and K. Wu, Substrate integrated waveguide (SIW) linear phase filter, *IEEE Microwave Wireless Components Lett.* **15**(11), 2005, 787–789.

[113] X.-P. Chen, K. Wu, and Z.-L. Li, Dual-band and triple-band substrate integrated waveguide filters with Chebyshev and quasi-elliptic responses, *IEEE Trans. Microwave Theory Techn.* **55**(12), Part 1, 2007, 2569–2578.

[114] B. Potelon, J.-F. Favennec, C. Quendo, E. Rius, C. Person, and J.-C. Bohorquez, Design of a substrate integrated waveguide (siw) filter using a novel topology of coupling, *IEEE Microwave Wireless Components Lett.* **18**(9), 2008, 596–598.

[115] X.-P. Chen, and K. Wu, Substrate integrated waveguide cross-coupled filter with negative coupling structure, *IEEE Trans. Microwave Theory Techn.* **56**(1), 2008, 142–149.

[116] Y. D. Dong, T. Yang, and T. Itoh, Substrate integrated waveguide loaded by complementary split-ring resonators and its applications to miniaturized waveguide filters, *IEEE Trans. Microwave Theory Techn.* **57**(9), 2009, 2211–2223.

[117] G.-H. Lee, C.-S. Yoo, J.-G. Yook, and J.-C. Kim, SIW (substrate integrated waveguide) quasi-elliptic filter based on LTCC for 60-GHz application, In *2009 European Microwave Integrated Circuits Conference, EuMIC 2009*, 28–29 Sept. 2009, 204–207.

[118] S. Alotaibi, and J.-S. Hong, Novel substrate integrated folded waveguide filter, *Microwave Optical Technol. Lett.* **50**(4), 2008, 1111–1114.

[119] S. K. Alotaibi, and J.-S. Hong, Substrate integrated folded-waveguide filter with asymmetrical frequency response, In *38th European Microwave Conference, EuMC 2008.* 27–31 Oct. 2008, 1002–1005.

[120] H.-Y. Chien, T.-M. Shen, T.-Y. Huang, W.-H. Wang, and R.-B. Wu, Miniaturized bandpass filters with double-folded substrate integrated waveguide resonators in LTCC, *IEEE Trans. Microwave Theory Techn* **57**(7), 2009, 1774–1782.

[121] J.-S. Hong, Compact folded-waveguide resonators, In *2004 IEEE MTT-S International Microwave Symposium Digest*, 1, 6-11 June 2004, 213–216.

[122] J.-S. Hong, Folded-waveguide resonator filters, In *2005 IEEE MTT-S International Microwave Symposium Digest*, 12–17 June 2005, 4.

[123] J.-S. Hong, Compact folded-waveguide resonators and filters, *IEE Proc. Microwaves Antennas Propagation* **153**(4), 2006, 325–329.

[124] S. Al-otaibi, and J.-S. Hong, Folded-waveguide resonator filter with asymmetric frequency response, In *36th European Microwave Conference*, 2006, 10–15 Sept. 2006, 646–648.

[125] S. Al-otaibi, and J.-S. Hong, Electromagnetic design of folded-waveguide resonator filter with single finite-frequency transmission zero, *Int. J. RF Microwave Computer-Aided Eng.* **18**(1), 2008, 1–7.

[126] S. K. Alotaibi, and J.-S. Hong, Miniature folded waveguide resonators, In *2007 European Microwave Conference*, 9–12 Oct. 2007, 286–289.

[127] S. K. Alotaibi, and J.-S. Hong, Novel folded waveguide resonator filter using slot technique, *IEEE Microwave Wireless Components Lett.* **18**(3), 2008, 182–184.

[128] S. K. Alotaibi, J.-S. Hong, and Z.-C. Hao, Multilayer folded-waveguide dual-band filter, In *2008 IEEE MTT-S International Microwave Symposium Digest*, 15–20, June 2008, 73–734.

[129] J. H. Lee, S. Sarkar, S. Pinel, J. Papapolymerou, J. Laskar, and M. M. Tentzeris, 3D-SOP millimeter-wave functions for high data rate wireless systems using LTCC and LCP technologies, In *Proceedings of 55th Electronic Components and Technology Conference*, 31 May–3 June, 2005, 764–768.

[130] D. C. Thompson, O. Tantot, H. Jallageas, G. E. Ponchak, M. Tentzeris, and J. Papapolymerou, Characterization of liquid crystal polymer (LCP) material and transmission lines on LCP substrate from 30-100 GHz, *IEEE Trans. Microwave Theory Tech.*, **52**(4), 2004, 1343–1352.

[131] L. K. Yeung, K.-L. Wu, and Y. E. Wang, Low-temperature cofired ceramic LC filters for RF applications, *IEEE Microwave Magazine*, **9**, Oct. 2008, 118–128.

[132] B. A. Kopp, and A. S. Francomacaro, Miniaturized stripline circuitry utilizing low temperature cofired ceramic (LTCC) technology, *IEEE MTT-S Dig.* 1992, 1513–1516.

[133] R. L. Brown, P. W. Polinski, and A. S. Shaikh, Manufacturing of microwave modules using low-temperature cofired ceramics, *IEEE MTT-S Dig.*, 1994, 1727–1730.

[134] R. E. Hayes, J. W. Gipprich, and M. G. Hershfeld, A stripline re-entrant coupler network for cofired multilayer microwave circuits, *IEEE MTT-S Dig.* 1996, 801–804.

[135] A. Bailey, W. Foley, M. Hageman, C. Murray, A. Piloto, K. Sparks, and K. Zaki, Miniature LTCC filters for digital receivers, *IEEE MTT-S Dig.* 1997, 999–1002.

[136] W. Eurskens, W. Wersing, S. Gohlke, V. Wannenmacher, P. Hild, and R.Weigel, Design and performance of UHF band inductors, capacitors and resonators using LTCC technology for mobile communication systems, *IEEE MTT-S Dig.* 1998, 1285–1288.

[137] A. Fathy, V. Pendrick, G. Ayers, B. Geller, Y. Narayan, B. Thaler, H. D. Chen, M. J. Liberatore, J. Prokop, K. L. Choi, and M. Swaminathan, Design of embedded passive components in low-temperature cofired ceramic on metal (LTCC-M) technology, *IEEE MTT-S Dig.* 1998, 1281–1284.

[138] H. Liang, A. Sutono, J. Laskar, and W. R. Smith, Material parameter characterization of multilayer LTCC and implementation of high Q resonators, *IEEE MTT-S Dig.* 1999, 1901–1904.

[139] T. Ishizaki, M. Fujita, H. Kagata, T. Uwano, and H. Miyake, A very small dielectric planar filter for portable telephones, *IEEE Trans.* **MTT-42**, 1994, 2017–2022.

[140] T. Ishizaki, H. Miyake, T. Yamada, H. Kagata, H. Kushitani, and K. Ogawa, A first practical model of very small and low insertion laminated duplexer using LTCC suitable for W-CDMA portable telephones, *IEEE MTT-S Dig.* 2000, 187–190.

[141] H. Miyake, S. Kitazawa, T. Ishizaki, M.Tsuchiyama, K. Ogawa, and I. Awai, A new circuit configuration to obtain large attenuation with a coupled-resonator band elimination filter using laminated LTCC, *IEEE MTT-S Dig.* 2000, 195–198.

[142] J.-W. Sheen, LTCC-MLC Duplexer for DCS-1800, *IEEE Trans.* **MTT-47**, 1999, 1883–1890.

[143] A. Sutono, J. Laskar, and W. R. Smith, Development of integrated three dimensional Bluetooth image reject filter, *IEEE MTT-S Dig.* 2000, 339–342.

[144] M. A. Skurski, M. A. Smith, R. R. Draudt, D. I. Amey, S. J. Horowitz, and M. J. Champ, Thick-film technology offers high packaging density, *Microwaves RF* **38**(2), 1999, 77–86.

[145] K. Kageyama, K. Saito, H. Murase, H. Utaki, and T. Yamamoto, Tunable active filters having multilayer structure using LTCC, *IEEE Trans. Microwave Theory Tech.* **49**(12), 2001, 2421–2424.

[146] W. Y. Leung, K. K. Cheng, and K.-L. Wu, Multilayer LTCC bandpass filter design with enhanced stopband characteristics, *IEEE Microwave Wireless Comp. Lett.* **12**(7), 2002, 240–242.

[147] L. K. Yeung, and K.-L. Wu, A compact second-order LTCC bandpass filter with two finite transmission zeros, *IEEE Trans. Microwave Theory Tech.* **51**(2), 2003, 337–341.

[148] V. Piatnitsa, E. Jakku, and S. Leppaevuori, Design of a 2-pole LTCC filter for wireless communications, *IEEE Trans. Wireless Commun.* **3**, 2004, 379–381.

[149] C. W. Tang, Harmonic-suppression LTCC filter with the step-impedance quarter-wavelength open stub, *IEEE Trans. Microwave Theory Tech.*, **52**(2), 2004, 617–624.

[150] J. H. Lee, S. Pinel, J. Papapolymerou, J. Laskar, and M. M. Tentzeris, Low-loss LTCC cavity filters using system-on-package technology at 60 GHz, *IEEE Trans. Microw. Theory Tech.* **53**(12), 2005, 3817–3824.

[151] P. Ferrand, D. Baillargeat, S. Verdeyme, J. Puech, M. Lahti, and T. Jaakola, LTCC reduced size bandpass filters based on capacitively loaded cavities for Q-band applications, In *IEEE MTT-S International Microwave Symposium Digest Long Beach, CA*, Jun. 2005, 2083–2086.

[152] G. Wang, M. Van, F. Barlow, and A. Elshabini, An interdigital bandpass filter embedded in LTCC for 5-GHz wireless LAN applications, *IEEE Microwave Wireless Components Lett.* **15**(5), 2005, 357–359.

[153] M. C. Park, B. H. Lee, and D. S. Park, A laminated balance filter using LTCC technology, In *Proceedings of IEEE Asia-Pacific Microwave Conference*, Dec. 2005, 1–4 [CD-ROM]

[154] K. Rambabu, and J. Bornemann, Simplified analysis technique for the initial design of LTCC filters with all-capacitive coupling, *IEEE Trans. Microwave Theory Tech.* **53**(5), 2005, 1787–1791.

[155] C. Chang, and S. Chung, Bandpass filter of serial configuration with two finite transmission zeros using LTCC technology, *IEEE Trans. Microwave Theory Tech.* **53**(7), 2005, 2383–2388.

[156] W. Tung, Y. Chiang, and J. Cheng, A new compact LTCC bandpass filter using negative coupling, *IEEE Microwave Wireless Comp. Lett.* **15**(10), 2005, 641–643.

[157] L. K. Yeung, and K.-L. Wu, An LTCC balanced-to-unbalanced extracted pole bandpass filter with complex load, *IEEE Trans. Microwave Theory Tech.* **54**(4), 2006, 1512–1518.

[158] Y. Jeng, S. Chang, and H. Lin, A high stopband-rejection LTCC filter with multiple transmission zeros, *IEEE Trans. Microwave Theory Tech.* **54**(2), 2006, 633–638.

[159] K. Lin, C. Chang, M. Wu, and S. Chung, Dual-bandpass filters with serial configuration using LTCC technology, *IEEE Trans. Microwave Theory Tech.* **54**(6), 2321–2328.

[160] C. W. Tang, and S. F. You, Design methodologies of LTCC bandpass filters, diplexer and triplexer with transmission zeros, *IEEE Trans. Microwave Theory Tech.* **54**(2), 2006, 717–723.

[161] V. Piatnitsa, D. Kholodnyak, T. Tick, A. Simin, P. Turalchuk, E. Zameshaeva, J. Jantti, H. Jantunen, and I. Vendik, Design and investigation of miniaturized high-performance LTCC filters for wireless communications. In *2007 European Microwave Conference*, 9–12 Oct. 2007; pp. 544–547.

[162] G. Brzezina, L. Roy, and L. MacEachern, Design enhancement of miniature lumped-element ltcc bandpass filters, *IEEE Trans. Microwave Theory Techn.* **57**(4), Part 1, 2009, 815–823.

[163] H.-Y. Chien, T.-M. Shen, T.-Y. Huang, W.-H. Wang, and R.-B. Wu, Miniaturized Bandpass filters with double-folded substrate integrated waveguide resonators in LTCC, *IEEE Trans. Microwave Theory Tech.* **57**(7), 2009, 1774–1782.

[164] S. Dalmia, V. Sundaram, G. White, and M. Swaminathan, "Liquid crystalline polymer (LCP) based lumped-element bandpass filters for multiple wireless applications, In *2004 IEEE MTT-S International Microwave Symposium Digest*, 3, 6–11 June 2004, 1991–1994.

[165] S. Mukherjee, B. Mutnury, S. Dalmia, and M. Swaminathan, Layout-level synthesis of RF inductors and filters in LCP substrates for Wi-Fi applications, *IEEE Trans. Microwave Theory Techn* **53**(6), Part 2, 2005, 2196–2210.

[166] V. Palazzari, S. Pinel, J. Laskar, L. Roselli, and M. M. Tentzeris, Design of an asymmetrical dual-band WLAN filter in liquid crystal polymer (LCP) system-on-package technology, *IEEE Microwave Wireless Components Lett.* **15**(3), 2005, 165–167.

[167] C. Quendo, E. Rius, C. Person, J.-F. Favennec, Y. Clavet, A. Manchec, R. Bairavasubramanian, S. Pinel, J. Papapolymerou, and J. Laskar, Wide band, high rejection and miniaturized fifth order bandpass filter on LCP low cost organic substrate, In *2005 IEEE MTT-S International Microwave Symposium Digest*, 12–17 June 2005, 3.

[168] R. Bairavasubramanian, S. Pinel, J. Papapolymerou, J. Laskar, C. Quendo, E. Rius, A. Manchec, and C. Person, Dual-band filters for WLAN applications on liquid crystal polymer technology, In *2005 IEEE MTT-S International Microwave Symposium Digest*, 12–17 June 2005, 4 pp.

[169] S. Mukherjee, M. Swaminathan, and E. Matoglu, Statistical diagnosis and parametric yield analysis of liquid crystalline polymer RF dual band filters, In *2005 European Microwave Conference*, Vol. 2, 4–6 Oct. 2005, 4 pp.

[170] S. Mukherjee, M. Swaminathan, and S. Dalmia, Synthesis and diagnosis of RF filters in liquid crystalline polymer (LCP) substrate, In *2005 IEEE MTT-S International Microwave Symposium Digest*, 12–17 June 2005, 4 pp.

[171] S. Pinel, R. Bairavasubramanian, J. Laskar, and J. Papapolymerou, Compact planar and vialess composite low-pass filters using folded stepped-impedance resonator on liquid-Crystal-polymer substrate, *IEEE Trans. Microwave Theory Techn.* **53**(5), 2005, 1707–1712.

[172] K. S. Yang, S. Pinel, I. Kim, and J. Laskar, Low-loss integrated-waveguide passive circuits using liquid-crystal polymer system-on-package (SOP) technology for millimeter-wave applications, *IEEE Trans. Microwave Theory Techn.* **54**(12), Part 2, 2006, 4572–4579.

[173] J. Dekosky, S. Lapushin, S. Dalmia, W. Czakon, and G. White, The stacking of integrated passive devices, substrates, and RFICs to make ultra-thin modules using organic liquid crystalline polymer (LCP) technology, In *2006 IEEE MTT-S International Microwave Symposium Digest*, 11–16 June 2006, 1257–1260.

[174] R. Bairavasubramanian, S. Pinel, J. Laskar, and J. Papapolymerou, Compact 60-GHz bandpass filters and duplexers on liquid crystal polymer technology, *IEEE Microwave Wireless Components Lett.* **16**(5), 2006, 237–239.

[175] R. Bairavasubramanian, and J. Papapolymerou, Multilayer Quasi-elliptic filters using dual mode resonators on liquid crystal polymer technology, In *2007 IEEE/MTT-S International Microwave Symposium*, 3–8 June 2007, 549–552.

[176] R. Wu, C. Mmasi, V. Govind, S. Dalmia, C. Ghiu, and G. White, High performance and compact balanced-filter design for wimax front-end modules (FEM) using LCP-based organic substrates, In *2007 IEEE/MTT-S International Microwave Symposium*, 3–8 June 2007, 1619–1622.

[177] A. Bavisi, M. Swaminathan, and E. Mina, Liquid crystal polymer-based planar lumped component dual-band filters for dual-band WLAN systems, In *2007 IEEE Radio and Wireless Symposium*, 9–11 Jan. 2007, 539–542.

[178] W. Yun, V. Sundaram, and M. Swaminathan, High-Q embedded passives on large panel multilayer liquid crystalline polymer-based substrate, *IEEE Trans. Adv. Packaging* **30**(3), 2007, 580–591.

[179] Y. X. Guo, H. J. Lu, M. F. Karim, L. Jin, C. K. Cheng, A. C. W. Lu, and L. C. Ong, Investigation and characterization of millimeter-wave transmission lines and bandpass filters on liquid crystal polymer, In *2008 Asia-Pacific Microwave Conference*, 16–20 Dec. 2008, p. 1–4.

[180] R. Bairavasubramanian, and J. Papapolymerou, Development of V-band integrated frond-ends on liquid crystal polymer (LCP) technology, *IEEE Antennas Wireless Propagation Lett.* **7**, 2008, 1–4.

[181] S. Dalmia, L. Carastro, R. Fathima, V. Govind, J. Dekosky, S. Lapushin, R. Wu, B. Bayruns, and G. White, A compact dual band 802.11n front-end module for MIMO applications using multi-layer organic technology, In *2008 IEEE MTT-S International Microwave Symposium Digest*, 15–20, June 2008, 89–92.

[182] A. Geise, U. Strohmaier, and A. F. Jacob, Investigations of transmission lines and resonant structures on flexed liquid crystal polymer (LCP) substrates up to 67 GHz, In *2009 European Microwave Conference*, Sept. 29–Oct. 1 2009, 735–738.

[183] Y.-H. Chun, R. Keller, J.-S. Hong, C. Fragkiadakis, R. V. Wright, and P. B. Kirby, Lead-Strontium-Titanate varactor-tuned CPW bandstop filter on liquid crystal polymer substrates, In *2009 European Microwave Conference*, Sept. 29–Oct. 1 2009, 1373–1376.

[184] Z.-C. Hao and J.-S. Hong, Ultra-wideband bandpass filter using multilayer liquid-crystal-polymer technology, *IEEE Trans. Microwave Theory Techn.* **56**(9), 2008, 2095–2100.

[185] Z.-C. Hao J.-S. Hong, J. P. Parry, and D. Hand, Miniature coupled resonator UWB filter using a multilayer structure on liquid crystal polymer, In *2008 Asia-Pacific Microwave Conference*, 16–20 Dec. 2008. p 1–4.

[186] Z.-C. Hao, and J.-S. Hong, Compact ultra-wideband bandpass filter using broadside coupled hairpin structures on multilayer liquid crystal polymer substrate, *Electron. Lett.* **44**(20), 2008, 1197–1198.

[187] Z.-C. Hao, J.-S. Hong, and S. K. Alotaibi, A novel ultra wideband bandpass filter using broadside coupled structures on multilayer organic liquid crystal polymer substrate, In *38th European Microwave Conference*, 27–31 Oct. 2008, 998–1001.

[188] Z.-C. Hao, and J.-S. Hong, High selective ultra-wideband (UWB) bandpass filter with wideband harmonic suppression, In *38th European Microwave Conference*, 27–31 Oct. 2008, 353–356.

[189] Z.-C. Hao and J.-S. Hong, Compact wide stopband ultra wideband bandpass filter using multilayer liquid crystal polymer technology, *IEEE Microwave Wireless Components Lett.* **19**(5), 2009, 290–292.

[190] Z.-C. Hao, J.-S Hong, S. K. Alotaibi, J. P. Parry, and D. P. Hand, Ultra-wideband bandpass filter with multiple notch-bands on multilayer liquid crystal polymer substrate, *IET Microwaves Antennas Propagation* **3**(5), 2009, 749–756.

[191] Z.-C. Hao, and J.-S. Hong, Compact UWB filter with double notch-bands using multilayer lcp technology, *IEEE Microwave Wireless Components Lett.* **19**(8), 2009, 500–502.

[192] Z.-C. Hao, and J.-S. Hong, Ultra wideband bandpass filter using embedded stepped impedance resonators on multilayer liquid crystal polymer substrate, *IEEE Microwave Wireless Components Lett.* **18**(9), 2008, 581–583.

[193] http://www.rogerscorp.com/documents/730/acm/ULTRALAM-3000-LCP-laminate-data-sheet-ULTRALAM-3850.aspx

[194] http://www.rogerscorp.com/documents/731/acm/ULTRALAM-3000-LCP-Prepreg-ULTRALAM-3908.aspx

[195] http://www.rogerscorp.com/documents/822/acm/Fabrication-Guidelines-ULTRALAM-3000-LCP-Materials.aspx

CHAPTER ELEVEN

Superconducting Filters

Since the discovery of high-temperature superconductors (HTS) in 1986 [1], high-temperature superconductivity has been at the forefront of advanced filter technology and has been changing the way we design communication systems, electronic systems, medical instrumentation, and military microwave systems [2–7]. Superconducting filters play an important role in many applications, especially those for mobile communication systems [8–28], satellite communications [29–40], and radio astronomy and radars [41–53]. Most superconducting filters are simply microstrip structures using HTS thin films [54–104]. Some typical superconducting filters are described in this chapter. In addition, this chapter will cover some important properties of superconductors and substrates for growing HTS films [105–110], which are essential for the design of HTS microstrip filters.

11.1 HIGH-TEMPERATURE SUPERCONDUCTING (HTS) MATERIALS

11.1.1 Typical HTS Materials

Superconductors are materials that exhibit a zero intrinsic resistance to direct current (dc) flow when cooled below a certain temperature. The temperature at which the intrinsic resistance undergoes an abrupt change is referred to as the critical temperature or transition temperature, denoted by T_c. For alternating current (ac) flow, the resistance does not go to zero below T_c, but increases with increasing frequency. However, at typical RF/microwave frequencies (in the cellular band, for example), the resistance of a superconductor is perhaps one thousandth of that in the best ordinary conductor. It is certainly low enough to significantly improve performances of RF/microwave microstrip filters.

Although superconductors were first discovered in 1911, for almost 75 years after the discovery, all known superconductors required a very low transition temperature,

---

Microstrip Filters for RF/Microwave Applications by Jia-Sheng Hong
Copyright © 2011 John Wiley & Sons, Inc.

TABLE 11.1 Typical HTS materials

Materials	$T_c(K)$
$YBa_2Cu_3O7\text{-}x$ (YBCO)	≈92
$Tl_2Ba_2Ca_1Cu_2Ox$ (TBCCO)	≈105

say 30 Kelvin (K) or lower; this limited the applications of these early superconductors. A revolution in the field of superconductivity occurred in 1986 with the discovery of the superconductors with transition temperature greater than 77K, the boiling point of liquid nitrogen. These superconductors are, therefore, referred to as the high-temperature superconductors (HTS). The discovery of the HTS made world headlines because it made possible many practical applications of superconductivity. Since then, the development of microwave applications has proceeded vary rapidly, particularly HTS microstrip filters.

The growth of HTS films and the fabrication of HTS microstrip filters are compatible with hybrid and monolithic microwave-integrated circuits. Although there are many hundreds of high-temperature superconductors with varying transition temperatures, yttrium barium copper oxide (YBCO) and thallium barium calcium copper oxide (TBCCO) are by far the two most popular and commercially available HTS materials. These are listed in Table 11.1 along with their typical transition temperatures [105].

11.1.2 Complex Conductivity of Superconductors

Superconductivity may be explained as a consequence of paired and unpaired electrons travelling with the lattice of a solid. The paired electrons travel, under the influence of an electric field, without resistive loss. In addition, because of the thermal energy present in the solid, some of the electron pairs are split, so that some normal electrons are always present at temperatures above absolute zero. It is, therefore, possible to model the superconductor in terms of a complex conductivity $\sigma_1 - j\sigma_2$; such a model is called the two-fluid model [2,3].

A simple equivalent circuit is depicted in Figure 11.1, which describes complex conductivity in a superconductor. J denotes the total current density and J_s and J_n are the current densities carried by the paired and normal electrons, respectively. The

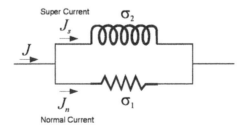

FIGURE 11.1 Simple circuit model depicting complex conductivity.

HIGH-TEMPERATURE SUPERCONDUCTING (HTS) MATERIALS   435

total current in the circuit is split between the reactive inductance and the resistance, which represents dissipation. As frequency decreases, the reactance becomes lower and more of the current flows through the inductance. When the current is constant, namely at dc, this inductance completely shots the resistance, allowing resistance-free current flow.

As a consequence of the two-fluid mode, the complex conductivity may be given by

$$\sigma = \sigma_1 - j\sigma_2$$
$$= \sigma_n \left(\frac{T}{T_c}\right)^4 - j\frac{1}{\omega\mu\lambda_0^2}\left[1 - \left(\frac{T}{T_c}\right)^4\right] \quad (11.1)$$

where σ_n is the normal state conductivity at T_c and λ_0 is a constant parameter, which will be explained in the next section. Note that the calculation of Eq. (11.1) is not strictly valid close to T_c.

11.1.3 Penetration Depth of Superconductors

Normally the approximation $\sigma_2 \gg \sigma_1$ can be made for good quality superconductors provided that the temperature is not too close to the transition temperature, where more normal electrons are present. Making this approximation, an important parameter called the penetration depth, based on the two-fluid model, is given by

$$\lambda = \frac{1}{\sqrt{\omega\mu\sigma_2}} \quad (11.2a)$$

Substituting σ_2 from Eqs. (11.1) into (11.2a) yields

$$\lambda = \frac{\lambda_0}{\sqrt{1 - \left(\frac{T}{T_c}\right)^4}} \quad (11.2b)$$

Thus, λ_0 is actually the penetration depth as the temperature approaches zero Kelvin. Depending on the quality of superconductors, a typical value of λ_0 is about 0.2 μm for HTS.

The penetration depth is actually defined as a characteristic depth at the surface of the superconductor such that an incident plane wave propagated into the superconductor is attenuated by e^{-1} of its initial value. It is analogous to the skin depth of normal conductors, representing a depth to which electromagnetic fields penetrate superconductors, and it defines the extent of a region near the surface of a superconductor in which current can be induced. The penetration depth λ is independent of frequency, but will depend on temperature, as can be seen from Eq. (11.2b). This dependence is different from that of the skin depth of normal conductors. Recall that the skin depth for normal conductors is

$$\delta = \sqrt{\frac{2}{\omega\mu\sigma_n}} \tag{11.3}$$

where σ_n is the conductivity of a normal conductor and is purely real. However, provided we are in the limit where σ_n is independent of frequency, the skin depth is a function of frequency.

Another distinguished feature of superconductors is that a dc current or field cannot fully penetrate them. This is, of course, quite unlike normal conductors, in which there is full penetration of the dc current into the material. As a matter of fact, a dc current decays from the surface of superconductors into the material in a very similar way to an ac current, namely, proportional to e^{-z/λ_L}, where z is the coordinate from the surface into the material and λ_L is the London penetration depth. Therefore, λ_L is a depth where the dc current decays by an amount e^{-1} compared to the magnitude at the surface of superconductors. In the two-fluid model, the value of the dc superconducting penetration depth λ_L will be the same as that of the ac penetration depth λ given in Eq. (11.2) with λ being independent of frequency.

11.1.4 Surface Impedance of Superconductors

Another important parameter for superconducting materials is the surface impedance. In general, solving Maxwell's equation for a uniform plane wave in a metal of conductivity σ yields a surface impedance given by

$$Z_s = \frac{E_t}{H_t} = \sqrt{\frac{j\omega\mu}{\sigma}} \tag{11.4}$$

where E_t and H_t are the tangential electric and magnetic fields at the surface. This definition of the surface impedance is general and applicable for superconductors as well. For superconductors, replacing σ by $\sigma_1 - j\sigma_2$ gives

$$Z_s = \sqrt{\frac{j\omega\mu}{(\sigma_1 - j\sigma_2)}} \tag{11.5a}$$

whose real and imaginary parts can be separated, resulting in

$$\begin{aligned} Z_s &= R_s + jX_s \\ &= \frac{\sqrt{\omega\mu}}{2}\left(\frac{\sqrt{k+\sigma_1} - \sqrt{k-\sigma_1}}{k} + j\frac{\sqrt{k-\sigma_1} + \sqrt{k+\sigma_1}}{k}\right) \end{aligned} \tag{11.5b}$$

with $k = \sqrt{\sigma_1^2 + \sigma_2^2}$. Using the approximations that $k \approx \sigma_2$ and $\sqrt{1 \pm \sigma_1/\sigma_2} \approx 1 \pm \sigma_1/(2\sigma_2)$ for $\sigma_2 \gg \sigma_1$, and replacing σ_2 with $(\omega\mu\lambda^2)^{-1}$, we arrive at

$$R_s = \frac{\omega^2 \mu^2 \sigma_1 \lambda^3}{2} \quad \text{and} \quad X_s = \omega\mu\lambda \tag{11.6}$$

It is important to note that for the two-fluid model, provided σ_1 and λ are independent of frequency, the surface resistance R_s will increase as ω^2. This is of practical significance for justifying the applicability of superconductors to microwave devices as compared with normal conductors, which will be discussed later. R_s will depend on temperature as well. Figure 11.2 illustrates typical temperature-dependent behaviors of R_s, where R_0 is a reference resistance. The surface reactance in Eq. (11.6) may also be expressed as $X_s = \omega L$, where the inductance $L = \mu\lambda$ is called the internal or kinetic inductance. The significance of this term lies in its temperature dependence, which will mainly account for frequency shifting of superconducting filters against temperature. Figure 11.3 demonstrates the typical temperature dependence of an HTS microstrip meander, open-loop resonator, obtained experimentally, where the resonant frequency f_0 is normalized by the resonant frequency at 60K. The temperature stability of cooling systems for HTS filters can be better than 0.5K; therefore, the frequency shifting would not be an issue for most applications.

Films of superconducting material are the main constituents of filter applications and it is crucial for these applications that a good understanding of the properties of these films be obtained. The surface impedance described above is actually for an infinite thick film; it can be modified in order to take the finite thickness of the film into account. If t is the thickness of the film then its surface impedance [3] is

$$Z_f = R_s \left\{ \coth\left(\frac{t}{\lambda}\right) + \frac{t}{\lambda} \frac{1}{\sinh^2\left(\frac{t}{\lambda}\right)} \right\} + jX_s \coth\left(\frac{t}{\lambda}\right) \tag{11.7}$$

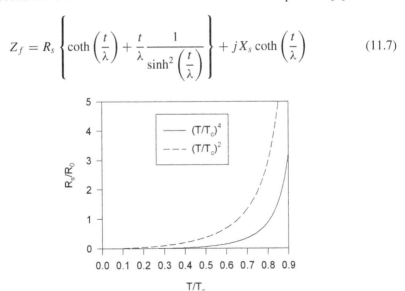

FIGURE 11.2 Temperature dependence of surface resistance of superconductor.

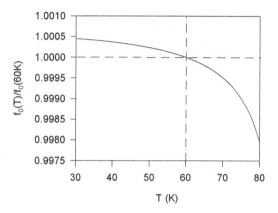

FIGURE 11.3 Temperature-dependent resonant frequency of a HTS microstrip resonator.

where R_s and X_s are given by Eq. (11.6). Again $\sigma_2 \gg \sigma_1$ is assumed in the derivation of the expression. The effect of the finite thickness of thin film tends to increase both the surface resistance and the surface reactance of thin film. Figure 11.4 plots the surface resistance of the thin film as a function of t/λ, indicating that in order to reduce the thin film surface resistance, the thin film thickness should be greater than three to four times the penetration depth. This is similar to the requirement for normal conductor thin film microwave devices, where the conductor thickness should at least three to four times thicker than the skin depth.

At this point, it is worthwhile comparing the surface resistance of HTS with that of normal conductors. For a normal conductor, the surface resistance and surface reactance are equal and are given by

$$R_s = X_s = \sqrt{\frac{\omega \mu}{2\sigma_n}} \tag{11.8}$$

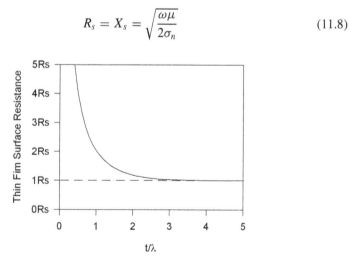

FIGURE 11.4 Surface resistance of superconducting thin films as a function of normalized thickness.

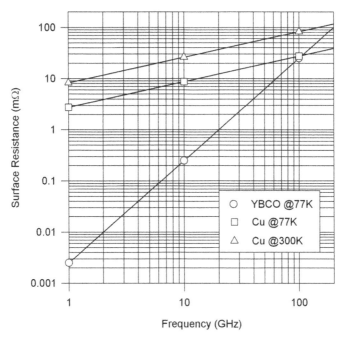

FIGURE 11.5 Comparison of the surface resistance of YBCO at 77K with copper as a function of frequency.

Both are proportional to the square root of frequency. Because the surface resistance of a superconductor increases more rapidly (as frequency squared), there is a frequency at which the surface resistance of normal conductors actually becomes lower than that of superconductors. This has become known as the crossover frequency. Figure 11.5 shows the comparison of the surface resistance of YBCO at 77K with copper, as a function of frequency. The typical values used to produce this plot are:

- YBCO thin film surface resistance (10 GHz and 77K) = 0.25 mΩ
- Copper surface resistance (10 GHz and 77K) = 8.7 mΩ
- Copper surface resistance (10 GHz and 300K) = 26.1 mΩ

In this case, the crossover frequency between copper and HTS films at 77K is about 100 GHz.

It can also be seen from Figure 11.5 that at 2 GHz the surface resistance of HTS thin film at 77K is a thousand times smaller than that of copper at 300K. Based on the discussion on microstrip resonator quality factors in Chapter 4, we may reasonably assume that a copper microstrip resonator has a conductor quality factor $Q_c = 250$ at 2 GHz and 300K. Since the conductor Q is inversely proportional to the surface resistance, if the same microstrip resonator is composed of HTS thin film, it follows immediately that the Q_c for the HTS microstrip resonator can be larger than 250×10^3.

11.1.5 Nonlinearity of Superconductors

Microwave materials exhibit nonlinearity when they are subject to an extreme electromagnetic field, namely, their material properties such as conductivity, permittivity, and permeability become dependent on the field. This is also true for HTS materials. It has been known that the surface resistance of an HTS film, which is related to the conductivity, as described above, will be degraded even when the RF peak magnetic field in the film is only moderately high [106–108]. In the limit when the peak magnetic field exceeds a critical value, the surface resistance rises sharply as the HTS film starts losing its superconducting properties. This critical value of the RF peak magnetic field is known as the critical field and may be denoted by $H_{rf,c}$. The $H_{rf,c}$ may be related to a dc current density by

$$J_c = \frac{H_{rf,c}}{\lambda_L} \qquad (11.9)$$

where λ_L is the London penetration depth, which has the same value as that of λ given by Eq. (11.2) and the J_c is called the critical current density. J_c is an important parameter for characterization of HTS materials. It is temperature-dependent and has a typical value of about 10^6 A/cm^2 at 77K for a good superconductor. Note that Eq. (11.9) is valid only when the HTS film is several times thicker than the penetration depth.

Nonlinearity in the surface resistance not only increases losses of HTS filters, but also causes intermodulation and harmonic generation problems. This, in general, limits the power handling of HTS filters. In many applications such as in a receiver, where HTS filters are operated at low powers, the nonlinear effects are either negligible or acceptable. For high-power applications of HTS filters, the power-handling capability of an HTS filter can, in general, be increased in two ways. The first method, from the HTS material viewpoint, is to increase the critical current density J_c by improving the material or to operate the filter at a lower temperature; J_c will increase as the temperature is decreased. The second method, from microwave design viewpoint, is to reduce the maximum current density in the filter by distributing the RF/microwave current more uniformly over a larger area. High-power HTS filters handling up to more than 100 W have been demonstrated [87–104].

11.1.6 Substrates for Superconductors

Superconducting films have to be grown on some sort of substrate that must be inert, compatible with both the growth of good quality film, and also have appropriate microwave properties for the application at hand. In order to achieve good epitaxial growth, the dimensions of the crystalline lattice at the surface of the substrate should match the dimensions of the lattices of the superconductors. If this is not the case, strain can be set up in the films, producing dislocations and defects. In some cases,

TABLE 11.2 Substrates for HTS Films

Substrate	ε_r (typical)	tan δ (typical)
LaAlO$_3$	24.2 @ 77K	7.6×10^{-6} @ 77K and 10 GHz
MgO	9.6 @ 77K	5.5×10^{-6} @ 77K and 10 GHz
Sapphire	11.6 ∥ c-axis @ 77K	1.5×10^{-8} @ 77K and 10 GHz
	9.4 ⊥ c-axis @ 77K	

the substrates can react chemically, causing impurity levels to rise and the quality of the film to fall. Cracks can be caused in the film if the thermal expansions of the substrate and film are not appropriately matched. Some of the above problems can be overcome by the application of a buffer layer between the films and the substrates. In addition, the surface of substrates should be smooth and free from defects and twinning, if possible. These cause unwanted growth and mechanisms that can lead to nonoptimal films. For microwave applications, it is of fundamental importance that the substrates have a low dielectric loss tangent (tan δ). If the loss tangent is not low enough, then the advantage of using a superconductor can be negated. It is also desirable, in most applications, that the dielectric constant, or ε_r, of a substrate not change much with temperature, improving the temperature stability of the final applications. Whatever the dielectric constant, it must be reproducible and not change appreciably from batch to batch. This is very important for mass production.

With all the above requirements, it is not surprising that an ideal substrate for HTS films has not been found yet. Nevertheless, a number of excellent substrates, producing high-quality films with good microwave properties, are in common use. Among these, the most widely used and commercially available substrates are the lanthanum aluminate (LaAlO$_3$ or LAO), magnesium oxide (MgO), and sapphire (Al$_2$O$_3$) [109–111]. LaAlO$_3$ has a higher dielectric constant than MgO and sapphire, but is generally twinned. Sapphire is a low-loss and low-cost substrate, but its dielectric constant is not isotropic and it requires a buffer layer for grow good HTS films. MgO is, in general, a very good substrate for applications, but is mechanically brittle. Table 11.2 lists some typical parameters of these substrates. For a sapphire substrate, the values of relative dielectric constant that are given are for both parallel and perpendicular to the c-axis (crystal axis) because of anisotropy.

11.2 HTS FILTERS FOR MOBILE COMMUNICATIONS

The technology and system challenges for the evolution of mobile communications have stimulated considerable interest in applications of high-temperature superconducting (HTS) technology [8–28]. The challenges for the cellular mobile base stations vary, but may focus on increasing sensitivity and selectivity, which are summarized as follows:

- Sensitivity — The benefits of increasing sensitivity in rural areas is obvious since the number of mobile base stations and thus the investment necessary to

442 SUPERCONDUCTING FILTERS

secure the radio coverage of a given area will be reduced as the range of each mobile base station is increased. Increasing sensitivity is also desirable in urban co-channel, interference-limited areas since it allows the mobile terminals to reduce the average radiated power and increase their autonomy.

- Selectivity — The soaring demand for mobile communications place severe demands on the frequency resources as the allocated bandwidth becomes increasingly congested. Interference is a growing, pervasive threat to the mobile communication industry, particularly in dense urban regions. Increasing selectivity to improve interference rejection will increase call clarity and reduce the number of dropped calls, which will lead to a general improvement in the quality of service (QoS).

For mobile communication applications, many advanced HTS filters have been developed; most of them are HTS microstrip filters. These filters are simply microstrip filters using HTS thin films instead of conventional conductor films. In general, because of very low conductor losses, the use of HTS thin films can lead to significant improvement of microstrip filter performance with regard to the passband insertion loss and selectivity. This is particularly substantial for narrowband filters, which are normally required in mobile base stations. Some typical high-performance HTS filters are described next.

11.2.1 HTS Filter with a Single Pair of Transmission Zeros

11.2.1.1 Design Consideration High-performance HTS filters are usually designed to have quasielliptic function. The reason for this is explained in light of the following study of three types of filters, namely Chebyshev, quasielliptic, and elliptic-function filters against these simplified specifications:

- Center frequency   1777.5 MHz
- Passband width   15 MHz
- Passband insertion loss ≤ 0.3 dB
- Passband return loss ≤ -20 dB
- 30-dB rejection bandwidth   17.5 MHz
- Transmit band rejection (1805–1880 MHz) ≥ 66 dB

Figure 11.6a shows typical transmission characteristics of the three types of filters. As can be seen, the distinguished differences among them are the locations of transmission zeros. Whereas the Chebyshev filter has all transmission zeros at dc and infinite frequencies, the elliptic-function filter has transmission zeros at finite frequencies and exhibits an equal ripple at the stopband. The quasielliptic-function filter shown here has only a single pair of transmission zeros at finite frequencies, with the remainders at dc and infinite frequencies. From this study, a number of conclusions can be drawn about the applicability of the filter types considered.

- Filters with a low number of poles require a lower Q to meet the passband insertion specification. It is estimated the Q that can be attained is 50,000.

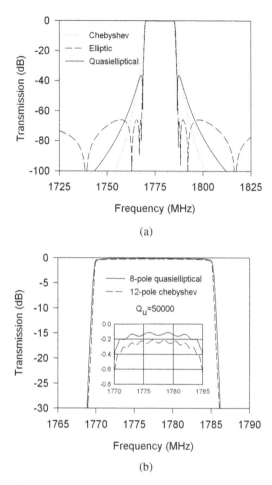

FIGURE 11.6 (a) Typical transmission characteristics of three types of eight-pole bandpass filters for a passband from 1770 to 1785 MHz. (b) Comparison of 12-pole Chebyshev and 8-pole quasielliptic function filters that meet the selectivity (a 30-dB rejection bandwidth of 17.5MHz). The Chebyshev filter has higher insertion losses in the passband (see the small insert).

Therefore, provided the required rejection and selectivity can be met, a filter with a lower number of poles as possible should be used.

- The insertion loss at the passband edge, as well as at midband, must be considered. The loss at the band edges is more predominant in the elliptic- and quasielliptic-function filters because of the effect of transmission zero near the cutoff frequency.
- The Chebyshev filter with the same number of poles as an elliptic- or quasielliptic-function filter has poorer selectivity close to the passband edge, although it has better rejection at the transmit band (see Figure 11.6a).

Increasing the number of poles can improve selectivity, but the penalty is an increased passband insertion loss (see Figure 11.6b). The Chebyshev filter is unsuitable for the requirement in which a very high selectivity is required.
- For a given number of poles, an elliptic-function filter has higher selectivity close to the passband edge, but poorer rejection in the transmit band compared with a quasielliptic-function filter (see Figure 11.6a).
- The elliptic-function filter is difficult to implement using distributed elements. An eight-pole quasielliptic-function filter with a single pair of finite-frequency attenuation poles, as described in Section 9.1, provides the best solution to meet the above specifications.

11.2.1.2 Filter Design and Realization The design of this type of filter with a single pair of transmission zeros has been detailed in Section 9.1. Referring to Section 9.1, the element values of the lowpass prototype for the eight-pole bandpass filter are

$$\begin{aligned} g_1 &= 1.02940 & g_2 &= 1.47007 \\ g_3 &= 1.99314 & g_4 &= 1.96885 \\ J_3 &= -0.40624 & J_4 &= 1.43484 \end{aligned} \quad (11.10)$$

Using the design equations given by Eq. (9.9) in Chapter 9, the coupling coefficients and external Q for this bandpass filter, with a fractional bandwidth $FBW = 0.00844$, are found

$$\begin{aligned} M_{12} &= M_{78} = 0.00686 & M_{23} &= M_{67} = 0.00493 \\ M_{34} &= M_{56} = 0.00426 & M_{45} &= 0.00615 \\ M_{36} &= -0.00172 & Q_{ei} &= Q_{eo} = 121.98281 \end{aligned} \quad (11.11)$$

The theoretical response of the filter can be computed using the method described in Section 9.1.

A novel HTS microstrip filter configuration as shown in Figure 11.7 was developed to fulfill the quasielliptic-function response. The filter is comprised of eight microstrip

FIGURE 11.7 Layout of the eight-pole quasielliptic function filter.

meander open-loop resonators on a 0.3-mm thick MgO substrate. The attractive features of this filter configuration are not only its small and compact size, but also its capability of allowing a cross coupling to be implemented, such that a pair of transmission zeros (or attenuation poles) at finite frequencies can be accomplished. The physical dimensions of the filter are determined using a full-wave EM simulator that simulated the coupling coefficients and external quality factors against physical structures, as described in Chapter 7.

11.2.1.3 Evaluation of Quality Factor It is important to evaluate the achievable unloaded quality factor Q_u of the HTS resonators used in the filter. This will serve as a justification whether or not that the required insertion loss of the bandpass filter can be met. As discussed in Chapter 4, three loss mechanisms should usually be considered for evaluation of Q_u. They are the losses associated with the microstrip conductor (the HTS thin film, in this case), the dielectric substrate, and the package housing made from a normal conductor, respectively. Recalling Eq. (4.63) and replacing the Q_r, the radiation quality factor, with the Q_h, the quality factor of the package housing, we have

$$\frac{1}{Q_u} = \frac{1}{Q_c} + \frac{1}{Q_d} + \frac{1}{Q_h}$$

Q_c, which is the quality factor of the HTS thin film microstrip resonator, and Q_d, which is the quality factor of the dielectric substrate. Direct calculations of these quality factors are nontrivial, because they require knowledge of electromagnetic field distributions that depend on the geometry of the microstrip resonator, the size of housing, and the boundary conditions imposed. However, estimation is possible following the discussions in Chapter 4. Recall Eq. (4.65) that

$$Q_c \approx \pi \left(\frac{h}{\lambda}\right)\left(\frac{\eta}{R_s}\right)$$

where h is the substrate thickness, λ and η ($\approx$377 Ω) are the wavelength and wave impedance in free space, and R_s the surface resistance of the HTS thin films for the microstrip and its ground plane (normally, a HTS ground is needed to achieve a higher Q_u). It is commonly accepted that the surface resistance of the HTS thin film is proportional to f^2, with f the frequency (refer to Eq. 11.6). Thus, the Q_c is actually proportional to wavelength, or inversely proportional to frequency. Having a thick substrate can increase the Q_c, but care must be taken, because this increases the radiation and unwanted couplings as well. At a frequency of 2 GHz and a temperature of 60K, $R_s \leq 10^{-5}$ Ω can be expected for a good YBCO thin film. If we let $h = 0.3$ mm, $\lambda = 150$ mm, and $R_s = 10$ $\mu\Omega$, we then find that $Q_c \approx 240{,}000$. Similarly, to estimate the dielectric loss, recall Eq. (4.66).

$$Q_d \geq \frac{\varepsilon'}{\varepsilon''} = \frac{1}{\tan\delta}$$

For MgO substrates, the typical value of tan δ is the order of 10^{-5} to 10^{-6} for a temperature ranging from 80 to 40K [109,110]. Therefore, at an operation temperature of 60K and frequency of about 2 GHz, we can expect that $Q_d > 10^5$ for MgO substrates.

The power loss of the housing because of nonperfect conducting walls at resonance, defined in Eq. (4.70) is

$$P_h = \frac{R_h}{2} \int |\underline{n} \times \underline{H}|^2 dS$$

Here R_h is the surface resistance of the housing walls, $\underline{n}$ is the unit normal to the housing walls, and $\underline{H}$ is the magnetic field at resonance. The housing walls are normally gold-plated to a thickness that is greater than the skin depth. Although the surface resistance of gold is about two orders larger than that of the HTS thin film, the fields intercepted by the housing walls are weaker, because they actually decay very fast away from the HTS thin film resonator, as discussed in Chapter 4. Therefore, the housing quality factor Q_h, which is inversely proportional to P_h, can generally reach a higher value with a larger housing size. Alternatively, one could optimize the shape of a resonator or use lumped resonators to confine the fields in the substrate to obtain a higher Q_h, which may, however, reduce Q_c. If we assume $Q_c = Q_d = 150,000$, we need a Q_h of 15,000, as well, to achieve a total unloaded quality factor Q_u of 50,000. For system cooling and assembling, a smaller filter housing is usually desirable. If the small housing causes a problem associated with the housing loss, an effective approach to increase the housing quality factor is to increase the housing size, particularly its height. By doing so, one must be careful about any housing mode that might be excited.

Probably, the best way to find the unloaded quality factor is to measure it experimentally. For this purpose, a single HTS microstrip, meander, open-loop resonator was fabricated using double-sided YBCO thin films on a MgO substrate with a size of $12 \times 10 \times 0.3$ mm. The resonator was fixed on a gold-plated titanium carrier and assembled in a package housing for the filter. The inner area of the housing is 39.2×25.6 mm with a distance of 3 mm from the substrate to the housing lid. Two copper microstrip feed lines were inserted between the HTS resonator and the input/output (I/O) ports on the housing to excite the HTS resonator. The loaded quality factor Q_L was measured using a network analyzer. The unloaded quality factor is extracted by

$$Q_u = \frac{Q_L}{1 - |S_{21}|} \tag{11.12}$$

with $|S_{21}|$ the absolute magnitude of S_{21} measured at the resonant frequency. The results of Q_u obtained at different temperature are $Q_u = 38,800$ at 70 K, $Q_u = 47,300$ at 60 K, and $Q_u = 50,766$ at 50 K.

11.2.1.4 Filter Fabrication and Test The superconducting filter was fabricated using $YBa_2Cu_3O_7$ thin film HTS material. This was deposited onto both sides of a

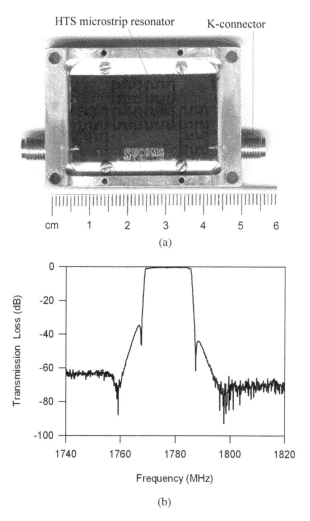

FIGURE 11.8 (a) Fabricated eight-pole HTS microstrip quasielliptic function filter in test housing. (b) Measured performance of the filter at 55K.

MgO substrate that was 0.3 × 39 × 22.5 mm and had a relative dielectric constant of 9.65. Figure 11.8a is a photograph of the fabricated HTS bandpass filter assembled inside the test housing. The packaged superconducting filter was cooled down to a temperature of 55K in a vacuum cooler and measured using a network analyzer. Excellent performance was measured, which is shown in Figure 11.8b. The filter showed the characteristic of the quasielliptic response with two diminishing transmission zeros near the passband edges, resulting in steeper skirts. One might notice that the filter frequency responses in Figure 11.8b shows asymmetric locations of

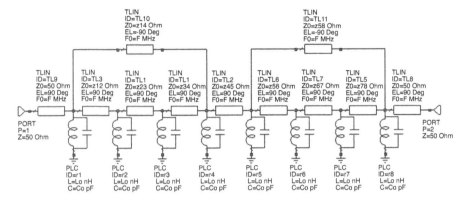

FIGURE 11.9 The circuit model showing the CQ coupling structure of the eight-pole HTS filter.

the transmission zeros. In this case, the asymmetric response is believed because of unwanted couplings [74].

11.2.2 HTS Filter with Two Pairs of Transmission Zeros

The next example is an eight-pole narrowband HTS microstrip bandpass filter on r-plane sapphire substrate [75]. The filter is designed to exhibit two pair transmission zeros with a cascaded quadruplet-(CQ) coupling structure, and to have a fractional bandwidth $FBW = 0.256\%$ and a center frequency $f_0 = 1968$ GHz.

11.2.2.1 Filter Circuit Model Figure 11.9 is the circuit model of the filter, which is created with Microwave Office, a commercially available software [112]. Following the discussion in Chapter 8, each quarter- wavelength line in this circuit model has an electrical length $EL = \pm 90°$ at the central frequency of the passband, which functions as an immittance inverter to represent the coupling between the associated pair of resonators. In this case, the positive sign is used for all the direct couplings, whereas the negative sign is for the cross couplings. All the resonators are supposed to be synchronously tuned at the central frequency $f_0 = 1/2\pi\sqrt{L_0 C_0}$.

The other circuit parameters are related, following the formulations given in Chapter 8, to a set of desired coupling coefficients M_{jk} and external quality factors $Q_{ei/o}$:

$$\begin{aligned}
Q_{ei} = Q_{eo} &= 359.6736 & M_{45} &= 0.001369 \\
M_{12} &= 0.002187 & M_{56} &= 0.001133 \\
M_{23} &= 0.001707 & M_{67} &= 0.002109 \\
M_{34} &= 0.001361 & M_{78} &= 0.001988 \\
M_{14} &= -0.000278 & M_{58} &= -0.000952
\end{aligned}$$
(11.13)

Thus, for the terminal impedance $Z = 50\ (\Omega)$, the remaining circuit parameters are found as

$$C_0 = \frac{Q_{ei}}{\omega_0 Z} = 581.74625345\ (\text{pF}) \quad L_0 = \frac{Z}{\omega_0 Q_{ei}} = 0.01124233\ (\text{nH})$$

$$z12 = \frac{Z}{Q_{ei} M_{12}} = 63.564\ (\Omega) \quad z23 = \frac{Z}{Q_{ei} M_{23}} = 81.438\ (\Omega)$$

$$z34 = \frac{Z}{Q_{ei} M_{34}} = 102.142\ (\Omega) \quad z45 = \frac{Z}{Q_{ei} M_{45}} = 101.545\ (\Omega)$$

$$z56 = \frac{Z}{Q_{ei} M_{56}} = 122.696\ (\Omega) \quad z67 = \frac{Z}{Q_{ei} M_{67}} = 65.915\ (\Omega)$$

$$z78 = \frac{Z}{Q_{ei} M_{78}} = 69.927\ (\Omega) \quad z14 = \frac{Z}{Q_{ei}\ |M_{14}|} = 500.054\ (\Omega)$$

$$z58 = \frac{Z}{Q_{ei}\ |M_{58}|} = 146.024\ (\Omega) \tag{11.14}$$

It is worth pointing out that a higher precision for the numerical results of C_0 and L_0 is necessary, because they determine the center frequency that is very sensitive for a narrowband filter, such as the one under discussion. The ideal performance of the filter, predicted by the circuit model, is plotted in Figure 11.10. As can be seen, the filter has a passband of 5 MHz centered at 1968 MHz and exhibits high selectivity resulting from the two pairs of transmission zeros.

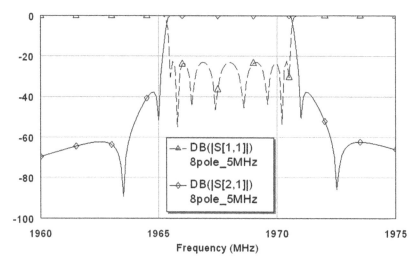

FIGURE 11.10 The ideal performance of the eight-pole HTS CQ filter with two pairs of transmission zeros.

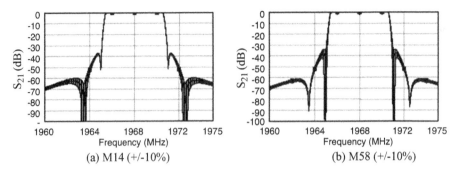

FIGURE 11.11 Cross-coupling sensitivity analysis of the eight-pole HTS CQ filter.

11.2.2.2 Sensitivity Analysis For the development of narrowband filter it is also important to carry out the sensitivity analysis since a narrowband filter tends to be more sensitive to the tolerances in both the design and fabrication. To this end, one can perform a series of sensitivity analyses based on the circuit model of Figure 11.9; the results are plotted in Figs. 11.11–11.13. The shading in Figure 11.11 indicates the changes in frequency response of the eight-pole HTS filter because of the plus and minus 10% variations of the two cross couplings, that is, M_{14} and M_{58}. It has been shown that varying M_{14} mainly affects the locations of the two outer transmission zeros, whereas the change in M_{58} has more effect on the locations of the two inner transmission zeros. Figure 11.12 shows the results of sensitivity analyses when the variation only occurs for each of the direct couplings. Again, all the variations are within ±10% of the desired couplings of Eq. (11.13). We notice that the variations of M_{12}, M_{23}, and M_{34} mainly cause the shifting of the two outer transmission zero locations as that of M_{14} does. This is because the first quadruplet section was designed to produce these two outer transmission zeros. In a similar pattern we see that, like the variation of M_{58}, the varying of M_{56}, M_{67}, and M_{78} has more effects on the frequency responses around to the inner pair of transmission zeros, which were designed to be controlled by the second quadruplet section. One can also observe that the coupling M_{45} is the least sensitive parameter for this type of filter. The analyzed results for the asynchronously tuned filter are given in Figure 11.13, where each diagram shows the sensitivity of frequency response against the variation of resonant frequency of each resonator. The resonant frequency was supposed to be changed within ±0.2% of the midband frequency, that is, 1968 ± 4 MHz. In contrast to the couplings, it is evident from the given results that the asynchronous tuning causes much more distortion in the desired filter response and, therefore, tuning resonators, in particular, resonators 2–7, is most important for this narrowband filter.

11.2.2.3 Filter Implementation Figure 11.14 shows the layout of the eight-pole CQ microstrip filter for the implementation using HTS thin films. The dielectric substrate used is the r-cut sapphire substrate, which not only allows high-quality YBCO films to be produced, but also is a low-cost standard industrial wafer. However, since

HTS FILTERS FOR MOBILE COMMUNICATIONS 451

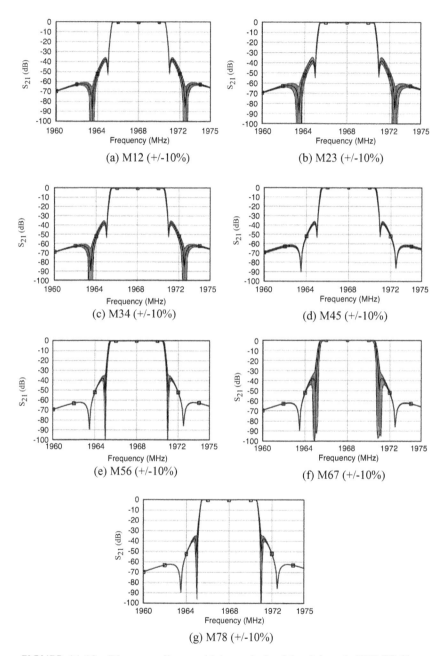

FIGURE 11.12 Direct-coupling sensitivity analysis of the eight-pole HTS CQ filter.

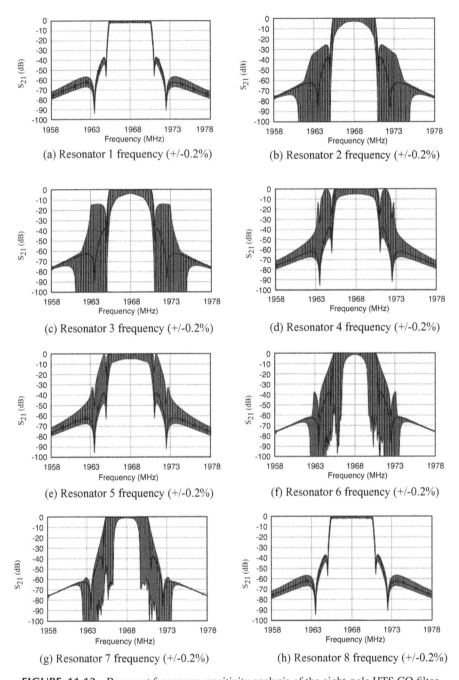

FIGURE 11.13 Resonant frequency sensitivity analysis of the eight-pole HTS CQ filter.

FIGURE 11.14 Eight-pole HTS microstrip CQ filter layout.

the dielectric properties of sapphire are anisotropic, the relative dielectric constant is not a single value, but a tensor given by

$$\hat{\varepsilon} = \begin{bmatrix} 9.4 & 0 & 0 \\ 0 & 9.4 & 0 \\ 0 & 0 & 11.6 \end{bmatrix} \quad (11.15)$$

in Cartesian coordinates with the z-axis being parallel to the principal crystal axis (c-axis) of sapphire. For a primary design, an effective dielectric constant of 10.0556 was used. The arrangements of resonators in Figure 11.14 not only allow the required couplings to be implemented, but also allow each resonator to experience the same permittivity tensor. This means that the frequency shift of each resonator because of the anisotropic permittivity of sapphire substrate is the same, which is very important for the synchronously tuned narrowband filter.

The design of this type of microstrip filter follows the similar procedure described in Chapter 9. Starting with the set of desired coupling coefficients and external quality factors of Eq. (11.13), full-wave EM simulations are performed to extract these desired design parameters so as to determine the physical layout with all the dimensions.

11.2.2.4 Fabrication and Experiment The filter was then fabricated on a 0.43-mm thick sapphire (Al_2O_3) wafer with double-sided YBCO films, which have a thickness of 300 nm. Both sides of the wafer are gold-plated with 200-nm thick gold (Au) for the RF contacts. The fabricated HTS filter used a wafer size of about 47 × 17 mm, which was assembled into a test housing, as shown in Figure 11.15a for measurements. For the experiments, some sapphire tuners were arranged on the lid to tune only the resonant frequencies of eight HTS resonators, which can be clearly seen from Figure 11.15b. This is because the frequency tuning is more important for this type of narrowband filter, as shown by the earlier sensitivity analysis. The RF measurement was done using a microwave network analyzer and in a cryogenic cooler. Figure 11.16 shows the measured results at 65K and after tuning the filter. The tuning is usually requested for a narrowband filter since there are tolerances in both the wafer thickness and fabrication.

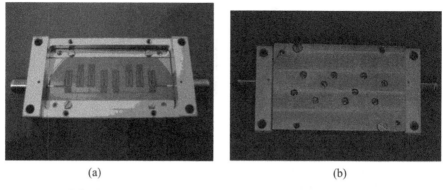

(a)   (b)

FIGURE 11.15 Assembled eight-pole HTS microstrip CQ filter.

11.2.3 HTS Filter with Group-Delay Equalization

As discussed in Chapter 9 for linear-phase filters, in many communications systems, flat group delay of a bandpass filter is also demanded in addition to its selectivity. The group delay represents the true signal delay between the input and output ports of a communication channel such as a filter and is defined in Eq. (2.12). The demand on a flat group delay is particularly true for high-capacity communication systems where it is essential for a bandpass filter to have a good linear-phase response or flat group delay over the central region of the passband, although the requirement of selectivity can allow deterioration in the phase performance at the edges of the

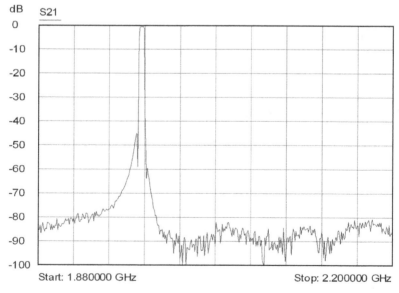

FIGURE 11.16 Measured response of the eight-pole HTS microstrip CQ filter.

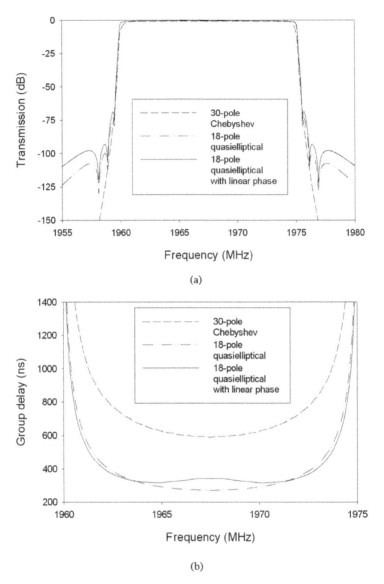

FIGURE 11.17 Performance comparison for three high-order filters. (**a**) Transmission response. (**b**) Group-delay response.

passband. Unfortunately, higher order selective-only filters tend to result in a poorer phase performance even over the band center. To demonstrate this, Figure 11.17 shows a comparison of performances for three different types of high-order filters, i.e., 30-pole Chebyshev, 18-pole quasielliptic function, and 18-pole quasielliptic function with linear phase.

All the three filters have a passband of 15 MHz from 1960 to 1975 MHz with the same ripple level and are supposed to meet a selectivity of 70-dB rejection bandwidth of about 16 MHz, as illustrated in Figure 11.17a. It is evident that, to meet this requirement, the Chebyshev filter requires much more resonators for practical implementation than that required by the quasielliptic-function filters, namely 30 against 18. The need for such a larger number of resonators for the Chebyshev filter not only leads to a larger size, but also results in larger insertion loss (because of finite resonator Q) and variation of group delay over the pass band. The group-delay response is shown in Figure 11.17b. Apparently, it is not a good practice to realize highly selective filters with the Chebyshev response. While the two quasielliptic-function filters are more superior in terms of achieving high selectivity with less number of resonators, the one with linear-phase response is eventually more attractive as it exhibits a flatter group delay over the band center, which can be seen from Figure 11.17b. It may also be worthwhile to compare the out-of-band rejection of the three filters, as shown in Figure 11.17a. The 30-pole Chebyshev filter has the best out-of-band rejection at the cost of the larger number of resonators included. For the other two 18-pole quasielliptic-function filters, the rejection band of the filter with linear delay is slightly worse than the other one without delay equalization, which is expected because of the different cross couplings introduced into the two filters. In practice, there will be a finite or minimum out-of-band rejection requirement, which, however, can easily be met by a high-order filter.

Hence, it is desirable to develop high-order HTS bandpass filters, not only to meet the selectivity requirement, but also with a capability of self-equalization of group delay over the central part of the passband [78]. The design, modeling, and experimental results of such a high-order HTS microstrip bandpass filter are presented next.

The filter model or coupling structure of an eighteen-pole filter is shown in Figure 11.18 for designing the high-order HTS microstrip filter. Each node with a number represents a resonator. Resonators 1 and 18 are coupled to the input and output ports, respectively, denoted by external quality factors, Q_{e1} and Q_{e2}. The full line between adjacent resonators indicates the direct coupling. There are four cross couplings, as indicated by the broken line, between the resonators 2 and 5, 6 and 9, 10 and 13, as well as 14 and 17. These cross couplings are denoted by coupling coefficients M_{25}, M_{69}, $M_{10,13}$, and $M_{14,17}$. As can be seen, each cross coupling is associated with a quadruplet section and, hence, the filter has basically a cascaded quadruplet (CQ) structure, as introduced in Chapter 9. The advantage of this coupling structure lies

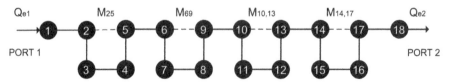

FIGURE 11.18 Coupling structure for 18-pole selective filter with group-delay equalization.

in that each of quadruplet sections can be such arranged either to produce a pair of transmission zeros (attenuation poles) at finite frequencies in order to achieve higher selectivity for the given number of resonators, or to result in a linear-phase performance to achieve a self-equalization of group delay. For this design, only one quadruple section, which consists of the resonators 10–13, will be used for the group-delay equalization, whereas the other three quadruplet sections are arranged for the high selectivity. Since the effect of the cross coupling in each quadruple can be tuned independently, this makes the tuning easier for such a high-order filter.

To design this type of filter with cross couplings, two main steps are involved. The first step is to obtain all desired parameters, such as coupling coefficients and external quality factors of Figure 11.18, which corresponds to a filter synthesis as discussed in Chapter 8. The second step is about the physical realization in a microwave transmission line medium, which, in this case, is the microstrip.

The 18-pole filter was designed to have a 15-MHz passband at a center frequency of 1967.5 MHz. The desired external quality factors and coupling coefficients are listed below

$$\begin{aligned}
Q_{e1} = Q_{e2} &= 122.899126 & M_{11,12} &= 0.002074 \\
M_{12} &= 0.006449 & M_{12,13} &= 0.003423 \\
M_{23} &= 0.004089 & M_{13,14} &= 0.003873 \\
M_{34} &= -0.005706 & M_{14,15} &= 0.002946 \\
M_{45} &= 0.003460 & M_{15,16} &= -0.006372 \\
M_{56} &= 0.003878 & M_{16,17} &= 0.003627 \\
M_{67} &= 0.003735 & M_{17,18} &= 0.006450 \\
M_{78} &= -0.004866 & M_{25} &= 0.001822 \\
M_{89} &= 0.003722 & M_{69} &= 0.001007 \\
M_{9,10} &= 0.003855 & M_{10,13} &= 0.001785 \\
M_{10,11} &= 0.003419 & M_{14,17} &= 0.002624
\end{aligned} \qquad (11.16)$$

Using the formulation given in Chapter 7 for the scattering parameters of the two-port network of a synchronously tuned n-coupled resonator filter, the filter response can be computed. Alternatively, a circuit model similar to that shown in Figure 11.9 may be created for the modeling [78]. Figure 11.19 shows the ideal or theoretical frequency responses of the filter. As can be seen, the filter exhibits three pairs of transmission zeros (attenuation poles) at finite frequencies to increase the selectivity, while has a flat group delay over about 50% of the passband in the middle.

To implement this type of filter in microstrip, two basic quadruplet sections of coupled microstrip resonators, as shown in Figure 11.20, have been investigated. Each quadruplet section is comprised of four meandered microstrip resonators as shown and the four resonators are arranged in such a way that the direct coupling between adjacent resonators results from the proximity of the resonators, whereas a

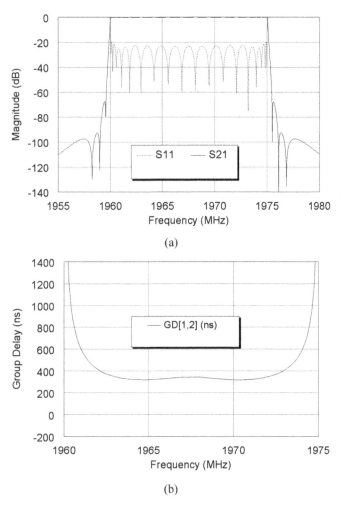

FIGURE 11.19 Theoretical responses of the 18-pole bandpass filter with self-equalization of group delay. (**a**) Magnitude; (**b**) group delay.

narrow line coupled to the first and last resonators is used for the cross coupling. By inspecting the quadruplet configurations or arrangements in Figures 11.20 a and b, we notice that they are almost the same, except for the spacing between two inner coupled resonators in each quadruplet section. This spacing in Figure 11.20a is much smaller than that in Figure 11.20b. As a matter of fact, it is this spacing difference that makes the characteristics of these two quadruplets totally different. It can be shown that the coupling between the two inner coupled resonators in Figure 11.20a is dominated by the electric coupling, whereas the magnetic coupling is dominant for the two inner coupled-resonators in Figure 11.20b. Because these two couplings have opposite effects, which tend to cancel out each other; we may denote the electric coupling as a

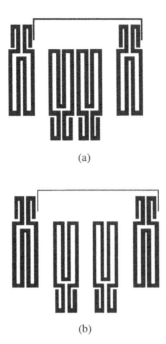

FIGURE 11.20 Two microstrip quadruplet sections. (**a**) For the realization of a pair of transmission zeros at finite frequencies. (**b**) For the realization of group-delay equalization.

negative coupling and the magnetic coupling a positive one. This then allows us to use the microstrip configuration of Figure 11.20a to implement the desired couplings for those quadruplet sections in Figure 11.18 that produce finite frequency transmission zeros. On the other hand, the microstrip configuration of Figure 11.20b is used for the realization of the required couplings for the quadruplet consisting of resonators 10–13 in Figure 11.18, which results in group-delay equalization.

For example, following Eq. (11.16), the desired coupling matrix for the first quadruplet, i.e., coupled resonators 2, 3, 4, and 5, is given by

$$\begin{bmatrix} 0 & M_{23} & 0 & M_{25} \\ M_{23} & 0 & M_{34} & 0 \\ 0 & M_{34} & 0 & M_{45} \\ M_{25} & 0 & M_{45} & 0 \end{bmatrix} = \begin{bmatrix} 0 & 0.004089 & 0 & 0.001822 \\ 0.004089 & 0 & -0.005706 & 0 \\ 0 & -0.005706 & 0 & 0.003460 \\ 0.001822 & 0 & 0.003460 & 0 \end{bmatrix}$$

(11.17)

where the coupling for M_{34} is negative. This quadruplet is supposed to be responsible for a pair of transmission zeros observable from the magnitude response in Figure 11.19a. The full-wave EM simulation has been performed for a microstrip quadruplet section in the form of Figure 11.20a to facilitate the coupling matrix of Eq. (11.17); the result is plotted (full line) in Figure 11.21a. Note that the response was obtained by weekly coupling the quadruplet to two input/output ports. The EM

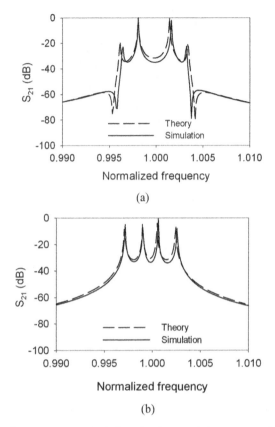

FIGURE 11.21 Frequency responses of quadruplet coupled resonators. (**a**) For the realization of a pair of transmission zeros at finite frequencies. (**b**) For the realization of group-delay equalization.

simulation was done using a commercially available simulator *em* [113]. A quadruplet resonator-circuit model, based on Eq. (11.17), can also be used to compute the theoretical response, as shown with the broken line in Figure 11.21a. As can be seen, there is a good agreement between the theory and simulation and the used microstrip quadruplet realization produces the desired transmission zeros.

Similarly, the EM simulation and circuit modeling can be done for the quadruplet that consists of coupled resonators 10, 11, 12, and 13 and has a coupling matrix

$$\begin{bmatrix} 0 & M_{10,11} & 0 & M_{10,13} \\ M_{10,11} & 0 & M_{11,12} & 0 \\ 0 & M_{11,12} & 0 & M_{12,13} \\ M_{10,13} & 0 & M_{12,13} & 0 \end{bmatrix} = \begin{bmatrix} 0 & 0.003419 & 0 & 0.001785 \\ 0.003419 & 0 & 0.002047 & 0 \\ 0 & 0.002047 & 0 & 0.003423 \\ 0.001785 & 0 & 0.003423 & 0 \end{bmatrix}$$

(11.18)

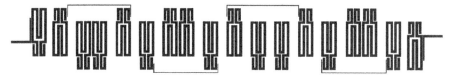

FIGURE 11.22 Layout of the designed 18-pole HTS microstrip filter on sapphire substrate.

In this case, a microstrip configuration of Figure 11.20b was used in the simulation. The simulated and theoretical results are plotted in Figure 11.21b, where no finite-frequency transmission zero is observable, which is expected as this quadruplet is only for the group-delay equalization. Again, the good agreement between the theory and simulation validates the microstrip realization of this type of quadruplet section.

Figure 11.22 shows the final layout of the designed 18-pole superconducting microstrip filter. The substrate is r-cut sapphire and the chip size is 74 ×17 mm. It can be recognized in Figure 11.22 that the third quadruplet from the left is for the self-equalization of group delay.

The filter was then fabricated on a 0.43-mm thick sapphire (Al_2O_3) wafer with double-sided YBCO films. The YBCO thin films have a thickness of 300 nm and a characteristic temperature of 87K. Both sides of the wafer are gold-plated with 200-nm thick gold (Au) for the RF contacts. The fabricated HTS filter was assembled into a test housing with two K connectors for measurements. The lid has accommodated a number of sapphire tuners, similar to that shown in Figure 11.15. Most of the tuners were used in the experiment to tune the resonator frequencies, as suggested by a sensitivity study, which is similar to that discussed in the previous section for the eight-pole HTS microstrip CQ filter.

The RF measurements were done using an HP network analyzer and in a cryogenic cooler. Figure 11.23 shows the measured results at 65K and after tuning the filter. The tuning is a must for such a high-order and narrowband filter, since there are tolerances in both the wafer thickness and fabrication. From Figure 11.23a, we can see that the measured band center frequency is about 1970 MHz, which is slightly higher than the designed one of 1967.5 MHz. This is because the center frequency is dependent on the orientation of the filter on the sapphire wafer as a result of anisotropic dielectric property. The measured bandwidth is close to 15 MHz. The insertion loss of 1.4 dB at the band center was measured, including the losses of the contacts. The resonator Q is estimated to be larger than 50,000. The measured return loss, shown in Figure 11.23a, is better than -10 dB across the passband. Further finer tuning could improve the return loss as well as shift the center frequency to the designed one. The transmissions zeros near the band edges for enhancing selectivity are observable. The measured group delay of the filter is plotted in Figure 11.23b, showing a flat group delay over the central region of the passband, which is in very good agreement with the theoretical response given in Figure 11.19b.

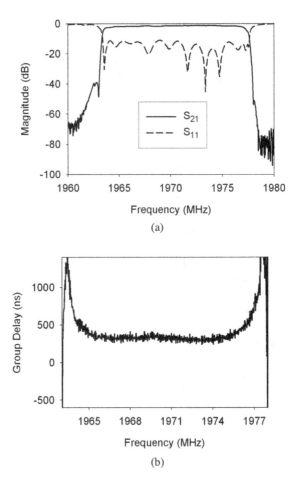

FIGURE 11.23 The measured responses of the 18-pole HTS filter with self-equalization of group delay. (a) Magnitude; (b) group delay.

11.3 HTS FILTERS FOR SATELLITE COMMUNICATIONS

In commercial satellite applications, superconducting filters offer the advantage of a reduction in mass and volume, combined with improved high-frequency performance. The mass and volume of payload electronic equipment is a significant contributor to the overall cost of space systems. Therefore, there has been an interest in the development of HTS planar filters for satellite applications [29–40].

11.3.1 C-Band HTS Filter

In an earlier work [33], HTS microstrip filters with Chebyshev response were developed for a satellite communication (SATCOM) system. In SATCOM applications,

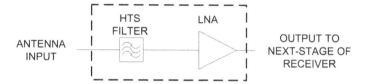

FIGURE 11.24 Block diagram of HTS filter and cryoLNA in cryogenic Dewar.

the power of the received signal is very low. In conventional systems, the first element after the antenna is usually a low-noise amplifier (LNA). The disadvantage of this is that the amplifier is a broad-band device and, thus, will process all the signals delivered to it, including those in the side lobes of the antenna. This could be a problem if there are large out-of-band signals present. These large signals could saturate the amplifier. Even if the signals are not strong enough to send the amplifier into compression, these large signals could produce intermodulation products in the band of interest. If one were to place a narrow filter in front of the LNA, then the interference can be greatly attenuated and only the desired in-band signals presented to the LNA. This would improve the selectivity of the system. This is not usually done because the added loss of a conventional filter can seriously degrade the receiver sensitivity, but because of the low loss of HTS materials, this approach becomes feasible.

Since the HTS filters need to be cooled, one can gain great advantage by placing as many components as possible in the Dewar. By placing the first-stage LNA in the Dewar, as shown in Figure 11.24, one gains a great advantage in noise temperature. A very good conventional -band SATCOM LNA has a noise temperature of about 30K, but by placing a cryoLNA (that is, an LNA designed to operate at 77K) in the Dewar with the filter, the noise temperature of the device is only about 10K. These two elements together provide a very high-performance front end.

The filter used above was designed and built to cover one transponder band with a bandwidth of 37 MHz at a center frequency of about 3.5 GHz. The filter is an eighth-order 0.01-dB equal-ripple Chebyshev design. This filter was fabricated on a 34 × 16 mm MgO substrate that was 0.5 mm thick. The physical realization is illustrated in Figure 11.25. From this physical layout, one can see that the resonator

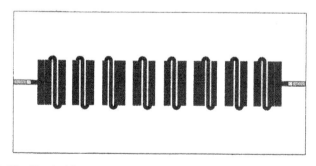

FIGURE 11.25 Physical layout of the eight-pole HTS filter with at a center frequency about 3.5 GHz on a MgO substrate with a size of 34 × 16 mm. (Taken from [33], ©2000 IEEE.)

used was a half-wavelength structure with two bends, which may be seen as a folded stepped-impedance resonator that has a higher impedance line in the middle and two lower impedance line sections on the both sides. Instead of optimizing this filter in the lumped domain, the filter performance was optimized directly in the distributed domain. This design approach was chosen because this structure closely resembles a half-wavelength resonator and, therefore, becomes practical to optimize directly as distributed elements. This allows a flexible implementation of inter-resonator couplings as well as a trade-off between the resonator size and unloaded Q. The measured passband insertion loss of this filter was about then 0.2 dB and had a return loss of approximately 25 dB. As this filter loss is a cold loss (at 77K), the added noise temperature because of the filter is only 3.6K. Thus, the advantage of HTS filters is in providing very high selectivity with only minimal increase in noise temperature that is not normally achievable with normal ambient temperature filters.

Another work, reported in [35], has successfully demonstrated the feasibility of building a C-band HTS multiplexer. The multiplexer consists of 60 channel filters that are self-equalized 10-pole HTS planar structures designed with drop-in cryogenic ferrite circulators and isolators. The multiplexer is integrated with two miniature pulse-tube cryocoolers to meet the reliability requirement of space applications. The whole package is designed to be in the form of "plug-and-play" to replace conventional C-band dielectric resonator input multiplexers inside the INTELSAT eight satellites. The results achieved in this work illustrate that an overall mass saving of 12 kg (approx. 50% saving) and a size reduction of over 50% can be achieved with the use of HTS technology. Although the system demonstrated has 60 channels, future multimedia satellites are expected to have a larger number of RF channels. As the number of channels increases, the benefits of inserting the HTS technology become more substantial.

More recently, a C-band eight-pole self-equalized quasielliptic HTS microstrip filter has been developed for satellite application [36]. This type of narrow-band filter ($FBW \approx 1.4\%$ at 4 GHz), is required to comply with stringent passband characteristics concerning the selectivity and the group delay. The filter design is based on a canonical coupling structure, as shown in Figure 11.26, where each black circle represents a resonator. The full lines indicate the direct couplings and, the broken lines the cross couplings. These cross couplings having positive and negative signs are implemented to obtain the good selectivity by placing finite-frequency transmission

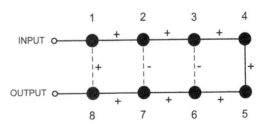

FIGURE 11.26 Canonical coupling structure of eight-pole quasielliptic filter with linear-phase response or group-delay self-equalization.

zeros close to the passband edges and to equalize the group delay in the channel band as well.

The normalized coupling matrix and scaled external quality factors, as discussed in Chapter 7, are given as follows:

$$m = \begin{bmatrix} 0 & 0.8251 & 0 & 0 & 0 & 0 & 0 & 0.0463 \\ 0.8251 & 0 & 0.5855 & 0 & 0 & 0 & -0.0562 & 0 \\ 0 & 0.5855 & 0 & 0.4935 & 0 & -0.2025 & 0 & 0 \\ 0 & 0 & 0.4935 & 0 & 0.7519 & 0 & 0 & 0 \\ 0 & 0 & 0 & 0.7519 & 0 & 0.4935 & 0 & 0 \\ 0 & 0 & -0.2025 & 0 & 0.4935 & 0 & 0.5855 & 0 \\ 0 & -0.0562 & 0 & 0 & 0 & 0.5855 & 0 & 0.8251 \\ 0.0463 & 0 & 0 & 0 & 0 & 0 & 0.8251 & 0 \end{bmatrix}$$

$$q_{e1} = q_{e2} = 0.9926$$

(11.19)

Using the formulation presented in Chapter 7, the ideal frequency response of the filter can be computed and the results plotted in Figure 11.27.

The physical realization of this filter presents some challenges. First, unlike a CQ filter, the effect of each cross coupling is not independent in a canonical filter, as discussed in Chapter 9. Second, there are parasitic couplings that can affect the filter performance. Figure 11.28a illustrates a realization of the eight-pole HTS canonical filter described in Seaux et al. [36], in which microstrip cross open-loop resonators are deployed. This type of resonator allows the coupling structure of Figure 11.26 in a flexible manner while limiting the problems of undesired couplings among resonators. It can be shown that only the parasitic couplings between resonators 1 and 7 and 2 and 8 have a dominative side effect on the frequency responses of the filter. To overcome this, one metallic patch is inserted between resonator 1, 2, 7, and 8, as shown in Figure 11.28a. The dimensions of the square patch are determined to suppress the parasitic couplings. In addition to this measure to minimize the unwanted cross couplings, one can also take into account the parasitic couplings in the design. To do so, first all parasitic couplings need to be extracted from EM simulation and then a new coupling matrix, which is different from the ideal one of Eq. (11.19) and includes the unwanted cross couplings, is optimized for the desired filter response [36]. This leads to an optimal design for which a true response can be predicted and a minimal effort for the filter final tuning can be achieved. Figure 11.28b shows the experimental performance of an eight-pole HTS filter that has a topology shown in Figure 11.28a and was built on a 520-μm thick $LaAlO_3$ substrate with a size of 21.16 × 12.08 mm and coated double-sided superconductor YBaCuO thin film. The measurement was taken at 77K without tuning [36].

11.3.2 X-Band HTS Filter

Most HTS microstrip filters developed for satellite applications require both high selectivity and flat group delay. An X-band (8–12 GHz) narrowband HTS filter

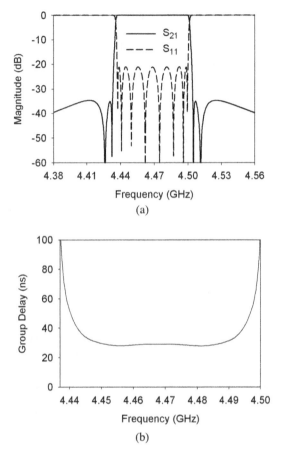

FIGURE 11.27 Ideal performance of eight-pole canonical filter with linear-phase response. (**a**) Magnitude; (**b**) group delay.

exhibiting such a stringent requirement has been developed [38]. The filter has a center frequency of 8.625 GHz and a 3-dB bandwidth of 42 MHz (a fractional bandwidth of 0.49%). This particular frequency is chosen since there are practical applications for microwave radio relay communication systems, satellite communications, and space research, etc., whose operation frequencies are around 8 GHz. The filter has a coupling structure of Figure 11.29, which is a cascaded quadruplet or CQ filter, as discussed in Chapter 9. In this filter design, the two quadruplet sections, which consist of resonators 2–5 and 6–9, were designed to create two pairs of real transmission zeros to improve group-delay flatness. The other quadruplet section, consisting of resonators 10–13, was designed to create one pair of imaginary transmission zeros to realize high selectivity. The three pairs of transmission zeros of the lowpass prototype filter were designed at $p_{1,2} = \pm 0.55$, $p_{3,4} = \pm 0.55$, and $p_{5,6} = \pm j1.30$ on the complex plane (see Chapter 3), respectively.

HTS FILTERS FOR SATELLITE COMMUNICATIONS   467

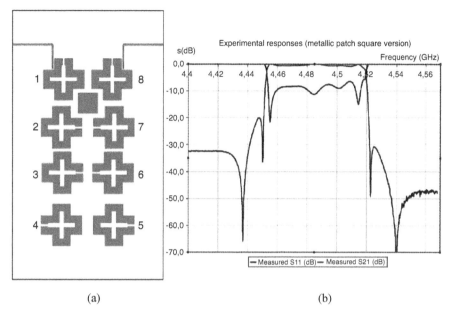

(a)   (b)

FIGURE 11.28 (a) Topology of the C-band eight-pole linear phase HTS microstrip filter. (b) Experimental response without tuning. (Taken from [36], ©2005 EuMA.)

Figure 11.30a illustrates the layout of the realized filter, which uses 14 double-folded resonators. Since the double-folded resonator is asymmetric, it is easy to realize different types of adjacent coupling, electric coupling, or magnetic coupling, by simply flipping over the resonators. As can be seen from Figure 11.30a, the arrangement of the inner pair of resonators of the first two CQ sections is different from that of the inner pair of resonators of the last CQ section on the right, which results in distinct coupling characteristics. As a result, the first two CQ sections are implemented for improving group-delay flatness, whereas the last CQ section is for improving selectivity.

The filter was fabricated on a 2-inch diameter 0.5-mm thick MgO wafer with double-sided YBCO films. It was then packaged into a housing box with inner dimensions of 40 × 8.4 × 5 mm. The frequencies of housing modes are higher than 17 GHz. The measurements were done in a cryogenic cooler by Agilent network

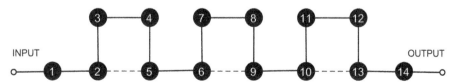

FIGURE 11.29 Coupling structure of 14-pole CQ filter for high selectivity and flat group delay.

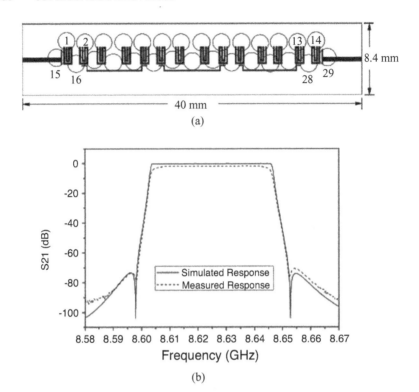

FIGURE 11.30 (a) Layout of the X-band 14-pole HTS microstrip filter. (b) Simulated and measured performance. (Taken from [38], ©2009 IEEE.)

analyzer N5230A at 77K. Calibration was done inside the cooler. The I/O cables and connectors inside the cooler were previously measured at both room and low temperatures. The effect of these cables and connectors was compensated for. The filter was tuned using sapphire rod screws, as denoted by circles in Figure 11.30a. Screws 1–14 were used to tune the frequency of each resonator, screws 16–28 were used to tune the coupling between adjacent resonators, and screws 15 and 29 were used to tune the I/O coupling. Figure 11.30b displays the measured and simulated performance of the filter. The measured midband insertion loss is 1.6 dB for a return loss better than 15.5 dB, which implies that the unloaded value of the double-folded resonator is higher than 19,000 at 77K. Band-edge steepness reaches over 11.7 dB/MHz showing a high selectivity. As reported [38], the variation of group delay is less than 23 ns over 33 MHz (78.5% of 3-dB bandwidth), and less than 30 ns over 34.5 MHz (82% of 3-dB bandwidth).

11.3.3 *Ka*-Band HTS Filter

The development of multimedia satellites and the saturation of the operational frequency bands necessitate an increase in frequency coverage. The use of high frequencies requires reconsidering the technologies and the design methods of the passive and active devices. An investigation has been carried out to study a superconducting front-end planar filter at *Ka*-band (26–40 GHz) for a potential application of the future geostationary orbit satellite communication system [39]. Since the theoretical model has shown the rapid increase of the superconductor surface resistance with frequency (Figure 11.5), care must be taken for microwave applications of HTS at such a high frequency. In this investigation, L-shaped microstrip resonators, similar to that shown in Figure 6.8, were used to design a four-pole Chebyshev bandpass filter having a fractional bandwidth of 6.2% centered at 29.1 GHz. The filter, which has a size of 6 × 4 mm only, was fabricated on a 0.25-mm thick $LaAlO_3$ substrate with double-sided YBaCuO films. Although the measured performance of the filter was encouraging, it was then difficult to find the insertion loss of the HTS filter accurately from the direct measurement. Thus, the unloaded quality factor (Q_u) of a L-shaped resonator with 50-Ω line characteristic impedance was measured to evaluate the insertion losses of the four-pole HTS filter at 29.1 GHz for a return loss of 10 dB. The experimental investigation [39] revealed that the HTS L-shaped resonator had a $Q_u = 1837$, which was about six time higher than that for a classical conductor resonator on alumina at 77K. Hence, the insertion loss of the HTS filter, which is estimated based on the measured Q_u of the HTS resonator, was found to be 0.66 dB. It was also found [39] that by reducing the line characteristic impedance from 50 to 30 Ω, the Q_u of the HTS L-shaped resonator at 29.1 GHz was increasing from 1834 to 4009 at 77K, indicating that the use of low-impedance line resonators can reduce the filter insertion loss further.

11.4 HTS FILTERS FOR RADIO ASTRONOMY AND RADAR

There has been an interest in radio astronomy or radar applications of high-temperature superconducting filters [41–53]. A radio astronomy observatory usually carries out important research into a wide range of astrophysical areas, including pulsars, the interstellar medium, circumstellar masers, active galactic nuclei, gravitational lensing, and the cosmic microwave background. However, proliferating traffic from mobile communications [global system for mobile communications (GSM) and third generation (3G)], global positioning systems (GPSs), digital audio broadcasting (DAB), and closed-circuit television (CCTV) is seriously contaminating the radio astronomical observations. The power levels of interference can be billions of times stronger than the studied radio astronomy signals, depending on the environment. The protected quiet bands are increasingly encroached on by interference. Efficient filtering is required to help alleviate this problem. In order to reduce the noise in the semiconductor front-end amplifiers, every radio telescope receiver system is cooled down to a temperature of around 20K. Under the cryogenic temperature environment, high-temperature superconducting (HTS) filters provide an effective solution to keep the strong interfering signals out of the passband and ensure negligible added noise. Depending on an operational system, either narrowband or wideband HTS filters is

11.4.1 Narrowband Miniature HTS Filter at UHF Band

For applications near 408 MHz in the ultrahigh-frequency (UHF) band, HTS microstrip bandpass filters with high selectivity in a physical size are required in the receivers. However, the maximum physical size of the HTS filters is limited by wafer size, cryostat space, and the maximum rate of heat extraction by the cryostat. To overcome this, a miniature HTS bandpass filter with microstrip quarter-wavelength spiral resonators has been developed [47]. The filter was designed as an eight-pole quasielliptic function filter that has a coupling structure shown in Figure 11.31. This coupling structure looks similar to the canonical one in Figure 11.26, but has only two cross couplings, that is, M_{27} and M_{36}. The normalized coupling matrix and scaled external quality factors are given in Eq. (11.20). The center frequency of the filter is 408 MHz and the bandwidth is 8.5 MHz, or $FBW = 2.083\%$. Using Eq. (11.20), the theoretical response of the filter can be calculated by the formulation introduced in Chapter 7; the result is plotted in Figure 11.32.

$$m = \begin{bmatrix} 0 & 0.8349 & 0 & 0 & 0 & 0 & 0 & 0 \\ 0.8349 & 0 & 0.5981 & 0 & 0 & 0 & -0.0145 & 0 \\ 0 & 0.5981 & 0 & 0.5574 & 0 & -0.0201 & 0 & 0 \\ 0 & 0 & 0.5574 & 0 & 0.5814 & 0 & 0 & 0 \\ 0 & 0 & 0 & 0.5814 & 0 & 0.5574 & 0 & 0 \\ 0 & 0 & -0.0201 & 0 & 0.5574 & 0 & 0.5981 & 0 \\ 0 & -0.0145 & 0 & 0 & 0 & 0.5981 & 0 & 0.8349 \\ 0 & 0 & 0 & 0 & 0 & 0 & 0.8349 & 0 \end{bmatrix}$$

$$q_{e1} = q_{e2} = 0.9969 \quad (11.20)$$

For the desired response, shown in Figure 11.32, the signs of the direct (or main) couplings in Eq. (11.20) do not affect the filter performance, except that the coupling (M_{45}) between resonators 4 and 5 has to be opposite the cross coupling, M_{36}, between

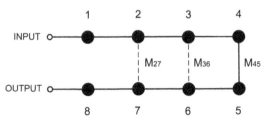

FIGURE 11.31 Coupling structure of an eight-pole quasielliptic function filter.

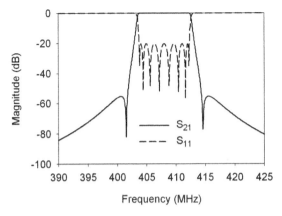

FIGURE 11.32 Theoretical response of the UHF-band eight-pole quasielliptic function filter.

resonators 3 and 6 and, M_{27}, between resonators 2 and 7. The determination of the sign of a coupling coefficient is rather relative and dependent on the physical coupling structure of the coupled resonators (refer to the discussion in Chapter 7).

Figure 11.33a illustrates the layout of the designed filter on 0.508-mm thick MgO substrate. The filter has an inline structure and is symmetrical with respect to the input and output. It consists of eight quarter-wavelength resonators, which were realized by HTS microstrip spirals shorted to an HTS grounding pad (the big black pad in the layout). Each quarter-wavelength spiral resonator resonating at 408 MHz occupies an area of 2.95 × 3.70 mm with a line width of 0.05 mm and a gap of 0.05 mm between tracks. The HTS grounding pad contacted the box wall with no gap in the simulation initially and then was bonded to the metal box wall by a number of bonding wires in the package. The grounding pad has a depth of 4.0 and a width of 36.0 mm making sure that they have only a small influence on the resonant frequencies.

The use of an inline structure as that of Figure 11.33a to implement the coupling structure of Figure 11.31 has an advantage of diminishing unwanted cross couplings, as compared to a folded structure. Consequently, these unwanted cross couplings appear sufficiently small and can be neglected in the filter design. On the other hand, for the two desired cross couplings, namely M_{36} and M_{27}, each is facilitated through a transmission strip capacitively coupled to the relevant resonators, as illustrated by the inset in Figure 11.33a. The smaller the gap and the longer the overlap length, the stronger the cross coupling. Note that the center frequency will slightly decrease when the transmission strip end is increased. Hence, the structures of these resonators involved need to be adjusted accordingly. External coupling is controlled by the tap position, as shown in the enlargement in Figure 11.33a. The closer the tapped line is to the shorted end of the resonator, the weaker the external coupling. By applying the technique described in Chapter 7, these design parameters can be extracted so as to determine the appropriate physical dimensions of the filter. The entire layout of the filter has a size of 36 × 12 mm, including the margin of 2.5 mm to the box wall.

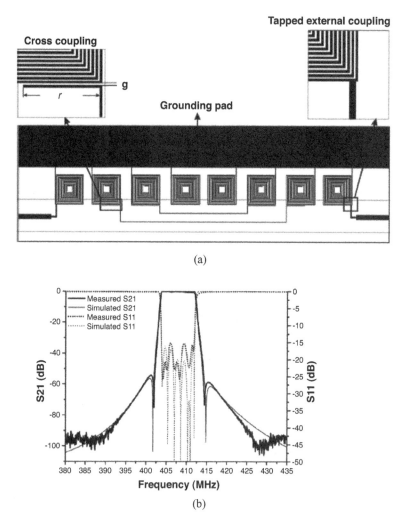

FIGURE 11.33 (a) Layout of the eight-pole microstrip HTS bandpass filter for the radio-astronomy applications at the UHF band with a center frequency of 408 MHz and a bandwidth of 8.5 MHz. (b) Measured and simulated performance of the filter. (Taken from [47], ©2006 IEEE.)

Because of the limitation of the structure, only negative cross couplings can be applied in the layout. Generally, one negative cross coupling using a transmission line between resonators 3 and 6 would be sufficient for two symmetrical transmission zeros in the stopband. However, near the shorted end of the spiral resonator, the electric field is too weak to produce enough electric coupling to achieve steep slope in the characteristics of the quasielliptic filter. One more negative cross coupling between resonators 2 and 7 was added in the layout, although theoretically a positive cross coupling can generate another two symmetrical transmission zeros.

The HTS bandpass filter was fabricated using double-sided YBCO thin film deposited on an 0.508-mm thick MgO substrate. A 1.0-mm wide and 0.1-μm thick gold strip was deposited along the edge of the ground pad to allow further wedge bonding. The circuit, with an overall dimension of 36 × 12 mm, was mounted in a gold-plated titanium box. More than a total of 80 aluminum bonding wires (25-μm thick) were then bonded to connect the grounding pad to the gold-plated box wall. The filter characteristics, measured at 20K, are shown in Figure 11.33b, where the simulation results are also shown. The measured minimum insertion loss is 0.01 dB, indicating that the aluminum bonding introduced virtually no loss. This is mainly because the high current at the shorted end of each spiral resonator has been spread out to a low level when reaching the other end of the large HTS grounding pad. The average unloaded quality factor of each resonator, estimated from the insertion loss, exceeds 300,000, although this value is obviously uncertain because of the difficulty in measuring such small loss values. It is also shown [47] that the first spurious passband of the filter appears at 1215 MHz, which is about three times the fundamental center frequency.

11.4.2 Wideband HTS Filter with Strong Coupling Resonators

A wideband superconducting filter using strong coupling resonators has been developed for a radio astronomy observatory whose observation band covers a quite wide range, from 1180 to 1730 MHz [53]. To reduce the out-of-band interferences, a wideband HTS filter with band edge-slope as steep as possible and out-of-band rejection as high as possible is required. The filter has a bandwidth of 550 MHz or a fractional bandwidth of 38% centered at 1445 MHz. There are at least two challenges for the design of such a wideband HTS microstrip filter.

The first challenge is the spurious peaks related to the second harmonic of the resonators. For an ordinary filter with ultranarrow bandwidth, the spurious peaks related to the second harmonic are sufficiently far away from the passband, which usually do not cause any problem unless, in some special cases, while much wider suppressed band above the passband are required. When the bandwidth of a filter becomes significantly large, for example, more than 2/3 of its central frequency or even larger, the spurious peaks might become very close to the passband edge, which can be a serious problem for the application under consideration.

The second challenge is the strong coupling strength between the resonators. It is well known that the bandwidth of a filter is closely related to the coupling strength between its basic elements, that is, the resonators. In general, the wider the passband, the stronger is the required coupling. For the design of a coupled resonator filter with extremely wide bandwidth, remarkably strong coupling between the resonators must be realized, which can be a difficult task as well.

To handle these challenges, two basic resonators shown in Figures 11.34a and b are deployed in the filter design [53]. The resonator of Figure 11.34a has a hairpin-like structure except an interdigital capacitor structure exists at the ends. The interdigital capacitor structure can lead to a wide stopband between the fundamental frequency (wanted) and the second harmonic (unwanted) because of the dispersion effects

474 SUPERCONDUCTING FILTERS

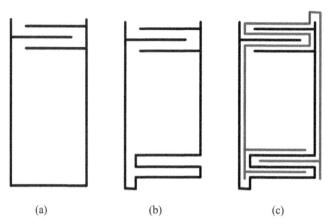

FIGURE 11.34 (a) Hairpin-like resonator with interdigital capacitor at the ends. (b) Modified resonator. (c) Interleaved coupled resonators.

similar to that discussed in Chapter 10 for the capacitive-loaded transmission-line resonator. The modified resonator of Figure 11.33b is for the facilitation of strong coupling resonators, as shown in Figure 11.33c, where the two modified resonators are interleaved together. The interleaved structure of Figure 11.33c is somehow like the psesudointerdigital structure described in Chapter 8, which, however, are developed for different purposes.

Using the resonators of Figure 11.34, a 12-pole filter with a bandwidth from 1180 to 1730 MHz was selected to meet the high selectivity demand. Furthermore, using the coupling structure of Figure 11.35a, a transmission zero was placed at the upper side of the passband to meet the attenuation specification at 1760 MHz, which makes the filter frequency response asymmetrical. The physical layout of the filter, however, can still be kept symmetrical by employing two identical cross couplings, as indicated in Figure 11.35b, to create the same transmission zero on filter frequency response. The symmetric physical configuration will efficiently simplify the EM simulation process. The topology in Figure 11.35 a has cascade trisectionlike (refer to Chapter 9) structure. Each node with a number represents a resonator, the full lines represent the main coupling, and the dotted lines between resonators 3 and 7, 6 and 10 represent the cross coupling, respectively.

The normal design procedure for cross coupled resonator filters using coupling matrix (refer to Chapter 7), which is based on narrowband approximation, is no longer suitable for this case because the fractional bandwidth here is wider than 35%. Nevertheless, a coupling matrix could be used as a starting point of the design, although the initial design is not very accurate. Optimization should then be performed by the full-wave EM simulation to finalize the filter design.

The filter was fabricated on a $36 \times 30 \times 0.5$ mm double-sided $YBa_2Cu_3O_7$ film deposited on MgO substrate [53]. The standard photolithography and ion-beam milling technology was used for the filter patterning. The fabricated filter was assembled into

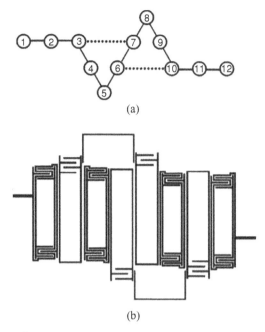

FIGURE 11.35 (a) Coupling structure of a 12-pole cross-coupled filter. (b) Physical realization (not to scale) of the 12-pole wideband HTS microstrip filter. (Taken from [53], ©2009 IEEE.)

testing housing, which was plated with gold. On the cover of the housing, tuning screws were provided for tuning the filter response. The filter was then measured in a cryogenic cooler using an Agilent 8510C network analyzer. The measured responses of the filter at 40K after tuning are plotted in Figure 11.36, compared with the simulated ones.

11.5 HIGH-POWER HTS FILTERS

The above-described HTS filters are primarily for low-power applications. However, HTS filters can also be designed for high-power applications [87–104]. In general, there are three main factors that may limit the power handling of an RF/microwave filter:

(i) RF breakdown;
(ii) heating in materials; and
(iii) nonlinearity in materials.

For a HTS filter, the power-handling limits are much the same. The RF breakdown or arcing occurs at very high electric fields. Using a thicker dielectric substrate with

476 SUPERCONDUCTING FILTERS

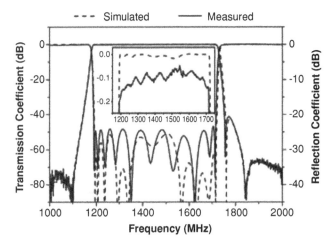

FIGURE 11.36 Measured and simulated performances of the 12-pole wideband HTS microstrip filter. (Taken from [53], ©2009 IEEE.)

a lower dielectric constant and avoiding very small coupling gaps can reduce the concentration of the electric field. Heating is associated with dissipation in materials, including dielectrics and conductors. This may play a minor role in limiting the power-handling capability of a high-Q HTS filter using high-quality HTS film and a low-loss tangent substrate. Nonlinearity in materials particular associated with the nonlinear surface resistance of superconductor appears to be the major concern for designing a high-power HTS filter.

Increasing input power of a HTS filter will raise the maximum current density at the surface of superconductor. When the maximum current density exceeds the critical current density of the HTS material, the surface resistance rises sharply, causing the transition from its superconducting state into nonsuperconducting state, and eventually the collapse of the HTS filter performance. However, before the maximum current density exceeds the critical current density, there is another effect because of the nonlinear surface resistance, which may limit the power handling of a HTS filter. This is the two-tone, third-order intermodulation (IMD).

For nonlinear impedance $Z = Z(I)$, the voltage will be a nonlinear function of current, $V(t) = I(t)Z(I) \approx a_1 I(t) + a_2 I^2(t) + a_3 I^3(t) + \cdots$. If we apply a two-tone fundamental signal $I(t) = I_1 \sin \omega_1 t + I_2 \sin \omega_2 t$, it will produce intermodulation products at frequencies $m\omega_1 \pm n\omega_2$, where m and n are integers. Among these products, the third-order IMD signals at $2\omega_1 - \omega_2$ and $2\omega_2 - \omega_1$ are of primary concern, because they may fall in the filter passband, causing interference with desired signals. To measure the two-tone third-order IMD in a filter, two-tone signals are usually adjusted to have the same power levels at the filter input and to have the frequencies such that the third-order IMD signals are in the passband of the filter. The power of the fundamental and the power of the third-order IMD at the output of the filter are measured and plotted as a function of the applied input power. In a log–log plot, the

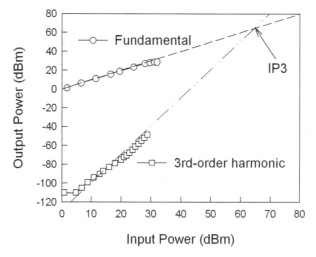

FIGURE 11.37 Determining the third-order intercept point (IP3) by linear extrapolation of measured data (plotted as symbols).

slope of the third-order IMD is about 3, compared to 1 for the fundamental. Consequently, the situation arises wherein the output power in the fundamental becomes equal to the output power in the third-order IMD. This intercept point, measured in dBm, is known as the third-order intercept point (TOI or IP3) and is used as a figure of merit for the nonlinearity present in the filter. A high intercept indicates a high power-handling capability of a filter. In practice, this intercept may not be measured directly, but can be measured by plotting the levels of the fundamental and two-tone intermodulation at lower power levels, and then using linear extrapolation to determine the intercept, as demonstrated in Figure 11.37.

From a microwave design viewpoint, an effective approach for reducing the nonlinear effects of a HTS filter is to reduce the current crowding in superconductor. A simple way to achieve a more uniform current distribution is to increase microstrip line widths. An example of this is a five-pole HTS pseudocombline filter with 1.2% fractional bandwidth centered on 2 GHz that uses half-wavelength resonators having a line characteristic impedance 10 Ω on 508-μm thick $LaAlO_3$ substrate and can handle 36 W of power at 45K [87].

However, the increase of resonator line width has its limitations. First, the resonant mode along the resonator width direction, i.e., the transverse mode, may become too close to the desired resonator mode, which, in turn, degrades the filter spurious performance. Second, its two outer edges always have high current concentration. To overcome these problems, the so-called split-resonators, shown in Figure 11.38, have been introduced [95–97]. All the elements in a split-resonator are resonating at the same frequency. The basic idea is to redistribute the current in a resonator onto the individual resonating elements such that the current density at edges of each element is reduced. With this approach, the peak current density in a resonator can be reduced significantly so that the overall power-handling capability of a filter using this type of

(a) (b)

FIGURE 11.38 (a) Four-element split-line resonator. (b) Two-element split-ring resonator.

resonator can be improved. There is another important advantage of applying the split resonator. Since the width of each resonating element is much smaller than its length, its transverse mode will resonate at much higher frequency than the fundamental frequency. This implies that a much wider resonator with a large number of elements can now be used without the potential interference of the unwanted transverse mode. Similarly, it is proposed [98] to form resonators having their incident power and resulting currents divided within interior arrays of "basic resonators." (A properly designed array of basic resonators acts as a single resonator.) Although other forms of basic resonators may also be attractive, the basic resonator used [98] is a "zig–zag resonator" to demonstrate that the use of parallel and cascade connections of basic resonators are analyzed and are found to both yield an increase in power handling proportional to the number of basic resonators used.

Using patch or 2D resonators for high-power filter designs is another common approach [89–94]. A two-pole, high-power HTS filter has been developed based on a circular-disk resonator [90]. This is a dual-mode filter that uses two orthogonal degenerate TM^z_{110} modes and, in order to couple the two modes, some sort of perturbation on the perfect circular disk is required (refer to Chapter 10). In this case, an elliptical deformation is used because of that smooth shape being free from the field concentration. The desired coupling can be obtained by suitable adjustment of the ellipticity of the disk shape so that the symmetric axes are oriented at 45° to the polarization directions of the modes. The filter is designed for a center frequency of 1.9 GHz with a passband about 15 MHz and, in this case, the diameter along to the major axis is 19.6 mm with an ellipticity being merely as little as 1%. The elliptic disk is capacitively coupled to the input/output feed lines of the filter. To avoid very narrow coupling gaps, which may cause electric discharge for high-power operation, the width of each feed line is expanded toward its open end. The filter was fabricated using double-sided YBCO thin films on 1-mm thick $LaAlO_3$ substrate. The two-tone, third-order intermodulation measurement was performed with the two fundamental input signals of the frequencies of 1.905 and 1.910 GHz and the input power up to 37.3 dBm. The generated third-order IMD signals were at 1.900 and 1.915 GHz. The IP3 value, obtained by linear extrapolation, is 73 dBm (20 kW). High-power tests indicate that this high-power HTS filter could have a power-handling capability beyond 100 W [90].

As mentioned in Chapter 4, the TM^z_{010} mode of a circular-disk resonator is of particular interest for design of high-power filters. This is because the disk resonator operating at this mode does not have current at the edge and has a fairly uniform current distribution along the azimuthal direction [91]. A ring resonator or a polygon-shape with sufficient number of sides can also operate at this mode. A four-pole, high-power HTS filter comprised of the edge current-free disk and ring resonators has been developed for extended C-band output multiplexers of communication satellites [92]. The filter has a 40-MHz bandwidth at about 4.06 GHz with a power handling of 60 W and a third-order intercept point of higher than 83 dBm. Another two-pole filter of this type using two octagon-shape resonators is reported [93]. The filter is composed of double-sided $Tl_2Ba_2CaCu_2O_8$ thin films on 0.508-mm thick $LaAlO_3$ with a size of 35 ×17 mm and is designed to have 1% fractional bandwidth at 6.04 GHz. The measurements show that the performance of this filter does not degrade up to 115 W of CW transmitting power at 77K.

11.6 CRYOGENIC PACKAGE

All known superconductors by far must be operated at temperatures well below room temperature and, hence, HTS filters needs to be encapsulated and operated with a cryogenic refrigerator (cooler). This is necessary for the HTS components, but the low-noise amplifiers (LANs) in a system can be also cooled, which gives the system an extra reduction in overall noise. It has been reported [14] that the noise figure of a LNA is reduced from < 0.8 dB at room temperature to < 0.2 dB at 77K. Cryogenic engineers have dramatically improved the reliability of cryogenic refrigerators, such as Stirling coolers, Gifford McMahon coolers, and pulse-tube coolers. With a large enough demand and careful design and manufacturing techniques, highly reliable cryogenic refrigerators can be made available on the commercial market place. For example, Superconductor Technologies Inc., (STI) has developed a cryogenic refrigerator suitable for a HTS wireless base station filter system applications that has

FIGURE 11.39 Inside views of the cryogenic rf housings (microenclosures) inside 2004 Dewar (left) and 2008 Dewar (right). The 2008 Dewar has been optimized to eliminate parts as compared to the 2004 design. (Taken from [114], ©2009 IEEE.)

480 SUPERCONDUCTING FILTERS

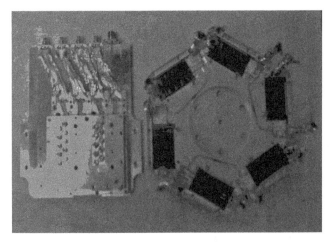

FIGURE 11.40 Expanded views of one of the three microenclosures in a 2004 Dewar (left) and a 2008 microenclosure. (Taken from [114], ©2009 IEEE.)

very impressive reliability statistics. As of late 2007, STI has deployed more than 6000 systems that have accumulated more than 200,000,000 h of operation. Each cryogenic refrigerator, on the average, has operated for over 4 years [7,114].

Design for low cost and high reliability implies eliminating parts and simplifying processes where possible, since much of the cost of a HTS filter system can lie in the cryopackaging [114]. As deployed in 1997, the STI Cryogenic Receiver Front End (CRFE), which typically contains six filter/LNA channels, was a very complicated assembly requiring many custom machined parts with features from multiple sides. While this is acceptable for satellites or other applications where these costs are dwarfed by other costs, it is not for widespread commercial applications. Thus, in 2004, STI undertook a major effort to redesign the CRFE to decrease costs. Much of this simplification can be seen in Figures. 11.39 and 11.40, showing pictures of the insides of a 2004 vintage Dewar and one from 2008. As can be seen, the assembly has been greatly simplified, with many parts being consolidated and/or altogether eliminated. In particular, many threaded parts, mainly screws and connectors, have been eliminated or replaced with spring clips. From Figure 11.40, one may see some tuning screws for two HTS filters on the top a filter housing in the 2004 microenclosure (left), while six HTS filters (in black) can be identified in the 2008 microenclosure (right).

REFERENCES

[1] I. G. Bednorz, and K. A. Müller, Possible high-Tc superconductivity in Be-La-Cu-O system, *Z. Phys.* **B64**(2), 1986, 189–191.

[2] Z.-Y. Shen, *High-Temperature Superconducting Microwave Circuits*, Artech House, Norwood, 1994.

[3] M. J. Lancaster, *Passive Microwave Device Applications of High-Temperature Superconductors*, Cambridge University Press, Cambridge, 1997.

[4] A. I. Braginski, Superconducting electronics coming to market, *IEEE Trans. Appl. Superconductivity* **9**, 1999, 2825–2836.

[5] Microwave and communication applications at low temperature, *IEEE Trans.* **MTT-48**, 2000, Part II [SI].

[6] R. R. Mansour, Microwave superconductivity, *IEEE Trans.* **MTT-50**, 2002, 750–759.

[7] M. Nisenoff, and J. M. Pond, Superconductors and microwaves, *IEEE Microwave Magazine* **10**, 2009, 84–95.

[8] D. Jedamzik, R. Menolascino, M. Pizarroso, and B. Salas, Evaluation of HTS subsystems for cellular basestations, *IEEE Trans. Appl. Superconductivity* **9**(2), 1999, 4022–4025.

[9] STI Inc. A receiver front end for wireless base stations. *Microwave J.* **39**(4), 1996, 116–120.

[10] S. H. Talisa, M. A. Robertson, B. J. Meler, and J. E. Sluz, Dynamic range considerations for high-temperature superconducting filter applications to receive front ends, *IEEE MTT-S Dig.* 1997, 997–1000.

[11] R. B. Hammond, HTS wireless filters: Past, present and future performance, *Microwave J.* **41**(10), 1998, 94–107.

[12] G. Koepf, Superconductors improve coverage in wireless networks, *Microwave RF* **37**(4), 1998, 63–72.

[13] Y. Vourc'h, G. Auger, H. J. Chaplopka, and D. Jedamzik, Architecture of future basestations using high temperature superconductors, In *ACTS Mobile Summit*, Aalbourg Demank, September 1997; p 802–807.

[14] R. B. Greed, D. C. Voyce, J.-S. Hong, M. J. Lancaster, M. Reppel, H. J. Chaloupka, J. C. Mage, R. Mistry, H. U. Häfner, G. Auger, and W. Rebernak, An HTS transceiver for third generation mobile communications—European UMTS, In *MTT-S European Wireless* Amsterdam 1998, 98–103.

[15] B. A. Willemsen, HTS filter subsystems for wireless telecommunications, *IEEE Trans. Appl. Superconductivity* **11**(1), Part 1, 2001, 60–67.

[16] M. V. Jacob, J. Mazierska, and S. Takeuchi, Miniaturized superconducting filter for mobile communications, In *TENCON 2003. Conference on Convergent Technologies for Asia-Pacific Region*, Vol. 2, 15–17 Oct. 2003. 631–634.

[17] G. Tsuzuki, Y. Shen, and S. Berkowitz, Ultra selective HTS bandpass filter for 3G wireless application, *IEEE Trans. Appl. Superconductivity* **13**(2), Part 1, 2003, 261–264.

[18] R. W. Simon, R. B. Hammond, S. J. Berkowitz, and B. A. Willemsen, Superconducting microwave filter systems for cellular telephone base stations, *Proc. IEEE* **92**(10), 2004, 1585–1596.

[19] E. P. McErlean, and J.-S. Hong, A high temperature superconducting filter for future mobile telecommunication systems, In *34th European Microwave Conference*, Vol. 2, 13 Oct. 2004, 1117–1119.

[20] E. Picard, V. Madrangeas, S. Bila, J. C. Mage, and B. Marcilhac, Very narrow band HTS filters without tuning for UMTS communications, In *34th European Microwave Conference*, Vol. 2, 13 Oct. 2004, 1113–1116.

[21] J.-S. Hong, E. P. McErlean, B. Karyamapudi, M. Cox, and M. Shiel, Superconducting filters for wireless communication applications, In *Proceedings of 4th International Conference on Microwave and Millimeter Wave Technology (ICMMT2004)*, 18–21 Aug. 2004, 264–267.

[22] J.-S. Hong, E. P. McErlean, and B. M. Karyamapudi, A high-temperature superconducting filter for future mobile telecommunication systems, *IEEE Trans. Microwave Theory Techn* **53**(6), Part 2, 2005, 1976–1981.

[23] A. P. Knack, S. Kolesov, and J. Mazierska, CDMA coverage capacity uplink model: an assessment of implementing superconducting technology in CDMA cellular networks, In *Proceedings of 2005 Asia-Pacific Microwave Conference (APMC 2005)*, Vol. 3, 4–7 Dec. 2005. 4 p.

[24] F. Ricci, V. Boffa, Guojun Dai, G. Grassano, R. Mele, R. Tebano, D. Arena, G. Bertin, N. P. Magnani, G. Zarba, A. Andreone, A. Cassinese, and R. Vaglio, Design and development of a prototype of hybrid superconducting receiver front-end for UMTS wireless network: first results and application perspectives, *IEEE Trans. Appl. Superconductivity* **15**(2), Part 1, 2005, 988–991.

[25] M. Shigaki, Y. Hagiwara, K. Yamanaka, and K. Kurihara, Frequency and power saving of the future generation wireless communications by superconducting filter, In *Proceedings of 2006 Asia-Pacific Microwave Conference (APMC 2006)*, 12–15 Dec. 2006, 835–838.

[26] M. I. Salkola, and D. J. Scalapino, Benefits of superconducting technology to wireless CDMA networks, *IEEE Trans. Vehicular Technol.*, **55**(3), 2006, 943–955.

[27] M. I. Salkola, Implications of third generation wireless standards for the design of superconducting transmit filters, *IEEE Commun. Lett.* **10**(5), 2006, 329–331.

[28] S. Narahashi, K. Satoh, K. Kawai, and D. Koizumi, Cryogenic receiver front-end for mobile base stations, In *2008 China-Japan Joint Microwave Conference*, 10–12 Sept. 2008, 619–622.

[29] A. Centeno, and J. Breeze, The potential use of HTS filters in satellite payloads, *IEE Colloq. Superconducting Microwave Circuits* **19**, 1996, 1–4.

[30] Z. M. Hejazi, P. S. Excell, and Z. Jiang, Compact dual-mode filters for HTS satellite communication systems, *IEEE Microwave Guided Wave Lett.* **8**(8), 1998, 275–277.

[31] T. Kasser, M. Klauda, C. Neumann, E. Guha, S. Kolesov, A. Baumfalk, and H. Chaloupka, A satellite repeater comprising superconducting filters, *IEEE MTT-S International Microwave Symposium Digest*, Vol. 1, 7–12 June 1998, :375–378.

[32] A. Baumfalk, H. Chaloupka, S. Kolesov, M. Klauda, and C. Newmann, HTS power filters for output multiplexers in satellite communications, *IEEE Trans. Appl. Superconductivity* **9**(2), Part 3, 1999, 2857–2861.

[33] E. R. Soares, J. D. Fuller, P. J. Morozick, and R. L. Alvarez, Applications of high-temperature-superconducting filters and cryoelectronics for satellite communication, *IEEE Trans. Microwave Theory Techn.* **48**(7), Part 2, 2000, 1190–1198.

[34] C. Lascaux, F. Rouchaud, V. Madrangeas, M. Aubourg, P. Guillon, B. Theron, and M. Maignan, Planar Ka-band high temperature superconducting filters for space applications, In *2001 IEEE MTT-S International Microwave Symposium Digest*, Vol. 1, 20–25 May 2001, 487–490.

[35] R. R. Mansour, T. Romano, S. Ye, S. Peik, T. Nast, D. Enlow, C. Wilker, and J. Warner, Development of space qualifiable HTS communication subsystems, *IEEE Trans. Appl. Superconductivity* **11**(1), Part 1, 2001, 806–811.

[36] J.-F. Seaux, S. Courreges, S. Bila, V. Madrangeas, M. Maignan, and C. Zanchi, An eight pole self-equalised quasi-elliptic superconductor planar filter for satellite applications, In *2005 European Microwave Conference*, Vol. 1, 4–6 Oct. 2005, 4 p.

[37] M. Nisenoff, and W. J. Meyers, On-orbit status of the High Temperature Superconductivity Space Experiment, *IEEE Trans. Appl. Superconductivity* **11**(1), Part 1, 2001, 799–805.

[38] L. Gao, L. Sun, F. Li, Q. Zhang, Y. Wang, T. Yu, J. Guo, Y. Bian, C. Li, X. Zhang, H. Li, J. Meng, and Y. He, 8-GHz narrowband high-temperature superconducting filter with high selectivity and flat group delay, *IEEE Trans. Microwave Theory Techn.* **57**, 2009, 1767–1773.

[39] J. F. Seaux, C. Lascaux, V. Madrangeas, S. Bila, and M. Maignan, Interest of the superconductivity at 30 GHz: application to the HTS preselect receive filters for satellite communications, In *2004 IEEE MTT-S International Microwave Symposium Digest*, Vol. 2, 6–11 June 2004, 1121–1124.

[40] I. V. Korotash, and E. M. Rudenko, High-temperature superconductor-films microstrip band-pass filters for extreme working conditions, In *Microwaves, Radar and Remote Sensing Symposium, MRRS 2008, 22–24 Sept.* 2008, 122–125.

[41] Y. Li, M. J. Lancaster, F. Huang, and N. Roddis, Superconducting microstrip wide band filter for radio astronomy, In *2003 IEEE MTT-S International Microwave Symposium Digest*, Vol. 1, 8–13 June 2003, 551–554.

[42] G. Zhang, M. J. Lancaster, F. Huang, M. Zhu, and B. Cao, Accurate design of high Tc superconducting microstrip filter at UHF band for radio astronomy front end, In *2004 IEEE MTT-S International Microwave Symposium Digest*, Vol.2, 6–11 June 2004, 1117–1120.

[43] D.-C. Niu, T.-W. Huang, H.-J. Lee, and C.-Y. Chang, An X-band front-end module using HTS technique for a commercial dual mode radar, *IEEE Trans. Appl. Superconductivity* **15**(2), 2005, 1008–1011.

[44] J. Zhou, M. J. Lancaster, F. Huang, N. Roddis, and D. Glynn, HTS narrow band filters at UHF band for radio astronomy applications, *IEEE Trans. Appl. Superconductivity* **15**(2), Part 1, 2005, 1004–1007.

[45] G. Zhang, F. Huang, and M. J. Lancaster, Superconducting spiral filters with quasi-elliptic characteristic for radio astronomy, *IEEE Trans. Microwave Theory Techn.* **53**(3), Part 1, 2005, 947–951.

[46] G. Zhang, M. J. Lancaster, F. Huang, and N. Roddis, A superconducting microstrip bandstop filter for an L-band radio telescope receiver. In *2005 European Microwave Conference*, Vol. 1, Oct. 4–6, 2005. p 4 p.

[47] G. Zhang, M. J. Lancaster, and F. Huang, A high-temperature superconducting bandpass filter with microstrip quarter-wavelength spiral resonators. *IEEE Trans Microwave Theory Techn* **54**, 2006, 559–563.

[48] G. Zhang, J. Zhou, N. Roddis, and M. J. Lancaster, Superconducting filters for radio astronomy, In *IEEE Mediterranean Electrotechnical Conference, MELECON 2006;* 16–19 May 2006, 553–556.

[49] G. Zhang, M. J. Lancaster, F. Huang, and N. Roddis, An HTS wideband microstrip bandpass filter for L band receivers in radio astronomy observatory, In *2007 European Microwave Conference*, 9–12 Oct. 2007, 450–453.

[50] L. Sun, Q. Zhang, F. Li, H. Li, S. Li, A. He, X. Zhang, C. Li, Q. Luo, C. Gu, and Y. He, High temperature superconducting microwave filters for applications in mobile

communication, satellite receiver and wind-profile radar, In *Sixth International Kharkov Symposium on Physics and Engineering of Microwaves, Millimeter and Submillimeter Waves (MSMW '07)*, Vol. 1. 25–30, June 2007, 67–72.

[51] Q. Zhang, C. Li, Y. Wang, L. Sun, F. Li, J. Huang, Q. Meng, X. Zhang, A. He, H. Li, J. Zhang, X. Jia, and Y. He A HTS bandpass filter for a meteorological radar system and its field tests, *IEEE Trans.Appl. Superconductivity* **17**(2):Part 1, 922–925.

[52] H. Kayano, T. Kawaguchi, N. Shiokawa, K. Nakayama, T. Watanabe, and T. Hashimoto, Narrow-band filter for transmitter of radar application, In *38th European Microwave ConferenceEuMC 2008*. 27–31 Oct. 2008, 853–856.

[53] T. Yu, C. Li, F. Li, Q. Zhang, L. Sun, L. Gao, Y. Wang, X. Zhang, H. Li, C. Jin, J. Li, H. Liu, C. Gao, J. Meng, and Y. He A wideband superconducting filter using strong coupling resonators for radio astronomy observatory. *IEEE Trans. Microwave Theory Techn.* **57**, 2009, 1783–1789.

[54] D. Zhang, G.-C. Liang, C. F. Shih, Z. H. Lu, and M. E. Johansson, A 19-pole cellular bandpass filter using 75mm diameter high-temperature superconducting film. *IEEE Microwave Guided-Wave Lett* **5**(11), 405–407.

[55] D. Zhang, G.-C. Liang, C. F. Shih, M. E. Johansson, and R. S. Withers, Narrow-band lumped-element microstrip filters using capacitively-loaded inductors, *IEEE Trans* **MTT-43**, 1995, 3030–3036.

[56] G. L. Matthaei, N. O. Fenzi, R. J. Forse, and S. M. Rohlfing, Hairpin-comb filters for HTS and other narrow-band applications, *IEEE Trans* **MTT-45**, 1997, 1226–1231.

[57] J.-S. Hong, M. J. Lancaster, D. Jedamzik, and R. B. Greed, 8-Pole superconducting quasi-elliptic function filter for mobile communicationsapplication, *IEEE MTT-S Dig.* 1998, 367–370.

[58] G. Tsuzuki, M. Suzuki, N. Sakakibara, and Y. Ueno, Novel superconducting ring filter. *IEEE MTT-S Dig.* 1998, 379–382.

[59] M. Reppel, H. Chaloupka, J. Holland, J.-S. Hong, D. Jedamzik, M. J. Lancaster, J.-C. Mage, and B. Marcilhac, Superconducting preselect filters for base transceiver stations, In *ACTS Mobile Communications Summit 98, Rhodes*, June 1998.

[60] E. R. Soares, K. F. Raihn, A. A. Davis, R. L. Alvarez, P. J. Marozick, and G. L. Hey-Shipton, HTS AMPS-A and AMPS-B filters for cellular receive base stations, *IEEE Trans. Appl. Superconductivity* **9**(2), 1999, 4018–4021.

[61] J.-S. Hong, M. J. Lancaster, D. Jedamzik, and R. B. Greed, On the development of superconducting microstrip filters for mobile communications application, *IEEE Trans.* **MTT-47**(9), 1999, 1656–1663.

[62] J.-S. Hong, M. J. Lancaster, D. Jedamzik, and R. B. Greed, Progress in superconducting preselect filters for mobile communications base stations. In *Proceedings of European Conference on Applied Superconductivity (EUCAS)*, Sept. 1999, 327–330.

[63] G. L. Hey-Shipton, Efficient computer design of compact planar band-pass filters using electrically short multiple coupled lines, *IEEE MTT-S Dig.* 1999, 1547–1550.

[64] J.-F. Liang, C.-F. Shih, Q. Huang, D. Zhang, and G.-C. Liang, HTS microstrip filters with multiple symmetric and asymmetric prescribed transmission zeros, *IEEE MTT-S Dig.* 1999, 1551–1554.

[65] E. R. Soares, Design and construction of high performance HTS pseudo-elliptic band-stop filters, *IEEE MTT-S Dig.* 1999, 1555–1558.

[66] J.-S. Hong, M. J. Lancaster, and J.-C. Mage Cross-coupled HTS microstrip open-loop resonator filter on LAO substrate, *IEEE MTT-S Dig.* 1999, 1559–1562.
[67] M. Reppel and H. Chaloupka, Novel approach for narrowband superconducting filters. *IEEE MTT-S Dig.* 1999, 1563–1566.
[68] J.-S. Hong, M. J. Lancaster, R. B. Greed, D. Voyce, D. Jedamzik, J. A. Holland, H. Chaloupka, and J.-C. Mage Thin film HTS passive microwave components for advanced communication systems. *IEEE Trans Appl Superconductivity* **9**, 1999, 3893–3896.
[69] H. T. Kim, B.-C. Min, Y. H. Choi, S.-H. Moon, S.-M. Lee, B. Oh, J.-T. Lee, I. Park, and C.-C. Shin, A compact narrowband HTS microstrip filter for PCS applications, *IEEE Trans. Appl. Superconductivity* **9**, 1999, 3909–3912.
[70] K. F. Raihn, R. Alvarez, J. Costa, and G. L. Hey-Shipton Highly selective HTS band pass filter with multiple resonator cross-couplings. *IEEE MTT-S Dig.* 2000, 661–664.
[71] G. Tsuzuki, M. Suzuki, and N. Sakakibara, Superconducting filter for IMT-2000 band. *IEEE MTT-S Dig.* 2000, 669–672.
[72] H. Kayano, H. Fuke, F. Aiga, Y. Terashima, H. Yoshino, R. Kato, and Y. Suzuki, Superconducting microstrip line band-pass filter for mobile applications, *IEEE MTT-S Dig.* 2000, 673–676.
[73] J.-S. Hong, M. J. Lancaster, R. B. Greed, D. Jedamzik, J.-C. Mage, and H. J. Chaloupka, A high-temperature superconducting duplexer for cellular base station applications. *IEEE Trans.* **MTT-48**, 2000, 1336–1343.
[74] J.-S. Hong, M. J. Lancaster, D. Jedamzik, R. B. Greed, and J.-C. Mage, On the performance of HTS microstrip quasi-elliptic function filters for mobile communications applications, *IEEE Trans.* **MTT-48**, 2000, 1240–1246.
[75] J.-S. Hong, E. P. McErlean, and B. Karyamapudi, Narrowband high temperature superconducting filter for mobile communication systems. *IEE Proc. Microwaves, Antennas Propagation* **151**(6), 2004, 491–496.
[76] J.-S. Hong and E. P. McErlean, Narrow-band HTS filter on sapphire substrate. *IEEE MTT-S Intern. Microwave Symp. Dig.* **2**, 2004, 1105–1108.
[77] J.-S. Hong, E. P. McErlean, and B. Karyamapudi, High-order superconducting filter with group delay equalization, *IEEE MTT-S Intern. Microwave Symp. Dig.* 2005, 4 p.
[78] J.-S. Hong, E. P. McErlean, and B. Karyamapudi, Eighteen-pole superconducting CQ filter for future wireless applications. *IEE Proc. Microwaves Antennas Propagation* **153**(2), 2006, 205–211.
[79] C. Li, Q. Zhang, L. Sun, F. Li, T. Yu, H. Li, X. Zhang, A. He, and Y. He, A high selective HTS filter with group delay self-equalization for 3G mobile telecommunication systems. In *Sixth International Kharkov Symposium on Physics and Engineering of Microwaves, Millimeter and Submillimeter Waves (MSMW '07)*, Vol. 1, 25–30 June 2007, 395–397.
[80] S. Li, J. Huang, Q. Meng, L. Sun, Q. Zhang, F. Li, A. He, X. Zhang, C. Li, H. Li, and Y. He, A 12-pole narrowband highly selective high-temperature superconducting filter for the application in the third-generation wireless communications, *IEEE Trans.Microwave Theory Techn.* **55**(4), 2007, 754–759.
[81] W. Hattori, T. Yoshitake, and K. Takahashi, An HTS 21-pole microstrip filter for IMT-2000 base stations with steep attenuation, *IEEE Trans. Appl Superconductivity* **11**(3), 2001, 4091–4094.

[82] G. Tsuzuki, Shen Ye, and S. Berkowitz, Ultra-selective 22-pole 10-transmission zero superconducting bandpass filter surpasses 50-pole Chebyshev filter, *IEEE Trans. Microwave Theory Techn.* **50**(12), 2002, 2924–2929.

[83] S. Pal, C. J. Stevens, and D. J. Edwards, Compact parallel coupled HTS microstrip bandpass filters for wireless communications, *IEEE Trans. Microwave Theory Techn.* **54**(2), Part 1, 2006, 768–775.

[84] S. Jin, B. Wei, B. Cao, X. Zhang, X. Guo, H. Peng, Y. Piao, and B. Gao, Design and performance of an ultra-narrowband superconducting filter at UHF band, *IEEE Microwave Wireless Components Lett.* **18**(6), 2008, 395–397.

[85] A. M. Abu-Hudrouss, A. B. Jayyousi, and M. J. Lancaster, Triple-band HTS filter using dual spiral resonators with capacitive-loading. *IEEE Trans. Appl. Superconductivity* **18**(3), 2008, 1728–1732.

[86] H.-F. Huang, and J.-F. Mao, Design of minimized electromagnetic bandgap structure high temperature superconducting filter, In *IEEE MTT-S International Microwave Workshop Series on Art of Miniaturizing RF and Microwave Passive Components (IMWS)* 2008 14–15 Dec. 2008, 23–25.

[87] G.-C. Liang, D. Zhang, C.-F. Shih, M. E. Johansson, R. S. Withers, D. E. Oates, A. C. Anderson, P. Mankiewich, E. de Obaldia, and R. E. Miller, High-power HTS microstrip filters for wireless communications. *IEEE Trans.* **MTT-43**, 1995, 3020–3029.

[88] S. Ye, and R. R. Mansour, A novel split-resonator high power HTS planar filter, *IEEE MTT-S Dig.* 1997, 299–302.

[89] Z.-Y. Shen, C. Wilker, P. Pang, and C. Carter, High-power HTS planar filters with novel back-side coupling, *IEEE Trans* **MTT-44**, 1996, 984–986.

[90] K. Setsune, and A. Enokihara, Elliptic-disc filters of high-T_c superconducting films for power-handling capability over 100 W, *IEEE Trans.* **MTT-48**, 2000, 1256–1264.

[91] H. Chaloupka, M. Jeck, B. Gurzinski, and S. Kolesov, Superconducting planar disk resonators and filters with high power handling capability *Electronics Lett.* **32**, 1996, 1735–1737.

[92] A. Baumfalk, H. Chaloupka, and S. Kolesov, HTS power filters for output multiplexers in satellite communications. *IEEE Trans. Appl. Superconductivity*, **9**, 1999, 2857–2861.

[93] Z.-Y. Shen, C. Wilker, P. Pang, D. W. Face, C. F. Carter, and C. M. Harrington, Power handling capability improvement on high-temperature superconducting microwave circuits. *IEEE Trans. Appl. Superconductivity* **7**, 1997, 2446–2453.

[94] A. P. Jenkins, D. Dew-Hughes, D. J. Edwards, D. Hyland, and C. R.M. Grovenor, Application of TBCCO based HTS devices to digital cellular communications, *IEEE Trans. Appl. Superconductivity* **9**, 1999, 2849–2852.

[95] R. R. Mansour, Y. Shen, V. Dokas, B. Jolley, W.-C. Tang, and C. M. Kudsia, Feasibility and commercial viability issues for high-power output multiplexers for space applications. *IEEE Trans. Microwave Theory Techn.* **48**(7), Part 2, 2000, 1199–1208.

[96] S. Futatsumori, T. Hikage, and T. Nojima, Microwave superconducting reaction-type transmitting filter using split open-ring resonator, *Electronics Lett.* **42**(7), 2006, 428–430.

[97] S. Futatsumori, T. Hikage, T. Nojima, A. Akasegawa, T. Nakanishi, and K. Yamanaka, HTS split open-ring resonators with improved power-handling capability for reaction-type transmitting filters, *Electronics Lett.* **43**(17), 2007, 956–957.

[98] G. L. Matthaei, B. A. Willemsen, E. M. Prophet, and G. Tsuzuki, Zig–zag-array superconducting resonators for relatively high-power applications, *IEEE Trans. Microwave Theory Techn.* **56**(4), 2008, 901–912.

[99] H. Sato, J. Kurian, and M. Naito, Third-order intermodulation measurements of microstrip bandpass filters based on high-temperature superconductors, *IEEE Trans. Microwave Theory Techn.* **52**(12), 2004, 2658–2663.

[100] K. Satoh, D. Koizumi, and S. Narahashi, New method to improve power handling capability for coplanar waveguide high-temperature superconducting filter, In *2005 European Microwave Conference*, Vol. 1, 4–6 Oct. 2005, 4 p.

[101] X. Guo, B. Wei, X. Zhang, B. Cao, S. Jin, H. Peng, L. Gao, and B. Gao, Design of a high-power superconducting filter using resonators with different linewidths, *IEEE Trans. Microwave Theory Techn.* **55**(12), Part 1, 2007, 2555–2561.

[102] H. Zhao, J. R. Dizon, R. Lu, W. Qiu, and J. Z. Wu, Fabrication of three-pole $HgBa_2CaCu_2O_{6+\delta}$ hairpin filter and characterization of its third order intermodulation. *IEEE Trans. Appl. Superconductivity* **17**(2), Part 1, 2007, 914–917.

[103] C. Collado, J. Mateu, O. Menendez, and J. M. O'Callaghan, Nonlinear distortion in a 8-pole quasi-elliptic bandpass HTS filter for CDMA system, *IEEE Trans. Appl. Superconductivity* **15**(2), Part 1, 2005, 992–995.

[104] J. Mateu, J. C. Booth, C. Collado, and J. M. O'Callaghan, Intermodulation distortion in coupled-resonator filters with nonuniformly distributed nonlinear properties—use in HTS IMD compensation. *IEEE Trans. Microwave Theory Techn.* **55**(4), 2007, 616–624.

[105] M. Zeisberger, M. Manzel, H. Bruchlos, M. Diegel, F. Thrum, M. Klinger, and A. Abramowicz, $Tl_2Ba_2Ca_1Cu_2O_x$ thin films for microstrip filters, *IEEE Trans. Appl. Superconductivity* **9**, 1999, 3897–3900.

[106] Z.-Y. Shen, and C. Wilker, Raising the power handling capacity of HTS circuits. *Microwave RF* **33**, 1994, 129–138.

[107] C. Wilker, Z.-Y. Shen, P. Pang, W. L. Holstein, and D. W. Face, Nonlinear effects in high temperature superconductors: 3rd order intercept from hamonic generation, *IEEE Trans. Appl. Superconductivity* **5**, 1995, 1665–1670.

[108] H. Xin, D. E. Oates, A. C. Anderson, R. L. Slattery, G. Dresslhaus, and M. S. Dresselhaus Comparison of power dependence of microwave surface resistance of unpatterned and patterned YBCO thin film, *IEEE Trans.* **MTT-48**, 2000, 1221–1226.

[109] J. Krupka, R. G. Geyer, M. Kuhn, and J. H. Hinken, Dielectric properties of single crystals of Al_2O_3, $LaAlO_3$, $NdGaO_3$, $SrTiO_3$ and MgO at cryogenic temperatures, *IEEE Trans.* **MTT-42**, 1994, 1886–1890.

[110] T. Konaka, M. Sato, H. Asano, and S. Kubo, Relative permittivity and loss tangents of substrate materials for high T_c superconducting thin films *J. Superconductivity* **4**, 1991, 283–288.

[111] V. B. Braginsky, V. S. Ilchenko, and K. S. Bagdassarov, Experimental observation of fundamental microwave absorption in high quality dielectric crystals, *Phys. Lett.* **A12**, 1987, 300–305.

[112] AWR Microwave Office/Analog Office V7.5, Applied Wave Research, Inc. 2007.

[113] *EM User's Manual*, Version 10, Sonnet Software Inc., Liverpool, NY, 2006.

[114] B. A. Willemsen, Practical cryogenic receiver front ends for commercial wireless applications, *IEEE MTT-S Intern. Microwave Symp. Dig.* 2009, 1457–1460.

CHAPTER TWELVE

Ultra-Wideband (UWB) Filters

Ultra-wideband (UWB) technology is being reinvented with many promising modern applications [1,2]. Ultra-wideband or broad-band microwave filters are essential components for these applications, such as UWB wireless and radar systems. The rapid growth in this field has prompted the development of various types of UWB filters [3–76]. In this chapter, typical types of UWB filters are described. This includes UWB filters comprised of short-circuit stubs, those using coupled single- or multimode resonators, or quasilumped elements, those based on cascaded high- and lowpass filters, and those with single- or multiple-notched bands.

12.1 UWB FILTERS WITH SHORT-CIRCUITED STUBS

12.1.1 Design of Stub UWB Filters

The optimum distributed highpass filter, which consists of short-circuited stubs, as introduced in Chapter 6, can be designed as an UWB bandpass filter [9]. Consider a filter design to have a passband covering the entire FCC-defined UWB band from 3.1 to 10.6 GHz [1]. Referring to Figure 6.3b, we can let the low cutoff frequency $f_c = f_{c1} = 3.1$ GHz and the high cutoff frequency $(\pi/\theta_c - 1)/f_c = f_{c2} = 10.6$ GHz. Thus, the electrical length θ_c, referenced at the low cutoff frequency, can be found to be

$$\theta_c = \frac{\pi}{(f_{c2}/f_{c1}) + 1} = 0.711 \text{ rad or } 40.73° \qquad (12.1)$$

Table 12.1 tabulates some typical element values of the transmission line circuit in Figure 6.3a, where n is the number of short-circuited stubs. These elements are handy for practical design of UWB bandpass filters with a degree from 5 to 11 and with a fractional bandwidth between 100% and 111.11%.

Microstrip Filters for RF/Microwave Applications by Jia-Sheng Hong
Copyright © 2011 John Wiley & Sons, Inc.

UWB FILTERS WITH SHORT-CIRCUITED STUBS   489

TABLE 12.1   Element Values of UWB Transmission Filters with Short-Circuited Stubs (Passband Ripple = 0.1 dB)

n	θ_c	y_1 y_n	$y_{1,2}$ $y_{n-1,n}$	y_2 y_{n-1}	$y_{2,3}$ $y_{n-2,n-1}$	y_3 y_{n-2}	$y_{3,4}$
3	40°	0.54659	1.00474	0.71896			
	41°	0.57995	0.99364	0.77446			
	42°	0.61487	0.98207	0.83294			
	43°	0.65141	0.97007	0.89456			
	44°	0.68966	0.95762	0.95944			
	45°	0.72970	0.94473	1.02773			
4	40°	0.60040	0.98420	0.87761	0.95561		
	41°	0.63530	0.97260	0.94034	0.94251		
	42°	0.67174	0.96061	1.00606	0.92904		
	43°	0.70978	0.94822	1.07488	0.91520		
	44°	0.74951	0.93543	1.14693	0.90102		
	45°	0.79101	0.92224	1.22236	0.88648		
5	40°	0.62508	0.97521	0.94191	0.93754	1.02709	
	41°	0.66044	0.96353	1.00621	0.92425	1.09583	
	42°	0.69733	0.95146	1.07347	0.91063	1.16753	
	43°	0.73583	0.93901	1.14379	0.89668	1.24230	
	44°	0.77602	0.92617	1.21731	0.88243	1.32026	
	45°	0.81797	0.91295	1.29417	0.86786	1.40156	
6	40°	0.63807	0.97063	0.97222	0.92979	1.08475	0.92140
	41°	0.67363	0.95892	1.03705	0.91650	1.15464	0.90798
	42°	0.71072	0.94683	1.10482	0.90290	1.22746	0.89428
	43°	0.74942	0.93436	1.17565	0.88898	1.30332	0.88028
	44°	0.78981	0.92152	1.24968	0.87476	1.38234	0.86598
	45°	0.83197	0.90830	1.32706	0.86025	1.46471	0.85143

For the UWB filter with a passband from 3.1 to 10.6 GHz, the fractional bandwidth $FBW = 109.49\%$, centered at $f_0 = 6.85$ GHz, and the desired electrical length is given by Eq. (12.1). Design a filter with five stubs or $n = 5$. By interpolation from the element values listed in Table 12.1, similar to the design example described in Section 6.1.2, we can find the desired normalized admittance parameters:

$$y_1 = 0.65089, \quad y_2 = 0.98885, \quad y_3 = 1.07727,$$
$$y_{1,2} = 0.96668, \quad y_{2,3} = 0.92784$$

The filter is a symmetrical circuit with the terminal impedance $Z_0 = 50\,\Omega$. Using Eq. (6.9), the characteristic impedances for all line elements are found to be

$$Z_1 = 76.81790\,\Omega \quad Z_2 = 50.56379\,\Omega \quad Z_3 = 46.41362\,\Omega$$
$$Z_{1,2} = 51.72342\,\Omega \quad Z_{2,3} = 53.88860\,\Omega \tag{12.2}$$

Using the design parameters given by Eq. (12.2), we can create a circuit model with Microwave Office, a commercial design tool for microwave CAD (refer to Chapter 8), as shown in Figure 12.1a. In this case, the electrical length of the short-circuited stubs and the electrical length of the connecting lines are set to be 90° and 180°, respectively, at the center frequency f_0, that is, 6.85 GHz. This is equivalent to 40.73° and 81.46°, respectively, at the low cutoff frequency of 3.1 GHz. The circuit modeling results are plotted in Figure 12.1b for the magnitude response and 12.1c for the group-delay response. The filter shows a ripple passband from 3.1 to 10.6 GHz with a return loss below −16.4 dB, as designed. One can see that the frequency response is equivalent to that of a nine-pole filter, although only five stubs are used. This is because the four connecting lines also contribute to the selectivity of the filter and, hence, add additional four poles to the filter order. Compared with the conventional stub bandpass filter described in Section 5.2.7.1, for achieving the same filtering characteristics, such as that shown in Figure 12.1b, this type of UWB bandpass filter not only uses less short-circuited stubs, but also has much more reasonable impedance levels for all the short-circuited stubs [9]. This makes the realization easy with a low-cost fabrication technology.

This circuit can be transferred to a microstrip filter and the result is shown in Figure 12.2a. The effects of T-junctions must be taken into account, which mostly affect the connecting lines. The microstrip UWB filter is realized on a substrate with a relative dielectric constant of 3.05 and a thickness of 0.508 mm. The layout is created using Sonnet software [79] for EM simulation. The arrowed patches indicate the via-hole grounding for the short-circuited stubs.

In the simulation, the conductor (17.5-μm copper), dielectric, and radiation losses are included. Figure 12.2b is the simulated performance of the filter, which shows a good response and verifies the design.

12.1.2 Stub UWB Filters with Improved Upper Stopband

Although the stub UWB filter of Figure 12.2a has a high cutoff frequency at 10.6 GHz, it exhibits periodic harmonic passbands because of the behaviors of the distributed transmission-line elements and the closet harmonic passband starting at about 16 GHz. To extend the upper stopband of the UWB bandpass filter, a simple method is to use a bandstop or lowpass filter to suppress the spurious response. An example of this is given in Figure 12.3a, where only the partial stub UWB filter is displayed in order to highlight the bandstop filter cascaded at the output of the filter. The bandstop filter is designed based on an optimum bandstop filter with three open-circuited stubs (refer to Section 6.2.3). Since the center frequency of the bandstop filter is much higher than that of the UWB bandpass filter, its size is quite small as compared to the entire UWB filter. The triangle stub in the middle of the bandstop filter is used as a low-impedance open-circuited stub in order to have a defined insertion point to the main signal path over a wide frequency range. Alternatively, a radial stub, as discussed in Section 6.2.4, can be used. The EM-simulated results of the entire stub UWB filter is shown in Figure 12.3b. As can be seen, the filter has an upper stopband extended to 20 GHz.

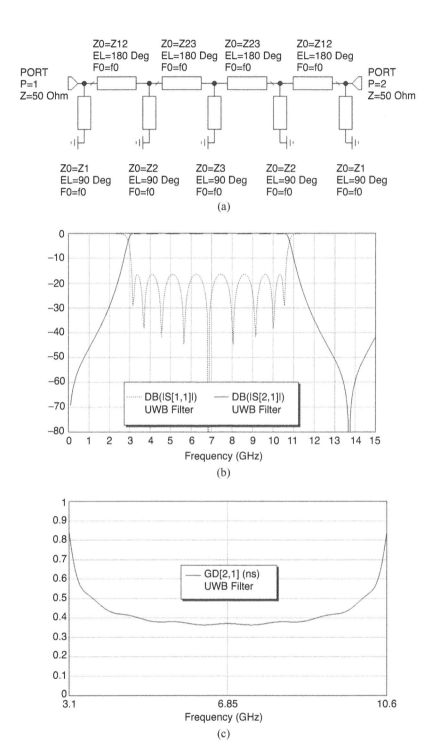

FIGURE 12.1 (a) Circuit model for a nine-pole UWB filter with five short-circuited stubs. (b) Magnitude response of the UWB filter. (c) Group-delay response over the passband of the UWB filter.

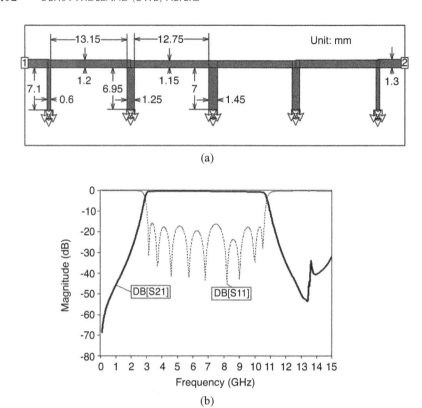

FIGURE 12.2 (a) Layout of a nine-pole microstrip UWB filter with five short-circuited stubs on a substrate with a relative dielectric constant of 3.05 and a thickness of 0.508 mm. (b) EM-simulated performance of the filter.

Another method to suppress the unwanted spurious response is to replace uniform connecting lines with stepped-impedance lines [36]. As a demonstration, Figure 12.4 shows two five-pole stub UWB filters, each of which consists of three short-circuited stubs. The filter in Figure 12.4a, or filter A, has two uniform connecting lines, which are replaced with two stepped-impedance lines for the filter (filter B), shown in Figure 12.4b. The stepped-impedance line can be seen as a stepped-impedance lowpass filter, introduced in Chapter 5. In principle, it should preserve the frequency characteristics, both magnitude and phase, of the uniform transmission line over the desired passband of the UWB bandpass filter. Furthermore, it should have a proper cutoff frequency to effectively suppress the unwanted spurious response. Both filters are designed on a substrate with a relative dielectric constant of 3.05 and a thickness of 0.508 mm. The final design of filter B, with the dimensions shown, has taken all discontinuities into account, which can normally be done in light of EM simulation. As a result, the length of the short-circuited stub in the middle has

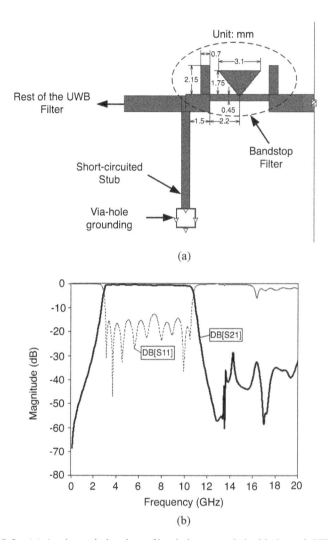

FIGURE 12.3 (a) A microstrip bandstop filter being cascaded with the stub UWB filter of Figure 12.2a to improve the upper stopband performance on a substrate with a relative dielectric constant of 3.05 and a thickness of 0.508 mm. (b) EM-simulated performance of the filter.

been adjusted. Nevertheless, it is evident, as compared to filter A, that the use of the stepped-impedance line reduces the filter size because of its slow-wave effect.

Figure 12.5 gives the comparison of performances of these two filters. We can see from Figure 12.5a that both filters have the same passband, but the spurious response of filter A is suppressed in filter B. In addition to an improved upper stopband, filter B also shows a better selectivity at the high side of the passband. This is because

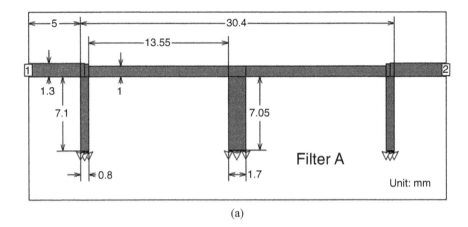

(a)

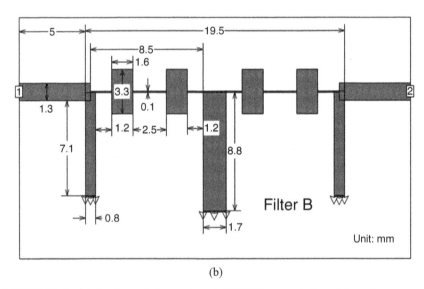

(b)

FIGURE 12.4 (a) A microstrip five-pole stub UWB filter consisting of three short-circuited stubs on a substrate with a relative dielectric constant of 3.05 and a thickness of 0.508 mm (filter A). (b) Its modification with the uniform connecting lines being replaced with the steeped-impedance lines (filter B).

the cutoff frequency of the stepped-impedance line or lowpass filter is near the edge of the passband of the UWB filter. The group-delay responses of the two filters are depicted in Figure 12.5b. As compared to the filter A, filter B has a large group delay, as well as the group- delay variation, over the passband. Particularly, its group delay is much larger at the edge of the high side of the passband because of the higher selectivity, which, however, causes a slightly increased in-band insertion loss toward this passband edge.

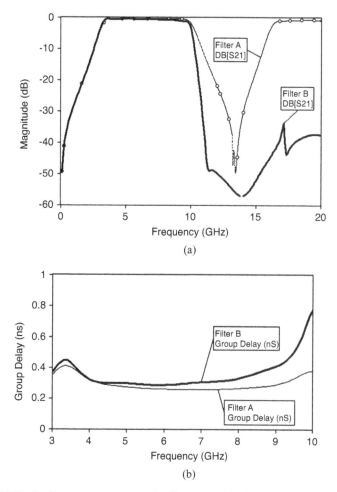

FIGURE 12.5 Performance comparison for filters A and B shown in Figure 12.4. (a) Magnitude response; (b) group-delay response.

To verify the design of Figure 12.4b, Figure 12.6 illustrates the experimental results, which are compared with the EM-simulated ones. The insert in Figure 12.6a is a photograph of the fabricated UWB filter.

12.2 UWB-COUPLED RESONATOR FILTERS

12.2.1 Interdigital UWB Filters with Microstrip/CPW-Coupled Resonators

In Chapter 5, we discussed a type of interdigital bandpass filter with tapped-line input and output. Figure 12.7 shows another type of interdigital filter. It differs from the filter in Figure 5.15 in that the first and last lines, which connect directly to the

496 ULTRA-WIDEBAND (UWB) FILTERS

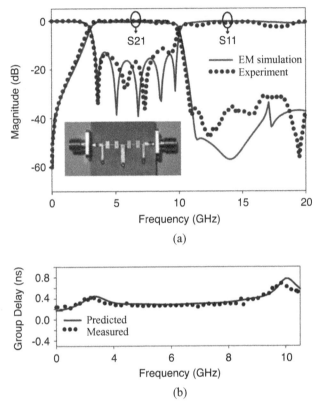

FIGURE 12.6 Experimental results of the filter B, as compared with the EM simulation. (**a**) Magnitude; (**b**) group delay.

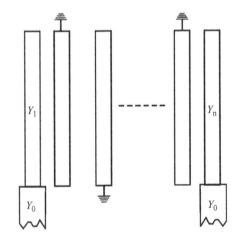

FIGURE 12.7 Interdigital filters with open-circuited lines at the input and output.

terminals (Y_0) at one end, and are open-circuited instead of shorted-circuited at the other end. These open-circuited lines at the input and output also count as resonators, so that there are n line elements for an n-pole filter [77]. Although this type of filter is most practical for designs having moderate or wide bandwidths, an implementation on microstrip with conventional parallel-coupled lines for an UWB filter can be difficult since very tight coupling gaps are required.

To overcome this difficulty, broadside coupling structures can be used. For example, Figure 12.8a illustrates a hybrid microstrip and coplanar waveguide (CPW) UWB filter structure, which is constructed on a single dielectric substrate with a thickness of h [33]. It can be seen as a three-pole interdigital filter having a form of Figure 12.7. On the top side of the substrate, two parallel microstrip open-circuited stub lines of W_1 wide and L_1 long are coupled to the CPW resonator through the dielectric substrate. The two open-circuited lines, separated by a gap of g, are connected to 50-Ω input/output (I/O) feed lines having a width of W_0. A top view of the microstrip is shown in Figure 12.8b. The filter has a symmetrical structure; in its middle, a quarter-wavelength CPW resonator with a width of W_2 and a length of L_2 is formed on the bottom side of the substrate. A bottom view for the CPW is given in Figure 12.8c.

Since no simple design formulation is available for microstrip to CPW coupled lines, EM-simulation tools can be used for designs of this type of filter. To investigate the operation of the filter, EM simulations have been carried out using commercially available software [79]. Figure 12.9 shows some typical frequency responses obtained by the simulations. For these results, the filter has fixed dimensions (refer to Figure 12.8a) where $W_0 = 1.2$ mm, $W_1 = 0.8$ mm, $L_1 = 6.2$ mm, $W_2 = 2.4$ mm, $L_2 = 8.6$ mm, and $s = 0.2$ mm on a 0.508-mm thick substrate with a relative dielectric

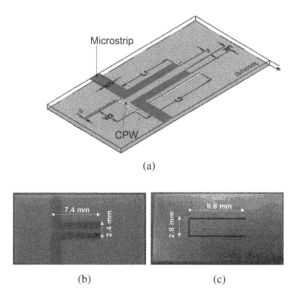

FIGURE 12.8 (a) 3D view of three-pole UWB microstrip/CPW filter structure; (b) top view (microstrip); (c) bottom view (CPW).

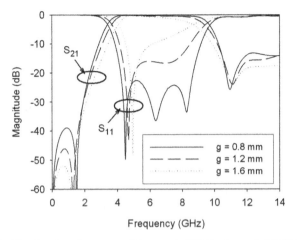

FIGURE 12.9 Simulated performance of three-pole UWB microstrip/CPW filter for different gap widths.

constant of 3.05. The gap (g) between the two parallel microstrip stub lines was allowed to vary in the simulations to optimize the performance of the filter, since it was found to be a critical design parameter.

As can be seen from the transmission (S_{21}) response in Figure 12.9, the filter exhibits a quasielliptic-function filtering characteristic with two transmission zeros closer to the passband. When the gap (g) between the two microstrip open-circuited lines is varied between 0.8 to 1.6 mm, two effects can be observed. The first effect is on the bandwidth of the filter. For a larger gap, a smaller bandwidth is observed. This is because the overlap between top microstrip stubs and the bottom CPW resonator is reduced, hence, reducing the resulting interstage coupling. The second effect of varying the gap is able to shift the locations of transmission zeros. The smaller gap moves the transmission zero, which is close to the low side of the passband, toward the passband, resulting in a higher selectivity. This strongly suggests that this finite-frequency transmission zero could result from the cross coupling within a trisection of microstrip–CPW–microstrip resonators (refer to Chapter 9). On the other hand, the changing of the gap hardly moves the transmission zero on the high side of the passband, which appears to result from the inherent attention pole at $2f_0$ of this type of quarter-wavelength resonator filter, where f_0 is the fundamental resonant frequency. In reality, this attenuation pole does not occur exactly at $2f_0$, because of the dispersion in the microstrip and CPW. For this three-pole UWB filter, the attenuation level on the upper stopband is limited because of the cross coupling between the input and output resonators.

Varying the gap between the two microstrip open-circuited stubs also affects the return loss of filter, as shown in Figure 12.9. When the gap is equal to 0.8 mm, the filter clearly shows three transmission poles with low return loss across the passband. This indicates the filter is a three-pole filtering structure where, in addition to the CPW quarter-wavelength resonator, the two microstrip open-circuited stub lines contribute to the degree of filter as well.

For an experimental demonstration, the designed filter with 0.8-mm gap between microstrip stub lines was fabricated using a standard photolithography process on GML1000 substrate, which has a relative dielectric constant of 3.05 and a thickness of 0.508 mm. The top and bottom view of the filter are shown in photographs Figure 12.8b and c. The two 50-Ω feed lines are extended for the purpose of measurement while the actual size of the filter is only 8.8 × 2.8 mm on the substrate used. This is a very compact filter only occupying a circuit area about 0.25 × 0.08 λ_{g0}, where λ_{g0} is the guided wavelength at the center frequency.

Figure 12.10a shows the measured and simulated S parameter response from the filter. The measured 3-dB bandwidth is from 3.5 to 9.3 GHz, representing a fractional bandwidth of 90% at the center frequency of 6.4 GHz. From the measurement, the two transmission zeros are observable at 1.95 and 10.36 GHz, respectively. The midband insertion loss was measured to be 0.6 dB, including the loss from two SMA connectors. The measured return loss performance was quite good (below −20 dB) around the midband, but deteriorated toward the passband edges, indicating

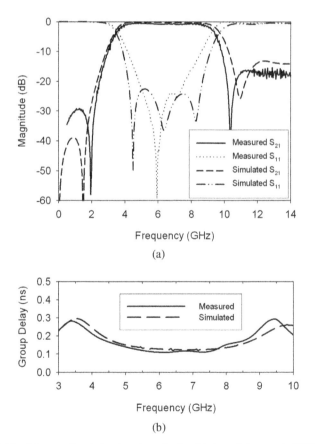

FIGURE 12.10 Measured and simulated results for fabricated three-pole UWB microstrip/CPW filter. (**a**) Magnitude; (**b**) group delay.

some mismatch of the fabricated filter. Because of the mismatching, the number of transmission poles observable in the measurement also appears to be different from that simulated. The mismatch could be attributed to a misalignment between the top and bottom patterned circuits during the fabrication. The measured and simulated group-delay responses of the filter are plotted in Figure 12.10b, where a good agreement between the two can be observed. The measured group delay at the midband was only about 0.1 ns and the maximum variation of the group delay in the whole passband was within 0.2 ns.

The design of higher order UWB filters based on the above compact filtering structure is also feasible. Figure 12.11 illustrates the top and bottom layouts of a

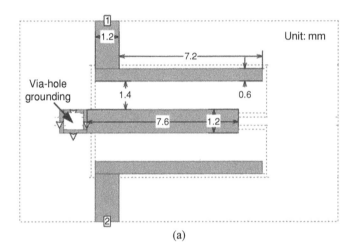

(a)

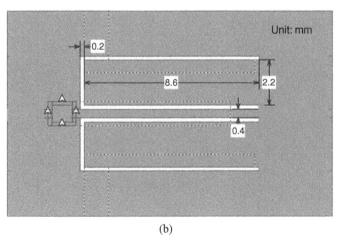

(b)

FIGURE 12.11 Five-pole UWB microstrip/CPW filter on a single substrate with a relative dielectric constant of 3.05 and a thickness of 0.508 mm. (**a**) Top view (microstrip); (b) bottom view (CPW).

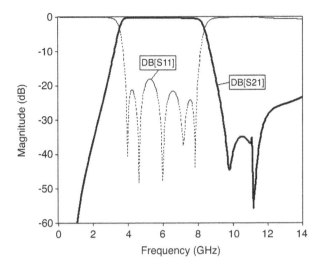

FIGURE 12.12 Simulated performance of five-pole UWB microstrip/CPW filter.

five-pole filter of this type. The filter consists of five quarter-wavelength resonators, three microstrip ones on the top and two CPW ones on the bottom side. From the input (port 1) to the output (port 2), the microstrip and CPW resonators are placed alternatively to achieve an interdigital filtering structure of Figure 12.7. For UWB designs, desired strong coupling between adjacent microstrip and CPW resonators can be facilitated via the broadside coupling. Coupling between the two CPW resonators is small, because they are edge-coupled and have the same orientation, which minimize the coupling. Figure 12.12 depicts the EM-simulated performance of the filter.

12.2.2 Broadside-Coupled Slow-Wave Resonator UWB Filters

To extend the upper stopband of a coupled resonator UWB filter, the slow-wave resonators or capacitive loaded transmission-line resonators, introduced in Chapter 10, can be deployed. However, these resonators need to be broadside coupled in order to achieve a high degree of coupling for UWB filter designs [66]. As a demonstration, Figure 12.13a is the 3D view of a five-pole broadside coupled slow-wave resonator UWB Filter. It consists of three metal layers, that is, the top metal layer having two resonators, the middle metal layer having three resonators, and the ground layer. The substrate between the top and middle metal layers has a thickness of h_1 and the substrate between the middle and the ground layers has a thickness of h_2.

The filter is designed on dielectric substrates with a relative dielectric constant of 3.15. The dielectric substrates have a thickness $h_1 = 50$ μm and $h_2 = 750$ μm, respectively. The design is based on EM simulation and optimization. In the design, the overlapping between top- and middle-layer resonators is adjusted to achieve desired coupling. Generally, larger overlapping results in tighter coupling so that a wider operating bandwidth is achieved. The final layouts and dimensions of the filter

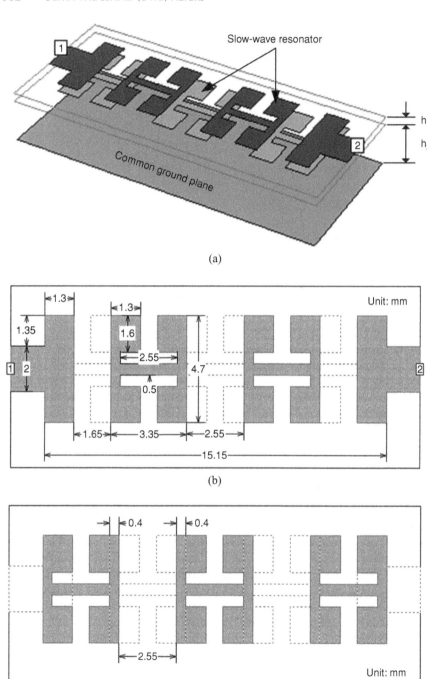

FIGURE 12.13 (a) 3D view of five-pole broadside coupled slow-wave resonator UWB filter; (b) top layer layout; (c) middle layer layout.

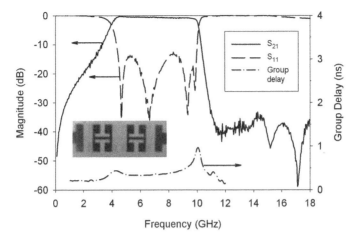

FIGURE 12.14 Experimental results of fabricated five-pole broadside coupled slow-wave resonator UWB filter.

are shown in Figure 12.13b and c. The input and output ports are designed on the top layer using 50-Ω microstrip lines.

The designed UWB filter is fabricated using multilayer LCP lamination technology. Because of the thickness limitation of available LCP films, three types LCP films are used in the fabrication, which have same electrical properties. One layer 50-μm thick and five layers 100-μm thick LCP films are used as core films, which have 315°C melting temperature. Five layer 50-μm thick LCP films are used as bonding film to bond core film in the fabrication process, which have 290°C melting temperature. The insert in Figure 12.14 shows the photograph of the fabricated vialess UWB filter, which has a size of 15.15 × 4.7 mm, (about 0.55 × 0.17 λ_{g0}, where λ_{g0} is the guide wavelength of 50-Ω microstrip line at 6.85 GHz).

The fabricated UWB bandpass filter is measured using vector network analyzer HP8720B, and the results are plotted in Figure 12.14. The measured 3-dB bandwidth is from 3.9 to 10.1 GHz. This filter has an insertion loss of 0.58 dB at center frequency 7.0 GHz, which includes the insertion loss from two SMA connectors. Moreover, the designed filter has excellent stopband performance. From 11 to 18 GHz, the rejection level is higher than 35 dB. It also can be seen in Figure 12.14, the measured group delay is 0.32 ns at 6.5 GHz and has a variation within 0.2 ns from 4.0 to 9.3 GHz, over 85% of the passband.

The above designed filter exhibits an asymmetrical selectivity, as can be seen from Figure 12.14. To improve the selectivity on the low side of the passband, two short-circuited stubs can be added at input/output ports, respectively, as shown in Figure 12.15. At low frequency, these short-circuited stubs act as shunt inductors, which function as additional highpass elements and are helpful to improve lower passband edge selectivity. To achieve a compact footprint, the short-circuit stubs are folded in the design, which, however, can degenerate the upper stopband rejection level because of the parasitic effects between folded short-circuited stubs.

504 ULTRA-WIDEBAND (UWB) FILTERS

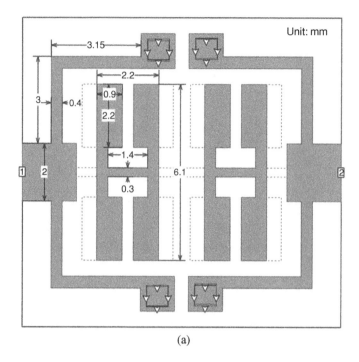

(a)

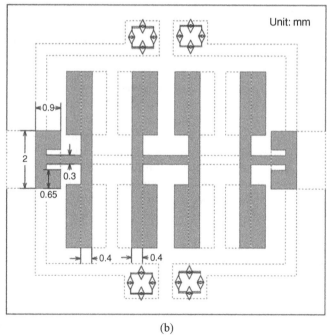

(b)

FIGURE 12.15 Layouts of a broadside coupled slow-wave resonator UWB Filter with short-circuited stubs at input/output ports; (**a**) top layer view; (**b**) middle layer view.

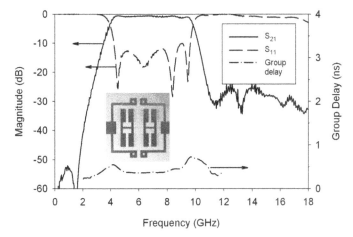

FIGURE 12.16 Experimental results of fabricated broadside coupled slow-wave resonator UWB filter with short-circuited stubs at input/output ports.

The designed filter is fabricated using the same multilayer LCP technology for the above vialess coupled resonator UWB filter. The insert in Figure 12.16 is the photograph of fabricated filter, which has a compact size of 9.6 × 9.2 mm (about 0.35 × 0.33 λ_{g0} at 6.85 GHz). The measured responses of the filter are plotted in Figure 12.16. The filter shows an improved selectivity at the low side of the passband, as expected. It has a 3-dB bandwidth from 3.92 to 9.8 GHz and a 10-dB insertion-loss bandwidth from 3.45 to 10.2 GHz. At center frequency 6.86 GHz, the measured filter has an insertion loss of 0.5 dB. Note that the measured filter also has a wide stopband because of the use of the slow-wave resonators. From 10.86 to 18.0 GHz, the rejection level is higher than 25 dB. It can be shown that without the folding of the short-circuited stubs, the rejection level is higher than 30 dB over the same upper stopband range. The measured group delay of the filter over the passband is also depicted in Figure 12.16, which shows a group delay of 0.36 ns at 6.8 GHz with a variation within 0.2 ns over the most of the passband, that is, from 4.0 to 9.2 GHz.

12.2.3 UWB Filters Using Coupled Stepped-Impedance Resonators

As shown in Figure 12.17a, UWB Filters can also be designed using coupled stepped-impedance resonators on a multilayer layer structure [48]. The multilayer structure includes three layer metals, namely, top, middle, and ground layer. The stepped-impedance resonator (SIR), which is similar to the slow-wave resonator discussed in Chapter 10, can be treated as a cascading of high- and low-impedance transmission lines. The high-impedance lines act as series inductors, whereas the low-impedance lines act as capacitors. This filtering structure may be represented using an equivalent circuit of Figure 12.17b. The SIRs embedded on the middle layer are adopted to realize series inductor L_1, L_3 and shunt capacitors C_2, C_3, C_6, and C_7. On the top

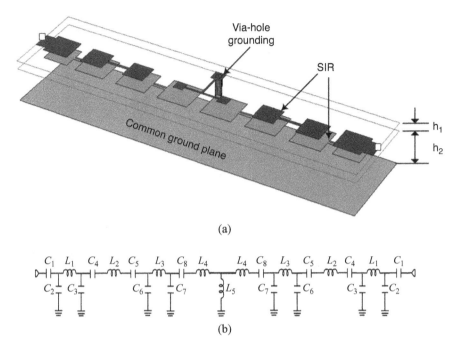

FIGURE 12.17 (a) 3D view of a broadside coupled stepped impedance resonator UWB filter; (b) its circuit model.

layer, 50-Ω microstrip lines are used for input and output; the series capacitors C_1, C_4, C_5, and C_8 are implemented using broadside coupled microstrip square patches; the series inductors L_2 and L_4 are the high-impedance microstrip lines connecting a pair of patches, respectively. The shunt inductor L_5 is implemented by a short-circuited high-impedance microstrip line.

Since it is difficult to achieve ideal lumped elements over a very wide frequency range for a UWB filter, the circuit model of Figure 12.17b may only serve as a prototype for an initial design, which, nevertheless, can provide some useful design information. For instance, sensitivity analysis can be carried out to identify the effect of each element on the filter performance, which helps to optimize the final filter design using EM simulation. Figures 12.18–12.20 show the results obtained by the sensitivity analysis based on the Monte Carlo method, which is available from a microwave design tool [78]. The dark lines represent the ideal circuit responses of S_{21} and S_{11} for a set of ideal lumped elements:

$$C_1 = 5.380 \text{ pF} \quad C_2 = 0.402 \text{ pF} \quad C_3 = 0.681 \text{ pF} \quad C_4 = 5.160 \text{ pF}$$
$$C_5 = 4.630 \text{ pF} \quad C_6 = 0.806 \text{ pF} \quad C_7 = 0.867 \text{ pF} \quad C_8 = 1.386 \text{ pF}$$
$$L_1 = 1.210 \text{ nH} \quad L_2 = 1.485 \text{ nH} \quad L_3 = 1.166 \text{ nH} \quad L_4 = 0.791 \text{ nH} \quad L_5 = 1.380 \text{ nH}$$

(12.3)

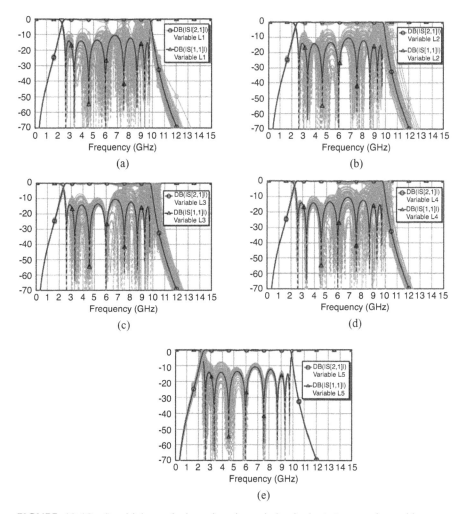

FIGURE 12.18 Sensitivity analysis against the variation in the *inductor* values with a standard deviation of 15% (assume a normal distribution). (**a**) For L_1 variable; (**b**) for L_2 variable; (**c**) for L_3 variable; (**d**) for L_4 variable; (**e**) for L_5 variable.

As can been seen, this is an eight-pole filter response. The grey shading in each diagram illustrates the sensitivity of the filter response against a variation of element value. The first set of the results, shown in Figure 12.18, is for the inductive elements. From these results, one can see that all the series inductors, from L_1 to L_4 have more influence on the passband response toward its high cutoff frequency, which is expected as these inductors function as a lowpass element. On the contrary, the shunt inductor L_5 mostly affects the low cutoff frequency, because it acts a highpass element. It can be shown that L_5 plays a very effective role in the improvement of the filter selectivity at its low edge of the passband.

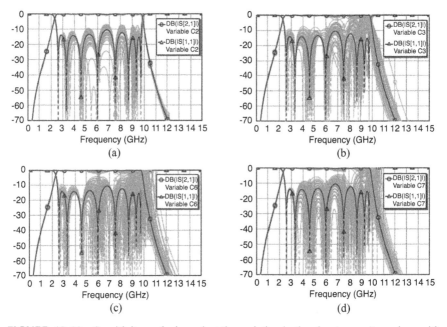

FIGURE 12.19 Sensitivity analysis against the variation in the *shunt capacitor* values with a standard deviation of 15% (assume a normal distribution). (**a**) For C_2 variable; (**b**) for C_3 variable; (**c**) for C_6 variable; (**d**) for C_7 variable.

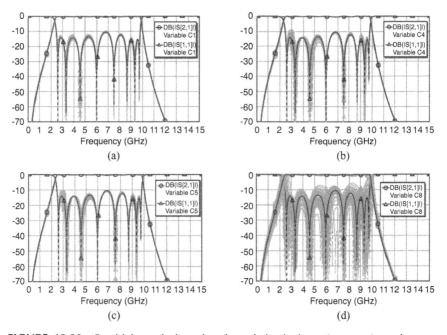

FIGURE 12.20 Sensitivity analysis against the variation in the *series capacitor* values with a standard deviation of 15% (assume a normal distribution). (**a**) For C_1 variable; (**b**) for C_4 variable; (**c**) for C_5 variable; (**d**) for C_8 variable.

The second set of the sensitivity analysis results shown in Figure 12.19 is for the shunt capacitive elements. All these elements show a greater influence on the passband response toward its high cutoff frequency. Again, this is expected, as they function as a lowpass element. However, the variation in C_2 is less sensitive as compared to that in C_3, C_6, and C_7.

From the remaining results of the sensitivity analysis given in Figure 12.20, it is evident that the series capacitors C_1, C_4, and C_5 have a much smaller effect on filter performance. Conversely, C_8 appears to be quite important, which can affect both the low and high cutoff frequencies.

For the given circuit parameters of Eq. (12.3), an initial realization of the filter can follow a similar path for the design of a stepped-impedance lowpass filter, as discussed in Chapter 5. All parasitic effects from the discontinuities have to be considered, which significantly affect the inductive elements. Note that it is difficult to extract an exact circuit parameter over a wide frequency range because, in reality, it can be frequency-dependent. Therefore, there needs to be some optimization in light of the EM simulation to finalize the filter design. Figure 12.21 shows the layouts of the designed multilayer UWB filter. The EM-simulated performance of the filter is plotted in Figure 12.22 along with the circuit modeling response, where good agreement between the two, over the passband, can be observed. However, the circuit modeling does not match the EM-simulated frequency response at the upper stopband, because the realized elements are frequency-dependent, as mentioned above. The simulated filter has a passband from 2.2 to 9.9 GHz with an in-band return loss better than 10.2 dB.

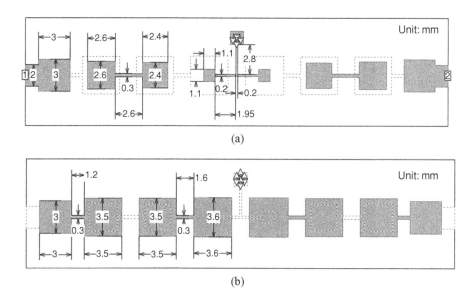

FIGURE 12.21 A designed eight-pole broadside coupled stepped impedance resonator UWB filter on dielectric substrates with a relative dielectric constant of 3.15 and thicknesses of $h_1 = 50$ μm and $h_2 = 750$ μm. (**a**) Top layer layout; (**b**) middle layer layout.

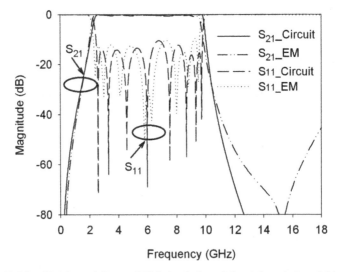

FIGURE 12.22 Circuit modeling and EM simulation of the eight-pole broadside coupled stepped-impedance resonator UWB filter.

The designed UWB filter is fabricated using a multilayer LCP lamination process similar to that described previously in Section 12.2.2. The fabricated filter is shown as an insert in Figure 12.23 together with the measured results. The fabricated filter, which is 37.6 × 5.6 mm, exhibits a desired passband from 2.2 to 9.9 GHz. An insertion loss of 0.18 dB was measured at center frequency (6.05 GHz). The measured return loss is greater than 10.0 dB over the passband. It also can be seen from Figure 12.23, the fabricated filter has a good rejection level, which is higher than 32.01 dB from

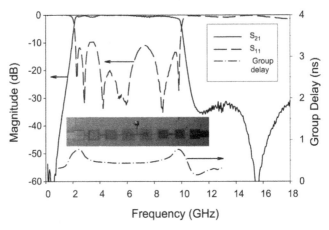

FIGURE 12.23 Experimental results of fabricated eight-pole broadside coupled stepped-impedance resonator UWB filter.

10.9 to 18.0 GHz. The measured group delay is 0.45 ns at 6.1 GHz within a variation of 0.35 ns in the passband.

12.2.4 Multimode-Resonator UWB Filters

Another development of UWB filters is based on the concept of a multimode resonator with a stepped-impedance configuration, which was originated in Zhu et al. [3,4] and extended to the implementation of a class of UWB bandpass filters [12–16]. To understand this concept, let us consider the multimode stepped-impedance resonator (SIR) in Figure 12.24a. In principle, this is a transmission-line resonator whose both ends are open circuited. It consists of a low-impedance line section in the middle and the two identical high-impedance line sections on the two sides. Z_1 and Z_2 denote the line characteristic impedances and θ_1 and θ_2 represent the electrical lengths, respectively. Note that this geometry is opposite to that of the SIR used for the filter designs described in the previous section, wherein the SIR has a high-impedance line section in the middle and two low-impedance line sections on the two sides. Because of this particular geometry, one can exploit the first several resonant modes of the resonator to design a UWB filter.

For a demonstration, consider such a multimode resonator being excited with an arrangement as shown in Figure 12.25a. Assume that $\theta_1 = \theta_2 = \pi/2$ or $90°$ at a nominal center frequency, for example, 5 GHz. The structure of Figure 12.25a is then first simulated with $W_1 = 2.7$ mm and $W_2 = 0.25$ mm on a 0.5-mm thick dielectric substrate having a relative dielectric constant of 3.15 for varying the input and output (I/O) coupling spacing s. The results are plotted in Figure 12.25. When the resonator is weakly excited with $s = 1.0$ mm, four sharp resonating peaks in the given frequency range can be clearly observed. These peaks correspond to the resonant frequencies of the first four resonant modes, denoted by f_1, f_2, f_3, and f_4. As the I/O coupling increases or s becomes smaller, a wide passband based on the first three resonant modes can be

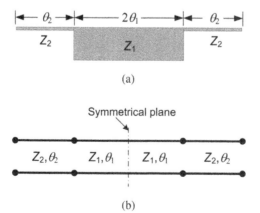

FIGURE 12.24 (a) Multimode stepped-impedance resonator (SIR). (b) Equivalent transmission-line model with the both ends open-circuited.

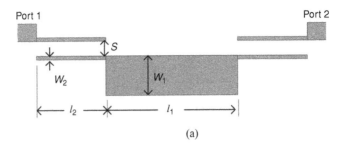

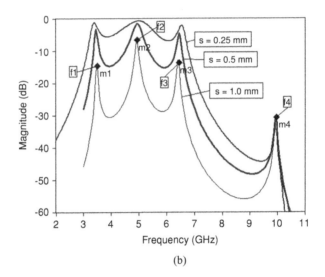

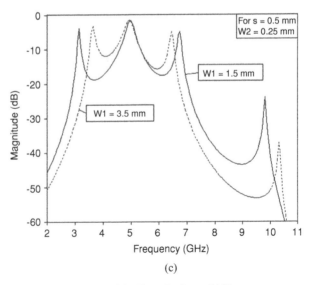

FIGURE 12.25 (a) Multimode SIR with I/O excitations. (b) Frequency responses against the I/O coupling spacing s. (c) Frequency responses against the width W_1 of the low-impedance line. (Assume $\theta_1 = \theta_2 = \pi/2$ at 5 GHz.)

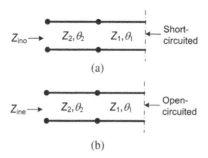

FIGURE 12.26 (a) Odd-mode equivalent circuit of the multimode SIR. (b) Even-mode equivalent circuit of the multimode SIR.

formed. This is the basic idea behind the so-called multimode resonator-based UWB filter. The bandwidth of such a UWB filter, centered at f_2, is largely dependent on the separation of these three resonant modes, which can be controlled by the impedance ratio or the width ratio of a multimode SIR and is demonstrated in Figure 12.25c. As can be seen, for the fixed high-impedance line with $W_2 = 0.25$ mm, changing the low-impedance line width W_1 alters the mode separation and an increase in W_1 tends to reduce the mode separation. Since Z_1 decreases as W_1 is increased, the impedance ratio $R = Z_2/Z_1$ will increase. This implies that, for $R > 1$, a smaller R leads to the realization of a wider bandwidth of a multimode resonator filter.

The property of the multimode SIR can also be studied in light of the transmission-line theory discussed in Chapter 2. By placing a short circuit at the symmetrical plane of the equivalent circuit mode in Figure 12.24b, we obtain an equivalent circuit for the odd modes, as shown in Figure 12.26a. Similarly, placing an open circuit at the symmetrical plane results in an equivalent circuit for the even modes; this is depicted in Figure 12.26b.

Using Eqs. (2.29) and (2.33), the input impedance Z_{ino} at the left end of the circuit, as indicated in Figure 12.26a, can be expressed as

$$Z_{ino} = Z_2 \frac{jZ_1 \tan\theta_1 + jZ_2 \tan\theta_2}{Z_2 - Z_1 \tan\theta_1 \tan\theta_2} \quad (12.4)$$

Since the boundary condition at the left end of the resonator is always an open circuit, at the resonances for the odd modes, we have

$$Z_{ino} = \infty \quad (12.5)$$

From Eq. (12.4), this requires

$$Z_2 - Z_1 \tan\theta_1 \tan\theta_2 = 0 \quad (12.6)$$

In the case of $\theta_1 = \theta_2 = \theta$, (12.6) becomes

$$\tan\theta = \pm\sqrt{R} \quad \text{for} \quad R = \frac{Z_2}{Z_1} \quad (12.7)$$

Two solutions can then be found for the fundamental mode frequency f_1 and the third resonant mode frequency f_3, respectively, that is,

$$\theta(f_1) = \tan^{-1}\sqrt{R} \tag{12.8a}$$

$$\theta(f_3) = \pi - \tan^{-1}\sqrt{R} \tag{12.8b}$$

In a similar way, using Eqs. (2.29) and (2.32), the input impedance Z_{ine} at the left end of the circuit, as indicated in Figure 12.26b, can be derived as

$$Z_{ine} = jZ_2\frac{Z_2\tan\theta_1\tan\theta_2 - Z_1}{Z_2\tan\theta_1 + Z_1\tan\theta_2} \tag{12.9}$$

At the resonances for the even modes, we have

$$Z_{ine} = \infty \tag{12.10}$$

For $\theta_1 = \theta_2 = \theta$, two solutions of Eq. (12.10) can be found from Eq. (12.9) for the second and fourth resonant mode frequencies f_2 and f_4, respectively, which are

$$\theta(f_2) = \frac{\pi}{2} \tag{12.11a}$$

$$\theta(f_4) = \pi \tag{12.11b}$$

Now, if we ignore the dispersion in microstrip and assume an equal phase velocity in the line sections with characteristic impedances Z_1 and Z_2, from Eqs. (12.8) and (12.11) for $\theta \propto f$, we can obtain

$$\frac{f_1}{f_2} = \frac{2\tan^{-1}\sqrt{R}}{\pi} \tag{12.12a}$$

$$\frac{f_3}{f_2} = \frac{2}{\pi}\left(\pi - \tan^{-1}\sqrt{R}\right) \tag{12.12b}$$

$$\frac{f_4}{f_2} = 2 \tag{12.12c}$$

Define the normalized separation between f_1/f_2 and f_3/f_2 as

$$\Delta f_{13} = \frac{f_3 - f_1}{f_2} = \frac{2}{\pi}\left(\pi - 2\tan^{-1}\sqrt{R}\right) \tag{12.13}$$

The two normalized frequencies, f_1/f_2 and f_3/f_2, and their normalized separation Δf_{13} are plotted in Figure 12.27, versus R. Note that, theoretically, the fourth resonant frequency f_4, is always twice the center frequency f_2, which can cause an unwanted spurious response near the desired passband when the bandwidth is wide.

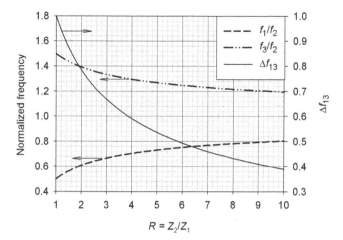

FIGURE 12.27 Normalized resonant frequencies and their separation versus impedance ratio of a multimode SIR.

Recalling the discussion in association with Figure 12.25, f_2 is set as the center frequency for a multimode resonator filter utilizing the first three resonant modes. Therefore, the separation between f_1 and f_3 is proportional to the filter bandwidth. From Figure 12.27, it can be seen that this separation ranges from 100% for $R = 1$ to 39% for $R = 10$; hence, it is evident that the larger the impedance ratio (R) of high- and low-impedance lines, the smaller is the filter bandwidth. In the following, we discuss an empirical method to determine an initial R for a filter design.

Assume that a UWB filter of this type is to be designed to match an n-pole Chebyshev filter response. From Eqs. (3.9) and (3.11), for $|S_{21}(\Omega)| = 1$, we have

$$\cos(n \cos^{-1} \Omega) = 0 \tag{12.14}$$

Solving Eq. (12.14) for Ω, we can find k transmission poles in the passband, that is,

$$\Omega_k = \cos\left[\frac{(2k-1)\pi}{2n}\right] \quad \text{for } k = 1, 2, \ldots n \tag{12.15}$$

By applying the frequency transformation for a wideband filter

$$\frac{\Omega}{\Omega_c} = \frac{2}{FBW}\left(\frac{f}{f_0} - 1\right) \tag{12.16}$$

with

$$FBW = \frac{f_2 - f_1}{f_0}$$

$$f_0 = \frac{f_2 + f_1}{2}$$

we obtain

$$f_p(k) = \frac{FBW \cdot \Omega_k}{2} + 1 \quad \text{for } k = 1, 2, \ldots n \quad (12.17)$$

where $f_p(k)$ is the kth transmission pole, which is normalized by the center frequency, and FBW is the fractional bandwidth of the UWB filter to be designed. These transmission poles can form a base for finding an initial R, which is demonstrated with two design examples below.

12.2.4.1 Five-Pole Filter Design with a Single Triple-Mode Resonator
Consider the design of a UWB filter using a single multimode or, in this case, triple-mode SIR to meet the following specifications

Center frequency f_0	5 GHz
Fractional bandwidth FBW	80%
Passband return loss L_R	-15 dB

It will be shown that with a single triple-mode SIR, a five-pole UWB filter can be designed. Thus, using Eq. (12.17) for $n = 5$ and $FBW = 0.8$, the normalized transmission poles are calculated (see tabulation below):

K	1	2	3	4	5
$f_p(k)$	1.38	1.24	1	0.76	0.62

Let

$$f_a = \frac{f_p(1) + f_p(2)}{2} = \frac{1.38 + 1.24}{2} = 1.31$$

$$f_b = \frac{f_p(4) + f_p(5)}{2} = \frac{0.76 + 0.62}{2} = 0.69$$

$$\Delta f_{13} = f_a - f_b = 0.62$$

Compared with Eq. (12.13), one can see that f_a corresponds to f_3/f_2, whereas f_b corresponds to f_1/f_2 for $f_2 = f_0$. The idea is to split the three resonant modes accordingly. Thus, from Figure 12.27 or Eq. (12.13), we can determine an impedance ratio $R = 3.6$ for the filter design.

On a substrate with a relative dielectric constant of 3.15 and a thickness of 0.5 mm, referring to Figure 12.25(a), choosing $W_2 = 0.25$ mm for the high-impedance line results in $Z_2 = 110\,\Omega$ and $l_2 = 9.8$ mm for $\theta_2 = 90°$ at 5 GHz. For $R = 3.6$, $Z_1 = Z_2/R = 30.6\,\Omega$. This gives $W_1 = 2.5$ mm and $l_1 = 18.1$ mm for $2\theta_1 = 180°$ at 5 GHz.

For the I/O coupled-line sections, the design equations from Section 5.2.2 may be utilized to find a pair of even- and odd-mode impedances. For $n = 5$ and $L_R = -15$ dB, we can find $g_0 = 1.0$ and $g_1 = 1.2328$ from (3.26). For a terminal impedance $Z_0 = 50\,\Omega$ ($Y_0 = 1/Z_0$) and $FBW = 0.8$, from Eq. (5.21a), we obtain $J_{01}/Y_0 = 1.0096$.

Substituting this result into Eqs. (5.22a) and (5.22b) yields $Z_e = 151.45 \, \Omega$ and $Z_o = 50.49 \, \Omega$. Since the width of couple lines is the same as W_2, we find $s = 0.05$ mm so as to arrive at a pair of even- and odd-mode impedances on the given substrate for matching to $(Z_e - Z_o)/(Z_e + Z_o)$, as discussed in association with Eq. (5.41).

The above-determined dimensions allow us to create an initial layout, similar to that of Figure 12.25a, for finalizing the design using EM simulation. The final layout of the design is shown in Figure 12.28a, where the final dimensions have

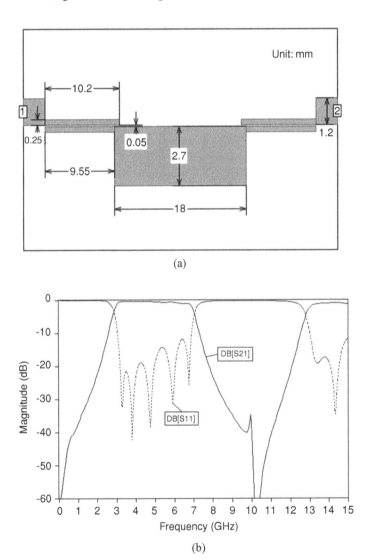

FIGURE 12.28 (a) Layout of five-pole UWB filter using a single triple-mode SIR on a substrate with a relative dielectric constant of 3.15 and a thickness of 0.5 mm. (b) Full-wave EM simulated frequency responses of the filter.

taken all the discontinuities into account. Note that the I/O feed line is slightly longer than the length of the high-impedance line, or the coupled-line section, for the suppression of the spurious response because of the fourth resonant mode. This spurious response is excited because of unequal even- and odd-mode phase velocities in microstrip, which is analogous to that occurring at $2f_0$ in a conventional parallel-coupled, half-wavelength resonator filter. Figure 12.28b shows the performance of the filter, obtained by the EM simulation. As can be seen, the spurious response at 10 GHz ($2f_0$) has been suppressed well below -30 dB. There are five transmission poles within the desired passband from 3 to 7 GHz. With a single triple-mode resonator, two extra poles result from the I/O coupled-line sections that have a very strong coupling. This transmission-line filter has its next periodic passband centered at $3f_0$, or 15 GHz. To remove this undesired passband, the technique discussed in Section 12.1 can be incorporated in the design.

12.2.4.2 Eight-Pole Filter Design with Two Triple-Mode Resonators

For this example, a higher-order UWB filter of this type using two triple-mode resonators is to be designed to have a fractional bandwidth $FBW = 0.8$ at a center frequency $f_0 = 5$ GHz. Assume the filter is to match a Chebyshev response with a passband return loss $L_R \leq -15$ dB.

The design starts with a determination of a pair of the even- and odd-mode impedances for the I/O coupled-line sections. For $n = 8$ and $L_R = -15$ dB, it can be found, from Eq. (3.26), that $g_0 = 1.0$ and $g_1 = 1.2747$. For $FBW = 0.8$, $J_{01}/Y_0 = 0.9926$ is calculated from Eq. (5.21a), where $Y_0 = 1/Z_0$ with the terminal impedance $Z_0 = 50\ \Omega$. By using Eqs. (5.22a) and (5.22b), we obtain

$$Z_e = 148.94\ \Omega \qquad Z_o = 49.65\ \Omega \qquad (12.18)$$

Next, applying Eq. (12.17), for $n = 8$ and $FBW = 0.8$, gives the normalized transmission poles (shown in the tabulation below):

k	1	2	3	4	5	6	7	8
$f_p(k)$	1.39	1.33	1.22	1.08	0.92	0.78	0.67	0.61

In this case, there is no transmission pole at the center frequency, but there are four transmission poles on each side of it. Let

$$f_a = \frac{f_p(2) + f_p(3)}{2} = \frac{1.33 + 1.22}{2} = 1.28$$

$$f_b = \frac{f_p(6) + f_p(7)}{2} = \frac{0.78 + 0.67}{2} = 0.73$$

$$\Delta f_{13} = f_a - f_b = 0.55$$

With $\Delta f_{13} = 0.55$, we can find, from Eq. (12.13) or Figure 12.27, an impedance ratio $R = 4.7$. Choosing $Z_2 = 130\,\Omega$, we obtain

$$Z_1 = Z_2/R = 27.66\,\Omega \tag{12.19}$$

To proceed with the filter design, it is useful to exploit an equivalent circuit as shown in Figure 12.29a. The filter is symmetrical with two identical transmission-line resonators that have a low-impedance section, denoted by Z_1. The two resonators couple to each other through the coupled-line section in the middle, characterized by a pair of even- and odd-mode impedances Z_{e2} and Z_{o2}. The input and output (I/O)

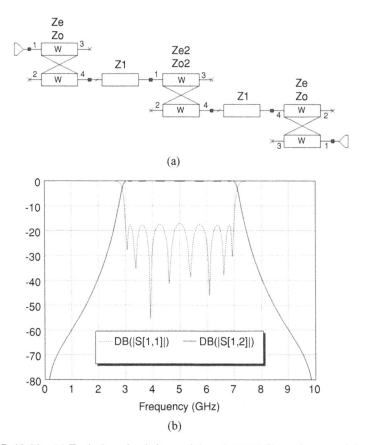

FIGURE 12.29 (a) Equivalent circuit for an eight-pole UWB filter using two triple-mode SIR. (b) Circuit response of the filter for $Z_e = 149\,\Omega$, $Z_o = 50\,\Omega$, $Z_1 = 27.25\,\Omega$, $Z_{e2} = 150.8\,\Omega$, $Z_{o2} = 65.7\,\Omega$. Each pair of the coupled lines has an electrical length of 90° and the low impedance (Z_1) transmission line section has an electrical length of 180° at 5 GHz.

are the coupled-line sections having the even-impedance of Z_e and the odd-mode impedance of Z_o. All pairs of the coupled lines have an electrical length of 90°, and the low-impedance (Z_1) transmission-line section has an electrical length of 180°, at the center frequency of 5 GHz. With the results obtained in Eqs. (12.18) and (12.19), only unknowns are Z_{e2} and Z_{o2}. Initially, set $Z_{e2} = Z_e$ and $Z_{o2} = Z_o$ as given in Eq. (12.18). The circuit parameters, including Z_1, are then finalized by optimizing the circuit to match to the Chebyshev response with the transmission poles shown above. The optimized circuit response is illustrated in Figure 12.29b, where the eight transmission poles can clearly indentified from the response of S_{11}. The final circuit parameters are determined as

$$Z_e = 149\ \Omega \quad Z_o = 50\ \Omega$$
$$Z_{e2} = 150.8\ \Omega \quad Z_{o2} = 65.7\ \Omega \quad (12.20)$$
$$Z_1 = 27.25\ \Omega$$

As can be seen, the final value of Z_1 has been slightly changed from the one given in Eq. (12.19). This is because the result of Eq. (12.19) is obtained rather empirically and to obtain the optimal values for Z_{e2} and Z_{o2} affects Z_1 as well.

Here we come to the final stage for the filter design, that is, the physical realization. On a substrate with a relative dielectric constant of 3.15 and a thickness of 0.5 mm, initial dimensions for the circuit parameters given in Eq. (12.20) can be found by using microstrip design equations, such as those described in Chapter 4. Referring to Figure 12.25a, while W_1 for the low-impedance line section of the triple-mode resonator can be decided from the Z_1 given in Eq. (12.20), W_2 for the high-impedance line section of the resonator can be found from the given impedance ratio R. For a determined W_2, the realization of each pair of coupled lines can then be facilitated by altering their coupling spacing for a pair of even- and odd-mode impedances that matches $(Z_e - Z_o)/(Z_e + Z_o)$ for the I/O coupling, or $(Z_{e2} - Z_{o2})/(Z_{e2} + Z_{o2})$ for the inter-resonator coupling. For the detail of this technique, one can refer to the discussion in association with Eq. (5.41) in Chapter 5. The final layout of the designed filter is shown in Figure 12.30a, which is obtained in light of EM simulation to take into account the effects of microstrip discontinuities. The EM-simulated performance of the filter is displayed in Figure 12.30b, which shows a desired ripple passband from 3 to 7 GHz.

It is feasible that the method described above can be used to design an even higher-order UWB filter with more than two triple-mode resonators.

12.3 QUASILUMPED ELEMENT UWB FILTERS

12.3.1 Six-Pole Filter Design Example

Figure 12.31a is a six-pole lumped-element bandpass filter prototype, which is adopted to develop compact UWB bandpass filters [49]. The input and output ports

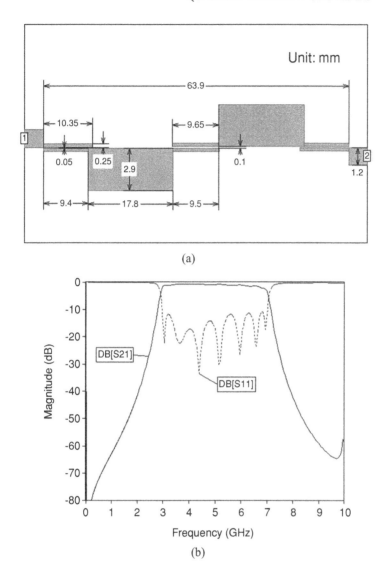

FIGURE 12.30 (a) Layout of eight-pole UWB filter using two triple-mode SIR on a substrate with a relative dielectric constant of 3.15 and a thickness of 0.5 mm. (b) Full-wave EM simulated frequency responses of the filter.

have 50 Ω impedance; the series inductors L_1, L_2, and L_3 and shunt capacitors C_2 and C_3 may be seen as lowpass elements, whereas the series capacitors C_1 and C_4 and shunt inductor L_4 act as highpass elements. Figure 12.31b shows the circuit response of the bandpass filter prototype. For the given element values, which are obtained by optimization, the prototype exhibits six transmission poles in the passband, which is equivalent to a six-pole bandpass filter.

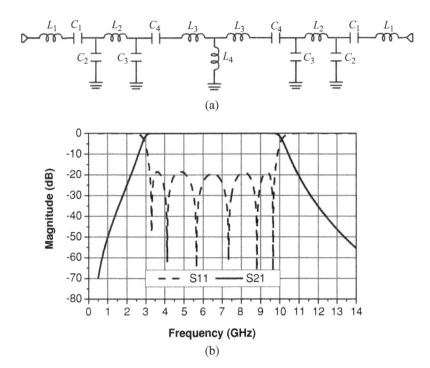

FIGURE 12.31 (a) Circuit schematic for a six-pole UWB bandpass filter prototype. (b) Circuit responses of the six-pole bandpass filter prototype with the elements: $L_1 = 1.28$ nH, $L_2 = 1.419$ nH, $L_3 = 0.976$ nH, $L_4 = 1.30$ nH, $C_1 = 0.936$ pF, $C_2 = 0.522$ pF, $C_3 = 0.687$ pF, $C_4 = 0.902$ pF.

Although other arrangements of the low- and highpass elements are possible to arrive at a different filter topology, the choice of the topology of Figure 12.31a is based on the consideration for an easy physical realization. As such, this lumped-element bandpass prototype is adopted for a physical implementation of multilayer structure illustrated in Figure 12.32. This multilayer structure has three metal layers, i.e., the top metal layer of Figure 12.32b, the middle metal layer of Figure 12.32c, and the bottom metal layer as the ground plane. Dielectric substrates are used to support these metal layers. On the top metal layer, the input and output use 50-Ω microstrip feed lines.

Since it is difficult to realize ideal and low-loss lumped elements to operate over a very wide frequency range, quasilumped elements are used for the implementation. The series capacitors with capacitances of C_1 and C_4 are realized by broadside-coupled radial stubs that are on two different metal layers, i.e., the top layer (see Figure 12.32b) and the middle layer (see Figure 12.32c). The radial stubs on the middle layer also produce shunt capacitances C_2 and C_3 with respect to the ground plane. The reason for choosing multilayer radial stubs to realize series and shunt

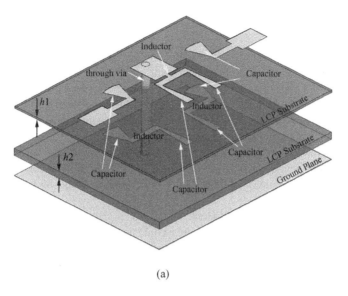

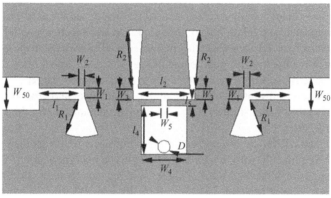

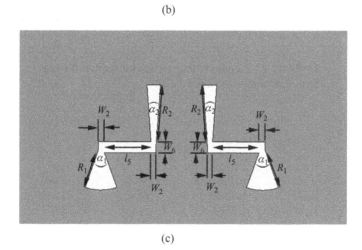

FIGURE 12.32 (a) 3D structure for a six-pole quasilumped element UWB filter; (b) top metal layer layout; (c) middle metal layer layout.

capacitors is because the radial stubs can provide very low input impedance over a large frequency range with small contacting point, which prevents excitation of transverse modes, as discussed in Section 6.2.4. Large series capacitances can also be conveniently implemented in small footprints by using overlapped radial stubs that take advantages of the multilayer structure. The desired capacitances can then be controlled by both radiuses and angles in the design.

The series inductors L_1, L_2, and L_3 are realized by the high-impedance microstrip line sections; the shunt inductor L_4 is implemented by using short-circuited microstrip stub. For the implementation, a large pad is used for connecting short-circuited stub to ground.

Even though the quasilumped elements have been chosen, there is a certain frequency range over which these elements could better match the desired lumped elements in Figure 12.31. Furthermore, since there are many discontinuities, such as steps and junctions in the physical implementation, the parasitic parameters associated with these discontinuities have not been considered in the filter equivalent circuit. In this case, the equivalence between the actual filter and equivalent circuit would be approximate mainly over the desired passband, and the actual filter design has been completed in light of EM simulation. Dielectric substrates with relative dielectric constant $\varepsilon_r = 3.15$ and loss tangent $\tan\delta = 0.0025$ are used in the design. The substrate between the top and middle metal layers has a thickness of $h_1 = 0.1$ mm. This thin dielectric substrate facilitates the implementation of larger series capacitances. The second substrate between the middle metal layer and ground plane has a thickness of $h_2 = 0.7$ mm. Table 12.2 gives geometry parameters for the UWB filter structure of Figure 12.32.

The designed UWB filter is fabricated using a multilayer LCP bonding technology. Thick bonding films (50 μm), which have a melting temperature around 290°C, are used to bond core films that have a higher melting temperature. Because of the thickness limitation of available LCP core film, 50- and 100-μm LCP core films are used in the fabrication. Figure 12.33 gives the experimental results of the fabricated UWB bandpass filter. The small insert in Figure 12.33a is a photograph of the

TABLE 12.2 Geometry Parameters for the UWB Filter Structure Shown in Figure 12.32

W_{50}	2.0 mm	l_1	2.23 mm
W_1	0.7 mm	l_2	3.05 mm
W_2	0.3 mm	l_3	0.45 mm
W_3	0.7 mm	l_4	3.0 mm
W_4	3.0 mm	l_5	3.0 mm
W_5	0.7 mm	R_1	2.15 mm
W_6	0.7 mm	R_2	3.43 mm
h_1	0.1 mm	α_1	40.0°
h_2	0.7 mm	α_2	8.0°
D	0.75 mm		

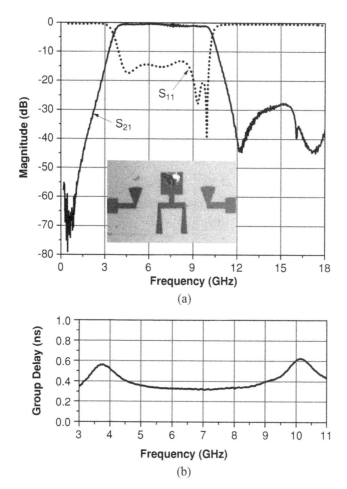

FIGURE 12.33 Experimental results of fabricated six-pole UWB filter. (**a**) Measured magnitude responses (the small insert is the photograph of the fabricated filter.) (**b**) Measured group-delay response.

fabricated UWB filter. The size of this six-pole filter (not including 50-Ω feed lines) is 15.4 × 7.6 mm, which is compact amounting to about 0.55 × 0.27 λ_{g0}, where λ_{g0} is the guide wavelength at the center frequency (6.9 GHz).

The measured 3-dB bandwidth is from 3.6 to 10.2 GHz and 10-dB bandwidth is from 3.2 to 10.62 GHz. This filter has a minimum insertion loss of 0.35 dB at 6.16 GHz. Although there is a ripple measured at about 15 GHz, the rejection level is higher than 28.1 dB from 11.5 to 18.0 GHz. Shown in Figure 12.33b is the measured group delay of the filter. It has a variation within 0.1 ns from 4.5 to 9 GHz, over 83% of the passband.

12.3.2 Eight-Pole Filter Design Example

To improve the selectivity of the above design, two short-circuited stubs can be integrated at the input and output ports, respectively [49]. These short-circuited stubs not only add two transmission poles in the passband, but also introduces a finite-frequency transmission zero at the high side of the passband. A circuit model developed from the above six-pole bandpass filter prototype is shown in Figure 12.34a, where two short-circuited transmission lines with electrical length θ and characteristic impedance Z_0 function as the aforementioned short-circuited stubs. The resultant eight-pole filter is designed to have an even wider bandwidth, as shown in Figure 12.34b, where a transmission zero at 11.0 GHz is clearly seen. In this case, the transmission line parameters for the short-circuited stubs are $Z_0 = 50\ \Omega$ and $\theta = 98.5°$ at 6.0 GHz. Thus, the short-circuited stubs will have an electrical length of 180° at 11.0 GHz and present a short circuit at the input and output, respectively, resulting in the transmission zero as observed.

Figure 12.35 shows the physical realization for the eight-pole UWB bandpass filter. Again, a three-metal layer structure on LCP substrates is used and quasilumped

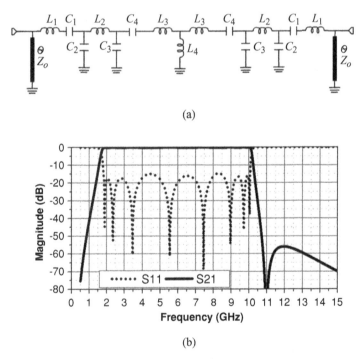

FIGURE 12.34 (a) Circuit schematic for an eight-pole UWB bandpass filter prototype. (b) Circuit responses of the eight-pole bandpass filter prototype with the elements: $Z_0 = 90\ \Omega$, $\theta = 98.5°$ at 6 GHz, $L_1 = 1.354$ nH, $L_2 = 1.356$ nH, $L_3 = 0.772$ nH, $L_4 = 2.058$ nH, $C_1 = 1.42$ pF, $C_2 = 0.6485$ pF, $C_3 = 0.746$ pF, $C_4 = 1.994$ pF.

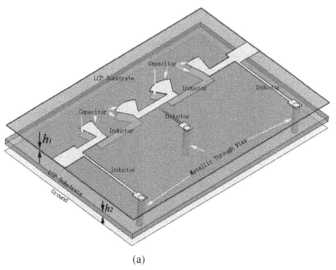

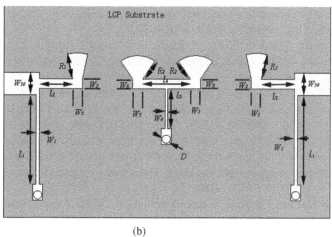

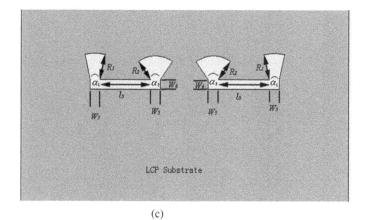

FIGURE 12.35 (a) 3D structure for an eight-pole quasi-lumped element UWB filter; (b) top metal layer layout; (c) middle metal layer layout.

TABLE 12.3 Geometry Parameters for the UWB Filter Structure Shown in Figure 12.35

W_{50}	2.0 mm	l_1	7.14 mm
W_1	0.33 mm	l_2	3.01 mm
W_2	0.9 mm	l_3	3.6 mm
W_3	0.9 mm	l_4	4.38 mm
W_4	0.2 mm	l_5	4.67 mm
W_5	0.9 mm	R_1	2.305 mm
W_6	0.9 mm	R_2	1.96 mm
h_1	0.1 mm	α_1	25.2°
h_2	0.7 mm	α_2	68.28°

elements are adopted for the implementation of *LC* elements in the ways similar to that described in the above section for the six-pole filter.

Full-wave EM simulations are used for determining the geometry parameters of the eight-pole filter, which are listed in Table 12.3.

The designed eight-pole filter is fabricated using the same multilayer LCP technology for the six-pole filter described previously. Figure 12.36 shows the photograph of fabricated filter. The physical size of the filter is $24 \times 10 \times 0.8$ mm^3. The measured responses of the fabricated eight-pole UWB filter are plotted in Figure 12.37.

The measured magnitude responses show that the fabricated filter has a center frequency of 5.925 GHz and a 3-dB fractional bandwidth of 139% from 1.8 to 10.05 GHz. The in-band return loss is greater than 10.0 dB. A transmission zero that is generated by the short-circuited stubs is found at 11.6 GHz, which causes a high selectivity at upper passband edge. The upper stopband performance is also improved. From 10.57 to 20.0 GHz, the rejection level is higher than 38.1 dB. In general, the filter exhibits good selectivity at both of lower and upper passband edges and high stopband rejection level in the stopband. The measured group-delay response of the filter is depicted in Figure 12.37b, which shows a group-delay response with a variation within 0.2 ns over the central portion, that is, from 3 to 9 GHz, of the passband.

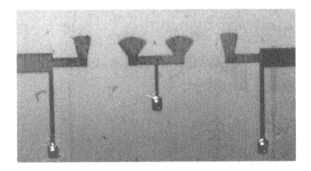

FIGURE 12.36 Photograph of fabricated eight-pole UWB bandpass filter.

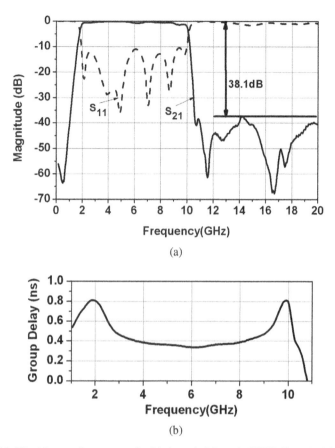

FIGURE 12.37 Measured responses for fabricated eight-pole UWB filter. (**a**) Magnitude; (**b**) group delay.

12.4 UWB FILTERS USING CASCADED MINIATURE HIGH- AND LOWPASS FILTERS

A UWB filter can also be formed by cascading high- and lowpass filters. This technique has a simple design procedure [75]. Unlike resonant-type bandpass filter, the high- and lowpass filters have small and flat group delay in their operation bands. Hence, by cascading them, small and flat group delay can be achieved for UWB bandpass filter. By properly choosing the cutoff frequencies of high- and lowpass filters, flexible passband for radar or communication systems can be achieved, such as 3.1–10.6 GHz for FCC-Defined UWB Indoor Limit and 2.0–18.0 GHz for a radar receiver. This approach also allows one to easily control the selectivity on each side of passband by independently adjusting the selectivity of high- and lowpass filters. A miniature UWB filter of this type is described in this section.

12.4.1 Miniature Wideband Highpass Filter

Figure 12.38 shows the 3D structure for a miniature highpass filter and its equivalent circuit. The filter structure has three metal layers, including a solid ground plane that is favorable for the system integration. To achieve a compact size, the middle layer shunt microstrip lines are short-circuited to the ground by sharing a common via. Broadside-coupled patches are used to implement series capacitors, whereas shunt short-circuited stubs are adopted to implement designed inductors. As a result, a desired highpass characteristic can be realized. Although only two stages are involved in this design, by cascading more highpass units, one can flexibly choose in- or out-of-plane input/output (I/O) ports in the design with improved selectivity.

To meet the lower stopband rejection level of FCC-defined UWB Indoor Limit, the 10-dB rejection level cutoff frequency is chosen as 3.1 GHz in the designing of highpass filter. The design can be started from the equivalent circuit, shown in Figure 12.38b, which includes two series capacitors, four shunt inductors, and two shunt capacitors. Because of the parasitic effects, in reality, this simplified lumped-element circuit model can not accurately model the structure shown in Figure 12.38a over an ultra-wide frequency band, for example, 1–20 GHz. Nevertheless, it is useful to explain the operation and to obtain the initial geometries for the filter.

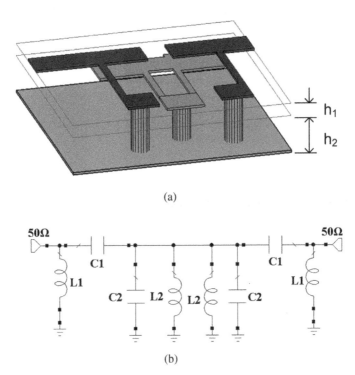

FIGURE 12.38 Miniature highpass filter. (**a**) 3D structure (two-stage with in-plane I/O ports). (**b**) Equivalent circuit ($C1 = 0.76$ pF, $C2 = 0.1$ pF, $L1 = 1.36$ nH, $L2 = 1.38$ nH).

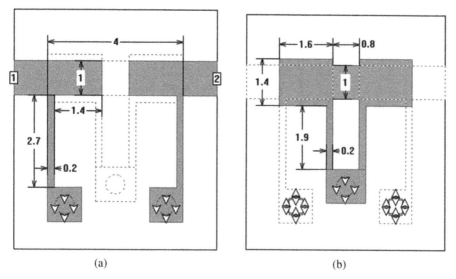

FIGURE 12.39 Layouts for designed two-stage highpass filter. (**a**) Top layer; (**b**) middle layer. All dimensions in millimeters.

By using a circuit optimization [78], a desired highpass response is obtained and the values of capacitors and inductors are listed in Figure 12.38. In the physical implementation, dielectric substrates with a relative dielectric constant of 3.15 and a loss tangent of 0.0025 are used. The substrate thicknesses, as referring to Figure 12.38a, are $h_1 = 0.1$ mm between the top and middle metal layers, and $h_2 = 0.3$ mm between the middle and ground metal layers, respectively. The corresponding microstrip components, such as the inductors and capacitors, can initially be determined based on the discussion in Chapter 4. A full-wave simulator [79] is used to finalize the physical implementation of the filter and the final layouts are shown Figure 12.39.

Figure 12.40 shows the full-wave simulated results compared with equivalent circuit response. As can be seen, the highpass filter, with a 10 dB cutoff at 3.1 GHz, has an ultra-wide 10 dB return loss bandwidth from 3.7 to 20.0 GHz. Note that the middle-layer copper has a thickness of 18 μm, which is comparable to the distance between top and middle layers. Hence, it has an impact on the performance of designed filter. Figure 12.40 also shows the full-wave simulation for the designed highpass filter with zero thickness copper, where a narrower operation bandwidth and worse return loss can be found. This indicates that the copper should be inserted into bonding films in the fabrication. The designed prototype also has a flat group delay, which is 0.05 ns at 14.0 GHz with a variation within 0.1 ns from 5.2 to 20.0 GHz.

The designed highpass filter has been fabricated by using multilayer liquid crystal polymer technology. To achieve designed thickness, three layers of 50-μm bonding films with a melting temperature of 290°C are used to bond four layers of liquid crystal polymer core films (three 50-and one 100-μm thick) together, which have a melting temperature of 315°C. Throughout the fabrication process, laser-drilled

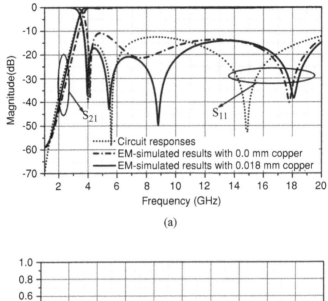

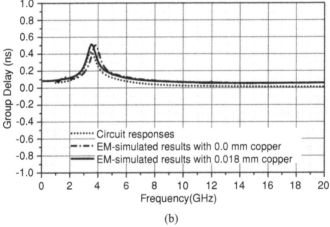

FIGURE 12.40 Full-wave simulation results and circuit responses of the two-stage highpass filter. (**a**) Sparameters; (**b**) group delay (ns).

alignment holes have been used to maintain the alignment between different layers. In the fabrication, pressure and temperature are carefully controlled in order to insert the copper into melting bonding films for obtaining a good return loss and to avoid squashing bonding films, which result in a changing of substrate thickness.

The experimental results are presented in Figure 12.41, where the small insert is a photograph of fabricated highpass filter. The fabricated filter is measured using HP8720 vector network analyzer (VNA). Anritsu 3680 universal test fixture is used to connect fabricated prototype with VNA and Thru-Reflect-Line calibration technique is adopted to remove the effects of the test fixture from the measurement. The measured filter has a 10-dB return loss bandwidth from 3.9 to 20.0 GHz with a 3-dB

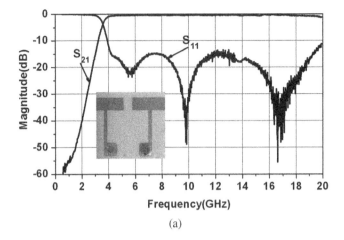

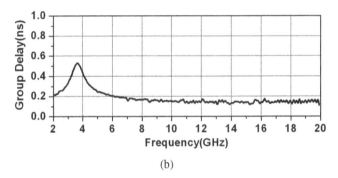

FIGURE 12.41 Experimental results for fabricated two-stage highpass filter (see insert). (a) Magnitude response; (b) group-delay response.

cutoff frequency of 3.6 GHz and a 10dB cutoff frequency of 3.24 GHz. From 3.9 to 19.90 GHz, the measured insertion loss is less than 1.0 dB. In Figure 12.41b, the measured result shows that the filter has a fairly small and flat group delay, which is 0.15 ns at 12.0 GHz with a variation within 0.05 ns from 6.0 to 20 GHz, over 87% of the its 10-dB return loss band.

12.4.2 Miniature Lowpass Filter

To achieve a small size for UWB bandpass filter, miniature lowpass filter has to be used in a cascaded high- and lowpass filter architecture. To this end, an open-stub lowpass filter is designed, as shown in Figure 12.42, which may be seen as a modification from that of an optimum bandstop filter, discussed in Chapter 6. However, the miniature lowpass filter allows an asymmetrical structure, in which all three stubs can have different characteristic impedances and electrical lengths. In addition, the connecting

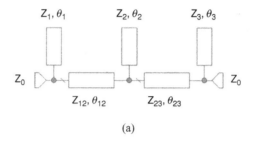

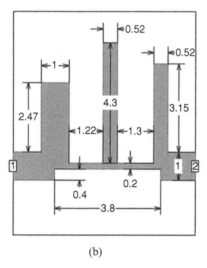

FIGURE 12.42 Miniature lowpass filter (a) Equivalent circuit ($Z_0 = 50\,\Omega$, $Z_1 = 59.93\,\Omega$, $Z_2 = 67.2\,\Omega$, $Z_3 = 68.34\,\Omega$, $Z_{12} = Z_{23} = 115\,\Omega$, $\theta_1 = 36.18°$, $\theta_2 = 51.86°$, $\theta_3 = 43.84°$, $\theta_{12} = \theta_{23} = 21.36°$, referenced at $f_0 = 6.85$ GHz.). (b) Layout (dimensions in millimeters) on a 0.4-mm thick dielectric substrate with a relative dielectric constant of 3.15.

lines between adjacent stubs are shorter. For this filter structure, three transmission zeros can be designed at desired frequencies by letting open stubs equal a quarter-wavelength at these frequencies. Since the sharper selectivity results in a larger variation of group delay, the first transmission zero is designed slightly far away from the cutoff frequency, that is, 12 GHz. The impedances of open stubs can be used to tune in-band match for a good return loss. With the goals of a small in-band return loss (say 15 dB), a 10-dB cutoff frequency of 10.6 GHz, the first transmission zero of 12 GHz, and a wide stopband with high rejection level of 20 dB, the lowpass filter circuit can be designed using a commercial available circuit simulator [78]. It can be shown that by tuning the lengths of open stubs, a wider stopband up to 30 GHz can be achieved at the cost of degraded rejection level. On the other side, by keeping the same cutoff frequency and rejection level, higher impedance, that is, Z_{12} and Z_{23}, of connecting transmission lines result in shorter electrical length θ_{12} and θ_{23}, leading

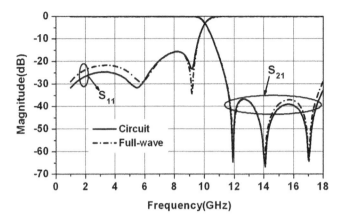

FIGURE 12.43 Circuit and full-wave simulation responses of miniature lowpass filter.

to a more compact filter. The final determined circuit parameters are given in the caption of Figure 12.42 and the circuit response is shown in Figure 12.43.

To be easily integrated with the designed highpass filter, the lowpass filter is implemented in microstrip on a 0.4-mm thick liquid crystal polymer substrate that has a dielectric constant of 3.15 and a loss tangent of 0.0025. Microstrip lines of 50 Ω impedance are adopted as input/output to keep consistent with that of the designed highpass filter's input/output. A full-wave simulator [79] is used to finalize the geometries for the lowpass filter; the layout is shown in Figure 12.42b. The filter has a very compact size of 4.56 × 4.9 mm (or about 0.19 × 0.2 λ_g, where λ_g is the guided wavelength at 6.85 GHz). The full-wave simulated response is also plotted in Figure 12.43, which agrees well with the circuit response.

12.4.3 Implementation of UWB Bandpass Filter

By cascading the high- and lowpass filters introduced above, a UWB bandpass filter has been developed. Generally, the integration of independently designed high- and lowpass filter may affect the return loss of resultant UWB bandpass filter because of interaction of phase at the joint. This phase effect can be compensated by a matched transmission-line section connecting the high- and lowpass filters. This is demonstrated in Figure 12.44. For this study, the highpass network is taken as that of Figure 12.38b and the lowpass network is that of Figure 12.42a. For a line characteristic impedance of $Z_0 = 50$ Ω, there exists an optimum electrical length θ_c for the inserted 50-Ω line that gives the best passband response of the resultant UWB bandpass filter. In this, case, it is found that $\theta_c = 67°$ at 6.85 GHz. For comparison, the responses for $\theta_c = 0°$ and $\theta_c = 67°$ are plotted in Figure 12.44b.

As can be seen, the length of the connecting line between the high- and lowpass filters is not very critical for the overall performance of the UWB bandpass filter. To achieve a smaller size, a short connecting line with a length of 0.3 mm is used to connect the high- and lowpass filters, without any tuning, for this UWB filter design.

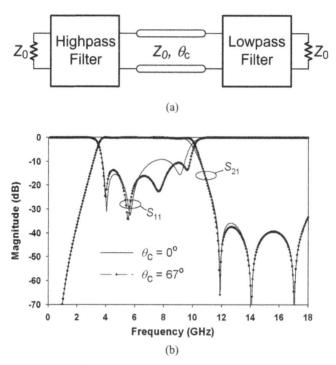

FIGURE 12.44 Theoretical study of the effect of connecting line between the high- and lowpass filters. (**a**) Circuit model; (**b**) magnitude response.

The insert in Figure 12.45 shows a photograph of the fabricated UWB filter, which has a size only amounting to 8.9 × 5.1 mm. Figure 12.45 also shows the measured results for the fabricated UWB bandpass filter, where FCC-defined UWB Indoor Limit is also plotted. It can be seen that this UWB bandpass filter can meet FCC-defined UWB Indoor Limit. The measured filter has a 3-dB insertion loss bandwidth from 3.5 to 10.0 GHz and a 10-dB insertion loss bandwidth from 3.14 to 10.6 GHz. From 3.9 to 9.3 GHz, small insertion loss is obtained, that is, 0.35 dB at 6.85 GHz with a variation within 0.15 dB. The measured UWB bandpass filter also exhibits good rejection level at upper stopband, which is higher than 29 dB from 11.5 to 16.0 GHz. Figure 12.45b shows the measured group delay, which is 0.18 ns at 6.85 GHz with a variation of 0.15 ns from 4.2 to 10.0 GHz, over 89% of its 3-dB bandwidth.

12.5 UWB FILTERS WITH NOTCH BAND(S)

For some practical applications, there is a need to introduce notch band(s) within a UWB passband to avoid the interference from existing wireless systems or to divide a UWB passband into a couple of subpassbands. Several techniques for the implementation of single or multiple notch bands are discussed in this section.

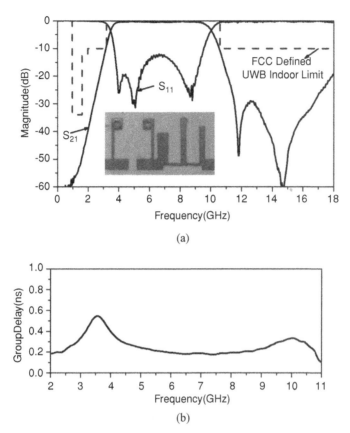

FIGURE 12.45 Experimental results for fabricated UWB bandpass filter, where the small insert shows a photograph of the fabricated filter. (**a**) Magnitude response; (**b**) group-delay response.

12.5.1 UWB Filters with Embedded Band Notch Stubs

In order to introduce a narrow notched band, let us first examine three different microstrip structures in Figure 12.46. These are a transmission line loaded with a conventional open-circuited stub, a spur line, and a transmission line with an embedded open-circuited stub. W_0 denotes the width of the main transmission line; W and L are the width and length of stub line, respectively, and G represents a gap. λ_g denotes the guided wavelength in the microstrip. In principle, to achieve a narrow notch the characteristic impedance of the conventional open-circuited stub will become extremely high, which may be difficult to fabricate. Alternatively, a spur-line or an embedded open-circuited stub [42] may be utilized.

For the investigation, assume that all the three structures are implemented on a microstrip substrate with a relative dielectric constant of 3.05 and a thickness of 0.508 mm. The main transmission line connecting to the I/O ports are the same for

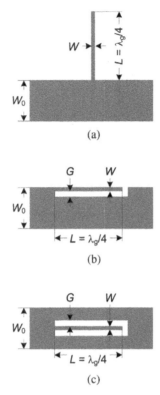

FIGURE 12.46 Schematic diagrams of (**a**) conventional open-circuited stub, (**b**) spur line, and (**c**) embedded open-circuited stub.

the three cases and has a width $W_0 = 1.3$ mm, which corresponds to a 50-Ω line on the substrate. The conventional quarter-wavelength ($\lambda_g/4$) microstrip open-circuited stub has a width $W = 0.1$ mm. The spur-line and embedded open-circuited stub have the same width $W = 0.1$ mm, as well as the same gap $G = 0.2$ mm. Full-wave EM simulations are carried out to obtain the transmission characteristics of these three structures, which are plotted in Figure 12.47 with frequencies normalized by the center frequency of notch band. As can be seen, the structure with the conventional open-circuited stub has a very wide notch bandwidth. On the contrary, the transmission line with embedded open-circuit stub results in a smallest notch. In this case, to achieve the same bandwidth of a notch produced by the embedded stub, the conventional quarter-wavelength open-circuited stub would need a high impedance of about 700 Ω, which is difficult to implement in practice. Therefore, the embedded open-circuited open stub in Figure 12.46c is favorable for implementing extremely narrow notched bands.

The narrowband characteristic of the embedded open-circuited stub is attributed primarily to its coupling to the main transmission line, and the notch bandwidth can easily be controlled by adjusting W and G. To demonstrate this, Figure 12.48 illustrates the simulated performance of the embedded stub with varying width and

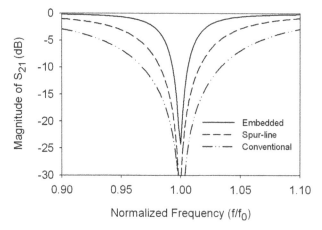

FIGURE 12.47 Simulated frequency responses for the three structures in Figure 12.46 with $W_0 = 1.3$ mm, $W = 0.1$ mm, and $G = 0.2$ mm on a microstrip substrate with a relative dielectric constant of 3.05 and a thickness of 0.508 mm.

gap. In all the cases, the width of main transmission line is fixed for $W_0 = 1.3$ mm. It can be seen that decreasing the gap or width reduces the bandwidth. This technique allows us to realize a narrow notch that would not be possible with a conventional open-circuited stub requiring extremely high impedance.

The achievement of extremely narrow bandwidth is one advantage of this technique. In addition, since the majority of the current flows around the edges of the microstrip line, the implementation of this embedded stub technique has less of an effect on the main transmission-line performance than the spur line over a wide frequency range. Hence, the embedded open-circuited stub structure is found to be

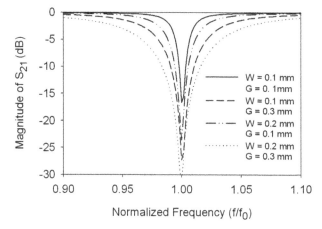

FIGURE 12.48 Notch characteristic of the embedded stub with varying width (W) and gap (G) for $W_0 = 1.3$ mm.

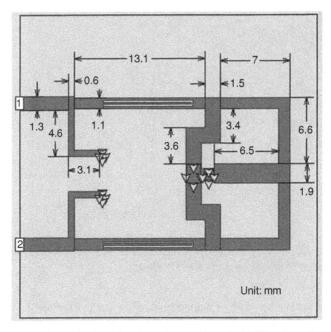

FIGURE 12.49 Microstrip UWB bandpass filter with embedded band notch stubs. The arrowed areas indicate the via-hole groundings. All the dimensions are in millimeters on a dielectric substrate with a relative dielectric constant of 3.05 and a thickness of 0.508 mm.

the best choice among the three types of structures that have been investigated for implementing a notch within the passband of a UWB bandpass filter.

Figure 12.49 demonstrates a microstrip UWB bandpass filter with embedded band notch stubs, which is realized on a microstrip substrate with a relative dielectric constant of 3.05 and a thickness of 0.508 mm. As a matter of fact, this UWB filter is based on a design described in Section 12.1.1, which consists of five short-circuited stubs and four nonredundant connecting lines. The filter is then folded to have a compact size. It has two embedded open stubs in the first and last connecting lines in order to introduce a very narrow notched (rejection) band in the UWB passband. The length of each of the two embedded open stubs should be $\lambda_g/4$ at the desired center frequency of the notched band to ensure that the second resonant harmonic of the embedded stub does not appear in the desired UWB passband.

Figure 12.50a is the photograph of an experimental filter of Figure 12.49. For this filter, its two embedded open-circuited stubs have a width $W = 0.2$ mm, and a gap $G = 0.2$ mm. Since the implementation of embedded stubs does not cause any increase in the filter size, the fabricated filter occupies a size of 22.2×15.1 mm on the substrate used. For the measurement, a microstrip feed line of 5-mm long is added at both the input and the output. Figure 12.50b illustrates the measured results, which show a good UWB bandpass performance with $FBW = 110\%$ at a midband frequency of 6.85 GHz and a designed narrow notched band centered at 5.8 GHz with

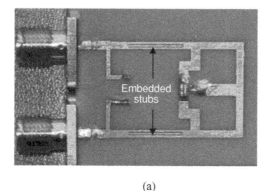

FIGURE 12.50 (a) Fabricated microstrip UWB bandpass filter with embedded band notch stubs. (b) Measured results.

a 10-dB fractional bandwidth of 6.5%. The attenuation at the center of two separated bands is more than 34 dB. At the midband frequency of each band, insertion loss of less than 0.7 dB was obtained. The filter also showed a flat group delay of about 0.5 ns at the midband frequency of each band.

Using the technique of embedded open stubs, it is also possible to produce two different notched bands when desired. A simple way is to allow the two open-circuited stubs in Figure 12.49 to have different lengths, which, however, may limit the rejection level of the notched bands. Alternatively, another pair of open-circuited stubs can be embedded in another two connecting lines, as depicted in Figure 12.51a. It should be noted that, as compared to that in Figure 12.49, the dimensions of the resultant filter are not changed except for the embedded stubs. In this case, for the first pair of the embedded open stubs, $W = 0.2$ mm, $G = 0.15$ mm, and $L = 9.2$ mm and for the second pair of the embedded open stubs, $W = 0.2$ mm, $G = 0.1$ mm,

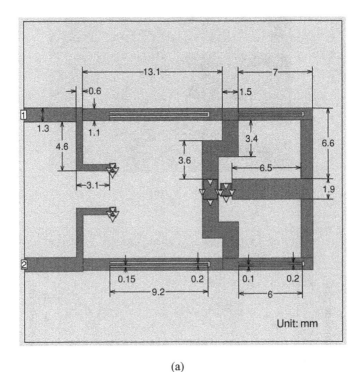

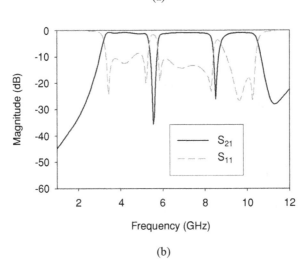

FIGURE 12.51 (a) Microstrip UWB bandpass filter with two notched bands on a dielectric substrate with a relative dielectric constant of 3.05 and a thickness of 0.508 mm. The arrowed areas indicate the via-hole groundings. (b) Full-wave EM-simulated performance.

and $L = 6$ mm. These changes will generate two desired notched bands without any significant influence on the wide performance of the filter, which is demonstrated in Figure 12.51b with the full-wave simulated results.

12.5.2 Notch Implementation Using Interdigital Coupled Lines

Another way to design UWB bandpass filters with notched bands is by using asymmetrical three parallel-coupled lines that can provide a tight coupling over a wide frequency band and, at the same time, generate a single narrow notch within the band [43]. Figure 12.52 illustrates a configuration of three coupled lines, introduced for this purpose. This is a two-port structure, where Z_0 denotes the terminal line characteristic impedance. There are three parallel-coupled lines in an interdigital form, which can enhance the coupling between the two ports, similar to the interdigital capacitor discussed in Chapter 4. All the three coupled lines have the same length of L, and their widths are denoted by W_1, W, and W_2, respectively. The spacings between the adjacent coupled lines are S_1 and S_2. The outer two lines, line 1 and line 2, are loaded with open stubs with the lengths and widths indicated by L_{s1}, L_{s2}, W_{s1}, and W_{s2}, respectively.

For an initial study, consider following cases: (1) symmetrical; (2) asymmetrical ($L_{s1} \neq L_{s2}$); (3) asymmetrical ($S_1 \neq S_2$); (4) asymmetrical ($W_1 \neq W_2$).

Figure 12.53 shows the transmission characteristics for these cases, obtained by full-wave EM simulations. One can find the values of the parameters for all these cases in the figure caption. It is evident that the symmetrical structure has good transmission over a wide frequency range without incurring any notch. However, for all the three asymmetrical cases, a notched band is produced.

By examining the phase responses as well as the field distributions for these cases, it can be shown that the symmetrical coupled-line structure does not resonate, whereas each of the asymmetrical coupled-line structures resonates at the notch frequency with

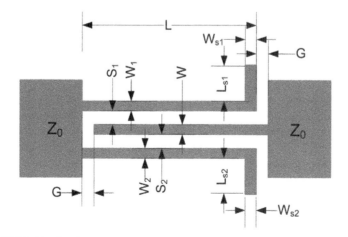

FIGURE 12.52 Configuration of three coupled line section with loading stubs.

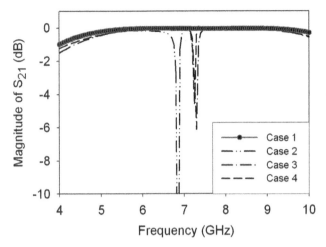

FIGURE 12.53 Full-wave EM simulated responses of the structure in Figure 12.52 on a microstrip substrate with a relative dielectric constant of 10.8 and a thickness of 1.27 mm, and a terminal line width $W_0 = 1.1$ mm for $Z_0 = 50\,\Omega$. For case 1: $W_1 = W_2 = W = 0.1$ mm, $S_1 = S_2 = 0.1$ mm, and $L_{s1} = L_{s2} = 0$. For case 2: $W_1 = W_2 = W = 0.1$ mm, $S_1 = S_2 = 0.1$ mm, $L_{s1} = 0.6$ mm, $W_{s1} = 0.25$ mm, and $L_{s2} = 0$. For case 3: $W_1 = W_2 = W = 0.1$ mm, $S_1 = 0.2$ mm, $S_2 = 0.1$ mm, and $L_{s1} = L_{s2} = 0$. For case 4: $W_1 = 0.3$ mm, $W_2 = W = 0.1$ mm, $S_1 = S_2 = 0.1$ mm, and $L_{s1} = L_{s2} = 0$. For all the cases: $L = 4$ mm and $G = 0.25$ mm.

a much higher current density and a phase transit. Under the asymmetrical condition, this resonance occurs along the outer lines when the total path length $2L + L_{s1} + L_{s2}$ is about half-guided wavelength at the center frequency of the notched band. This leads to a short circuit at the input port and no coupling exists with the central line at the notch frequency. This indicates that an asymmetrical structure of Figure 12.52 not only exhibits an enhanced degree of coupling, but it also generates a notch at a desired frequency. Furthermore, it appears that the frequency and bandwidth of notched band can conveniently be controlled by the parameters of the two added open stubs, which is discussed next.

Figure 12.54a displays the full-wave EM-simulated transmission property of the asymmetrical structure for varying L_{s1}, while the other parameters are kept the same as those in the case (2) discussed above. As expected, increasing L_{s1} moves the notch to lower frequencies. However, the bandwidth increases because of the increase in the degree of the asymmetric loading stubs. Figure 12.54b shows another case with varying W_{s1} for $L_{s1} = 1.0$ mm. As can be seen, increasing W_{s1} also shifts the notch frequency down because of an increase in the equivalent loading capacitance from the open-circuit stub, which leads to an increase in the bandwidth as well.

The above discussed two cases assuming $L_{s2} = 0$. Although using only a single loaded open stub may meet most applications, adding the second open stub with $L_{s2} \neq 0$, as a matter of fact, allows more flexibly in controlling both the frequency and the bandwidth of notched band. The controlling of the notch frequency is

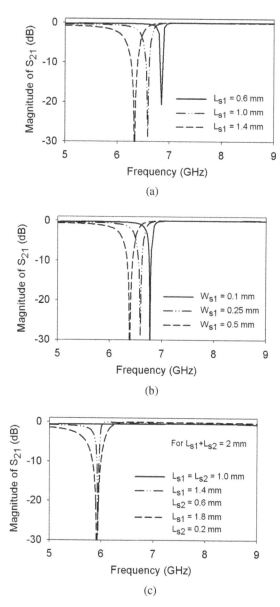

FIGURE 12.54 Transmission property of asymmetric structure for varying open stub parameters. (a) For varying L_{s1} ($W_1 = W_2 = W = 0.1$ mm, $S_1 = S_2 = 0.1$ mm, $W_{s1} = 0.25$ mm, and $L_{s2} = 0$). (b) For varying W_{s1} ($W_1 = W_2 = W = 0.1$ mm, $S_1 = S_2 = 0.1$ mm, $L_{s1} = 1.0$ mm, and $L_{s2} = 0$). (c) For different combinations of L_{s1} and L_{s2} with $L_{s1} + L_{s2} = 2$ mm ($W_1 = W_2 = W = 0.1$ mm, $S_1 = S_2 = 0.1$ mm, $W_{s1} = W_{s2} = 0.25$ mm).

obvious, as this resonant frequency is proportional to the length of $2L + L_{s1} + L_{s2}$. The bandwidth of the notched band can be easily controlled by adjusting the difference of lengths between the two stubs ($\Delta \ell = |L_{s1} - L_{s2}|$) where the bandwidth is proportional to $\Delta \ell$ as demonstrated in Figure 12.54c. For example, when $\Delta \ell = 0$, no notch exists, which corresponds to a symmetrical case. When increasing $\Delta \ell$, the bandwidth of the notch increases effectively, but the notch frequency is hardly changed as $L_{s1} + L_{s2} =$ constant. Therefore, in order to shift the notch band to a desired frequency without changing the bandwidth of the notch, $\Delta \ell$ can be kept constant for the determined bandwidth.

The asymmetrical three-coupled-line structure discussed above can easily be applied to a UWB bandpass filter, such as the one based on multimode resonators described in Section 12.2.4. Thus, the filter can exhibit an extremely narrow notched band in the UWB passband. Figure 12.55a illustrates the layout for such a filter,

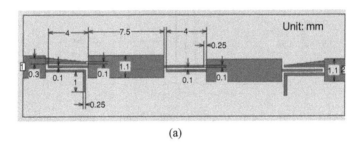

(a)

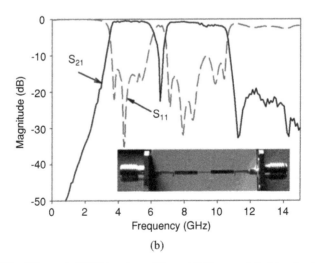

(b)

FIGURE 12.55 Microstrip UWB bandpass filter exhibiting a notched band using asymmetric three-coupled lines. (**a**) Layout of the filter on a dielectric substrate with a relative dielectric constant of 10.8 and a thickness of 1.27 mm. (**b**) Experimental results.

designed on a microstrip substrate with a relative dielectric constant of 10.8 and a thickness of 1.27 mm. The filter is comprised of two multimode stepped-impedance resonators. The input and the output sections use the asymmetrical three-coupled lines with a single-loaded open stub, which has a length of 1.0 mm and a width of 0.25 mm, as shown in the figure. The top outer lines of the I/O coupled lines are tapered to enhance the filter performance. The experimental results of the filter are shown in Figure 12.55b, where the small insert shows the fabricated filter with two SMA connectors for the measurement. As can be seen, a notched band is implemented, which is centered at 6.5 GHz with a measured rejection of greater than 22 dB. The measured 3-dB bandwidth of the filter, without the notch, is from 3.5 to 10.5 GHz. Thus, the two subbands, separated by the notch, have the measured 3-dB bandwidths from 3.5 to 6.1 GHz and from 6.85 to 10.5 GHz, respectively.

In order to implement more than one notched band, one may consider another two-port wideband coupling structure based on four or five interdigital coupled lines. The structure of Figure 12.56a consists of four interdigital coupled lines, which may also be seen to consist of a pair of coupled pseudointerdigital resonators (see Chapter 8), which resonate when the paths L_{r1} and L_{r2} amount to a half-guided wavelength at their fundamental resonant frequencies. Although the interdigital structure can provide a strong coupling, between the two terminal lines with a characteristic impedance of Z_0, over a wide frequency range, the structure inherently exhibits two notched bands resulting from the pair of coupled resonators. This is demonstrated with the EM-simulated results plotted in Figure 12.56b.

In the simulation, the terminal line has a width of 1.1 mm on a 1.27-mm thick microstrip substrate with a relative dielectric constant of 10.8. The widths for all the interdigital lines stay the same, namely, $W_{r1} = W_{r2} = 0.1$ mm. For case (1), $L_{r1} = L_{r2} = 8.4$ mm, $S_1 = 0.1$ mm, and $S_2 = 0.1$ mm. In this instance, two notched bands are centered at 6.38 and 8.3 GHz, respectively. In fact, for $L_{r1} = L_{r2}$, these two frequencies result from a modal split because of the coupling between the two identical pseudointerdigital resonators, which is similar to that for the synchronously tuned coupled resonators (see Chapter 7).

To close these two notch frequencies, we can reduce the coupling between the two resonators by increasing S_1, as discussed in Chapter 8 for the pseudointerdigital filter; in the meantime, a small S_2 maintains a tight coupling between the two ports. This is shown for case (2), when $L_{r1} = L_{r2} = 8.45$ mm, $S_1 = 0.15$ mm, and $S_2 = 0.1$ mm. As can be seen from Figure 12.56b, the two notched bands are now closer, centered at 6.64 and 7.96 GHz, respectively. Reducing the coupling also increases the notch-band bandwidths.

In order to be more flexible to control two notch frequencies, we can allow $L_{r1} \neq L_{r2}$, namely, the two pseudointerdigital resonators to resonate at different frequencies. This is shown in Figure 12.56b for case (3), when $L_{r1} = 8.45$ mm, $L_{r2} = 7.45$ mm, $S_1 = 0.15$ mm, and $S_2 = 0.1$ mm. As compared to the case (2), only with L_{r2} being changed from 8.45 to 7.45 mm, can one see that both notches are shifted to higher frequencies, but at different rates. The separation of the two notched bands, centered at 6.92 and 8.64 GHz, respectively, is also wider because of the different resonant frequencies of the coupled pseudointerdigital resonators. For

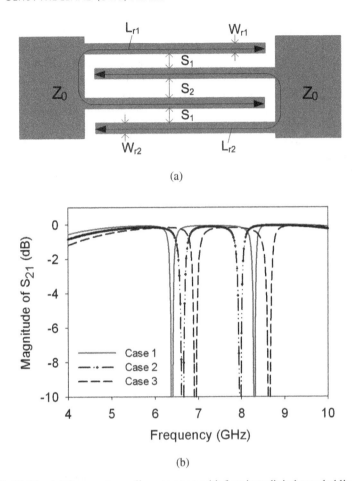

FIGURE 12.56 (a) Two-port coupling structure with four interdigital coupled lines. (b) Double notches generated by the structure simulated on a 1.27-mm thick microstrip substrate with a relative dielectric constant of 10.8. The 50-Ω terminal line has a width of 1.1 mm. $W_{r1} = W_{r2} = 0.1$ mm. For case 1, $L_{r1} = L_{r2} = 8.4$ mm, $S_1 = 0.1$ mm, $S_2 = 0.1$ mm. For case 2, $L_{r1} = L_{r2} = 8.45$ mm, $S_1 = 0.15$ mm, $S_2 = 0.1$ mm. For case 3, $L_{r1} = 8.45$ mm, $L_{r2} = 7.45$ mm, $S_1 = 0.15$ mm, $S_2 = 0.1$ mm.

the given spacings, asynchronously tuned resonators also reduce the coupling and, hence, result in an increase in the notch-band bandwidths as well.

Shown in Figure 12.57a is another two-port wideband coupling structure based on five interdigital coupled lines. The lengths of these coupled lines are denoted by $L_1, L_2 \ldots L_5$, and their spacings are indicated by $S_1, S_2 \ldots S_4$. Assume an equal line width for our discussion, although not necessarily for a practical implementation. This coupling structure is capable of producing three notched bands within its wide transmission band. When the structure is symmetrical, it exhibits only one notched band. To generate three notched bands, the structure has to be asymmetrical,

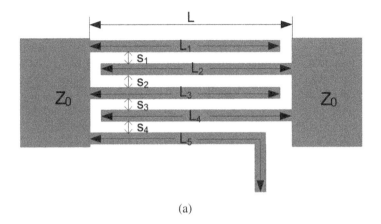

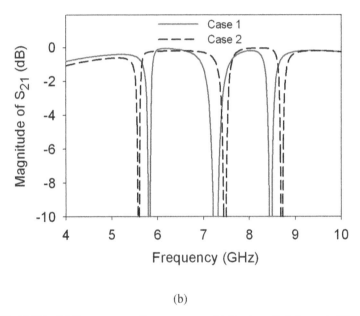

FIGURE 12.57 (a) Two-port coupling structure with five interdigital coupled lines. (b) Notched band characteristics of the structure simulated on a 1.27-mm thick microstrip substrate with a relative dielectric constant of 10.8. The 50-Ω terminal line has a width of 1.1 mm. For all cases, the interdigital lines have a width of 0.1 mm, $L = 4.25$ mm, $L_1 = 3.5$ mm, $L_5 = 5$ mm (including a 1 mm long open stub), $L_2 = L_4 = 4$ mm. For case 1, $L_3 = 4.15$ mm, $S_1 = S_4 = 0.2$ mm, $S_2 = S_3 = 0.1$ mm. For case 2, $L_3 = 3.85$ mm, $S_1 = S_4 = 0.15$ mm, $S_2 = S_3 = 0.15$ mm.

particularly $L_1 \neq L_5$. The difference between the two outer lines L_1 and L_5 defines an asymmetrical degree, which affects the bandwidths of notched bands, similar to that discussed above for the asymmetrical three-coupled lines. Varying the spacings S_i ($i = 1$ to 4) not only changes the coupling, but also affects the characteristics of the notched bands. Figure 12.57b shows the two typical cases. In general, although any parameter of the structure has an effect on the overall frequency response of the coupling structure, including the notched band frequencies and bandwidths, some parameter can be more effective in changing a particular characteristic. For example, it can be shown that changing L_3 is more effective in shifting the middle notched band. Since so many parameters, including interdigital line widths, are involved in this coupling structure, it is envisaged that a design for three desired notched bands needs to perform a series of parametric studies in light of full-wave EM simulation.

To demonstrate applications of the above discussed two-port coupling structures with multiple interdigital coupled line, Figure 12.58 illustrates a UWB filter of this type, whose input and output sections use a coupling structure that is based on four interdigital coupled lines. The middle portion of the filter consists of a high-impedance transmission section loaded with open stubs, which functions like a lowpass filtering unit to improve the high-side selectivity, as well as the upper stopband performance of the UWB filter. As expected, according to the above discussion, the filter exhibits two notched bands, which can clearly be seen from its EM-simulated performance, shown in Figure 12.58b.

Although the above example has shown what can be achieved with multiple interdigital coupled lines, in general, it is nontrivial to control the characteristics, namely the bandwidths and frequencies of notched bands. This is because these characteristics depend on the coupling of interdigital coupled lines, which also strongly affects the main passband response. Hence, there are some limitations in the design of this type of UWB filter with multiple notched bands. A more flexible approach for controlling notched band characteristics is introduced next.

12.5.3 UWB Filters with Notched Bands Using Vertically Coupled Resonators

In principle, to introduce a single-notch or multiple-notched bands into the passband of a UWB bandpass filter, narrowband bandstop filter(s), as discussed in Chapter 6, may be designed separately and then cascaded with the UWB filter. However, this approach will lead to a large size for the resultant UWB filter having notched band(s). Alternatively, an integrated design approach can be employed, which can not only make the filter compact, but also make the function implementation flexible, although it may somehow complicate the design because a multilayer structure [67]. This approach is explained with a design example shown in Figure 12.59.

The filtering structure of Figure 12.59 consists of four metal layers including the bottom one used as the common ground plane. The first and second layers, separated with a substrate thickness of h_1, accommodate the main signal track for the UWB

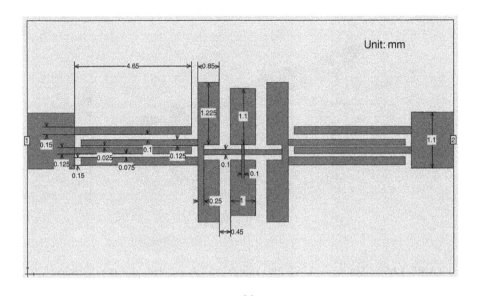

(a)

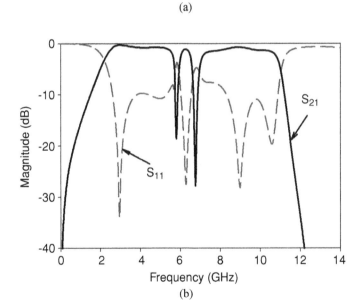

(b)

FIGURE 12.58 Microstrip UWB bandpass filter exhibiting two notched bands using four interdigital coupled lines. (**a**) Layout of the filter on a dielectric substrate with a relative dielectric constant of 10.8 and a thickness of 1.27 mm. (**b**) EM-simulated performance.

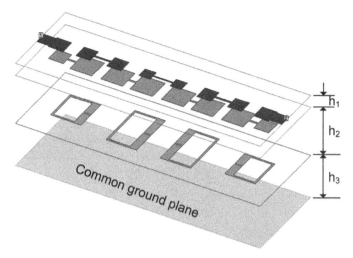

FIGURE 12.59 Multilayer UWB filter using vertically coupled resonators to realize notched bands.

bandpass filter, which is similar to the design shown in Figure 12.17a, based on coupled stepped-impedance resonators. For the implementation of notched bands, a third metal layer is inserted between the main UWB bandpass filter and its ground. On the third metal layer, as shown in Figure 12.59, there is an array of slow-wave open-loop resonators (see Chapter 10), which are vertically coupled to the main signal track above it. These resonators are designed to have their fundamental resonant frequencies centered at the desired notched bands. Although other resonators may be deployed, here, the reason to use the slow-wave, open-loop resonator is twofold: (i) to make the resonator size small, and (ii) to move its second resonant frequency away from the desired wide passband. The bandwidths of the notched bands can be controlled in the design by adjusting the substrate thicknesses of h_2 and h_3, while keeping $h_2 + h_3 =$ constant to maintain the UWB bandpass response. This is demonstrated with three design cases, which have the exact same layouts as shown in Figure 12.60, but different combinations of h_2 and h_3.

For the design, an initial UWB bandpass filter without notch can be designed first, based on the method discussed above in Section 12.2.3. After that, the third circuit layer with slow-wave open-loop resonators are inserted and full-wave EM simulation is used to finalize the design. The final layouts, shown in Figure 12.60, are symmetrical with respect to the input and the output ports for double notched bands. The performances of these three design cases, obtained using full-wave EM simulation, are shown in Figure 12.61. As can be seen, case 1 has the largest bandwidths for the two implemented notched bands because it has the smallest of h_2 and the largest of h_3 among the three designs. Without changing the layouts, all the three designs, with different h_2 and h_3, maintain good return loss over three passbands divided by the two notched bands. This shows that h_2 and h_3 can be adjusted jointly in design to

UWB FILTERS WITH NOTCH BAND(S) 553

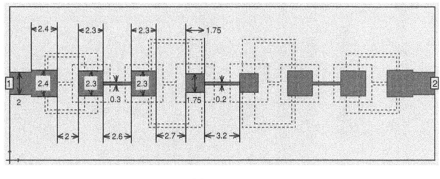

(a)

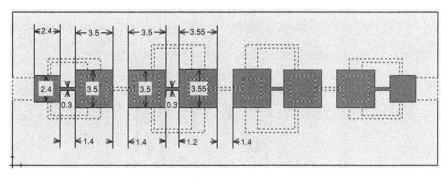

(b)

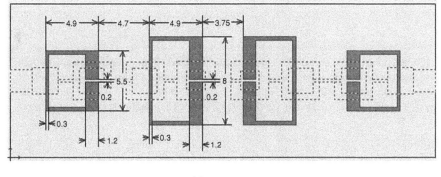

(c)

FIGURE 12.60 Layouts for multilayer UWB filter using vertically coupled resonators on the third layer for notched band implementation. (**a**) Top layer; (**b**) second layer; (**c**) third layer (all dimensions are in millimeters on dielectric substrates with a relative dielectric constant of 3.15 with thicknesses of $h_1 = 0.05$ mm and $h_2 + h_3 = 0.75$ mm).

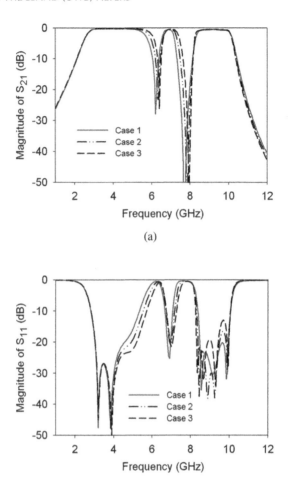

FIGURE 12.61 Full-wave EM-simulated performances of three design cases with the same layouts of Figure 12.60, but the different combinations of h_2 and h_3. For case 1: $h_2 = 0.4$ mm and $h_3 = 0.35$ mm. For case 2: $h_2 = 0.45$ mm and $h_3 = 0.3$ mm. For case 3: $h_2 = 0.5$ mm and $h_3 = 0.25$ mm. (a) Magnitude of S_{21}. (b) Magnitude of S_{11}.

vary the notch bandwidths effectively with a minimum effort to modify the design. The slight shift in notch frequencies because of the change of h_2 and h_3 can easily be compensated by tuning the gap of the open-loop resonators.

For an experimental demonstration, the designed case 2 filter is fabricated using multilayer liquid crystal polymer (LCP) lamination technology. Three types of LCP films are used in the fabrication, which are 100- and 50-μm thick LCP core film and 50-μm thick LCP bonding film. Shown in Figure 12.62 are the fabricated filter and the schematic of layered LCP films in the fabrication. Eleven layers LCP films

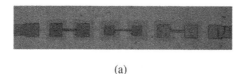

(a)

50 µm Core Film	18 µm Copper
	18 µm Copper
50 µm Bonding Film	
100 µm Core Film	
50 µm Bonding Film	
100 µm Core Film	
50 µm Bonding Film	
100 µm Core Film	
	18 µm Copper
50 µm Bonding Film	
100 µm Core Film	
50 µm Bonding Film	
100 µm Core Film	
	18 µm Copper

(b)

FIGURE 12.62 (a) Photograph of fabricated multilayer UWB filter with notched bands. (b) Schematic of layered LCP films used in the fabrication.

are used in the implementation. Bonding films are inserted between core films and bonded together in fabrication.

The fabricated filter is measured using an HP8720 vector network analyzer and the results are plotted in Figure 12.63. Two notched bands are achieved, which have center frequencies at 6.4 and 8.0 GHz and 3-dB bandwidths of 9.5% and 13.4%, respectively. The first notched band has a rejection level of 26.4 dB at 6.4 GHz; the second notched band has a rejection level of 43.7 dB at 8.0 GHz. The measured filter has three passbands with center frequencies at 4.28, 7.08, and 9.53 GHz and 3-dB bandwidths of 78.5%, 11.5% and 21.2%, respectively. The insertion losses at center frequencies of passbands are 0.16, 0.54, and 0.75 dB. The measured return losses are higher than 15 dB over the passbands. The fabricated filter exhibits a wide upper stopband with good rejection level, which is higher than 30 dB from 11.4 to 18.0 GHz. Figure 12.63 also shows the measured group delays, which are 0.4 ns at 4.28 GHz with a variation within 0.14 ns for the first passband, 0.75 ns at 7.08 GHz with a variation within 0.15 ns for the second passband, and 0.54 ns at 9.53 GHz with a variation within 0.4 ns for the third passband, respectively.

To demonstrate the flexibility of this multilayer filtering structure further, replacing the circuit on the third layer in Figure 12.60c with the one shown in Figure 12.64a,

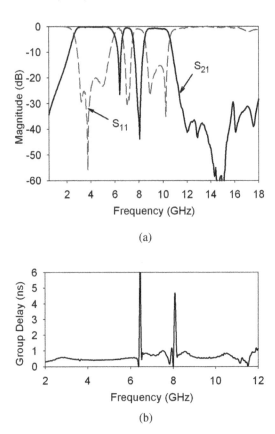

FIGURE 12.63 Measured responses for the fabricated multilayer UWB filter with notched bands. (**a**) Magnitude; (**b**) group delay.

we obtain, without any changes in the top and second layers, another filtering characteristic depicted in Figure 12.64b. As compared with the responses shown in Figure 12.61, not only the center frequencies of the two notched bands can shift, but their bandwidths also become narrow. These changes in the filtering characteristic are all attributed to the meander, slow-wave, open-loop resonators implemented. As can be seen from Figure 12.64a, the sizes of the meandered resonators are smaller than their counterparts in Figure 12.60c. It is this miniaturization that reduces the radiations of the resonators, as the fields are more confined within the substrate with the thickness of h_3; hence, their couplings to the main signal track become weak, resulting in the narrower notched bands. As a matter of fact, to be more flexible to control the bandwidths of multiple notched bands, in addition to the reshaping of the resonators, as demonstrated, the resonators themselves may also be placed on different circuit layers, although in any case, the main UWB bandpass filter design on the top two layers may or may not need to be modified to obtain a satisfactory performance.

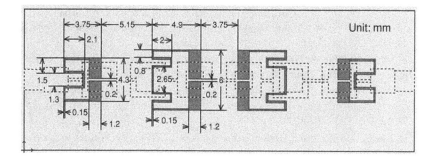

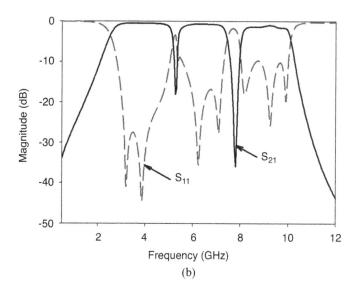

FIGURE 12.64 Modified design with meander slow-wave open-loop resonators. (a) Layout to replace the third layer of Figure 12.60c; (b) full-wave EM-simulated performances of the design with $h_2 = 0.45$ mm and $h_3 = 0.3$ mm.

REFERENCES

[1] Federal Communications Commission (FCC), Revision of part 15 of the commission's rules regarding ultra-wideband transmission system. FCC, Washington, Tech. Rep. ET-Docket 98-153 FCC02-48, Apr. 2002.

[2] Mini-Special Issue on Ultra-Wideband, *IEEE Trans Microwave Theory Tech* **52**, 2004.

[3] L. Zhu, H. Bu, and K. Wu, Aperture compensation technique for innovative design of ultra-broadband microstrip bandpass filter, *IEEE MTT-S Int. Dig.* **1**, 2000, 315–318.

[4] L. Zhu, H. Bu, and K. Wu, Broadband and compact multi-mode microstrip bandpass filters using ground plane aperture technique, *IEE Proc. Microwave Antennas Propagation* **149**(1), 2002, 71–77.

[5] W. Menzel, L. Zhu, K. Wu, and F. Bogelsack, On the design of novel compact broad-band planar filters, *IEEE Trans. Microwave Theory Tech.* **51**(2), 2003, 364–370.

[6] A. Saito, H. Harada, and A. Nishikata, Development of bandpass filter for ultra wideband (UWB) communication, In *Proceedings of IEEE Conference In Ultra Wideband System Technology* 2003, 76–80.

[7] H. Ishida, and K. Araki, Design and analysis of UWB bandpass filter with ring filter, *IEEE MTT-S Microwave Symp. Dig.* 2004, 1307–1310.

[8] S.-G. Kim, and K. Chang, Ultrawide-band transitions and new microwave components using double-sided parallel-strip lines, *IEEE Trans. Microwave Theory Tech.* **52**, 2004, 2148–2152.

[9] J.-S. Hong, and H. N. Shaman, An optimum ultra-wideband microstrip filter, *Microwave Opt. Technol. Lett.* **47**, 2005, 230–233.

[10] W. Menzel, M. S. Rahman, and L. Zhu, Low-loss ultra-wideband (UWB) filters using suspended stripline, In *Asia-Pacific Microwave Conference, Suzhou*, 2005, 2148–2151.

[11] C.-L. Hsu, F.-C. Hsu, and J.-T. Kuo, Microstrip bandpass filters for ultra-wideband (UWB) wireless communications, *IEEE MTT-S Microwave Symp. Dig.* 2005, 679–682.

[12] L. Zhu, S. Sun, and W. Menzel, Ultra-wideband (UWB) bandpass filters using multiple-mode resonator, *IEEE Microwave Wireless Components Lett.* **15**, 2005, 796–798.

[13] Y.-C. Chiou, J.-T. Kuo, and E. Cheng, Broadband quasi-Chebyshev bandpass filters with multimode stepped-impedance resonators (SIRs), *IEEE Trans. Microwave Theory Tech.* **54**(8), 2006, 3352–3358.

[14] H., Wang, L. Zhu, and W. Menzel, Ultra-wideband bandpass filter with hybrid microstrip/CPW structure, *IEEE Microwave Wireless Components Lett.* **15**, 2005, 844–846.

[15] K. Li, D. Kurita, and T. Matsui, An ultra-wideband bandpass filter using broadside-coupled microstrip-coplanar waveguide structure, *IEEE MTT-S Microwave Symp. Dig.* 2005, 675–678.

[16] L. Zhu, and H. Wang, Ultra-wideband bandpass filter on aperture-backed microstrip line, *Electron. Lett.* **41**(18), 2005, 1015–1016.

[17] H. N. Shaman, and J.-S. Hong, A compact ultra-wideband (UWB) bandpass filter with transmission zero, In *Proceedings of the 36th European Microwave Conference*, 2006, 603–605.

[18] H. Shaman, and J.-S. Hong, An optimum ultra-wideband (UWB) bandpass filter with spurious response suppression, In *Proceedings of the IEEE Wimcon*, FL, Dec. 2006, 1–5.

[19] K. Li and J.-S. Hong, Modeling of an ultra-wideband bandpass filtering structure, In *Proceeding of Asia-Pacific Microwave Conference*, APMC2006, Japan, WE1B-4, 12–15 December, Vol. 1, 2006, 37–40.

[20] Sun, S., and L. Zhu, Capacitive-ended interdigital coupled lines for UWB bandpass filters with improved out-of-band performances, *IEEE Microwave Wireless Components Lett.* **16**, 2006, 440–442.

[21] R. Li, and L. Zhu, Compact UWB bandpass filter using stub loaded multiple-mode resonator, *IEEE Microwave Wireless Components Lett.* **16**(8), 2006, 440–442.

[22] H. L. Hu, X. D. Huang, and C. H. Cheng, Ultra-wideband bandpass filter using CPW-microstrip coupling structure, *Electronics Lett.* **42**, 2006, 586–587.

[23] T.-N. Kuo, S.-C. Lin, and C. H. Chen, Compact ultra-wideband bandpass filters using composite microstrip–coplanar-waveguide structure, *IEEE Trans. Microwave Theory Tech.* **54**, 2006, 3772–3778.

[24] W. T. Wong, Y. S. Lin, C. H. Wang, and C. H. Chen, Highly selective microstrip bandpass filters for ultra-wideband applications, In *Proceedings of the Asia-Pacific Microwave Conference*, Nov. 2005, 2850–2853.

[25] K. Li, D. Kurita, and T. Matsui, UWB bandpass filters with multi-notched bands, In *Proceedings of the 36th European Microwave Conference*, 2006, 591–594.

[26] W. Menzel, and P. Feil, Ultra-wideband (UWB) filter with WLAN notch, In *Proceedings of the 36th European Microwave Conference*, 2006, 595–598.

[27] J. Garcia-Garcia, J. Bonache, F. Martin, Application of electromagnetic bandgaps to the design of ultra-wide bandpass filters with good out-of-band performance, *IEEE Trans. Microwave Theory Tech.* **54**, 2006, 4136–4140.

[28] R. Gomez-Garcia and J. I. Alonso, Systematic method for the exact synthesis of ultra-wideband filtering responses using highpass and low-pass sections, *IEEE Trans. Microwave Theory Tech.* **54**(10), 2006, 3751–3764.

[29] P. Cai, Z. Ma, X. Guan, Y. Kobayashi, T. Anada, and G. Hagiwara, Synthesis and realization of novel ultra-wideband bandpass filters using 3/4 wavelength parallel-coupled line resonators, In *Proceedings of the Asia-Pacific Microwave Conference*, Dec. 2006, 159–162.

[30] P. Cai, Z. Ma, X. Guan, Y. Kobayashi, T. Anada, and G. Hagiwara; "A novel compact ultra-wideband bandpass filter using a microstrip stepped-impedance four-modes resonator," *IEEE MTT-S Int. Microwave Symp. Dig.* 2007, 751–754.

[31] S. W. Wong, and L. Zhu, EBG-Embedded multiple-mode resonator for UWB bandpass filter with improved upper-stopband performance, *IEEE Microwave Wireless Components Lett.* **17**, 2007, 421–423.

[32] R. Li, and L. Zhu, Ultra-wideband microstrip-slotline bandpass filter with enhanced rejection skirts and widened upper stopband, *Electron. Lett.* **43**(24), 2007, 1368–1369.

[33] N. Thomson, and J.-S. Hong, Compact ultra-wideband microstrip/coplanar waveguide bandpass filter, *IEEE Microwave Wireless Components Lett.* **17**(3), 2007, 184–186.

[34] J.-S. Hong, and K. Li, Recent development of ultra-wideband (UWB) filters, In *Proceedings of IEEE Microwave, Antenna Propagation and EMC Technologies for Wireless Communication International Symposium*, Aug. 2007, 442–445.

[35] H. Shaman, and J.-S. Hong, Wideband bandpass microstrip filters with triple coupled lines and open/short stubs, *Asia-Pacific Microwave Conference*, Bangkok, Dec. 2007, 2293–2296.

[36] H. Shaman, and J.-S. Hong, Compact wideband bandpass filter with high performance and harmonic suppression, In *37th European Microwave Conference*, Munich, Oct. 2007, 528–531.

[37] H. Shaman, and J.-S. Hong, Compact wideband bandpass filter with multiple coupled lines, In *IEEE MMS2007*, Budapest, May 2007, 319–322.

[38] T.-N. Kuo, C.-H. Wang, and C. H. Chen, A compact ultra-wideband bandpass filter based on split-mode resonator, *IEEE Microwave Wireless Components Lett.* **17**(12), 2007, 852–854.

[39] J.-W. Baik, T.-H. Lee, and Y.-S. Kim, UWB bandpass filter using microstrip-to-CPW transition with broadband balun, *IEEE Microwave Wireless Components Lett.* **17**(12), 2007, 846–848.

[40] A. M. Abbosh, Planar bandpass filters for ultra-wideband applications, *IEEE Trans. Microwave Theory Tech.* **55**, 2007, 2262–2269.

[41] H. Shaman, and J.-S. Hong, A novel ultra-wideband (UWB) bandpass filter (BPF) with pairs of transmission zeroes, *IEEE Microwave Wireless Components Lett.* **17**(2), 2007, 121–123.

[42] H. Shaman, and J.-S. Hong, Ultra-wideband (UWB) bandpass filter with embedded band notch structures, *IEEE Microwave Wireless Components Lett.* **17**(3), 2007, 193–195.

[43] H. Shaman, and J.-S. Hong, Asymmetric parallel-coupled lines for notch implementation in UWB filters, *IEEE Microwave Wireless Components Lett.* **17**(7), 2007, 516–518.

[44] K. Li, D. Kurita, and T. Matsui, Dual-band ultra-wideband bandpass filter, In *IEEE MTT-S Int. Dig.* **2007**, 1193–1196.

[45] C.-W. Tang, and M. G. Chen, A Microstrip ultra-wideband bandpass filter with cascaded broadband bandpass and bandstop filters, *IEEE Trans. Microwave Theory Tech.* **55**, 2007, 2412–2418.

[46] C. H. Wu, Y. S. Lin, C. H. Wang, and C. H. Chen, A compact LTCC ultra-wideband bandpass filter using semi-lumped parallel-resonance circuits for spurious suppression, In *2007 European Microwave Conference*, 9–12 Oct. 2007, 532–535.

[47] C.-W. Tang, and D.-L. Yang, Realization of multilayered wide-passband banspass filter with low-temperature co-fired ceramic technology, *IEEE Trans. Microwave Theory Tech.* **56**, 2008, 1668–1674.

[48] Z.-C. Hao, and J.-S. Hong, Ultra wideband bandpass filter using embedded stepped impedance resonators on multilayer liquid crystal polymer substrate, *IEEE Microwave Wireless Components Lett.* **18**, 2008, 581–583.

[49] Z.-C. Hao, and J.-S. Hong, Ultra-wideband bandpass filter using multilayer liquid-crystal-polymer technology, *IEEE Trans. Microwave Theory Tech.* **56**(9), 2008, 2095–2100.

[50] Z.-C. Hao, and J.-S. Hong, Compact ultra-wideband bandpass filter using broadside coupled hairpin structures on multilayer liquid crystal polymer substrate, *IET Electronic Lett.* **44**, 2008, 1197–1198.

[51] Z.-C. Hao, and J.-S. Hong, A novel ultra wideband bandpass filter using broadside coupled structures on multilayer organic liquid crystal polymer substrate, In *38th European Microwave Conference, EuMC2008*, 998–1001.

[52] Z.-C. Hao, and J.-S. Hong, High selective ultra-wideband (UWB) bandpass filter with wideband harmonic suppression, In *38th European Microwave Conference, EuMC2008*, 353–356.

[53] H. Shaman, and J.-S. Hong, Ultra-wideband (UWB) microstrip bandpass filter with narrow notched band, In *38th European Microwave Conference*, Amsterdam, Oct. 2008, 857–860.

[54] A. Balalem, W. Menzel, J. Machac, and A. Omar, A simple ultra-wideband suspended stripline bandpass filter with very wide stop-band, *IEEE Microwave Wireless Components Lett.* **18**(3), 2008, 170–172.

[55] S. W. Wong, and L. Zhu, Implementation of compact UWB bandpass filter with a notch-band, *IEEE Microwave Wireless Components Lett.* **18**(1), 2008, 10–12.

[56] G.-M. Yang, R. Jin, C. Vittoria, V. G. Harris, and N. X. Sun, Small ultra-wideband (UWB) bandpass filter with notched band, *IEEE Microwave Wireless Components Lett.* **18**(3), 2008, 176–178.

[57] M. Mokhtaari and J. Bornemann; Ultra-wideband and notched wideband filters with grounded vias in microstrip technology, In *Asia-Pacific Microwave Conference*, APMC 2008, 16–20 Dec. 2008, 1–4.

[58] G.-M. Yang, R. Jin, C. Vittoria, V. G. Harris, and N. X. Sun, Small ultra-wideband (UWB) bandpass filter with notched band, *IEEE Microwave Wireless Components Lett.* **18**(3), 2008, 176–178.

[59] Z.-C. Hao, J.-S. Hong, J. P. Parry, and D. Hand, Miniature coupled resonator UWB filter using a multilayer structure on liquid crystal polymer, In *Asia-Pacific Microwave Conference, APMC 2008*, 16–20 Dec. 2008, 1–4.

[60] Y. H. Chun, H. Shaman, and J.-S. Hong, Switchable embedded notch structure for UWB bandpass filter, *IEEE Microwave Wireless Components Lett.* **18**(9), 2008, 590–592.

[61] H. R. Arachchige, J.-S. Hong, Z.-C. Hao, UWB bandpass filter with tunable notch on liquid crystal polymer substrate, In *Asia-Pacific Microwave Conference, APMC2008*, 16–20 Dec. 2008, 1–4.

[62] C.-P. Chen, Z. Ma, and T. Anada, Synthesis of ultra-wideband bandpass filter employing parallel-coupled stepped-impedance resonators, *IET Microwaves Antennas Propagation* **2**(8), 2008, 766–772.

[63] W. M. Fathelbab, and H. M. Jaradat, New stepped-impedance parallel-coupled line (PCL) filters for ultra-wide band (UWB) applications, In *38th European Microwave Conference, EuMC 2008*, 27–31 Oct. 2008, 496–499.

[64] Z. Ma, W. He, C.-P. Chen, Y. Kobayashi, and T. Anada, A novel compact ultra-wideband bandpass filter using microstrip stub-loaded dual-mode resonator doublets, In *2008 IEEE MTT-S International Microwave Symposium Digest*, 15–20 June 2008, 435–438.

[65] M. Uhm, K. Kichul and D. S. Filipovic, Ultra-wideband bandpass filters using quarter-wave short-circuited shunt stubs and quarter-wave series transformers, *IEEE Microwave Wireless Components Lett.* **18**(10), 2008, 668–670.

[66] Z.-C. Hao, and J.-S. Hong, Compact wide stopband ultra wideband bandpass filter using multilayer liquid crystal polymer technology, *IEEE Microwave Wireless Components Lett.* **19**(5), 2009, 290–292.

[67] Z.-C. Hao and J.-S. Hong, Compact UWB filter with double notch-bands using multilayer LCP technology, *IEEE Microwave Wireless Components Lett.* **19**(8), 2009, 500–502.

[68] Z.-C. Hao, J.-S. Hong, S. K. Alotaibi, J. P. Parry, and D. Hand, Ultra-wideband bandpass filter with multi notch-bands on multilayer liquid crystal polymer substrate, *IET Proc. Microwaves Antennas Propagation* **3**, 2009, 749–756.

[69] K. Song and Y. Fan, Compact ultra-wideband bandpass filter using dual-line coupling structure, *IEEE Microwave Wireless Components Lett.* **19**, 2009, 30–32.

[70] K. Liu, R. C. Frye, and R. Emigh, Miniaturized ultra-wideband band-pass-filter from silicon integrated passive device technology, *IEEE MTT-S Int. Microwave Symp. Dig.* 7–12 June 2009, 1057–1060.

[71] S. W. Wong, and L. Zhu, Quadruple-mode UWB bandpass filter with improved out-of-band rejection, *IEEE Microwave Wireless Components Lett.* **19**(3) 2009, 152–154.

[72] Y.-T. Lee, K. Liu, R. Frye, H.-T. Kim, G. Kim, and B. Ahn, Ultra-wide-band (UWB) band-pass-filter using integrated passive device (IPD) technology for wireless applications, In

59th Electronic Components and Technology Conference, ECTC 2009, 26–29 May 2009, 1994–1999.
[73] S. P. Lim, and J. Y. Park, Fully embedded UWB filter into organic packaging substrate, In *59th Electronic Components and Technology Conference, ECTC 2009*, 26–29 May 2009, 1585–1589.
[74] Z.-C. Hao, and J.-S. Hong, UWB bandpass filter with a multilayer non-uniform periodical structure on LCP substrates, *IEEE MTT-S International Microwave Symposium Digest*, 7–12 June 2009, 853–856.
[75] Z.-C. Hao, and J.-S. Hong, UWB bandpass filter using cascaded miniature highpass and lowpass filters with multilayer liquid crystal polymer technology, *IEEE Trans. Microwave Theory Techniques*, MTT-58, April 2010, 941–948.
[76] Z. C. Hao, J.-S. Hong, J. P. Parry, and D. Hand, Ultra wideband bandpass filter with multiple notch bands using nonuniform periodical slotted ground structure, *IEEE Trans. Microwave Theory Techniques MTT-57,* Dec. 2009, 3080–3088.
[77] G. L. Matthaei, Interdigital band-pass filters, *IEEE Trans.* **MTT-10**, 1962, 479–492.
[78] AWR Microwave Office/Analog Office V7.5, Applied Wave Research, Inc., El Segundo, CA, 2007.
[79] *EM User's Manual*, Version 10.: Sonnet Software Inc. Liverpool, New York, 2006.

CHAPTER THIRTEEN

Tunable and Reconfigurable Filters

Electronically tunable or reconfigurable microwave filters have attracted more and more attention for research and development, because of their increasing importance in improving the capability of current and future wireless systems [1]. Thinking ahead to future cognitive radio and radar applications, it is certain that electronically reconfigurable microwave filters will play an even more important role in wireless systems.

In general, to develop an electronically reconfigurable filter, active switching or tuning elements such as semiconductor p-i-n and varactor diodes, RF MEMS, or other functional material-based components including ferroelectric varactors (see Table 13.1) need to be integrated within a passive filtering structure. Since microstrip filters can conveniently facilitate this kind of small size integration, there has been increasing interest in developing tunable or reconfigurable filters based on microstrips [2–36]. These filters may be classified as tunable combline bandpass filters [2–9]; RF MEMS tunable filters [10–16]; piezoelectric transducer (PET) tunable filters [17–19]; tunable high-temperature superconductor (HTS) filters [20–22]; reconfigurable UWB filters [23,24]; tunable dual-band filters [25]; tunable bandstop filters [26–30]; reconfigurable/tunable dual-mode filters [31–36]; reconfigurable bandpass filters based on a switched delay-line approach [37]; and wideband bandpass filter with reconfigurable bandwidth [38,39].

Generally speaking, bandwidth tuning or controlling is more challenging than frequency tuning and the design of an electronically tunable filter with a wider bandwidth is more difficult than a narrow bandwidth in terms of tuning range and bandwidth control. Some techniques for the bandwidth control in a tunable filter were reported [40–43]. The nonlinear behavior of an electronically reconfigurable filter is very dependent on the tuning element used. The use of RF MEMS and a piezoelectric

---

Microstrip Filters for RF/Microwave Applications by Jia-Sheng Hong
Copyright © 2011 John Wiley & Sons, Inc.

TABLE 13.1 Comparison of Tuning Device Technologies[a]

Tuning Technology	Tuning Speed	Q or R_s at 10 GHz	Bias	Power Consumption
Semiconductor p-i-n diode	ns	$R_s \approx 1-4 \; \Omega$	10–30 mA	High
Semiconductor varactor	ns	$Q \approx 30-40$	< 30 V	Low
MEMS	μs	$Q \approx 50-400$	30–90 V	Negligible
Ferroelectric thin film	ns	$Q \approx 30-150$	< 30 V	Negligible
Pizotransducer	>100 μs	$Q > 500$	> 100 V	Negligible

[a]From reference [16] and [44].

transducer usually results in better linearity. Innovation in reconfigurable filter design can also improve the performance and increase the functionality. The relatively low Q of tuning elements can restrict the implementation of higher-order and narrowband tunable filters. This is because, for a given Q of the tuning elements and other losses associated with the circuit, the insertion loss of a filter is increased with an increase of its order and a decrease of its bandwidth. As a result, the insertion loss of a high-order tunable narrowband filter can be much larger for practical application. In addition, the tuning range is also more limited for a higher-order than for a lower-order filter [4]. The development of reconfigurable filters involves some trade-offs, such as the filter size and the complexity of the bias circuit, which add to challenges. Several typical electronically tunable or reconfigurable microstrip filters are described in the rest of this chapter.

13.1 TUNABLE COMBLINE FILTERS

A microstrip combline filter, such as that shown in Figure 5.19, is a favorite structure for developing tunable or reconfigurable bandpass filters [2–9]. To have a simple dc bias circuit, it is desired that all loading capacitances have the same value, denoted by C_L. Hence, all resonators have the same electrical length of θ_0 as well as the same characteristic admittance of Y_a. This results in a general schematic, as illustrated in Figure 13.1, for designing tunable combline filters, where Y_A denotes the characteristic admittance of the terminal line. The two short-circuited lines at the input and output are not resonators, but simply part of impedance-transforming sections that facilitate the input and output couplings. Coupling between the resonators is controlled by their separation as discussed in Chapter 5.

To design a bandpass filter of this type with tunable center frequency, we need to specify a nominal center frequency f_0, which is usually at the middle of a desired frequency tuning range. For a wide tuning range, the electric length θ_0 at this frequency may be chosen as 45°. Thus, the resonators are $\lambda_{g0}/8$ long, where λ_{g0} is the guided wavelength at f_0. Next, we need to determine the loading capacitance C_L at f_0. Following Eq. (5.43), C_L is given by

$$C_L = \frac{Y_a \cot\theta_0}{2\pi f_0} \qquad (13.1)$$

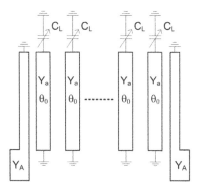

FIGURE 13.1 Schematic of tunable combline filter.

which can be determined by choosing a proper Y_a. A choice could consider having C_L around the middle of capacitance range of a varactor, although other factors, such as resonator line width, may be taken into account.

After determining θ_0, Y_a, and C_L, we design the filter for a given fractional bandwidth FBW, centered at f_0, and for a desired filtering characteristic, such as a Chebyshev response. For our demonstration, consider a varactor-tuned combline filter design to meet the following specifications:

Nominal center frequency f_0	1.0 GHz
Bandwidth of passband	70 MHz
Degree of filter	3
Filtering characteristic	Chebyshev with 0.04321-dB passband ripple
Terminal impedance	50 Ω

Thus, the fractional bandwidth $FBW = 0.07$. Choosing $\theta_0 = 45°$ and $Y_a = 1/100$ mhos, we find $C_L = 1.6$ pF from Eq. (13.1). The lowpass prototype parameters for the desired Chebyshev response can be found from Table 3.2, which are $g_0 = g_4 = 1.0$, $g_1 = g_3 = 0.8516$, and $g_2 = 1.1032$. From Eq. (5.45), we can determine the external quality factor and coupling coefficient as

$$Q_{e1} = Q_{e3} = Q_e = 12.166$$
$$M_{12} = M_{23} = M = 0.072 \qquad (13.2)$$

The filter is to be implemented on a dielectric substrate with a relative dielectric constant of 3.0 and a thickness of 1.016 mm. The next step in the design is to find the physical dimensions of the filter on the given substrate, which can be done by using the parameter extraction technique described in Chapter 7. Figure 13.2 displays the arrangements for extracting Q_e and M, using a commercially available EM-simulation tool, Sonnet Suites (see Chapter 8). The holes indicate via-hole groundings. In Figure 13.2a, port 2 is very weakly coupled to the resonator. This is an additional practical method allowing one to extract the single loaded Q_e from

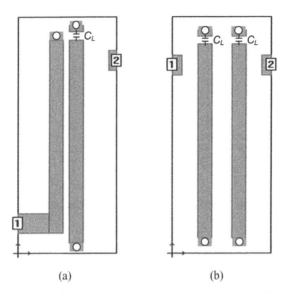

FIGURE 13.2 (a) Arrangement for extracting external quality factor. (b) Arrangement for extracting coupling coefficient.

an S_{21} amplitude response (see Figure 7.14), as long as any parasitic effects from port 2 are negligible. In this case, the single-loaded external quality can simply be extracted using $Q_e = \omega_0/\Delta\omega_{3dB}$. Similarly, to extract the inter-resonator coupling coefficient M from the EM simulation, the input and output ports, which function like a probe, as shown in Figure 13.2b, just excite the coupled resonators to make two resonant peaks observable for the parameter extraction following the discussion in Chapter 7. In the simulation, the loading capacitance is set to $C_L = 1.6$ pF, and the resonators have an electrical length of $\theta_0 = 45°$ and a characteristic admittance of $Y_a = 1/100$ mhos.

After extracting the desired parameters of Eq. (13.2), a layout of the filter is determined, which is illustrated in Figure 13.3a, where resonators 1 and 3 are slightly longer than resonator 2 to compensate for the effect from the input/output (I/O) feed structure on the resonant frequency. Figure 13.3b plots the EM-simulated performance of the filter for different loading capacitances, that is, $C_L = 0.8, 1.0, 1.6, 2.5,$ and 3.5 pF, respectively. It can be seen that the designed filter can be tuned over a wide frequency range from 0.7 to 1.3 GHz with a fairly well-preserved passband as the varactor capacitance C_L varies from 3.5 down to 0.8 pF. When $C_L = 1.6$ pF, which is the loading capacitance used in the filter design, the filter exhibits the most desired response with a return loss below -20 dB over the passband centered at the nominal center frequency of $f_0 = 1.0$ GHz. Tuning away from this frequency, the passband return loss deteriorates gradually and the fractional bandwidth changes slightly as well. This is because the I/O coupling (Q_e) and the inter-resonator coupling (M) change against the tuning or loading capacitance, which is demonstrated in Figure 13.4. Since for different C_L, these two parameters are extracted at different frequencies,

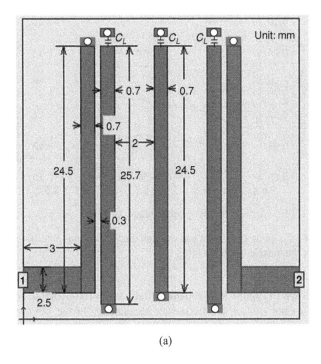

(a)

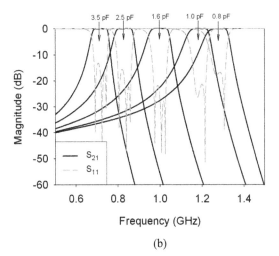

(b)

FIGURE 13.3 (a) Layout of the tunable combline filter on a dielectric substrate with a relative dielectric constant of 3.0 and a thickness of 1.016 mm. (b) EM-simulated performance for five different loading capacitances.

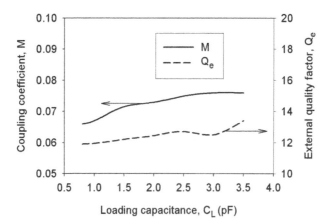

FIGURE 13.4 Coupling coefficient and external quality factor as a function of loading capacitance for the tunable combline filter.

the results shown in Figure 13.4 somehow indicates a frequency dependence of Q_e and M. As a matter of fact, the frequency dependence of the I/O coupling and the inter-resonator coupling is relatively small over a wide frequency range. This is a reason for the combline filter structure of Figure 13.1 to remain popular for designing a tunable filter with a wide frequency tuning range.

The designed tunable combline filter is fabricated on a Rogers RO3003 substrate with a relative dielectric constant of 3.0 and a thickness of 1.016 mm. Figure 13.5a is a photograph of the fabricated filter. Each resonator is associated with an M/A-COM surface mount GaAs high-Q varactor diode (MA46H202-1088), a bypass capacitor of 33 pF, and a chip resistor of 10K Ω, as indicated. In addition, two shunt capacitors of 100 pF are used in the dc bias circuit. The filter is measured using a microwave network analyzer and the results are plotted in Figure 13.5b, which only shows the responses when the applied dc voltage is 3, 7.6, and 20 V, respectively. In fact, by varying dc bias voltage from 3 to 20 V, the center frequency of the filter can be continuously tuned from 0.65 to 1.27 GHz and a typical middle band insertion loss is around 3.5 dB.

Variation in bandwidth as center frequency is tuned is a well-known problem. In general, to maintain a constant absolute passband bandwidth independent of tuned frequency, coupling coefficients must vary inversely with the tuning frequency. There have been some techniques that address this issue, for instance, those reported in Vendik et al. [4] and Kim et al. [8]. In Kim et al. [8], a varactor-tuned combline bandpass filter using step-impedance microstrip resonators is considered so that the inter-resonator coupling can be better controlled to meet the constant bandwidth requirement with shorter electrical lengths of line elements. In the meantime, lumped inductors are used for input and output coupling networks to enable the external Q (quality factor) to vary directly with the tuning frequency. A four-pole filter of this type has been demonstrated, showing that the 3-dB passband bandwidth varies less than 3.2% within the 250-MHz tuning range at 2 GHz.

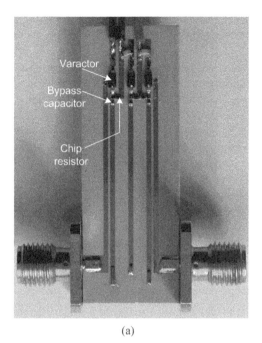

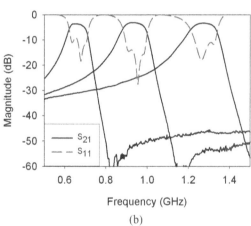

FIGURE 13.5 (a) Fabricated tunable combline filter on a dielectric substrate with a relative dielectric constant of 3.0 and a thickness of 1.016 mm. (b) Measured performance.

In Vendik et al. [4], a theoretical study reveals that, for a tunable n-pole filter, the figure of merit defined as the ratio of the shift of the center frequency of the passband of the filter to the average passband is dependent on the loss or Q of the tuning varactors and the filter order. As the quality factor is inversely proportional to the power loss, the larger the loss, the smaller the Q. Generally speaking, the figure of merit (or tuning range) is reduced for small Q and large n.

13.2 TUNABLE OPEN-LOOP FILTERS WITHOUT VIA-HOLE GROUNDING

Following the concept introduced for the pseudocombline filter in Chapter 5, a tunable open-loop filter without using any via-hole connections can be developed. To this end, a tunable open-loop resonator is developed in light of Figure 13.6. As can be seen, the varactor-tuned transmission-line resonator 1 (R1), which has an electrical length of θ_0 with a characteristic admittance of Y_a and loaded with a capacitance of C_L, can be converted to the tunable resonator 2 (R2), having equivalent characteristics in the vicinity of the fundamental-resonant frequency. This is because there is virtual grounding at resonance in the middle of resonator 2. The resonator 2 can then be folded to form the resonator 3 (R3). Since the two loaded capacitors in resonator 3 can be seen, from the two open ends of the folded transmission section, as two series-connected capacitors, their physical grounding can be removed to result in the tunable open-loop or hairpin resonator, as illustrated in Figure 13.6. The loading capacitance in the resultant resonator 4 (R4) is only one-half of that in the resonator 1, used in the tunable combline filter above. The tunable resonator 4 is particularly useful when via-hole grounding is undesired for some substrate and associated manufacturing process. An RF bypass or dc-blocking capacitor can easily be inserted into the loop to facilitate a dc bias for the varactor.

Figure 13.7 shows a schematic for designing tunable open-loop or hairpin filters without via-hole grounding. Since there is an electrical equivalence between the

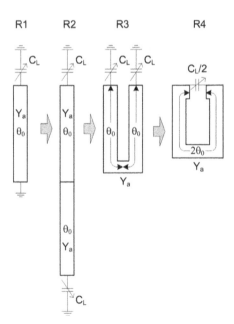

FIGURE 13.6 Evolution of tunable open-loop or hairpin resonator without via-hole grounding.

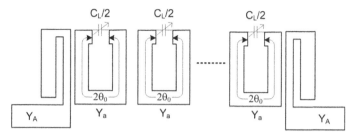

FIGURE 13.7 Schematic of tunable open-loop or hairpin filter without via-hole grounding.

tunable open-loop resonator, discussed here, and the resonator used in the tunable combline filter of Figure 13.1, the design of this type of tunable filter can follow the similar procedures described for the design of tunable combline filter in the last section. This is demonstrated with a design example below.

Consider the following specifications for this design:

Nominal center frequency f_0	1.5 GHz
Passband fractional bandwidth FBW	0.058
Degree of filter	3
Filtering characteristic	Chebyshev with 0.04321-dB passband ripple
Terminal impedance $1/Y_A$	50 Ω

First, choose $\theta_0 = 45°$ and $Y_a = 1/130$ mhos. The first choice of the electrical length is to achieve a wide tuning range, as discussed above for the tunable combline filter, whereas the second choice of the line admittance is to consider having a stronger inductive coupling effect with a narrower or higher impedance line. From Eq. (13.1), we obtain $C_L = 0.816$ pF. Thus, at f_0, the loading capacitance is actually $C_L/2 = 0.408$ pF and the electrical length for the open loop is $2\theta_0 = 90°$. Next, find the design parameters for the bandpass filter, that is, the external quality factor and coupling coefficient, from the desired Chebyshev lowpass prototype parameters of $g_0 = g_4 = 1.0$, $g_1 = g_3 = 0.8516$, and $g_2 = 1.1032$. Using Eq. (5.47), we obtain

$$Q_{e1} = Q_{e3} = Q_e = 14.683$$
$$M_{12} = M_{23} = M = 0.06 \qquad (13.3)$$

The filter is achieved on a dielectric substrate with a relative dielectric constant of 3.0 and a thickness of 1.52 mm. The line width for the open-loop resonator is found to be 0.5 mm for the chosen characteristic impedance ($1/Y_a$) of 130 Ω. The open-loop resonator can then be implemented with a loading capacitance of 0.408 pF to resonate at $f_0 = 1.5$ GHz. The desired external quality factor and coupling coefficient of Eq. (13.3) can be extracted from EM simulation with the formulations introduced in Chapter 7, or in a similar manner as described in association with Figure 13.2. The layout of the designed filter is shown in Figure 13.8a. The input

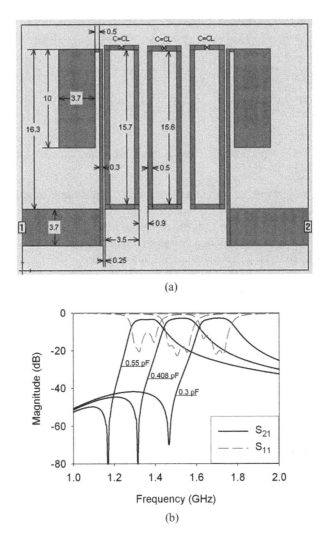

FIGURE 13.8 (a) Layout of the tunable open loop filter on a dielectric substrate with a relative dielectric constant of 3.0 and a thickness of 1.52 mm. All dimensions are in millimeters. (b) EM-simulated performance for three different loading capacitances.

and output (I/O) resonators are slightly longer than the middle one to compensate for the coupling effect of the I/O feed line on their fundamental resonant frequency. The I/O feed line has a high characteristic impedance with a line width of 0.3 mm, which is connected to an open-circuited stub with a lower characteristic impedance. The open stub is part of the I/O impedance transformer, which aims to generate large capacitance to ground, like an RF bypass capacitor, so that the narrow feed line may somehow mimic a short-circuited line transformer described in the last section for the tunable combline filter design. It can be shown that the lower the characteristic impedance of the open stub, the wider is the operating frequency of the impedance transformer. The length of the open stub can then be adjusted to achieve a maximum coupling at $f_0 = 1.5$ GHz for the loading capacitance equal to 0.408 pF.

The EM-simulated performance of the designed tunable filter is shown in Figure 13.8b with three different loading capacitances of $C_L = 0.55$, 0.408, and 0.3 pF, respectively. The losses are considered in the simulation by assuming a varactor Q in the range of 90 to 110, wherein a larger loading capacitance tends to have a lower Q or higher loss. As can be seen from the filter frequency responses plotted, for $C_L = 0.408$ pF, the passband is centered at 1.5 GHz with a desired bandwidth having the return loss almost below −20 dB. The filter also exhibits a finite-frequency transmission zero near the low side of the passband, which improves the selectivity on this side. The resultant asymmetrical frequency response, in this case, is contrary to that of the tunable combline filter shown in Figure 13.3b.

From Figure 13.8b, it can also be seen that, for given capacitance tuning range from 0.3 to 0.55 pF, the filter shows a reasonable frequency tuning range of more than 300 MHz, centered at 1.5 GHz. However, for the 0.3 and 0.55 pF loadings, the passband performance somewhat degrades the return loss, in particular. This is because the external quality factor and coupling coefficient, which are extracted at the nominal frequency of 1.5 GHz when the filter is designed, vary against frequency or loading capacitance, as shown in Figure 13.9. As compared with Figure 13.4, the external quality factor for the I/O feed structure used in this design seems to be much more strongly frequency dependent than for the one used in the combline filter design. This implies that the I/O impedance transformer, used in the design of Figure 13.8a, is inherently narrowband, which somewhat limits the tuning range of the filter, although it eliminates the need for via-hole grounding. In order to achieve a wider tuning range, a wideband I/O feed structure would be required.

Figure 13.10 shows a wideband frequency response of the designed tunable open-loop filter, obtained by EM simulation. The spurious response occurs at about $3f_0$

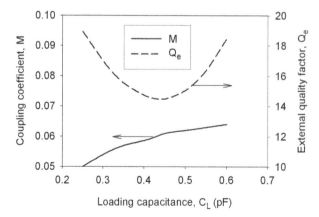

FIGURE 13.9 Coupling coefficient and external quality factor as a function of loading capacitance for the tunable open-loop filter.

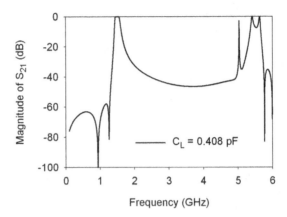

FIGURE 13.10 EM-simulated wideband frequency response of the tunable open-loop filter.

because of the loading capacitance. Two finite-frequency transmission zeros appear in the low stopband, which would seem to be inherent in the filter topology resulting from a combined effect of the I/O feed structure and the coupled open-loop resonators.

The above designed tunable open-loop filter is fabricated on a dielectric substrate with a relative dielectric constant of 3.0 and a thickness of 1.52 mm. The photograph of the fabricated filter is illustrated in Figure 13.11a. Each resonator is implemented with an M/A-COM gallium arsenide flip chip hyperabrupt varactor diode (MA46H120) and a bypass capacitor of 33 pF. The bypass capacitor is used in order to apply a dc voltage across the varactor. Two chip resistors of 10K Ω and a shunt capacitor of 100 pF are also used to form the dc bias circuit for each resonator. All the three tunable open-loop resonators are tuned with the same dc bias. The measured responses of the filter at three different dc bias voltages of 2.6, 4.0, and 5.8 V are shown in Figure 13.11b. In general, this experimental filter demonstrates a reasonable frequency tuning range of this type of tunable filter.

13.3 RECONFIGURABLE DUAL-MODE BANDPASS FILTERS

Reconfigurable or tunable bandpass filters may be designed based on single- or dual-mode resonators. This section demonstrates several examples of reconfigurable dual-mode filters developed based on the open-loop dual-mode filter introduced in Chapter 10.

13.3.1 Reconfigurable Dual-Mode Filter with Two dc Biases

Figure 13.12 shows the layout of a two-pole reconfigurable dual-mode microstrip open-loop resonator bandpass filter. C_1 and C_2 are variable capacitances of loaded varactors. On the opposite arm of the open-loop, there are two dc-blocking capacitors

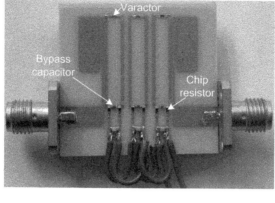

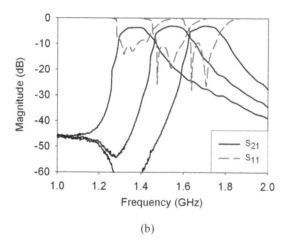

FIGURE 13.11 (a) Fabricated tunable open-loop filter on a substrate with a relative dielectric constant of 3.0 and a thickness of 1.52 mm. (b) Measured performance.

(C_{bypass}) to separate dc biases for the varactors C_1 and C_2. Without the capacitors, the filter is similar to a fixed-frequency filter described in Section 10.3.4. Hence, it has the same coupling scheme as shown in Figure 10.31, in which there is no coupling between the two operating resonant modes, that is, the even and odd modes.

Because there is no coupling between the two operating resonant modes, it is envisaged that the center frequency of the passband can be tuned if the resonant frequencies of the odd mode and even mode are proportionally shifted. To this end, C_1 in Figure 13.12 is used to vary the odd-mode frequency, while C_2 is used to change the even-mode frequency. Depending on the combinations of C_1 and C_2, not only the passband frequency can be tuned, but also the filtering characteristic can be reconfigured to exhibit a higher selectivity on the high side of the passband (case 1)

576 TUNABLE AND RECONFIGURABLE FILTERS

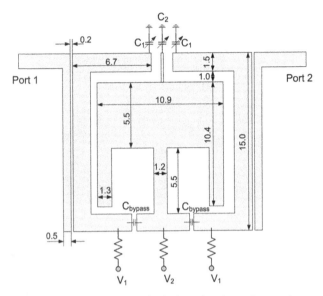

FIGURE 13.12 Layout of a reconfigurable dual-mode microstrip open-loop resonator bandpass filter on a substrate with a relative dielectric constant of 10.8 and a thickness of 1.27 mm. All dimensions are in millimeters.

or on the low side of the passband (case 2). These two cases are demonstrated experimentally with a fabricated filter shown in Figure 13.13. Infineon BB857 varactors, which have typical variable capacitances from 0.5 to 6.6 pF, are implemented for electronic tuning. In order to reconfigure the filter characteristic, two dc biases are deployed. The first dc bias V_1 is used to vary the odd-mode frequency and the second dc bias V_2 is used to change the even-mode frequency. The varactors are connected to the dc voltages via 10K Ω chip resistors.

The measured frequency responses are plotted in Figure 13.14, which shows that, depending on the combinations of the two dc biases, not only the passband frequency can be tuned, but also the filtering characteristic can be reconfigured from the high

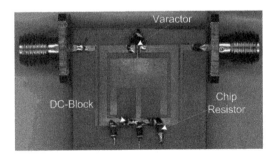

FIGURE 13.13 Fabricated reconfigurable dual-mode microstrip open-loop resonator bandpass filter with two dc biases.

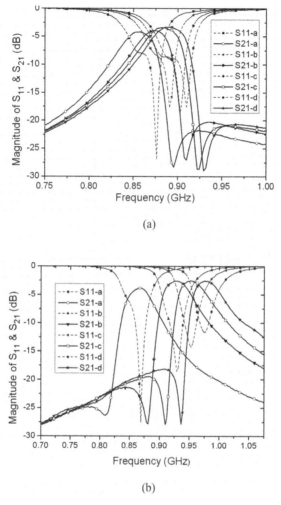

FIGURE 13.14 Measured performance of the reconfigurable dual-mode filter with two dc biases. (**a**) Case 1; (**b**) case 2.

selectivity on the high side of the passband for case 1 to the high selectivity on the low side of the passband for case 2. Table 13.2 lists the measured center frequencies against different bias voltage combinations, for cases 1 and 2, respectively.

13.3.2 Tunable Dual-Mode Filters Using a Single dc Bias

By modifying the dual-mode microstrip open-loop resonator, the center frequency of the dual-mode filter can be tuned electronically using a single dc bias [34–36]. In other words, the change of the even- and odd-mode resonant frequencies is facilitated with the same bias voltage. Figure 13.15 shows the schematic of a two-pole tunable

TABLE 13.2 Center Frequencies Against dc Bias Voltages

Case 1	V_1 (V)	V_2 (V)	f_0 (MHz)	Case 2	V_1 (V)	V_2 (V)	f_0 (MHz)
a	8.0	9.5	859.6	a	9.0	4.0	866.8
b	9.0	12.0	876.4	b	15.0	7.5	926.8
c	9.5	20.0	888.4	c	20.0	12.0	950.8
d	10.0	34.0	898.0	d	35.0	34.0	977.2

filter of this type, where all the loaded varactors have the same capacitance, denoted by C_v. As compared with the filter structure in Figure 13.12, the filter in Figure 13.15 uses a different input/output coupling arrangement, which, in fact, is a wideband transformer, for achieving a wide tuning range with a constant absolute bandwidth. This is discussed in detail next.

13.3.2.1 Ideal Tunable Dual-Mode Filter Requirement To tune this kind of dual-mode filter while keeping filter characteristics, including bandwidth, unchanged, ideally, two factors need to be considered. First, the resonant frequencies of odd mode (f_0^o) and even mode (f_0^e) need to be shifted proportionally. Second, the shape and bandwidth of odd- and even-mode frequency response must keep constant over the entire tuning range. This would require that the external quality factor of odd mode (Q_{exo}) and even mode (Q_{exe}) vary directly with the tuning frequency. These parameters may be represented by

$$|f_0^e - f_0^o| = A \tag{13.4}$$

$$Q_{exo} = \frac{f_0^o}{\Delta f_{3dB}^o} \tag{13.5}$$

$$Q_{exe} = \frac{f_0^e}{\Delta f_{3dB}^e} \tag{13.6}$$

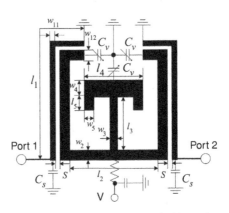

FIGURE 13.15 Schematic of tunable dual-mode filter using a single dc bias.

where A denotes the separation between the even- and odd-mode resonant frequencies and Δf_{3dB}^{o}, Δf_{3dB}^{e} are the 3-dB bandwidths of the odd and even modes, respectively. Note that Δf_{3dB}^{e}, Δf_{3dB}^{o} and A are ideally to be constant over the entire tuning range for a constant absolute bandwidth tuning.

13.3.2.2 Odd- and Even-Mode Tuning Rates

To shift the odd- and even-mode resonant frequencies proportionally by loading the same capacitance, the tuning rates of the odd- and even-modes need to be characterized. The tuning rate indicates how much a modal frequency is shifted by varying capacitance C_v. Assume that when C_v varies from C_{v1} to C_{v2}, the odd-mode frequency shifts from f_{01}^{o} to f_{02}^{o}, and the even-mode frequency shifts from f_{01}^{e} to f_{02}^{e}. Thus, the tuning rate can be defined as

$$\text{Tuning rate} = \frac{\left|f_{02}^{o/e} - f_{01}^{o/e}\right|}{|C_{v2} - C_{v1}|} \text{ (GHz/pF)} \tag{13.7}$$

where the superscript o/e denotes the odd or even mode.

By placing a short or open circuit at the symmetrical plane of the circuit in Figure 13.15, we obtain the circuit model for the odd or even mode, without the input/output coupling, as shown in Figure 13.16a and c. To demonstrate how to control the modal tuning rate, the circuit models of Figure 13.16a and c may be modified to that of Figure 13.16 b and d, respectively. A reference port is added for deriving an input admittance. For the odd mode, this admittance is given by

$$Y_{ino} = j\left(\omega C_v - \frac{Y_o}{\tan\theta_o}\right) \tag{13.8}$$

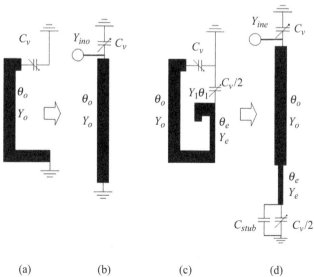

(a)   (b)   (c)   (d)

FIGURE 13.16 Circuit models. (**a**) and (**b**) Odd mode. (**c**) and (**d**) Even mode.

580 TUNABLE AND RECONFIGURABLE FILTERS

and for the even mode,

$$Y_{ine} = j\left\{Y_o \frac{Y_L + Y_o \tan\theta_o}{Y_o - Y_L \tan\theta_o} + \omega C_v\right\} \quad (13.9)$$

with

$$Y_L = Y_e \frac{\omega(C_v/2 + C_{stub}) + Y_e \tan\theta_e}{Y_e - [\omega(C_v/2 + C_{stub})]\tan\theta_e}$$

where C_v is the loading capacitance; C_{stub} represents the bended short open-circuited stub, which may be estimated from $C_{stub} = (Y_1 \tan\theta_1)/\omega$. $Y_o, Y_e, Y_1, \theta_o, \theta_e$, and θ_1 are the admittances and electrical lengths for the transmission line sections shown in Figure 13.6. The resonant frequencies of the odd and even mode can be found by solving

$$\text{Im}[Y_{ino}] = 0 \qquad \text{Im}[Y_{ine}] = 0 \quad (13.10)$$

From Eq. (13.8), it can be seen that the odd-mode resonant frequency depends on the parameters C_v, Y_o, and θ_o. Assume that the loading capacitance C_v varies from C_{v1} (0.6 pF) to C_{v2} (5 pF). For $\theta_o = 80°$ at a nominal frequency of 1 GHz, the modal tuning rate can be calculated by Eq. (13.7) with different values Y_o, and the results are plotted in Figure 13.17a. Similarly, for $Y_o = 0.02$S, the modal tuning rate can also be calculated against different values of θ_o; the results are present in Figure 13.17b. These two sets of results show that the smaller the admittance Y_o or the shorter the electrical length θ_o, the larger the modal tuning rate.

With the same loading capacitance range, that is, C_v varying from C_{v1} (0.6 pF) to C_{v2} (5 pF), the tuning rate of the even mode may be solely adjusted by C_{stub}, Y_e, and θ_e to match the tuning rate of the odd mode. Thus, for given separations between the odd- and even-mode resonant frequencies over a loading capacitance range, the values of Y_e and θ_e may be determined from Eqs. (13.11) and (13.12) for a given C_{stub},

$$\left|f_{01}^e - f_{01}^o\right|_{C_v=C_{v1}} = A \quad (13.11)$$

$$\left|f_{02}^e - f_{02}^o\right|_{C_v=C_{v2}} = B \quad (13.12)$$

where A and B are the separations between the odd- and even-mode resonant frequencies for the loading capacitances C_{v1} and C_{v2}, respectively. Ideally, they are also proportional to the bandwidth. For the constant absolute bandwidth tuning, the resonant frequencies of the odd and even modes need to be shifted proportionally. Hence, B should be equal to A. To demonstrate how to achieve this, assume that $A = 100$ MHz; $Y_o = 0.02$ S; $\theta_o = 80°$ at 1 GHz, which is the nominal high frequency of a given tuning range; $C_{v1} = 0.6$ pF; $C_{v2} = 5.0$ pF. Also, for the demonstration, C_{stub} is chosen for three values: $C_{stub} = 0$, $C_{stub} = 0.3$ pF, and $C_{stub} = 1.3$ pF to represent

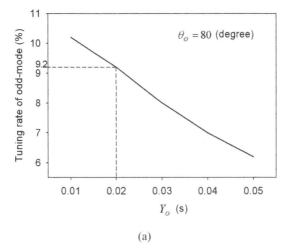

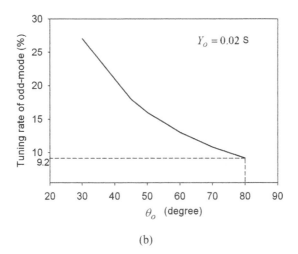

FIGURE 13.17 The tuning rate of odd mode varies with (a) Y_o and (b) θ_o (For $C_{v1} = 0.6$ pF and $C_{v2} = 5$ pF).

different cases of the short open-circuited stub. By applying Eqs. (13.8–13.12), Y_e and θ_e can be determined for $B = A$ as follows: $Y_e = 0.018$ S and $\theta_e = 68.54°$ when $C_{stub} = 0$; $Y_e = 0.016$ S and $\theta_e = 62.17°$ when $C_{stub} = 0.3$ pF; and $Y_e = 0.006$ S and $\theta_e = 23.43°$ when $C_{stub} = 1.3$ pF. After Y_e and θ_e are determined, the tuning rate of the even mode can then be calculated by Eq. (13.7). For the above three cases, the same tuning rate of 9.2% is obtained, which is equal to that of the odd mode for $Y_o = 0.02$ S and $\theta_o = 80°$ (see Figure 13.17). The resonant frequencies of the even and odd modes against the loading capacitance C_v, varying from 0.6 to 5.0 pF, are plotted in Figure 13.18 for a comparison. Note that, for the even mode, the curves for

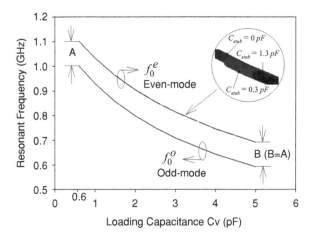

FIGURE 13.18 The modal resonant frequencies against loading capacitance. For the odd mode: $Y_o = 0.02$ S and $\theta_o = 80°$. For the even mode: case 1 ($Y_e = 0.018$ S and $\theta_e = 68.54°$ when $C_{stub} = 0$), case 2 ($Y_e = 0.016$ S and $\theta_e = 62.17°$ when $C_{stub} = 0.3$ pF), and case 3 ($Y_e = 0.006$ S and $\theta_e = 23.43°$ when $C_{stub} = 1.3$ pF).

the three different values of C_{stub} coincide, which implies that the effect of the short open-circuited stub can easily be compensated by Y_e and θ_e with negligible influence to the even-mode tuning rate. This, however, allows a more flexible design for the loading element inside the open loop for the even mode. From the analysis above, it is clear that the tuning rate of the even mode is controllable to be equal to the tuning rate of the odd mode. As such, the resonant frequencies of the two operating modes can be controlled to shift proportionally.

13.3.2.3 External Coupling of the Filter Referring to Figure 13.15, for a given geometry of the dual-mode resonator with loading capacitance, the external quality factors of the odd and even modes, that is, Q_{exo} and Q_{exe}, can be extracted using Eq. (10.28). To maintain the shape and bandwidth of the odd- and even-mode frequency responses over the entire tuning range, the Q_{exo} and Q_{exe} of the tunable filter must satisfy Eqs. (13.5) and (13.6), respectively. This may be achieved by properly choosing the transformer parameters w_{11}, l_1, s, and C_s shown in Figure 13.15. Following the discussion on dual-mode open-loop filters in Chapter 10, in general, reducing w_{11} and s can increase the I/O coupling leading to a smaller Q_{exo} or Q_{exe}. In addition, employing a large value of C_s can also enhance the I/O coupling.

13.3.2.4 Design Procedure and Examples Based on the previous discussions, a design procedure for this type of tunable filter may be developed as follows.

Step1: Determine the requirements of the ideal tunable dual-mode filter
The requirements of the ideal tunable filter may be derived from its fixed-frequency response centering at the high-frequency edge of a given tuning range. The separation

(A) of the odd- and even-mode resonant frequencies and the required Q_{exo} and Q_{exe} for the ideal tunable filter then can be determined from Eqs. (13.4–13.6). Using a prescribed $n + 2$ coupling matrix [see the example associated with Eq. (10.29) given in Chapter 10], these required parameters are ready to be calculated.

Step 2: Design of the even-mode tuning rate
The tuning rate of the even mode may be designed to match the tuning rate of the odd mode by using Eqs. (13.11) and (13.12) with $B = A$ for a constant absolute bandwidth tuning. As there are six degrees of freedom to determine the tuning rate of the even mode, that is, w_3, l_3, w_4, l_4, w_5, and l_5, four parameters may be chosen first, and the last two can then be determined by Eqs. (13.11) and (13.12). Usually, w_4, l_4, w_5, and l_5 can be chosen first, because their main purpose is to shorten l_3 so as to have a compact loading element for the even mode inside the open loop (see Figure 13.15) and their effect on the even-mode tuning rate can easily be compensated by l_3 and w_3 (see the above discussion associated with Figure 13.18).

Step3: Design of the I/O transformer
To maintain the shape and bandwidth of odd- and even-mode frequency responses over the tuning range, the simulated Q_{exo} and Q_{exe} of the tunable filter must satisfy the required ones by properly choosing the transformer parameters w_{11}, l_1, s, and C_s.

Two design examples are described below. The first one (filter A) is designed with the following specifications:

Tunable range:	$0.6 \sim 1.07$ GHz
Fractional bandwidth (*FBW*)	2.9% at 1.07 GHz
Number of poles:	2

In addition, the filter is to have a finite-frequency transmission zero located at the high side of the passband. To derive the required design parameters for the tunable filter, a filter centering at 1.07 GHz can be designed, first, with a finite-frequency transmission zero located at 1.12 GHz, a fractional bandwidth of 2.9%, and passband return loss of -20 dB. The desired parameters may be derived from a target-coupling matrix corresponding to the prescribed response, which are $f_0^o = 1.053$ GHz, $f_0^e = 1.098$ GHz, $Q_{exo} = 27.2$, and $Q_{exe} = 105.2$. A microstrip substrate with a relative dielectric constant of 10.2 and a thickness of 1.27 mm is used for realizing the filter. The odd-mode resonator is simply formed using a 50-Ω line to have a tuning rate similar to that shown in Figure 13.17b. Although it is possible to use other characteristic impedance line, the resultant tuning rate will be different. The loading capacitance C_v is chosen to be 0.6 pF, and the electrical length of the odd-mode resonator can then be determined initially for $f_0^o = 1.053$ GHz. Steps 2 and 3 of the above design procedure are then repeated until some best-fit results for the even-mode tuning rate and the external quality factors are obtained. Figure 13.19 shows the extracted design curves (dash line) compared with that of the desired ones (full line). It can be seen that the Q_{exe} does not meet the ideal values after the even-mode

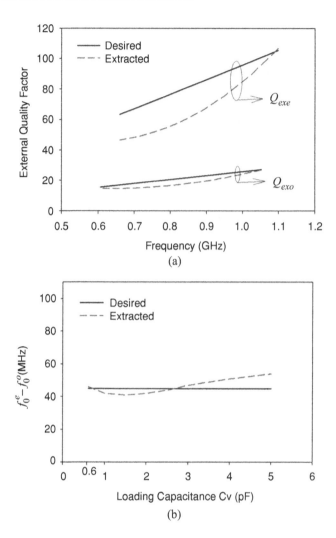

FIGURE 13.19 Design curves of filter A. (**a**) External quality factors (Q_{exo}, Q_{exe}). (**b**) The separation of the modal frequencies ($|f_0^e - f_0^o|$).

tuning rate being modified. This is because the parameters of the even mode have effects on both Q_{exe} and the tuning rate. Therefore, there is a trade-off in the design and, as a result, the designed bandwidth would be slightly different from the desired specification.

Referring to Figure 13.15, the parameters for the final layout of this tunable filter design are given in Table 13.3. Figure 13.20 shows the fabricated filter. The capacitance C_s is realized by AVX chip capacitor [80], while the variable capacitance C_v is implemented by M/A COM MA46H202 varactor [81]. In this case,

TABLE 13.3 Parameters of the Tunable Dual-Mode Filter (Filter A) (in millimeters) (Referring to Figure 13.15)

w_{11}	w_2	w_3	w_4	w_5	w_{12}
0.2	0.7	0.5	0.3	1.2	1.0
l_1	l_2	l_3	l_4	l_5	s
17.7	12.7	13	8.5	2	0.2
C_s	2.8 pF		C_v	0.6 to 5.0 pF	

three varactors are used, which are applied with a single dc bias circuit. The measured results for a dc bias ranging from 2.2 to 22.0 V are plotted in Figure 13.21 along with the EM-simulated ones. The EM simulations, performed using Sonnet *em* [82], assume the capacitance C_v of the loading varactors changing from 5.0 to 0.6 pF. One can observe that the measured tunable characteristics are in good agreement with the simulated ones. The experimental varactor-tuned bandpass filter shows a high selectivity on the high side of the passband with less than 1.8-dB insertion loss and more than 10 dB return loss over a tuning range of 41% from 0.6 to 1.03 GHz. The measured 3-dB bandwidth is 85± 5 MHz.

The second filter, namely, filter B, is designed to exhibit a finite-frequency transmission zero located at the low side of the passband. Two possible layouts for this filter design are illustrated in Figure 13.22, where the one on the left has a symmetrical varactor loading scheme for the even modeand the one on the right has an asymmetrical varactor loading scheme for the even mode. In either case, the location of loaded varactor for the even mode is different from that of filter A. This is because the even-mode resonant frequency of the filter B is required to be lower than that of the odd-mode in order to produce a finite-frequency transmission zero located at

FIGURE 13.20 Photograph of the fabricated tunable dual-mode filter (filter A).

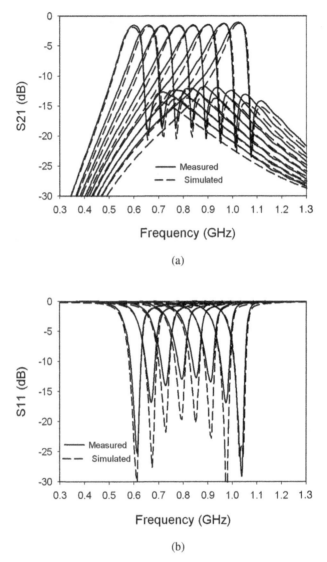

FIGURE 13.21 Measured and EM simulated S parameters of the tunable dual-mode filter (filter A). (**a**) S_{21} and (**b**) S_{11}. The bias voltage is between 2.2–22 V for the measurement.

low side of the passband. By employing the similar design procedure, filter B with either a symmetrical or asymmetrical varactor loading scheme can be designed. The designed filter parameters are listed in Table 13.4. Both designs result in the same filtering responses as shown in Figure 13.23. In this case, the only difference between the two designs, that is, the symmetrical and asymmetrical, is the loading capacitance for the even mode, as indicated in Table 13.4. This implies that the two layouts of

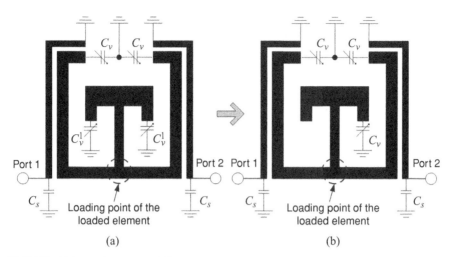

FIGURE 13.22 The layout of filter B (**a**) with symmetrical structure and (**b**) with asymmetrical structure. (Dimensions refer to Figure 13.15.)

Figure 13.22 can be equivalent as long as the loading point, as indicated, for the even-mode element is symmetrical with respect to the I/O ports, which allows a more flexibility in the design.

The experimental filter for filter B, fabricated using the same substrate as that for filter A, is shown in Figure 13.24. It's EM-simulated and measured responses are displayed in Figure 13.25. The experimental varactor-tuned bandpass filter shows a high selectivity on the low side of the passband with less than 2.2-dB insertion loss

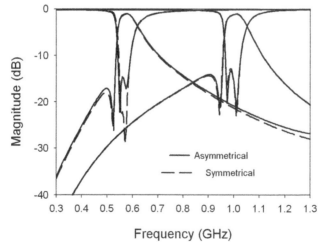

FIGURE 13.23 The EM-simulated performance of filter B with symmetrical and asymmetrical varactor loading for the even mode.

TABLE 13.4 Parameters of the Tunable Dual-Mode Filter (Filter B) (in millimeters) (Referring to Figures 13.15 and 13.22)

w_{11}	w_2	w_3	w_4	w_5	w_{12}
0.2	0.7	0.3	0.5	0.3	1.0
l_1	l_2	l_3	l_4	l_5	s
17.7	12.7	14	6.5	3.9	0.2
C_s	2.8 pF		C_v^1	0.41 to 2.95 pF (sym)	
			C_v	0.6 to 5.0 pF (asym.)	

and more than 10-dB return loss over a tuning range of 41% from 0.57 to 0.98 GHz. The applied bias voltage is between 2.2 and 22.0 V. The measured 3-dB bandwidth is 91 ± 6 MHz over the tuning range.

13.3.3 Tunable Four-Pole Dual-Mode Filter

Two or more dual-mode open-loop resonators can be cascaded to build up a higher-order tunable filter. For example, Figure 13.26 displays a fabricated four-pole tunable filter of this type [35]. Each dual-mode, open-loop resonator is loaded with three Infineon BB857 varactors. The whole filter is tuned using a single dc bias circuit. The measured results for a dc bias ranging from 10.6 −34.0 V are plotted in Figure 13.27, compared with the simulation for different loading varactor capacitances.

Figure 13.27 shows that the experimental four-pole tunable bandpass filter exhibits a quasielliptic-function response with a finite-frequency transmission zero on each side of the passband. The transmission zero close to the low side of the passband is inherently associated with the even mode of the first dual-mode resonator, whereas the transmission zero close to the high side of the passband is inherently associated

FIGURE 13.24 Photograph of the fabricated tunable dual-mode filter (filter B).

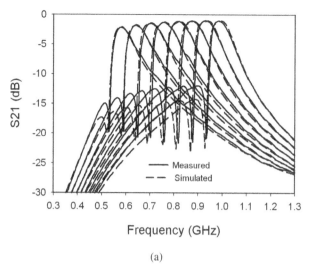

(a)

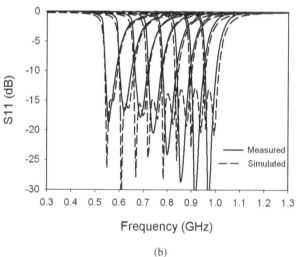

(b)

FIGURE 13.25 Measured and EM-simulated S parameters of the tunable dual-mode filter (filter B). (**a**) S_{21} and (**b**) S_{11}. The bias voltage is between 2.2–22 V for the measurement.

FIGURE 13.26 Photograph of the fabricated four-pole tunable dual-mode filter.

590 TUNABLE AND RECONFIGURABLE FILTERS

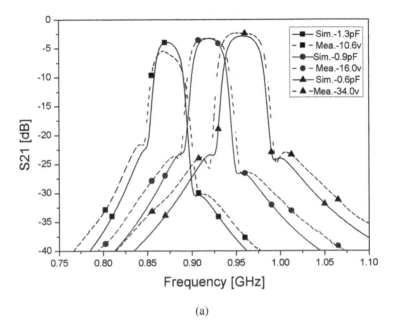

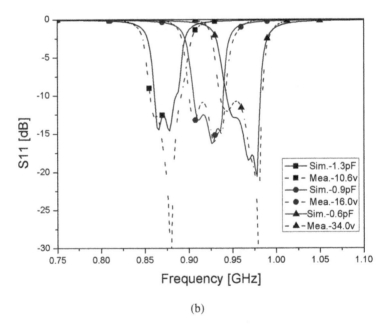

FIGURE 13.27 Measured tunable filter performance as compared to that simulated. (**a**) S_{21}. (**b**) S_{11}.

with the even mode of the second dual-mode resonator. Therefore, when the even-mode frequency is tuned, the associated transmission zero shifts accordingly. For the given dc bias voltage range, the filter is tuned over a tuning range from 0.86 to 0.96 GHz. The tubing range can be improved by employing a wideband transformer for I/O coupling [84], as described in the last section. The insertion loss in the passband is mainly because of the loss from the varactors used, which can be reduced by using alternative varactors with lower loss or higher Q.

13.4 WIDEBAND FILTERS WITH RECONFIGURABLE BANDWIDTH

In comparison to tunability of center frequency, there has been little effort made in tunability of bandwidth. A reason for this is the lack of methods to vary the inter-resonator coupling electronically essential for bandwidth control. There have been some papers which tackle this subject in a variety of ways, some concentrating on the tunability of bandwidth at a fixed center frequency [45,46] and others dealing with simultaneous control of both bandwidth and center frequency [47–49]. Upon inspection of the literature, it is clear that most reported filter topologies and methods used only apply to narrowband applications. However, there is increasing demand for tunable wideband filters. Typical applications would be receiver modules and RF converters, where filter banks can be replaced by such a filter in order to save board space. A new topology has been developed to achieve such a filter with a reconfigurable bandwidth [38,39]. Shown in Figure 13.28 is the basic filter building block, which mainly consists of short-circuited coupled lines and stubs that have an electrical length of 90° at the center frequency of the passband. The tuning elements used are p-i-n diodes, as they are much less lossy than the semiconductor varactor diodes, thus limiting the degradation in passband width and filter operation.

The tunability of the filter is based on the concept that, for a given pair of coupled lines with fixed even- and odd-mode impedance, altering the impedance of the stubs

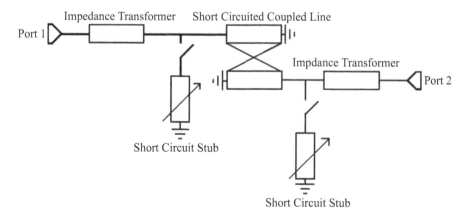

FIGURE 13.28 Filter building block for reconfigurable bandwidth.

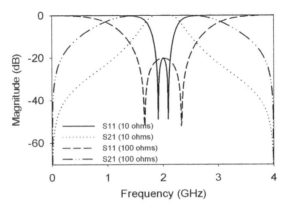

FIGURE 13.29 Theoretical response showing change of bandwidth with change of impedance of stub ($Z_{stubs} = 10\Omega$ - $100\ \Omega$, $\theta_{stub} = \theta_{coupled\ line} = 90°$, $Z_e = 110.4\ \Omega$, $Z_o = 18.8\ \Omega$).

alters the coupling and, in turn, alters the bandwidth, hence, by switching the stubs in and out of the circuit alters the passband width. It can be shown that the return loss in the passband only depends on the pair of even- and odd-mode impedances and is independent of the characteristic impedance of the other stubs (see Figure 13.29). This makes bandwidth tuning easy. In addition to this advantage, by simply cascading several same filter blocks of this type, a higher-order filter that increases the selectivity, while theoretically maintaining the same return loss over the same bandwidth, can be built. This trait is very desirable while designing wideband filters with a reconfigurable bandwidth (see Figure 13.30).

To obtain more reconfigurable bandwidth states, more short-circuited stubs can be deployed. Figure 13.31 shows the circuit model for the filter building block with four reconfigurable bandwidth states. Figure 13.32 depicts the circuit modeling results for four reconfigurable bandwidth states when (i) all short circuit stubs off; (ii) only stubs "A" with characteristic impedance Z_a on; (iii) only stubs "B" with characteristic impedance Z_b on; (iv) all the stubs on. As can be seen, the bandwidth of the filter is changed effectively over a wide frequency band, while keeping the same return loss level in the passband. This is because the theoretical return loss in the passband only depends on the pair of even- and odd-mode impedances and is independent of the characteristic impedance of the other stubs.

To validate the circuit concept described in the above section, a microstrip reconfigurable filter, based on the circuit model of Figure 13.31, is demonstrated through experimentally measured results, which are described as follows.

Figure 13.33 shows the reconfigurable microstrip filter developed. The filter is implemented on a substrate with $\varepsilon_r = 3$ and thickness = 1.02 mm. Semiconductor p-i-n diodes were used to switch on/off the short-circuit stubs. Each stub was designed to include two p-i-n diodes to improve isolation. The p-i-n diodes were modeled by using a capacitor of value 0.025 pF for isolation and a resistor of value of $4\ \Omega$ for connection (these values being obtained from the data sheet for the diode used). The initial dimensions of the circuits were obtained using Microwave Office

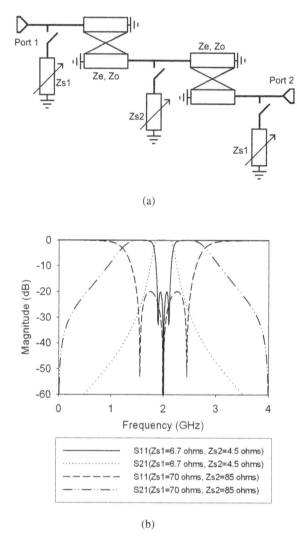

FIGURE 13.30 (a) Filter with two blocks cascaded. (b) Theoretical response with all stubs switched on (Z_{s1} = 6.7 Ω - 70 Ω, $Zs2$ = 4.5 Ω - 85 Ω, Z_e = 110.4 Ω, and Z_o = 18.8 Ω).

[83] and finalized using EM software Sonnet [82]. The dimensions of the filter layout needed to be optimized in order to compensate for the effects because of the loading elements and the transmission-line discontinuities. The final dimensions are as follows: $w(width)$ = 5.6 mm and $l(length)$ = 20.1 mm (narrowband stubs); $w(width)$ = 0.2 mm and $l(length)$ = 25.8 mm (wideband stubs). The coupled lines have a gap s = 0.2 mm, $l\,(length)$ = 26.85 mm, and $w\,(width)$ = 0.5 mm. In this case, the impedance transformer has a width and length of 0.6 and 25 mm, respectively.

The filter was fabricated using a Rogers Duroid 3003 substrate in compliance with the design. Figure 13.34 illustrates the fabricated filter with bias lines to apply bias

594 TUNABLE AND RECONFIGURABLE FILTERS

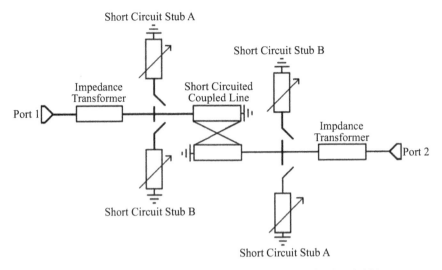

FIGURE 13.31 Filter building block with four reconfigurable bandwidth states.

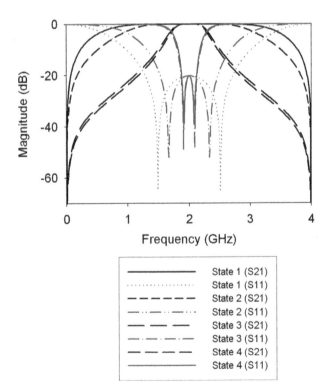

FIGURE 13.32 Theoretical response showing four bandwidth states for $Z_a = 100\ \Omega$, $Z_b = 10\ \Omega$, $Z_e = 110.4\ \Omega$, and $Z_o = 18.8\ \Omega$.

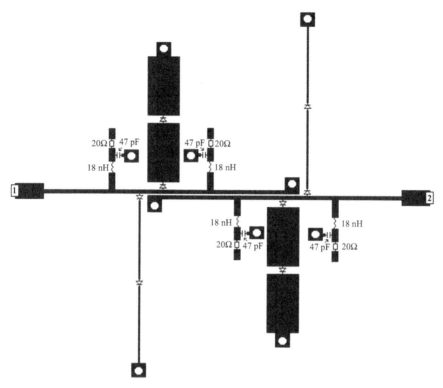

FIGURE 13.33 Layout of the filter with four reconfigurable bandwidth states.

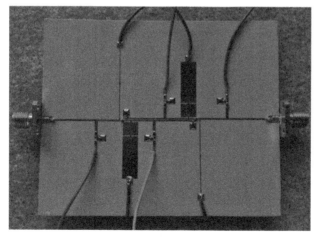

FIGURE 13.34 Fabricated filter for four reconfigurable bandwidths.

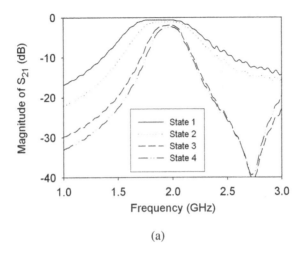

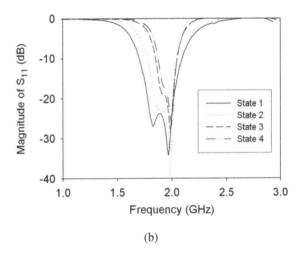

FIGURE 13.35 Measured responses of the reconfigurable filter. (**a**) Insertion Losses (S21). (**b**) Return losses (S11). State 1: all diodes "off;" state 2: wideband stubs "on" and narrowband stubs "off;" state 3: narrowband stubs "on" and wideband stubs "off;" state 4: all diodes "on."

voltages to the circuit using MACOM PIN diodes (MA4AGSBP907). Figure 13.35 shows the measured responses.

The fabricated filter shows four reconfigurable 3-dB fractional bandwidths of 33.5%, 23.23%, 14.65%, and 13.06% with the measured insertion losses of 0.62, 1.07, 1.19, and 2.5 dB at around 1.9 GHz, respectively. The insertion losses associated with the narrowband state are higher, as expected. Note that each short-circuited stub uses two p-i-n diodes to improve the isolation, which, however, increases the insertion loss. It was found that moving the diode that intersects the stub away from its short-circuited end can decrease the insertion loss. However, a spike begins to appear closer to the passband. Furthermore, because of unequal even- and odd-mode phase velocities, a spurious response is caused around 3.8 GHz ($2f_0$). This is because of the fact that microstrips are not pure TEM transmission lines. The noise figure of the filter may be considered equal to its insertion loss since the p-i-n diodes are basically a resistive element.

13.5 RECONFIGURABLE UWB FILTERS

Emerging ultra-wideband (UWB) technology requires the use of a wide radio spectrum. However, the frequency spectrum as a resource is valuable and limited, so the spectrum is always being used for several purposes, which means it is full of unwanted signals when an operation such as a UWB wireless system is concerned. In this case, existing undesired narrowband radio signals, which vary from place to place and from time to time, may interfere with the UWB system's range. A solution for this is to introduce an electronically switchable or tunable narrow rejection band (notch) within the passband of a UWB bandpass filter [23,24]. Such an electronically reconfigurable filter is also desired for wideband radar or electronic warfare systems.

13.5.1 UWB Filter with Switchable Notch

Figure 13.36 depicts the schematic for a UWB bandpass filter with a switchable notched band. The UWB bandpass filter is basically an optimum distributed highpass filter consisting of five short-circuited stubs and four connecting lines on microstrip, as described in Chapter 12. For the filter design, all the short-circuited stubs have an electrical length of 90° and all the connecting lines have an electrical length of 180° at a center frequency of 6.85 GHz. Stubs 1 and 5 have a line impedance of 78.02 Ω; stubs 2 and 4 have a line impedance of 43.8 Ω; stub 3 has a line impedance of 39.6 Ω. The connecting lines between stubs 1 and 2 and 4 and 5 have an impedance of 56.94 Ω; the connecting lines between stubs 2 and 3 and 3 and 4 have an impedance of 60.37 Ω. Two identical switchable notch structures are integrated into the two connecting lines as indicated. A wideband dc bias circuit is added. Three bypass capacitors, which have low impedance over the desired UWB frequency band, are inserted as shown to prevent the dc bias from being short circuited to ground.

Two different schemes of switchable notch structure are illustrated in Figure 13.37 for a comparison study. In principle, they are a section of transmission line with an embedded stub (see Section 12.5.1). The transmission line has a width of W and the embedded stub has a width of W_1 and a length of l_1. The electrical length of l_1 is about 90° at the desired notch frequency.

For scheme A of the switchable notch structure shown in Figure 13.37a, a p-i-n diode is attached at the open end of the embedded stub. Two bypass capacitors

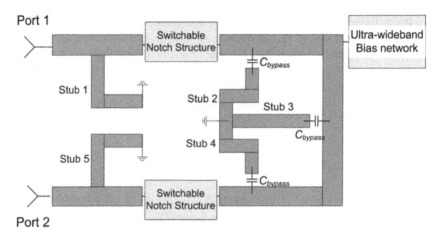

FIGURE 13.36 The schematic diagram of a reconfigurable UWB bandpass filter with switchable notch structures.

are used to avoid a short circuit for the dc bias applied to the p-i-n diode. When the p-i-n is at the zero bias, it presents large impedance because of its very small junction capacitance and, hence, the embedded stub functions as an open-circuited stub that resonates. Thus, at its fundamental- resonant frequency, the embedded open-circuited stub presents a short circuit at the main transmission line, resulting in a narrow reject band or notch in the frequency response. This case corresponds to the notch on. To switch the notch off, a forward bias is applied to the p-i-n diode. At the forward bias, the p-i-n diode behaves like a very small resistance so that the open end of the embedded stub is basically joined to the main transmission line. Thus there is no resonance from this embedded stub. As a consequence, the notch disappears.

On the contrary, for scheme B of the switchable notch structure shown in Figure 13.37b, an applied forward bias will switch the notch on and a zero bias will switch the notch off. In principle, both schemes should work the same; however,

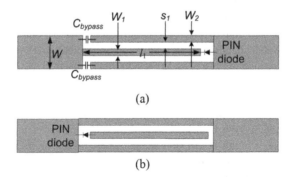

FIGURE 13.37 Switchable notch structures. (**a**) Scheme A; (**b**) scheme B.

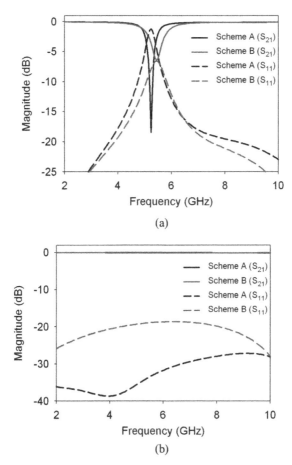

FIGURE 13.38 EM-simulated results for the switchable notch structures of Figure 13.37. (a) Notch switched on; (b) notch switched off.

because of the loss associated with the p-i-n diode, their performances are quite different in regard to the rejection level at the notched band. This is demonstrated with the EM-simulated results shown in Figure 13.38.

In the simulation, both structures in Figure 13.37 have dimensions $W = 1.3$ mm for a 50-Ω line, $W_1 = 0.2$ mm, $l_1 = 9.4$ mm, $W_2 = 0.35$ mm, and $s_1 = 0.2$ mm on a dielectric substrate with a relative dielectric constant $\varepsilon_r = 3.05$ and a thickness of 0.508 mm. The p-i-n diode is assumed to have a capacitance of 0.025 pF at the zero bias and to have a 4-Ω resistance under a forward bias. For scheme A, the bypass capacitance is 33 pF. It is evident, from Figure 13.38, that scheme A is superior to scheme B for two reasons: (i) the rejection at the notched band is much higher; and (ii) the return loss performance is better when the notch is switched off. This is because, for scheme A, the p-i-n diode is at off state when the embedded quarter-wavelength resonator resonates (the notch on) so that little loss is introduced by the p-i-n diode to degrade the quality factor of the resonator leading to a higher rejection. On the other hand, when the notch is switched off in the scheme A, the p-i-n diode is on, and, thus, the embedded line becomes an integrated part of the 50 Ω line causing less physical distortion of the 50 Ω line and resulting in a better matching. To demonstrate this type of filter experimentally, Figure 13.39a displays a fabricated reconfigurable UWB bandpass filter with a switchable notch band, where two identical notch structures (scheme A) are implemented. M/A-COM p-i-n diode (MA4AGSBP907) is implemented for a switch, which has a typical forward resistance of 4 Ω (determined by measuring its insertion loss in a 50-Ω line at 10 GHz when switched on) and 25 fF of low capacitance (determined by measuring its isolation in a 50-Ω line at 10 GHz when switched off). The filter is fabricated on a microstrip substrate with a relative dielectric constant of 3.05 and thickness of 0.508 mm by using a printed-circuit technology. Scattering parameter measurements are performed using Agilent 8720B network analyzer. Figure 13.39b plots the measured S_{21} responses for a notch band being switched on/off at the center frequency of around 5.1 GHz and, when it is turned on, its rejection ratio is more than 35 dB. To switch the notch on, the p-i-n diode is at a 0 V bias. To switch the notch off, it is found that the filter performance is almost the same when the p-i-n diode is subject to a forward bias of

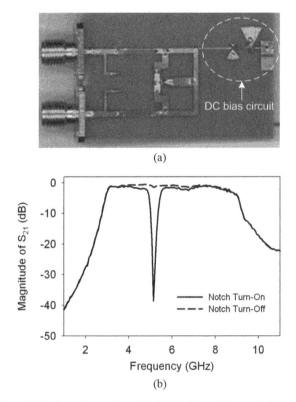

FIGURE 13.39 (a) Fabricated reconfigurable UWB filter with a switchable notch band. (b) Measured performance.

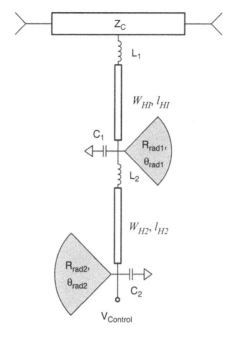

FIGURE 13.40 Schematic of a UWB dc bias circuit.

2.5 to 5 mA. The measured minimum insertion loss is 0.5 dB and the measured 3-dB bandwidth is 5.92 GHz.

A bias network can be critical for a wide operating frequency band. Since the p-i-n diodes are used, resistors should be avoided to minimize dc power consumption. Although a dc bias circuit with all lumped-element LC components may be used, the self-resonance frequency of lumped-element LC components could cause a problem for wideband and high-frequency operation. To overcome this, a mixed lumped- and distributed-element dc bias circuit, as depicted in Figure 13.40, is used in the experimental filter, as indicated in Figure 13.39a. The dc bias circuit parameters/dimensions (on the same aforementioned substrate) are as follows: $L_1 = 8.2$ nH, $L_2 = 33$ nH, $C_1 = C_2 = 33$ pF, $W_{H1} = W_{H2} = 0.2$ mm with a line impedance of 121 Ω, $l_{H1} = 7.4$ mm with an electrical length of 90° at 6.8 GHz, $l_{H2} = 4.5$ mm with an electrical length of 90° at 11.2 GHz, $R_{rad1} = 3.6$ mm, $R_{rad2} = 5.0$ mm, $\theta_{rad1} = 76.7°$, and $\theta_{rad2} = 88.3°$. It is shown that this dc bias circuit has little effect on RF transmission over the desired UWB from 3.1 to 10.6 GHz [23].

13.5.2 UWB Filter with Tunable Notch

In Figure 13.39a, replacing the p-i-n diode with a varactor diode leads to a tunable notch structure. Hence, the reconfigurable UWB filter described previously can be modified to have an electronically tunable notch band. For the experimental demonstrations, the reconfigurable UWB filter with a tunable notch band was implemented on a liquid crystal polymer (LCP) substrate with a relative dielectric constant of 3.0

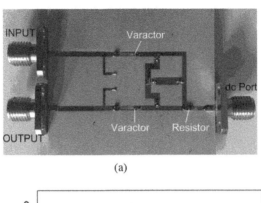

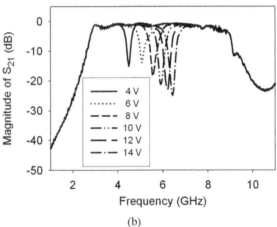

FIGURE 13.41 (a) The fabricated reconfigurable UWB filter with a tunable notch using varactor diodes. (b) Measured performance.

and thickness 0.5 mm. Figure 13.41a illustrates the fabricated filter. Its layout is similar to that shown in Figure 13.39a, except that varactors are used instead of the p-i-n diodes. In this case, GaAs constant gamma flip-chip varactor diodes (MA46H120) from M/A-COM were used for the electronic tuning [24]. The varactor is simply connected to the dc voltage via a 10-kΩ resistor.

Figure 13.41b gives the measured responses of the reconfigurable UWB filter. As can be seen, when the dc bias changes from 4 to 14 V, the notch band is tuned from 4.5 to 6.5 GHz within the UWB passband. It can be shown that, when there is no dc bias (0 V), there is no notch in the passband [24]. This is because the varactor capacitance is so large at the 0 V bias that it shifts the notch frequency well below the passband.

13.5.3 Miniature Reconfigurable UWB Filter

By deploying a multilayer structure, miniature reconfigurable UWB filters, such as that with a switchable notch band can be realized [85]. Figure 13.42 shows a 3D

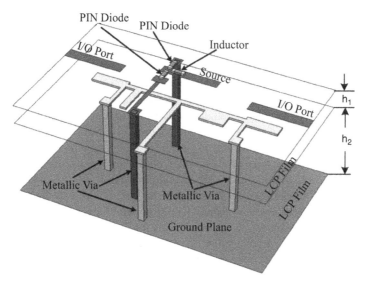

FIGURE 13.42 3D structure of a multilayer UWB bandpass filter with a switchable notch.

structure of this type of filter, which includes three metal layers, namely, top and middle layer and a solid ground plane. On the middle layer, a UWB bandpass filter without notching response is comprised of embedded short- and open-circuited stubs with embedded microstrip connecting lines. On the top layer, 50-Ω microstrip feed lines are used for input/output ports and broadside coupling is adopted to achieve the tight coupling from input/output feed lines to the middle layer circuit. To realize a switchable notch, additional arrangements including a notch structure and dc bias circuit need to be integrated. For this purpose, a short-circuited quarter-wavelength resonator, which resonates at the desired notch frequency, is designed on the top layer and is coupled to the middle layer UWB bandpass filter with a broadside coupling.

The operation of this reconfigurable UWB filter is described as follows. In order to make the notched band switchable, the open end of the quarter-wavelength resonator is connected to a pair of p-i-n diodes, as indicated in Figure 13.42. The two p-i-n diodes are head-to-head connected, and a dc bias is fed at their connection node. Thus, the two p-i-n diodes can be switched on/off at the same time with a single dc bias. When the dc bias is off, there is no dc current to drive the p-i-n diodes and the p-i-n diodes are off as well. In this case, the quarter-wavelength resonator functions properly and results in a notch at the desired frequency within the UWB passband. When a dc current is applied to drive the two p-i-n diodes, the p-i-n diodes are switched on, leading to an additional short circuit of the quarter-wavelength resonator on the other end. As a result, the quarter-wavelength resonator becomes a half-wavelength resonator with both ends short-circuited, which is resonating at a frequency that is higher than the highest frequency of the passband of the designed UWB bandpass filter and, hence, the notch band is switched off.

As illustrated in Figure 13.42, to control the p-i-n diodes, a single source is used for switching. Since the notching response has a narrow frequency band and the dc bias circuit on the top layer can be isolated from the UWB bandpass filter on the middle layer, a complicated wideband dc bias circuit, such as the one described in Section 13.5.1, is not required in this design. As a result, a simple dc bias circuit consisting of only a chip inductor is used for the designed filter, which is advantageous.

The design of this reconfigurable UWB filter has three steps. In the first step, the middle layer UWB bandpass filter without a notch is designed, based on an equivalent transmission-line circuit, which includes short- and open-circuited transmission lines. In the second step, the quarter wavelength resonator is integrated with the designed middle layer UWB filter. With the p-i-n diodes being switched on, the designed middle layer UWB filter needs to be tuned with the influence of broadside-coupled circuit on the top layer so as to realize a good passband response without notch. In the third step, the p-i-n diodes are switched off, and the quarter-wavelength resonator is resonating at desired notch frequency. The structure of the designed filter needs to be further adjusted to obtain a good return loss for the UWB bandpass filter with a notch. In the design, the second and third steps may need to be repeated until a good return loss is achieved for both cases, that is, with and without a notch.

A full-wave EM simulator from Sonnet Software [82] is employed for the filter design. The liquid crystal polymer film, which has a relative dielectric constant $\varepsilon_r = 3.15$ and a loss tangent $\tan \sigma = 0.0025$, is used to support the metal layers. The thickness between the top and middle layers, that is, h_1 referred to in Figure 13.42, is 0.15 mm and the thickness between the middle layer and ground plane $h_2 = 0.35$ mm. Figure 13.43 shows the final physical layouts for the designed filter, where all dimensions are in millimeters.

The designed filter is fabricated using a multilayer liquid crystal polymer technology similar to that described in Chapter 10. Figure 13.44a is a photograph of the fabricated filter, where only the top layout circuit can be seen. The filter is 19.7 × 11.3 mm. In the implementation, a 39 nH chip inductor is used for the dc bias circuit. Two p-i-n diodes (MA4AGSBP907 from M/A-COM), which have a 4.2 Ω resistance when switched on, are used in the filter fabrication. The measured results of the reconfigurable UWB filter with and without a notch are plotted in Figure 13.44b. When the notch is turned off, which is the case when the two p-i-n diodes are switched on under a forward dc bias current, the measured filter has a 10-dB insertion loss bandwidth from 3.1 to 10.6 GHz with a measured insertion loss of 0.61 dB at the center frequency 6.8 GHz. When the dc bias for the p-i-n diodes is switched off, the notch is turned on. In this case, as can be seen from Figure 13.44b, a notched band is measured with a 3-dB bandwidth from 5.3 to 6.0 GHz and the measured rejection at 5.58 GHz is 17.8 dB.

13.6 RF MEMS RECONFIGURABLE FILTERS

13.6.1 MEMS and Micromachining

Microelectromechanical system (MEMS) provides a class of new devices and components which display superior high-frequency performance and which enable new

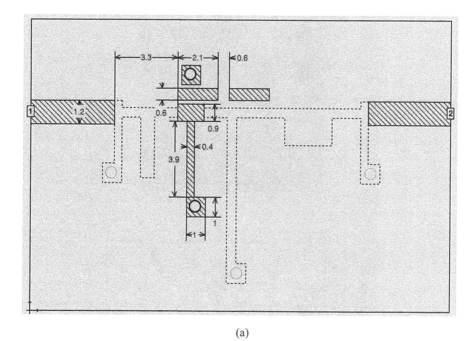

FIGURE 13.43 Layouts for the designed multilayer reconfigurable UWB filter. (**a**) Top layer; (**b**) middle layer. All dimensions are in millimeters.

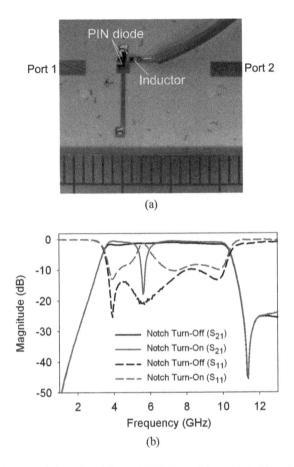

FIGURE 13.44 (a) Fabricated multilayer UWB filter with a switchable notch. (b) Measured responses.

system capabilities. For a general definition, a MEMS is a miniature device or an array of devices combining electrical and mechanical components and fabricated with integrated circuit (IC) batch-processing techniques [50–52]. There are several MEMS fabrication techniques including bulk micromachining and surface micromachining.

Bulk micromachining involves the creation of mechanical structures directly in silicon, gallium arsenide (GaAs), or other substrates by selectively removing the substrate materials. The process includes the steps of wet chemical etching, reactive-ion etching (RIE), or both to form the released or stationary microstructures. With wet etching, the resulting structures depend on the directionality of the etch, which is a function of the crystallinity of the substrate and the etching chemistry. The shape of the resulting microstructures becomes a convolution between the etch-mask pattern and the etching directionality. Therefore, the narrow deep microstructures generally pursued in bulk micromachining are difficult to achieve and better results are often achieved by dry etching with the RIE technique.

Surface micromachining consists of the deposition and lithographic patterning of various thin films, usually on silicon substrates. It may be intended to make one or more of the ("release") films freestanding over a selected part of the substrate, thereby able to undergo the mechanical motion or actuation characteristic of all MEMS. This can be done by depositing a "sacrificial" film (or films) below the released one(s); these are removed in the last steps of the process by selective etchings. The variety of materials for the release and sacrificial layers is great, including many metals (Au, Al, etc.), ceramics (SiO_2 and Si_3N_4), and plastics such as photoresist, polymethyl methacrylate (PMMA), and others. Depending on the details of the MEMS process and other materials in the thin-film stack, the release and sacrificial layers can be deposited by evaporation, sputtering, electrodeposition, or other methods.

Figure 13.45a demonstrates a structure of MEMS switch, which has been developed for improving RF power handling [53–56]. This MEMS switch consists of two pairs of bridge beams for a good stress balance. The bridges comprise four cantilever switching elements, pairs of which are joined head-to-head (as shown). Each of the cantilever switching elements is 600 × 150 μm, and the whole high-power switch chip, including a built-in dc bias circuit, is 1.05 × 1.75 mm on a 250-μm thick high-resistivity, silicon wafer. Figure 13.45b is the SEM image of a fabricated 2 × 2 MEMS switch.

The fabrication processes for this RF MEMS switch used a seven-layer mask set, illustrated in Figure 13.46, to produce a sandwiched- or balanced-bridge structure

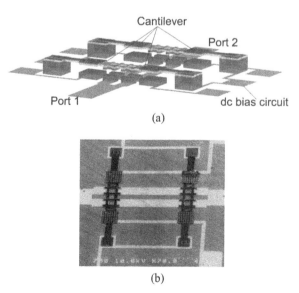

FIGURE 13.45 (a) Structure of a 2 × 2 MEMS switch. (b) SEM image of a fabricated 2 × 2 MEMS switch.

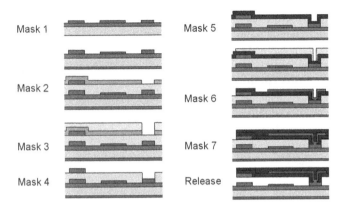

FIGURE 13.46 Fabrication process.

of dielectric-metal-dielectric. The mechanical nature of the balanced structure counteracts stresses in the beam, which are process induced or because of the inherent differences in the material properties of the dielectric and metal. The substrate chosen was 250 μm FZ-Silicon with 10 kΩ/cm resistively. The first stage involved initially depositing SiO_2 as an insulating layer followed by patterning the resist (mask 1); then, 5 nm of Cr was deposited followed by 2 μm Au. The resist was then removed using a lift-off process. The second stage involved raising the height level of the pillar and RF contact to be above that of the dc electrode to provide increased switch contact force. To achieve this, a layer of resist was deposited, followed by approximately 1 μm of Au being added to the pillar and RF lower contact region (mask 2). This was followed by another lift-off stage. A sacrificial layer was deposited next and the top contact metal Au was defined (mask 3). A thin Al layer was deposited on the front side surface as an etch-stop layer and a gold layer was deposited on the backside of wafer as a ground plane. The fourth stage defined the post hole by etching this in the sacrificial layer (mask 4). The photoresist and Al layer were then removed, followed by 0.5-μm deposition of SiN. In the next stage, photoresist was patterned (mask 5) and then the SiN layer was dry etched by reactive ion etch to provide the through hole to the electrode layer. The resist was then removed and approximately 1-μm Au was deposited for the top electrode (mask 6). Finally another 0.5-μm layer of SiN was deposited and defined (mask 7). After this the sacrificial layer was removed by plasma etching thus released the cantilever structure.

13.6.2 Reconfigurable Filters Using RF MEMS Switches

Using RF MEMS switches instead of the varactor to change the resonator length or its circuit parameters in a resonator filter forms a class of reconfigurable filters [10–16]. In this case, electronic tuning is normally carried out digitally, which can achieve a large tuning range with good performance, including low loss and high linearity. Figure 13.47 illustrates an example of this type of filter. In this case, as reported

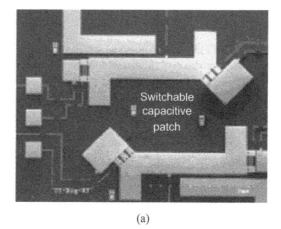

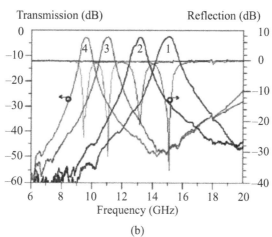

FIGURE 13.47 (a) Photograph of a two-bit RF MEMS reconfigurable filter and (b) its measured responses. (Taken from [11], © 2007 IEEE.)

in [11], the topology is based on distributed half-wavelength microstrip resonators, fabricated on a high-resistivity silicon substrate. This is a two-resonator bandpass filter and capacitive patches are added at the end of each resonator, which allow a pseudo two-bit center frequency shift with low loss. The measured response of the filter is shown in Figure 13.47b, where the passband is reconfigured at four different center frequencies.

A pioneering work on a differential four-bit RF MEMS tunable filter has been reported [13]. This filter demonstrates a wideband tunable filter over 6.5 to 10 GHz with 16 different filter responses adjacent to each other in frequency with very fine tuning resolution like a continuous tunable filter. More examples of RF MEMS reconfigurable filters can be found in a recent overview article [16].

610   TUNABLE AND RECONFIGURABLE FILTERS

13.7   PIEZOELECTRIC TRANSDUCER TUNABLE FILTERS

Piezoelectric transducers (PET) have also been used to develop electronically tunable microstrip filters [17–19]. Figure 13.48 illustrates the configuration of a PET tunable microstrip bandpass filter. As reported in [17], the tunable filter circuit consists of the filter using cascaded microstrip open-loop resonators [57], a PET, and an attached dielectric perturber above the filter. The PET is a composition of lead, zirconate, and titanate. The PET shown in Figure 13.48 consists of two piezoelectric layers and one shim layer. The center shim laminated between the two same polarization piezoelectric layers adds mechanical strength and stiffness. The shim is connected to one polarity of a dc voltage to deflect the PET and move it up or down vertically. As we can see from the structure in Figure 13.48, when the perturber moves up or down the effective dielectric constant of the filter decreases or increases, respectively, allowing the passband of the filter to shift toward higher or lower frequencies.

13.8   FERROELECTRIC TUNABLE FILTERS

Ferroelectrics have been studied since the early 1960s for application in microwave devices [58] and their properties have been studied extensively in the intervening

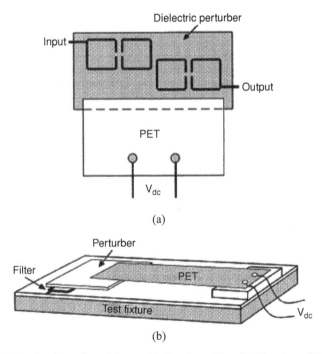

FIGURE 13.48 Configuration of the tunable bandpass filter. (**a**) Top view; (**b**) 3Dview. (Taken from [17], © 2003 IEEE.)

years. However, it is only relatively recently that applications are beginning to emerge [2–4,7,29,31,44,59–70]. The change in permittivity as a function of electric field is key to a wide range of applications.

13.8.1 Ferroelectric Materials

A ferroelectric material exhibits spontaneous polarization. Such a crystal can be seen to contain positive and negative ions. In a certain temperature range, the positive and negative ions are displaced. The displacement results in a net dipole moment. The orientation of the dipole moment in a ferroelectric can be shifted from one state to another by the application of an electric field. The appearance of the spontaneous polarization is highly temperature dependent and, in general, ferroelectric crystals have phase transitions, where the crystal undergoes structural changes [71]. This transition temperature is known as the Curie temperature (T_c) at which the material properties change abruptly.

Because of the nature of the crystal structure close to the Curie temperature, thermodynamic properties show large anomalies. This is usually the case with the dielectric constant, which increases to a large value close to the Curie temperature, as demonstrated in Figure 13.49; it is also the point where there is the largest sensitivity of the dielectric constant to the application of an electric field. Below the Curie temperature, the crystal is in polar phase, characterized by a hysteresis loop and the piezoelectric effect; in this phase, the dc-dependent permittivity has a butterfly shape [44]. Above the Curie temperature, the crystal is in paraelectric phase, with a bell-shaped dc-dependence of the permittivity; this phase is preferable for most agile microwave applications, because it has no hysteresis and, hence, the permittivity of the crystal may be determined unambiguously via the applied dc field.

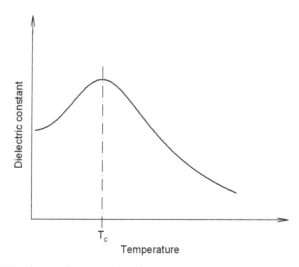

FIGURE 13.49 Curve of dielectric constant as a function of temperature.

Some materials which have shown a variable permittivity with electric field are: $SrTiO_3$, $(Ba,Sr)TiO_3$, $(Pb,Sr)TiO_3$, $(Pb,Ca)TiO_3$, $Ba(Ti,Sn)O_3$, $Ba(Ti,Zr)O_3$, and $KTaO_3$ dopents [72,73]. However, strontium titanate $SrTiO_3$ (STO) and barium strontium titanate $Ba_x Sr_{1-x}TiO_3$ (BST), where x can vary from 0 to 1, are two of the most popular ferroelectric materials current being studied for frequency-agile components and circuits. $SrTiO_3$ is of particular interest because of its crystalline compatibility with high-temperature superconductors (HTS) and its properties at low temperature. Pure STO is not supposed to have Curie temperature above 0K. Some thin films and amorphous ceramic forms show a low-temperature peak in the dielectric constant, implying that the Curie temperature is above 0K, probably due to stresses or impurities in the films. For BST, as the value of x varies from 0 to 1, the Curie temperature varies from the value of STO to about 400K, the Curie temperature of $BaTiO_3$ (BTO). This allows tailoring of the Curie temperature. Generally a value of $x = 0.5$ is used to optimize for room temperature and a value of around 0.1 used when the material is to be used in conjunction with HTS films.

There are a number of different forms of these materials which are of interest for applications. Single crystals have been studied for many years [74]. More recently, thin films of the materials have been studied. These films are almost exclusively made by laser ablation and are usually less than 1-μm thick. However, trilayer films have also been produced, forming an metal/ferroelectric/metal structure. Films on sapphire have also been produced with a CeO_2 buffer layer to compensate for the lattice and thermal expansion mismatch [75]. The Sol-Gel technique [76] for producing BST has also been developed. This technique is able to produce material that is of the order of 0.1-mm thick. More recently, lead strontium titanate $(Pb_{0.4}Sr_{0.6}(Ti_{1-x}Mn_x)O_3)$ or PST thin film deposited by Sol-Gel technique on a high-resistivity Si substrate has been reported [69,70].

13.8.2 Ferroelectric Varactors

Capacitive tunability of a ferroelectric varactor may be defined as

$$\text{Capacitive\_Tunability} = \frac{C_{\max} - C_{\min}}{C_{\max}} \times 100(\%) \tag{13.13}$$

where $C_{\max}$ is the capacitance of the varactor at 0 V bias and $C_{\min}$ is the capacitance obtained at a nonzero dc bias. $C_{\min}$ decreases as the absolute value of dc bias voltage increases. This is because the relative permittivity of a ferroelectric material in paraelectric phase is reduced by the applied dc field [44].

In general, a ferroelectric varactor can be designed in the form of a metal-insulator-metal (MIM) capacitor or an interdigital capacitor (IDC). The MIM capacitor usually exhibits a high capacitive tunability and small dc voltage is required. On the other hand, the interdigital capacitor may exhibit a small capacitive tunability because of additional fringing capacitance, which is less sensitive to the applied dc voltages. However, the IDC can be a more attractive option when a lower capacitance and a simpler fabrication process are desired.

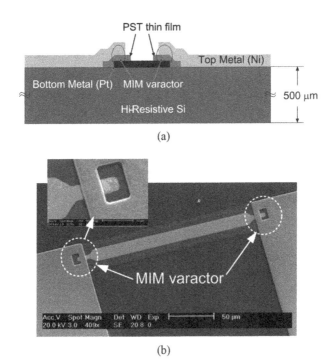

FIGURE 13.50 Ferroelectric MIM varactor. (**a**) Profile of layers. (**b**) SEM image of a fabricated MIM varactor.

Figure 13.50 demonstrates an example of MIM varactor, where a 0.3- to 0.5-μm thick PST thin film was deposited onto a high-resistivity Si substrate [69]. The use of high-resistivity silicon (>10 KΩ/cm) can minimize substrate conduction losses, which heavily occur in low-resistivity silicon. The silicon substrate is 525-μm thick, coated with a 1-μm SiO_2 polished buffer layer. Platinum was used for the bottom electrode, being a popular choice in thin-film ferroelectrics, mainly because of its excellent thermal properties, resistance to oxidation, and ability to promote PST growth. Standard photolithography and a lift-off process were used to define the bottom electrode. To form the specific capacitor areas and provide access to the bottom electrode, the PST film was then patterned using deionized water-diluted HF and HCl etch solution. To protect the relevant PST areas, a Shipley S18181 photoresist mask was used, again using standard photolithography. After the etching, the resist layer was lifted with acetone. In order to reduce the ohmic losses from a thin electrode, a nickel electroplating facility was used to deposit 1.7-μm thick nickel top electrodes.

For the given dimensions, that is, the thickness of the PST thin film and the overlapped area between top and bottom electrodes, the capacitance of MIM varactor is determined by applied dc voltage. In Figure 13.50, the varactor is actually formed by cascading two small MIM capacitors, each of which, as indicated, has an overlapped

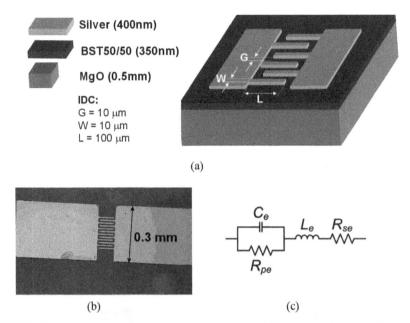

FIGURE 13.51 Ferroelectric interdigital varactor. (a) Profile of layers (not scaled). (b) Fabricated interdigital varactor chip. (c) The equivalent circuit of interdigital varactor.

area of 5×5 μm². The experiment at GHz frequencies reveals that the relative dielectric constant for the fabricated 0.3-μm thick PST thin film varies from 1100 at 0 V bias down to 500 at 6 V bias, which corresponds to a capacitive tunability of about 55% for the varactor. Larger tunability can be achieved by using a higher dc bias voltage. A typical fabricated PST MIM capacitor exhibits tunability at 1 MHz of 70 % with the change of the electric field strength from 0 to 30 kV/m. The loss tan δ of the PST film is measured from 0.004 to 0.02 at 1 MHz [69].

Figure 13.51 illustrates a typical ferroelectric varactor having a form of an IDC. BST varactors of this type have been developed based on the structure of Figure 13.51a [68]. In the fabrication, a 0.35-μm thick $Ba_{0.5}Sr_{0.5}TiO_3$ (BST) thin films were deposited onto (001) MgO substrates (of a thickness 0.5 mm) by pulsed laser deposition, using a laser fluence of 1.5 Jcm^{-2} at 5-Hz pulse rate in the presence of an oxygen gas background (0.1 mbar) [67]. The top silver metal on the structure of Figure 13.51a is used for patterning the desired IDC and its electrodes. An IDC has 12 fingers as shown in Figure 13.51b. The fingers, with a gap of 10 μm between them, are 100 μm long and 10 μm wide, respectively. As shown in Figure 13.52, a typical fabricated BST IDC exhibits a capacitance at 1 MHz varying from 0.48 to 0.69 pF with the change of the electric field strength from 4.0 to 0 V/μm. The loss tan δ of the BST film is measured from 0.005 to 0.01 at 1 MHz. Normally, for a ferroelectric material in paraelectric phase, the C-V curve should be symmetrical with respect to 0 V. The slightly asymmetric C-V curve of Figure 13.52 may originate from the

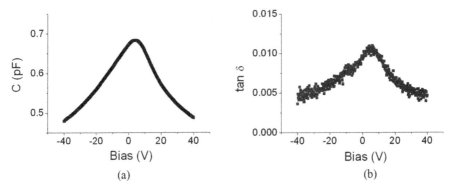

FIGURE 13.52 Typical properties of fabricated BST IDC against bias voltage. (**a**) Capacitance; (**b**) loss tangent.

movement of charged defects, such as oxygen vacancies, under the bias electric field, as discussed [77].

The capacitance and the loss vary as the operating frequency increases and the interdigital BST varactor can be modeled by an equivalent circuit of Figure 13.51c by means of RF measurements and electromagnetic simulations. The extracted values from the measurements are given in Table 13.5. As can be seen, this basic BST varactor cell with $C_{max} = 0.70$ pF at 0 V bias and $C_{min} = 0.42$ pF at 70 V bias has a capacitive tunability of 40% for the given range of the dc bias voltages according to Eq. (13.13).

13.8.3 Frequency Agile Filters Using Ferroelectrics

Frequency agile filters are among many other device applications of ferroelectrics. Such components have wide range applications in many communications and radar systems. Frequency agility in microwave circuits can be realized using ferroelectric thin films incorporated into conventional microstrip circuits. Electronically tunable filters can be produced with applications of interference suppression, secure communications, dynamic channel allocation, signal jamming, and satellite and ground-based communications switching. Many new systems concepts will appear as

TABLE 13.5 Equivalent Values with Bias Voltages

Bias Voltage [V]	Equivalent Values			
	C_e [pF]	L_e [nH]	R_{pe} [Ω]	R_{sc} [Ω]
0	0.70	0.5	16500	5
40	0.50	0.5	25000	5
70	0.42	0.5	30000	5

616 TUNABLE AND RECONFIGURABLE FILTERS

high-performance materials emerge; these systems will have considerably improved performance over conventional systems.

Ferroelectric tunable filters are fast, small, lightweight, and, because they work on electric field, have low power consumption. The range of tuning is quite large and devices are relatively simple in nature. The main problems currently being addressed are the relatively high-loss tangents of the practical ferroelectric materials and the large bias voltages required. This may be tackled by novel device structures.

For example, a microstrip defected or slotted ground structure, which is similar to that of Figure 10.52a, has been deployed to develop tunable bandstop filters using BST varactor chips [29,68]. Figure 13.53 illustrates a fabricated three-pole tunable bandstop filter of this type on a substrate having a relative dielectric constant of 3.05 and a thickness of 0.5 mm. The used tuning device is the BST interdigital varactor discussed above (see Figure 13.51). Its capacitance value varies from 0.70 down to 0.42 pF, according to the applied dc voltages from 0 to 70 V, as illustrated in Table 13.5. In order to practically apply this technology, a large BST varactor chip was produced, which can be easily attached on a conventional microwave circuit board. Its dimension is 2×1 mm^2 and consists of two large electrodes for attachment onto a microwave substrate circuit.

The bandstop filter includes bias circuits that provide proper dc voltages to the BST varactor and keep the RF responses similar to a bandstop filter without bias circuits. All the dc bias circuits are arranged on the ground plane of the microstrip filter, as

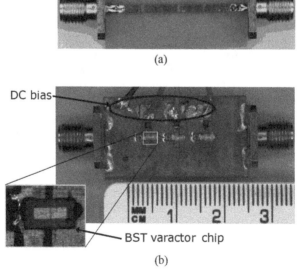

FIGURE 13.53 The fabricated tunable microstrip bandstop filter with BST varactor chips. (**a**) Top and (**b**) bottom.

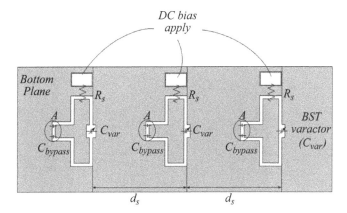

FIGURE 13.54 The schematic on the ground plane of the three-pole microstrip tunable bandstop filter with bias circuits.

shown in Figure 13.54. A high-resistance resistor is used instead of RF choke inductor because an ideal BST varactor does not consume dc currents and no voltage drop across the resistor occurs. A part of ground metal, (A in Figure 13.54), is cut to apply dc voltage to a BST varactor. Bypass capacitors, C_{bypass}, make short circuit at RF frequencies. Therefore, the overall structures operate equivalently to one without bias circuits.

The identical slot resonators on the ground plane are separated with a distance of d_s. This type of bandstop filter can also be considered as an electromagnetic bandgap (EBG) structure [78] having a general circuit model as in Figure 13.55, where P denotes the period, Z_c and β are the characteristic impedance and the propagation constant of the microstrip transmission line, respectively, and Z is the lumped element (resonator) loaded periodically. The parallel LC resonant circuit has an angular resonant frequency, $\omega_0 = 2\pi f_0 = 1/\sqrt{L \cdot C}$. The susceptance slope parameter of the resonator, given by Eq. (6.22), is proportional to C. If the susceptance slope parameter is larger, the coupling to the main signal line becomes weak leading to a narrower bandwidth. For the physical realization, $p = d_s$ is about a one-quarter

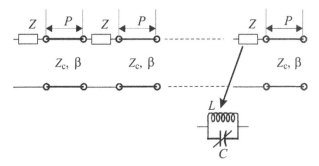

FIGURE 13.55 Circuit model for a microstrip transmission line coupled to periodic slotted ground structure.

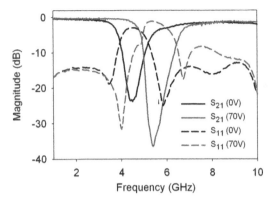

FIGURE 13.56 Measured results of a BST varactor tunable microstrip bandstop filter when the control voltage is adjusted from 0 to 70 V.

guided wavelength at f_0, and $Z_c = 50\ \Omega$. When the varactor is tuned, which is equivalent to a change of the capacitance value C, it is envisaged that the filter bandwidth will also change. At a lower frequency band, when the BST varactor has a larger capacitance, the fractional bandwidth is also smaller. This has been verified by measurement shown in Figure 13.56.

Figure 13.56 gives the measured responses of the bandstop filter in which we can observe that it operates at the center frequency from 4.585 to 5.52 GHz and has the 20-dB rejection bandwidth of 530 MHz at 4.585 GHz and 880 MHz at 5.52 GHz. Hence, the measured tuning range is 20%, and the fractional bandwidth is 11.6% and 15.9%, respectively. The rejection is smaller at 0 V bias. The reason for this is twofold. First, the fractional bandwidth is narrower leading to a higher loss; and second, the BST varactor itself has a higher loss at 0 V as shown in Figure 13.52b.

Other recently developed ferroelectric tunable filters include a fabricated tunable combline filter with the monolithic integration of BST thin films [3]. As reported in [3], for the dc bias, a resistive tantalum nitride (TaN) film is deposited and patterned above the BST films. The TaN film has a sheet resistance of approximately 1–10 kΩ/square and is used to route dc biasing signals to the circuits while minimally impacting the RF performance of the circuits. The integration of BST thin-film capacitors with alumina substrates, copper electrodes, and through-substrate vias provides several advances in tunable dielectric technologies, which allow for the development of more complex RF and microwave circuits. Under a 0–100 V bias, the X-band filter changes center frequency from 8.75 to 10.96 GHz with 4–8 dB of loss, while the Ku-band filter changes center frequency from 11.7 to 14.3 GHz with 6–10 dB of loss.

More recently, tunable ferroelectric filters with constant fractional bandwidths and constant return losses have been developed for X-band active electronically steered antenna (AESA) arrays used in radar systems [79]. The center frequency, the fractional bandwidth, and the return loss of the filters are simultaneously tuned through high-Q barium–strontium–titanate (BST) varactors with fast switching speed. The constant fractional bandwidth is achieved while tuning the center frequency and keeping the constant return loss. Two two-pole filters of this type have been demonstrated; each filter consists of two half-wavelength U-shaped resonators loaded with ferroelectric BST tunable capacitors at the open ends (one BST varactor for each resonator), which are used to tune the center frequency. Additional BST varactors are also used to control the inter-resonator coupling or I/O coupling. The filter designs are optimized for all the BST varactors having a tunable capacitance in the range of 20–60 fF. The filters are fabricated on a sapphire substrate with a thickness of 430 μm. The fabrication was carried out by using standard photolithographical techniques. Epitaxial $Ba_{0.45}Sr_{0.55}TiO_3$ thin films, 0.25 μm thick, were grown on c-plane sapphire substrates by using the combustion chemical vapor-deposition (CCVD) process. BST was first patterned and etched, following those resistor films that were deposited using CCVD and patterned. Metals were then evaporated and patterned by using a liftoff process and the typical metal stack was Ti/Cu/Au with a total thickness of 2 μm. An optional benzocyclobutene (BCB) layer was finally applied to passivate the filters. As reported in [79], a center frequency of 10 GHz and a frequency tuning range of 740 MHz (7.4%) are achieved with a low dc bias voltage of 30 V. The first tunable filter with constant fractional bandwidth has an equal-ripple bandwidth of 8.45%. The second filter with a constant fractional bandwidth and a constant return loss exhibits an equal-ripple bandwidth of 8% and a return loss of 17 dB. The measured insertion loss is between 2.7–2.2 dB for both filters, with a dc bias range of 0 to 30 V.

REFERENCES

[1] J.-S. Hong, Reconfigurable planar filters, *IEEE Microwave Magazine* **10**(6), 2009, 73–83.

[2] J. Nath, D. Ghosh, J.-P. Maria, A. I. Kingon, W. Fathelbab, P. D. Franzon, M. B. Steer, An electronically tunable microstrip bandpass filter using thin-film barium-strontium-titanate (BST) varactors, *IEEE Trans. Microwave Theory Techn.* **53**(9), 2005, 2707–2712.

[3] J. Sigman, C. D. Nordquist, P. G. Clem, G. M. Kraus, P. S. Finnegan, Voltage-controlled Ku-band and X-band tunable combline filters using barium-strontium-titanate, *IEEE Microwave Wireless Components Lett.* **18**(9), 2008, 593–595.

[4] I. Vendik, O. Vendik, V. Pleskachev, A. Svishchev, and R. Wordenweber, Design of tunable ferroelectric filters with a constant fractional band width, In *2001 IEEE MTT-S International Microwave Symposium Digest*, Vol. 3, 20–25 May 2001, 1461–1464.

[5] W. M. Fathelbab, and M. B. Steer, A reconfigurable bandpass filter for RF/microwave multifunctional systems, *IEEE Transactions Microwave Theory Techn.* **53**(3), Part 2, 2005, 1111–1116.

[6] G. Torregrosa-Penalva, G. Lopez-Risueno, and J. I. Alonso, A simple method to design wide-band electronically tunable combline filters, *IEEE Trans. Microwave Theory Techn.* **50**(1), Part 1, 2002, 172–177.

[7] I. Vendik, O. Vendik, V. Pleskachev, and M. Nikol'ski, Tunable microwave filters using ferroelectric materials, *IEEE Trans. Appl. Superconductivity*, **13**(2), Part 1, 2003, 716–719.

[8] B.-W. Kim, and S.-W. Yun, Varactor-tuned combline bandpass filter using step-impedance microstrip lines, *IEEE Trans. Microwave Theory Tech.* **52**(4), 2004, 1279–1283.

[9] M. Sanchez-Renedo, High-selectivity tunable planar combline filter with source/load-multiresonator coupling, *IEEE Microwave Wireless Components Lett.* **17**(7): 2007, 513–515.

[10] G. M. Kraus, C. L. Goldsmith, C. D. Nordquist, C. W. Dyck, P. S. Finnegan, F. Austin, IV, A. Muyshondt, and C. T. Sullivan, "A widely tunable RF MEMS end-coupled filter, In *IEEE MTT-S International Microwave Symposium Digest*, Vol. 26-11 June 2004, 429–432.

[11] P. Blondy, C. Palego, M. Houssini, A. Pothier, and A. Crunteanu, RF-MEMS reconfigurable filters on low loss substrates for flexible front ends, In *Asia-Pacific Microwave Conference APMC 2007*, 11–14 Dec. 2007, 1–3.

[12] A. Pothier, J.-C. Orlianges, G. Zheng, C. Champeaux, A. Catherinot, D. Cros, P. Blondy, and J. Papapolymerou, Low-loss 2-bit tunable bandpass filters using MEMS DC contact switches, *IEEE Trans. Microwave Theory Techn.* **53**(1) 2005, 354–360.

[13] K. Entesari, and G. M. Rebeiz, A differential 4-bit 6.5-10-GHz RF MEMS tunable filter. *IEEE Trans. Microwave Theory Techn.* **53**(3), Part 2, 2005, 1103–1110.

[14] I. C. Reines, C. L. Goldsmith, C. D. Nordquist, C. W. Dyck, G. M. Kraus, T. A. Plut, P. S. Finnegan, F. Austin, IV, and C. T. Sullivan, A low loss RF MEMS Ku-band integrated switched filter bank, *IEEE Microwave Wireless Components Lett.* **15**(2) 2005, 74–76.

[15] R. Zhang, and R. R. Mansour, Novel tunable lowpass filters using folded slots etched in the ground plane, In *IEEE MTT-S International Microwave Symposium Digest*, 12–17 June 2005, 4 p.

[16] G. M. Rebeiz, K. Entesari, I. C. Reines, S.-J. Park, M. A. El-Tanani, A. Grichener, and A. R. Brown, Tuning into RF MEMS, *IEEE Microwave Magazine* **10**(6), 2009, 55–72.

[17] L.-H. Hsieh, and K. Chang, Tunable microstrip bandpass filters with two transmission zeros, *IEEE Trans.Microwave Theory Techn.* Volume **51**(2) Part 1, 2003, 520–525.

[18] W.-T. Tu, and K. Chang, Piezoelectric transducer-controlled dual-mode switchable bandpass filter, *IEEE Microwave Wireless Components Lett.* **17**(3), 2007, 199–201.

[19] Y. Poplavko, D. Schmigin, V. Pashkov, M. Jeong, and S. Baik, Tunable microstrip filter with piezo-moved ground electrode, In *2005 European Microwave Conference*, Vol. 2, 4–6 Oct. 2005, 3p.

[20] S. Pal, C. Stevens, and D. Edwards, Tunable HTS microstrip filters for microwave electronics, *Electronics Lett.* **41**(5), 286–288.

[21] G. L. Matthaei, Narrow-band, fixed-tuned, and tunable bandpass filters with zig-zag hairpin-comb resonators. *IEEE Trans. Microwave Theory Techn.* **51**(4), Part 1, April 2003, 1214–1219.

[22] G. Subramanyam, F. W. Van Keuls, and F. A. Miranda, A K-band-frequency agile microstrip bandpass filter using a thin-film HTS/ferroelectric/dielectric multilayer configuration. *IEEE Trans. Microwave Theory Techn.* **48**(4), Part 1, 2000, 525–530.

[23] Y.-H. Chun, H. Shaman, and J.-S. Hong; Switchable embedded notch structure for UWB bandpass filter, *IEEE Microwave Wireless Components Lett.* **18**(9), 2008, 590–592.

[24] H. R. Arachchige, J.-S. Hong, and Z.-C. Hao, UWB bandpass filter with tunable notch on liquid crystal polymer substrate, *2008 Asia-Pacific Microwave Conference, APMC 2008*, 16–20 Dec. 2008, 4p.

[25] E. E. Djoumessi, M. Chaker, and K. Wu, Varactor-tuned quarter-wavelength dual-bandpass filter, *IET Microwaves Antennas Propagation* **3**(1) 117–124.

[26] D. R. Jachowski, "Compact, frequency-agile, absorptive bandstop filters," In *IEEE MTT-S International Microwave Symposium Digest*, 2005, 4 p.

[27] Y.-H. Chun, J.-S. Hong, P. Bao, T. J. Jackson, and M. J. Lancaster, BST varactor tuned bandstop filter with slotted ground structure, *IEEE MTT-S Int. Microwave Symp.* 2008, 1115–1118.

[28] S. Y. Huang, and Y. H. Lee, A compact E-shaped patterned ground structure and its applications to tunable bandstop resonator, *IEEE Trans. Microwave Theory Techn.* **57**(3) 2009, 657–666.

[29] Y.-H. Chun, J.-S.Hong, P. Bao, T. J. Jackson, and M. J. Lancaster, An electronically tuned bandstop filter using bst varactors, In *38th European Microwave Conference, EuMC 2008*, 27–31 Oct. 2008, 1699–1702.

[30] W. D. Yan, and R. R. Mansour, Compact tunable bandstop filter integrated with large deflected actuators, In *IEEE/MTT-S International Microwave Symposium*, 3–8 June 2007, 1611–1614.

[31] Y.-H. Chun, J.-S. Hong, P. Bao, T. J. Jackson, and M. J. Lancaster, BST-varactor tunable dual-mode filter using variable Z_C transmission line, *IEEE Microwave Wireless Components Lett.* **18**(3), 2008, 167–169.

[32] M. R. Al Mutairi, A. F. Sheta, and M. A. AlKanhal, A novel reconfigurable dual-mode microstrip meander loop filter, In *38th European Microwave Conference, EuMC 2008*, 27–31 Oct. 2008, 51–54.

[33] Y.-H. Chun, and J.-S. Hong, Electronically reconfigurable dual-mode microstrip open-loop resonator filter, *IEEE Microwave Wireless Components Lett.* **18**(7), 2008, 449–451.

[34] W. Tang, and J.-S. Hong, Compact tunable microstrip bandpass filters with asymmetrical frequency response, In *38th European Microwave Conference, EuMC2008*, 599–602.

[35] W. Tang, and J.-S. Hong, Tunable microstrip quasi-elliptic function bandpass filters, In *Proceedings of the 39th European Microwave Conference*, 2009, 767–770.

[36] W. Tang, and J.-S. Hong, Varactor-tuned dual-mode bandpass filters, *IEEE Trans. Microwave Theory Techn.* **58**(8), 2010, 2213–2219.

[37] P. W. Wong, and I. C. Hunter, A new class of low-loss high-linearity electronically reconfigurable microwave filter, *IEEE Trans. Microwave Theory Techn.* **56**(8), 2008, 1945–1953.

[38] A. Miller, and J.-S. Hong, Wideband bandpass filter with reconfigurable bandwidth, *IEEE Microwave Wireless Components Lett.* **20**(1), 2010, 28–30.

[39] A. Miller, and J.-S. Hong, Wideband bandpass filter with multiple reconfigurable bandwidth states, In *40th European Microwave Conference, EuMC 2010*.

[40] W. L. Jones, Design of tunable combline filters of near-constant bandwidth, *Electronics Lett.* **1**(6), 1965, 156–158.

[41] B. E. Carey-Smith, P. A. Warr, M. A. Beach, and T. Nesimoglu, Wide tuning-range planar filters using lumped-distributed coupled resonators, *IEEE Trans. Microwave Theory Techn.* **53**(2), 2005, 777–785.

[42] M. Sanchez-Renedo, R. Gomez-Garcia, J. I. Alonso, and C. Briso-Rodriguez, Tunable combline filter with continuous control of center frequency and bandwidth, *IEEE Trans.Microwave Theory Techn.* **53**(1), 2005, 191–199.

[43] M. Sanchez-Renedo, and R. Gomez-Garcia, Small-size planar tunable combline filter using decoupling walls, *Electronics Lett.* **43**(9), 2007, 532–534.

[44] S. Gevorgian, Agile microwave devices, *IEEE Microwave Magazine* **10**(5), 2009, 93–98.

[45] C. Rauscher, Reconfigurable bandpass filter with a three-to-one switchable passband width, *IEEE Trans. Microwave Theory Techn.* **51**, 2003, 573–577.

[46] L. Zhu, V. Devabhaktuni, C. Wang, and M. Yu, Adjustable bandwidth filter design based on interdigital capacitors, *IEEE Microwave Wireless Components Lett.* **18**, 2008, 16–18.

[47] M. S. Renedo, R. G. García, J. I. Alonso, and C. B. - Rodríguez, Tunable combline filter with continuous control of center frequency and bandwidth, *IEEE Trans. Microwave Theory Techn.* **53**(1), 2003, 191–199.

[48] C.-K. Liao, C.-Y. Chang, and J. Lin, A reconfigurable filter based on doublet configuration, *IEEE MTT-S International Microwave Symposium Digest*, July 2007, 1607–1610.

[49] R. V. Snyder, A wide-band tunable filter technique based on double-diplexing and low-Q tuning elements *IEEE MTT-S International Microwave Symposium Digest*, June 2000; 1759–1762.

[50] J.-Bryzek, K. Petersen, and W. McCulley, Micromachines on the march, *IEEE Spectrum Magazine*, May 1994, 20.

[51] E. R. Brown, RF-MEMS switches for reconfigurable integrated circuits, *IEEE Trans.* **MTT-46**, 1998, 1868–1880.

[52] G. M. Rebeiz, *RF MEMS: Theory, Design, and Technology*. Hoboken, NJ: Wiley, 2003.

[53] J.-S. Hong, E. P. McErlean, S. G. Tan, Y.-H. Chun, Z. Cui, L. Wang, R. B. Greed, and D. C. Voyce, Challenge and progress in high power MEMS switches for reconfigurable RF front-ends, In *2005 IEEE International Symposium on Microwave, Antenna, Propagation and EMC Technologies for Wireless Communications Proceedings*, 523–526.

[54] S. G. Tan, E. P. McErlean, J.-S. Hong, Z. Cui, L. Wang, R. B. Greed, and D. C. Voyce, Electromechanical modelling of high power RF MEMS switches with ohmic contact, In *35th European Microwave Conference Proceedings*, Paris, 2005, 1451–1454.

[55] E. P. McErlean, J.-S. Hong, S. G. Tan, Z. Cui, L. Wang, R. B. Greed, and D. C. Voyce, An 2 × 2 RF MEMS switch matrix, *IEE Proc. H Microwaves Antennas Propagation*, **152**(6), 2005, 449–454.

[56] L. Wang, Z. Cui, J.-S. Hong, E. P. McErlean, R. B. Greed, and D. C. Voyce, Fabrication of high power RF MEMS switches, *Microelectronic Eng.*, **83**(4–9), 2006, 1418–1420.

[57] J.-S. Hong, and M. J. Lancaster, Couplings of microstrip square open-loop resonators for cross-coupled planar microwave filters, *IEEE Trans. Microwave Theory Techn.* **44**, 1996, 2099–2109.

[58] S. N. Das, Quality of a ferroelectric material, *IEEE Trans.* **MTT-12**, 1964, 440–445.

[59] A. T. Findikoglu, Q. X. Jia, X. D. Wu, G. J. Chan, T. Venkatesan, and D. W. Reagour Tunable and adaptive bandpass filter using non-linear dielectric thin film of $SrTiO_3$, *Appl. Phys. Lett.* **68**(12), 1996, 1651–1653.

[60] A. Kozyrev, A. Ivanov, V. Keis, M. Khazov, V. Osadchy, T. Samoilova, O. Soldatenkov, A. Pavlov, G. Koepf, C. Mueller, D. Galt, and T. Rivkin, "Ferroelectric films: nonlinear properties and applications in microwave devices" IEEE MTT-S, Digest, 1998, 985–988.

[61] G. Subramanyam, F. W. Van Keuls, and F. A. Miranda, A K-band tunable microstrip band pass filter using a thin film conductor/ferroelectric/dielectric multilayer configuration, *IEEE Microwave Guided-Wave Lett.* **8**, 1998, 78–80.

[62] M. J. Lancaster, J. Powell, and A. Porch, Thin-film ferroelectric microwave devices, *Superconductivity Sci. Technol.* **11**, 1998, 1323–1334.

[63] R. Romanofsky, J. Bernhard, G. Washington, F. VanKeuls, F. Miranda, and C. Cannedy, A K-band linear phased array antenna based on $Ba_{0.60}Sr_{0.40}TiO_3$ thin film phase shifters, *IEEE MTT-S Dig.* 2000, 1351–1354.

[64] I. Vendik, O. Vendik, V. Sherman, A. Svishchev, V. Pleskachev, A. Kurbanov, and R. Wordenweber, Performance limitation of a tunable resonator with a ferroelectric capacitor, *IEEE MTT-S Dig.* 2000, 1371–1374.

[65] F. A. Miranda, G. Subramanyam, F. W. Van Keuls, R. R. Romanofsky, J. D. Warner, and C. H. Mueller, Design and development of ferroelectric tunable microwave components for Ku- and K-band satellite communication systems, *IEEE Trans.* **MTT-48**, 2000, 1181–1189.

[66] A. Tombak, J.-P. Maria, F. Ayguavives, Zhang Jin, G. T. Stauf, A. I. Kingon, and A. Mortazawi, Tunable barium strontium titanate thin film capacitors for RF and microwave applications, *IEEE Microwave Wireless Components Lett.* **12**(1), 2002, 3–5.

[67] P. M. Suherman, T. J. Jackson, Y. Y. Tse, I. P. Jones, R. I. Chakalova, and M. J. Lancaster, Microwave properties of $Ba_{0.5}Sr_{0.5}TiO_3$ thin film coplanar phase shifters, *J. Appl, Phys.* **99**(104101), 2006, 1–7.

[68] Y.-H. Chun, J.-S. Hong, P. Bao, T. J. Jackson, and M. J. Lancaster, Tunable slotted ground structured bandstop filter with bst varactors, *IET Proc. Microwaves Antennas Propagation* **3**(5), 2009, 870–876.

[69] Y.-H. Chun, C. Fragkiadakis, P. Bao, A. Luker, R. V. Wright, J.-S. Hong, P. B. Kirby, Q. Zhang, T. J. Jackson, and M. L. Lancaster, Tunable bandstop resonator and filter on Si-substrate with PST thin film by sol-gel deposition, In *38th European Microwave Conference, EuMC2008*, 13–16.

[70] Y.-H. Chun, R. Keller, J.-S. Hong, C. Fragkiadakis, R. V. Wright, and P. B. Kirby, Lead-strontium-titanate varactor-tuned cpw bandstop filter on liquid crystal polymer substrates, In *39th European Microwave Conference, EuMC2009*, 1373–1376.

[71] Y. Xu, *Ferroelectric Materials*, Elsevier Science, New York, 1991.

[72] S. B. Herner, F. A. Selmi, V. V. Varadan, and V. K. Varadan, The effect of various dopants on the dielectric properties of barium strontium titanate *Mater. Lett.* **15**, 1993, 317–324.

[73] C. M. Jackson, J. H. Kobayashi, A. Lee, C. Prentice-Hall, J. F. Burch, and R. Hu Novel monolithic phase shifter combining ferroelectrics and high temperature superconductors, *Microwave Opt. Technol. Lett.* **5**, 1992, 722–726.

[74] R. C. Neville, B. Hoeneisen, and C. A. Mead, Permittivity of strontium titanate, *J. Appl. Phys.* **43**(5), 1972, 2124–2131.

[75] Y. A. Boikov, and T. Claeson, High tunability of the permittivity of $YBa_2Cu_3O_{7-\delta}/SrTiO_3$ hetrostructures on sapphire substrates, *J. Appl. Phys.* **81**(7), 1997, 3232–3236.

[76] F. De Flavis, D. Chang, N. G. Alexopoulos, and O. M. Stafsudd, High purity ferroelectric materials by sol-gel process for microwave applications, *IEEE MTT-S Dig.* 1996, 99–102.

[77] X. H. Zhu, L. P. Yong, H. F. Tian, W. Peng, J. Q. Li, and D. N. Zheng, The origin of the weak ferroelectric-like hysteresis effect in paraelectric $Ba_{0.5}Sr_{0.5}TiO_3$ thin films grown epitaxially on $LaAlO_3$, *J. Phys.Condens. Matter* **18**(19), 2006, 4709–4718.

[78] B. M. Karyamapudi, and J.-S. Hong, Characterization and applications of a compact CPW defected ground structures, *Microwave Opt. Technol.Lett.* **47**(1), 2005, 26–31.

[79] S. Courreges, Y. Li, Z. Zhao, K. Choi, A. Hunt, and J. Papapolymerou, Two-pole X-band-tunable ferroelectric filters with tunable center frequency, fractional bandwidth, and return loss, *IEEE Trans.* **MTT-57**, 2009, 2872–2881.

[80] AVX Accu-P data sheet, AVX Corporation, Myrtle Beach, SC, 2006.

[81] M/A COM MA46H202 data sheet, M/A COM, Lowell, MA, 2006.

[82] *EM User's Manual*, Version 12. Sonnet Software Inc., Liverpool, New York, 2009.

[83] AWR Microwave office V7.5. Applied Wave Research, Inc., El Segundo, CA, 2007.

[84] W. Tang, and J.-S. Hong, Microstrip quasi-elliptic function bandpass filter with improved tuning range, *the 40th European Microwave Conference, EuMC 2010*.

[85] Z.-C. Hao, and J.-S. Hong, UWB bandpass filter with switchable notching band using multilayer LCP technology, In *40th European Microwave Conference, EuMC 2010*.

APPENDIX

Useful Constants and Data

TABLE A.1 Physical Constants

Speed of light in vacuum	$c = 2.99792458 \times 10^8$ m/s
Permittivity of vacuum	$\varepsilon_0 = 8.85418782 \times 10^{-12} \approx (1/36\pi) \times 10^{-9}$ F/m
Permeability of vacuum	$\mu_0 = 4\pi \times 10^{-7}$ H/m
Impedance of free space	$\eta_0 = 376.7303 \approx 120\pi\,\Omega$
Boltzmann's constant	$k = 1.3806 \times 10^{-23}$ J/K
Planck's constant	$h = 6.626 \times 10^{-34}$ J·s
Charge of electron	$e = 1.602177 \times 10^{-19}$ C
Electron rest mass	$m = 9.10938 \times 10^{-31}$ kg

TABLE A.2 Conductivity of Metals at 25°C (298K)

Material	Conductivity σ(S/m)	Material	Conductivity σ(S/m)
Silver	6.18×10^7	Zinc	1.66×10^7
Copper	5.84×10^7	Nickel	1.40×10^7
Gold	4.43×10^7	Platinum	0.97×10^7
Aluminium	3.69×10^7	Chromium	0.79×10^7

Microstrip Filters for RF/Microwave Applications by Jia-Sheng Hong
Copyright © 2011 John Wiley & Sons, Inc.

TABLE A.3 Electrical Resistivity ρ in 10^{-8} Ωm of Metals[a]

T/K	Aluminium	Copper	Gold	Silver
40	0.0181	0.0239	0.141	0.0539
60	0.0959	0.0971	0.308	0.162
100	0.442	0.348	0.650	0.418
200	1.587	1.046	1.462	1.029
273	2.417	1.543	2.051	1.467
300	2.733	1.725	2.271	1.629
400	3.87	2.402	3.107	2.241

[a]Conductivity $\sigma = 1/\rho$

TABLE A.4 Properties of Dielectric Substrates

Material	Relative Dielectric Constant at 10 GHz	Loss Tangent at 10 GHz	Thermal Conductivity (W/m K)
Alumina	9.7	0.0002	30
Fused quartz	3.8	0.0001	1
Polystyrene	2.53	0.00047	0.15
Beryllium oxide	6.6	0.0001	250
GaAs	12.9	0.0016	30
Si	11.7	0.005	90
RT/duroid 5880	2.20±0.015	0.0009	0.26
RT/duroid 6002	2.94±0.04	0.0012	0.44
RT/duroid 6006	6.15±0.15	0.0027	0.49
RT/duroid 6010LM	10.2±0.25	0.0023	0.78
RO3003	3.00±0.04	0.0013	0.50
RO3006	6.15±0.15	0.0020	0.61
RO3010	10.2±0.30	0.0023	0.66

Index

ABCD matrix, 11
  admittance inverter, 56
  impedance inverter, 56
  normalized, 306
  unit element, 65
ABCD Parameters, 11
  coupled-line subnetwork, 236
  two-port networks, 12
Adjacent resonators, 276
Admittance inverter, 55–56
Admittance matrix, 10
  normalized, 199
All-pass network, 36
  C-type sections, 37
  D-type sections, 37
  lowpass prototype, 48
  external delay equalizer, 295
All-pole filter(s), 32, 39
Amplitude-squared transfer function, 28
Anisotropic permittivity, 335, 453
Anisotropy, 441
Aperture couplings, 387
Approximation
  lumped-element capacitor, 113
  lumped-element inductor, 113
  problem, 29
Arithmetic mean, 214
Asymmetric coupled lines, 137
Asymmetrical frequency response, 276
Asynchronous tuning, 197

Asynchronously tuned coupled-resonator circuits, 209
Asynchronously tuned filter, 197, 209, 244
Attenuation constant, 81
  conductor loss, 81
  dielectric loss, 82

Bandpass filter(s), 53, 123
  using immittance inverters, 58–59
Bandpass transformation, 52
Bandstop filter(s), 54, 169
  with open-circuited stubs, 177
  optimum, 183
Bandstop transformation, 54
Barium strontium titanate (BST), 612
BST varactor, 614
Bessel filters, 35
Bias circuit, 568, 601, 604, 607
Bias network, 188
Bias T, 188
Biquadratic equation, 210, 214
Buffer layer, 441, 612
Bulk micromachining, 606
Butterworth (Maximally flat) response, 30
Butterworth lowpass prototype, 40

CAA, 233
CAD
  examples, 248
  tools, 233

Call clarity, 442
Canonical
  filters, 316
  coupling structure, 318, 464
Capacitively loaded transmission line resonator, 338
Capacitive tunability, 612
Cascade connection, 15
Cascaded couplings, 317
Cascaded network, 235
Cascaded quadruplet (CQ) filters, 271
Cascaded trisection (CT) filters, 275
Cauer filters, 35
Cavity mode, 100, 350, 357
Cell size, 239
Cellular mobile, 441
Chain matrix, 11
Characteristic admittance, 56, 124
Characteristic impedance, 13, 56, 76, 104, 108
Chebyshev
  filter, 32
  function, 32
  lowpass prototype, 40–42
  response, 31
Circular
  disk, 351, 478
  patch resonator, 99
  spiral inductor, 92
C-mode, 137
Combline filters, 144, 564
Commensurate
  length, 62
  network, 166
Complex amplitude, 7
Complex conductivity, 434
Complex conjugate symmetry, 306
Complex frequency variable, 29
Complex permittivity, 102
Complex plane zeros, 297
Complex plane, 29
Complex propagation constant, 13, 97
Computer aided analysis, 233
Computer aided design, 232
Computer optimization, 242
Conductor loss, 81
Conductor Q, 101
Convergence analysis, 238
Coplanar waveguide (CPW), 104
Coupled microstrip lines, 83

Coupled resonator circuits, 193–230
Coupled resonator filters, 194, 386, 495
Coupled-line
  network, 68
  subnetworks, 236
Coupling capacitances, 126
Coupling coefficient(s), 132, 151, 196–230, 294, 302, 389
  normalized, 196
Coupling gaps, 126
Coupling matrix, 201, 228, 243
Critical current density, 440
Critical field, 440
Critical temperature, 433
Cross coupling, 243, 246, 252, 263, 271, 296, 329, 448
Cryogenic
  cooler, 453, 461
  dewar, 463
  package, 479
Crystal axis, 441, 453
Crystalline lattice, 440
Curie temperature, 611
Current law, 198
Cutoff frequency, 28

dc
  bias, 188, 568, 575, 601, 607, 612
  block, 162
  blocking, 570, 574
Decibels (dB), 8
  per unit length, 81
De-embedding, 126
Definite nodal admittance matrix, 212
Defected ground structure, 393
Degenerate modes, 100, 351
Delay, 8
Dielectric
  loss, 82
  loss tangent, 95, 102
  properties, 413, 453
  substrate, 75, 104, 107
Dielectric Q, 95
Dispersion
  effect, 341, 473
  equation, 341
  in microstrip, 80
Dissipation, 69
  effects, 70
Distributed circuits, 59

Distributed element, 62
Distributed filter, 62
Distributed line resonators, 98
Dominant mode, 76, 104
Doubly loaded resonator, 219
Dual networks, 24
Dual-mode filter(s), 100, 352, 368
Dual-mode resonators, 101, 350, 366

Edge-coupled half-wavelength resonator filter, 249
Edge-coupled microstrip bandpass filters, 128
Effect of strip thickness, 79
Effective conductivity, 102
Effective dielectric constant, 76
Effective dielectric Permittivity, 76
Eigenequation, 210
Eigenfrequencies, 209
Eigenvalues, 210
Electric aperture coupling, 391
Electric coupling, 198, 203
Electric coupling coefficient, 205, 211
Electric discharge, 478
Electric wall, 83, 204, 350, 361
Electrical length(s), 63, 78
Electromagnetic (EM) simulation(s), 138, 232, 367
Electromagnetic (EM) waves, 1
Electromagnetic analysis, 241
Electromagnetic bandgap, 617
Electromagnetic coupling, 237
Electromagnetic waves in free space, 77
Electronic packaging, 412
Electronically tunable filters, 615
Element transformation, 49–54
Element values, 39
　Butterworth, 40
　Chebyshev, 41–42
　elliptic, 45
　Gaussian, 47
Elliptic function(s), 85, 105, 108
　lowpass filter, 123
　lowpass prototype, 43
　response, 33
EM simulation(s), *see* Electromagnetic simulation(s)
EM solver, 233
End-coupled half-wavelength resonator filters, 123

End-coupled slow-wave resonators filters, 341
Envelope delay, 9
Equal-ripple
　passband, 31
　stopband, 33
Equivalent networks, 24
Error function, 243
Even- and odd-mode
　capacitances, 84
　characteristic impedances, 85, 129
　networks, 19
Even excitation, 19
External delay equalizer, 295
External quality factor(s), 132, 193, 216, 229
Extracted pole filters, 304
Extracting coupling coefficient, 215
Extracting external quality factor, 216

Ferroelectric
　materials, 611
　varactors, 612
Filters,
　bandpass, 123–153
　bandstop, 169–190
　compact, 334–384
　canonical, 316
　cascaded quadruplet (CQ), 271
　cascaded trisection (CT), 275
　extracted pole, 304
　highpass, 162–168
　linear phase, 295
　lowpass, 112–119
　multiband, 320
　multilayer, 386–412
　selective, 261
　superconducting, 433–475
　tunable and reconfigurable, 563–618
　ultra-wideband (UWB), 488–550
Filtering characteristic(s), 166, 178, 289
Filtering function, 166, 184
Finite frequency attenuation poles, 119, 243, 444
Finite frequency zeros, 266, 298
Finite-frequency transmission zeros, 30
Flat group delay, 35, 295, 454, 529
Floquet's theorem, 340
Fractal dual-mode resonator, 354
Fractional bandwidth, 53, 170, 195

Frequency
  agile filters, 615
  agility, 615
  bands, 1
  invariant immittance inverter(s), 296
  invariant susceptance, 277
  mapping, 49, 63, 162, 170, 178
  spectrums, 1
  transformation, 28, 49–54, 515
  tuning, 177, 453, 563
Frequency-dependent couplings, 393
Fringe capacitance, 84
Fringe fields, 221
Full-wave EM simulation(s), *see* Electromagnetic simulation(s)

Gallium arsenide (GaAs), 606
Gaussian response, 35
Gaussian lowpass prototype, 44
General coupling matrix, 194
  including source and load, 228
Generalized scattering matrix, 26
Generalized Y and Z matrices, 26
Group delay, 9, 218
  equalization, 304, 454 (*See also* All-pass network)
  equalizers, 36
  response(s), 29, 227, 455
  self-equalization, 272, 318
Guided wavelength, 78
Guided-wave media, 76
g values, 39

Hairpin
  filter, 133, 570
  resonator, 131
Half-wavelength line resonators, 91, 128
High temperature superconductors (HTS), 433
Higher-order modes, 82
Highpass
  filters, 162–168
  transformation, 51
High-power applications, 440, 475
High-resistivity silicon, 613
Housing loss, 82
Housing Q, 102
HTS films, 434

HTS filters, 337
  for mobile communications, 441
  for satellite communications, 462
  for Radio Astronomy and Radar, 469
  high-power, 475
Hurwitz polynomial, 30

I/O coupling structures, 216
Ideal admittance inverter, 56
Ideal impedance inverters, 56
Identity matrix, 201
Immittance, 3 (*see also* Admittance, Impedance)
  inverter(s), 55
Impedance
  Inverter(s), 55
  matching, 9
  matrix, 15
  scaling, 50
Incident waves, 7
Induced voltages, 237
Inductance, 25, 39, 56, 68–69
Input/output (I/O), 136
Insertion loss, 8
  response, 29
Interdigital bandpass filter, 135, 238, 495
Interdigital capacitor(s), 93, 164, 473, 543, 612
Intermodulation, 440, 463, 476
Intrinsic resistance, 433
Isolation, 420, 592

J inverters, 56, 124

K inverters, 56
Kuroda identities, 62–66

Lanthanum aluminate (LaAlO$_3$ or LAO), 441
L-C ladder type, 112
LCP technology and filters, 414
Linear circuit simulator, 233
Linear phase, *see also* Group delay
  characteristics, 290
  filters, 295, 392
Liquid Crystal Polymer (LCP) Films, 416
London penetration depth, 436
Loop, *see also* Coupled resonator circuits
  current, 194
  equations, 194

INDEX **631**

Loop inductors, 94
Loss tangent(s), 82, 626
Lossless network, 10
Lossy
 capacitor, 69
 elements, 69
 inductor, 69
 resonators, 70
Lowpass
 filters, 112–122
 prototype filter, 39–48
 transformation, 50
Low-temperature cofired ceramic (LTCC), 5
L-shape resonators, 175
LTCC
 filters, 414–415
 materials, 413–414
Lumped inductors and capacitors, 91
Lumped L-C elements, 116
Lumped LC resonators, 59
Lumped-element
 immittance inverters, 60
 filters, 379

Magnesium oxide (MgO), 441
Magnetic aperture coupling, 391
Magnetic coupling, 198, 205
 coefficient, 212
Magnetic loss, 81
Magnetic substrates, 81
Magnetic wall, 83, 100, 205, 350, 361, 389, 408
Maximally flat group delay, 35. *See also* Group delay
Maximally flat stopband, 31
Meander open-loop resonators, 334
MEMS, 563, 604
Mesh sizes, 238
Metal-insulator-metal (MIM) capacitor, 93
Metallic enclosure, 82
Metallization, 107, 412
Metals, 625
Microelectromechanical system (MEMS), 604
Microstrip lines, 75
MIM capacitor, 93, 612
MIM varactor, 613
Minimum phase two-port network, 296
Mixed coupling coefficient, 207, 214

Multiband filters, 320
Multilayer filters, 386–420
Multilayered microstrip, 103
Multimode networks, 25
Multimode-resonator UWB filters, 511
Multiple coupled resonator filters, 243
Multiplexers, 464
Multiport networks, 20
Mutual capacitance, 68, 198
Mutual inductance, 68, 206

Narrow band approximation, 196, 200
Natural frequencies, 30
Natural resonance, 210–213
Natural resonant frequencies, 209
Negative coupling, 215
Negative electrical lengths, 124
Negative polarity, 298
Neper frequency, 29
Nepers per unit length, 81
Network
 analysis, 6
 connections, 13
 identities, 65
 parameter conversions, 16
 variables, 6
Node equations, 198
Node voltage, 198
Noise figure, 479
Nonadjacent resonators, 263, 296
Nonlinearity, 440
Nonminimum phase network, 296
Nonresonating node, 375–377
Normalized impedance matrix, 195
Normalized impedances/admittances, 20
Normalized reactance slope parameter, 174
Normalized susceptance slope parameter, 175
Notched band(s), 537, 543, 550
Numerical methods, 238

Objective function, 242
Odd mode, 83
 excitation, 19, 138
 networks, 19, 297
One-port networks, 13
Open-circuit impedance, 10
Open-circuited stub(s), 97, 115, 150, 177, 183, 400, 497, 537, 572, 598

632 INDEX

Open-end effect, 116
Open-ends, 89
Open-loop resonator(s), 216, 221, 267, 322, 335, 343, 366, 386, 437, 552, 571, 577
Optimization-based filter synthesis, 243
Optimum bandstop filters, 183
Optimum distributed highpass filters, 162
Optimum transfer function, 183

Parallel plate capacitance, 84
Parallel connection, 14, 23
Parallel-coupled half-wavelength resonator filters, 128
Parameter extraction, 138, 148, 364, 566
Passband ripple, 31
Patch resonators, 100
Penetration depth, 435
Periodic frequency response, 64
Periodically loaded transmission line, 340
Permeability, 75
Permittivity tensor, 335, 453
Permittivity, 75, 102
Phase delay, 9. *See also* Group delay
Phase linearity, 298
Phase response, 29
Phase shifter(s), 306, 313
Phase transitions, 611
Phase velocity, 78. *See also* Slow-wave
Phases, 8
p-i-n diode, 564, 591, 598, 603
Pizotransducer, 564
Piezoelectric transducers (PET), 610
  tunable filters, 610
π-mode impedances, 138
π-network, 127
Polarization, 138, 478, 610–611
Pole locations, 32
Poles and zeros, 29
Pole-zero patterns, 30
Polymethyl methacrylate, 607
Polynomials, 29
Positive coupling, 215
Power
  consumption, 564
  handling, 100, 405, 440, 475
  conservation, 9
p-plane, 29
Propagation constant, 78

Pseudohighpass filters, 166
Pseudocombline filter, 150
Pseudointerdigital bandpass filter, 238
Pseudointerdigital resonators, 239, 547
PST thin film, 612
Pure TEM mode, 84

Q factor, 95. *See also* Quality factor
Quadrantal symmetry, 37, 297
Quality factor, 69, 101
Quarter-wavelength
  line, 60
  resonators, 98
Quasielliptic function
  filter, 243, 336, 442, 444, 456, 470, 588
  response, 344, 378
Quasilumped elements, 91, 95, 162, 522
Quasilumped element resonators, 98
Quasilumped highpass filters, 162
Quasistatic analysis, 76
Quasi-TEM approximation, 76
Quasi-TEM-mode, 78, 83, 106

Radial stubs, 189
Radian frequency variable, 28
Radiation loss, 81
Radiation Q, 101
Rational function, 29, 297
Reactance slope parameters, 172
Reactance slope, 60
Reactances, 60
Reciprocal matrix, 202, 228
Reciprocal network(s), 9–11, 21
Reconfigurable filters, 563–618
  RF MEMS, 604
Redundant unit elements, 183
Reflected waves, 7
Reflection coefficients, 8
Resonator
  bandstop, 174
  capacitively Loaded Transmission-Line, 338
  coupled, 193
  CPW, 497
  dual-band, 321, 325
  dual-mode, 349
  dual-mode open-loop, 366
  doubly loaded, 219
  folded-waveguide, 408

INDEX 633

hairpin, 131
HTS thin film, 446
I/O, 216
lossy, 70
microstrip, 98–99
mixed, 321
multimode, 511
open-loop, 221
patch, 100
pseudointerdigital, 547
series, 59
shunt-parallel, 58
singly loaded, 216
SIW, 405
slot, 395
slow-wave, 336
split, 478
stepped-impedance, 505
stripline, 414
triangular-microstrip patch, 356
Return loss response, 29
Return loss, 8
RF
  breakdown, 475
  chokes, 188
  MEMS, 604
RF MEMS switch, 607
Richards transform variable, 185
Richards variable, 62
Richards' transformation, 62
Right-half plane zeros, 296
Ring resonator, 99
Ripple constant, 28
r-plane sapphire substrate, 335, 448

S matrix, 8
S parameters, 8
Sapphire (Al_2O_3), 441
Scaled external quality factors, 196
Scattering matrix, 8
Scattering Parameters, 8
Scattering transfer matrices, 26
Scattering transmission or transfer parameters, 26
Selective filters, 261
Selective linear phase response, 319
Selectivity, 44, 70, 72, 244, 262, 301, 442–443

Self- and mutual capacitances per unit length, 137
Self-capacitance, 68
Self-inductance, 68
Sensitivity analysis, 450, 506
Sensitivity, 384, 441
Series connection, 23
Series inductors, 116
Series-resonant shunt branch, 313
Short line sections, 91
Short-circuit admittance, 10
Short-circuited stub, 65, 97, 153, 166, 417, 488, 492, 530, 592, 597
Shunt capacitors, 113
Shunt short-circuited stubs, 153
Si substrate, 612
Sign of coupling, 215
Signal delay, 8
Signal jamming, 615
Silicon
  substrates, 607, 609, 613
  wafers, 607
Single crystal, 612
Single mode, 25, 321, 353,
Singly Loaded Resonator, 216
Skin depth, 436
Slotlines, 107
Slow-wave
  open-loop resonator filters, 343
  propagation, 336
  resonators, 338, 341, 501
Software defined radio, 272
Software, 233
Source or generator, 6
Source/load couplings, 228
Spiral inductor, 92
Spurious passband, 160, 337
Spurious resonance, 339
Spurious response, 136, 338, 492, 573
Spurious stop bands, 179
Spurious-free, 327
Square and meander loops, 100
Standing wave, 339, 357, 408
Stepped impedance
  resonators, 505, 547
  lowpass, 112, 492
Stored energy, 202
Straight-line inductor, 94
Stripline, 385, 410, 413

634 INDEX

Strontium titanate, 612
Stub bandpass filters, 153
Subnetworks, 16, 234
Substrates, 441, 626
Substrate-integrated waveguide filters, 404
Superconducting
 filters, 433
 materials, 433
Superconductors, 433
Superposition, 209, 358
Surface
 impedance, 436
 micromachining, 607
 reactance, 437
 resistance, 82
 wave, 82
Susceptance, 60, 122
Susceptance slope, 60
 parameter(s), 60, 145
Suspended and inverted microstrip lines, 103
Suspended microstrip line, 103
Switchable notch, 597
Symmetrical interface, 19
Symmetrical matrix, 21
Symmetrical network, 9, 11
 analysis, 19
Synchronously tuned coupled-resonator circuits, 203
Synchronously tuned filter, 195, 199
Synthesis,
 example, 244, 311
 procedure, 242

Tapped line, 133, 254
 coupling, 216
 I/O, 136
TEM wave, 75
TEM-mode, 135
Temperature
 coefficient, 414
 dependence, 437
 stability, 384, 437
Terminal impedances, 6
Thallium barium calcium copper oxide (TBCCO), 434
3-dB bandwidth, 35, 173
Theory of couplings, 193
Thermal expansions, 441

Thick film, 412, 437
Thin film, 79, 95, 438, 564, 607, 613
 microstrip (TFM), 103
Third-order intercept point (TOI or IP3), 477
Third-order intermodulation (IMD), 476
Thomson filters, 35
Topology matrix, 243
t-plane, 63
Transfer function (s), 28
 all-pass, 36
 amplitude squared, 28
 Butterworth, 30
 Chebyshev, 31
 CT filter, 276
 elliptic, 33
 Gaussian, 35
 linear phase filter, 296
 rational, 29
 single pair of transmission zero, 261
Transfer matrix, 11, 26
Transformers, 65
Transition temperature, 433, 611
Transmission
 coefficient(s), 8, 236
 zeros, 30, 43, 228
Transmission line
 elements, 61
 filter(s), 65, 179, 294
 inserted inverters, 287
 networks, 12
Trisection filters, 276
Tubular lumped-element bandpass filter, 379
Tunable capacitors, 619. *See also* Varactor
Tunable filters, 563
 ferroelectric, 610
 piezoelectric transducer, 610
Tunable combline filters, 564
Tunable notch, 601
Tuning device technologies, 564
Tuning element, 563
Tuning range, 564, 591
Tuning rate, 579
Two-fluid model, 434
Two-port network, 6

UE, 65. *See also* Unit element
Ultra-wideband (UWB) filters, 488–557
 reconfigurable, 597

UMTS filter, 245
Unit element(s), 65, 166, 178, 183
Unity immittance inverter, 310
Universal mobile telecommunication system (UMTS), 245
Unloaded quality factor, 69, 101
Unwanted couplings, 102, 242, 445

Varactor,
   semiconductor, 564
   ferroelectric, 612
Varactor-tuned combline filter, 565
Valley microstrip, 103
Velocity of light, 78
Via hole(s), 136, 386, 405, 414
Via-hole grounding, 104, 142, 170, 254, 490, 540, 565
Voltage and current variables, 6
Voltage law, 194
Voltage standing wave ratio ($VSWR$), 8

Wave(s)
   in CPW, 104
   in microstrip, 75
   in slotline, 107
   impedance, 77
   variables, 7
Waveguide
   cavities, 100
   filter, 406
   folded, 408
   substrate-integrated, 404
Wideband
   bandstop filters, 183
   filters with reconfigurable bandwidth, 591

Y matrix, 10
Y parameters, 10
Yttrium barium copper oxide (YBCO), 434

Z matrix, 10
Z parameters, 10

WILEY SERIES IN MICROWAVE AND OPTICAL ENGINEERING

KAI CHANG, Editor
Texas A&M University

FIBER-OPTIC COMMUNICATION SYSTEMS, Fourth Edition • *Govind P. Agrawal*

ASYMMETRIC PASSIVE COMPONENTS IN MICROWAVE INTEGRATED CIRCUITS • *Hee-Ran Ahn*

COHERENT OPTICAL COMMUNICATIONS SYSTEMS • *Silvello Betti, Giancarlo De Marchis, and Eugenio Iannone*

PHASED ARRAY ANTENNAS: FLOQUET ANALYSIS, SYNTHESIS, BFNs, AND ACTIVE ARRAY SYSTEMS • *Arun K. Bhattacharyya*

HIGH-FREQUENCY ELECTROMAGNETIC TECHNIQUES: RECENT ADVANCES AND APPLICATIONS • *Asoke K. Bhattacharyya*

RADIO PROPAGATION AND ADAPTIVE ANTENNAS FOR WIRELESS COMMUNICATION LINKS: TERRESTRIAL, ATMOSPHERIC, AND IONOSPHERIC • *Nathan Blaunstein and Christos G. Christodoulou*

COMPUTATIONAL METHODS FOR ELECTROMAGNETICS AND MICROWAVES • *Richard C. Booton, Jr.*

ELECTROMAGNETIC SHIELDING • *Salvatore Celozzi, Rodolfo Araneo, and Giampiero Lovat*

MICROWAVE RING CIRCUITS AND ANTENNAS • *Kai Chang*

MICROWAVE SOLID-STATE CIRCUITS AND APPLICATIONS • *Kai Chang*

RF AND MICROWAVE WIRELESS SYSTEMS • *Kai Chang*

RF AND MICROWAVE CIRCUIT AND COMPONENT DESIGN FOR WIRELESS SYSTEMS • *Kai Chang, Inder Bahl, and Vijay Nair*

MICROWAVE RING CIRCUITS AND RELATED STRUCTURES, Second Edition • *Kai Chang and Lung-Hwa Hsieh*

MULTIRESOLUTION TIME DOMAIN SCHEME FOR ELECTROMAGNETIC ENGINEERING • *Yinchao Chen, Qunsheng Cao, and Raj Mittra*

DIODE LASERS AND PHOTONIC INTEGRATED CIRCUITS • *Larry Coldren and Scott Corzine*

EM DETECTION OF CONCEALED TARGETS • *David J. Daniels*

RADIO FREQUENCY CIRCUIT DESIGN • *W. Alan Davis and Krishna Agarwal*

RADIO FREQUENCY CIRCUIT DESIGN, Second Edition • *W. Alan Davis*

MULTICONDUCTOR TRANSMISSION-LINE STRUCTURES: MODAL ANALYSIS TECHNIQUES • *J. A. Brandão Faria*

PHASED ARRAY-BASED SYSTEMS AND APPLICATIONS • *Nick Fourikis*

SOLAR CELLS AND THEIR APPLICATIONS, Second Edition • *Lewis M. Fraas and Larry D. Partain*

FUNDAMENTALS OF MICROWAVE TRANSMISSION LINES • *Jon C. Freeman*

OPTICAL SEMICONDUCTOR DEVICES • *Mitsuo Fukuda*

MICROSTRIP CIRCUITS • *Fred Gardiol*

HIGH-SPEED VLSI INTERCONNECTIONS, Second Edition • *Ashok K. Goel*

FUNDAMENTALS OF WAVELETS: THEORY, ALGORITHMS, AND APPLICATIONS, Second Edition • *Jaideva C. Goswami and Andrew K. Chan*

HIGH-FREQUENCY ANALOG INTEGRATED CIRCUIT DESIGN • *Ravender Goyal (ed.)*

ANALYSIS AND DESIGN OF INTEGRATED CIRCUIT ANTENNA MODULES • *K. C. Gupta and Peter S. Hall*

PHASED ARRAY ANTENNAS, Second Edition • *R. C. Hansen*

STRIPLINE CIRCULATORS • *Joseph Helszajn*

THE STRIPLINE CIRCULATOR: THEORY AND PRACTICE • *Joseph Helszajn*

LOCALIZED WAVES • *Hugo E. Hernández-Figueroa, Michel Zamboni-Rached, and Erasmo Recami (eds.)*

MICROSTRIP FILTERS FOR RF/MICROWAVE APPLICATIONS, Second Edition • *Jia-Sheng Hong*

MICROWAVE APPROACH TO HIGHLY IRREGULAR FIBER OPTICS • *Huang Hung-Chia*

NONLINEAR OPTICAL COMMUNICATION NETWORKS • *Eugenio Iannone, Francesco Matera, Antonio Mecozzi, and Marina Settembre*

FINITE ELEMENT SOFTWARE FOR MICROWAVE ENGINEERING • *Tatsuo Itoh, Giuseppe Pelosi, and Peter P. Silvester (eds.)*

INFRARED TECHNOLOGY: APPLICATIONS TO ELECTROOPTICS, PHOTONIC DEVICES, AND SENSORS • *A. R. Jha*

SUPERCONDUCTOR TECHNOLOGY: APPLICATIONS TO MICROWAVE, ELECTRO-OPTICS, ELECTRICAL MACHINES, AND PROPULSION SYSTEMS • *A. R. Jha*

TIME AND FREQUENCY DOMAIN SOLUTIONS OF EM PROBLEMS USING INTEGTRAL EQUATIONS AND A HYBRID METHODOLOGY • *B. H. Jung, T. K. Sarkar, S. W. Ting, Y. Zhang, Z. Mei, Z. Ji, M. Yuan, A. De, M. Salazar-Palma, and S. M. Rao*

OPTICAL COMPUTING: AN INTRODUCTION • *M. A. Karim and A. S. S. Awwal*

INTRODUCTION TO ELECTROMAGNETIC AND MICROWAVE ENGINEERING • *Paul R. Karmel, Gabriel D. Colef, and Raymond L. Camisa*

MILLIMETER WAVE OPTICAL DIELECTRIC INTEGRATED GUIDES AND CIRCUITS • *Shiban K. Koul*

ADVANCED INTEGRATED COMMUNICATION MICROSYSTEMS • *Joy Laskar, Sudipto Chakraborty, Manos Tentzeris, Franklin Bien, and Anh-Vu Pham*

MICROWAVE DEVICES, CIRCUITS AND THEIR INTERACTION • *Charles A. Lee and G. Conrad Dalman*

ADVANCES IN MICROSTRIP AND PRINTED ANTENNAS • *Kai-Fong Lee and Wei Chen (eds.)*

SPHEROIDAL WAVE FUNCTIONS IN ELECTROMAGNETIC THEORY • *Le-Wei Li, Xiao-Kang Kang, and Mook-Seng Leong*

ARITHMETIC AND LOGIC IN COMPUTER SYSTEMS • *Mi Lu*

OPTICAL FILTER DESIGN AND ANALYSIS: A SIGNAL PROCESSING APPROACH • *Christi K. Madsen and Jian H. Zhao*

THEORY AND PRACTICE OF INFRARED TECHNOLOGY FOR NONDESTRUCTIVE TESTING • *Xavier P. V. Maldague*

METAMATERIALS WITH NEGATIVE PARAMETERS: THEORY, DESIGN, AND MICROWAVE APPLICATIONS • *Ricardo Marqués, Ferran Martín, and Mario Sorolla*

OPTOELECTRONIC PACKAGING • *A. R. Mickelson, N. R. Basavanhally, and Y. C. Lee (eds.)*

OPTICAL CHARACTER RECOGNITION • *Shunji Mori, Hirobumi Nishida, and Hiromitsu Yamada*

ANTENNAS FOR RADAR AND COMMUNICATIONS: A POLARIMETRIC APPROACH • *Harold Mott*

INTEGRATED ACTIVE ANTENNAS AND SPATIAL POWER COMBINING • *Julio A. Navarro and Kai Chang*

ANALYSIS METHODS FOR RF, MICROWAVE, AND MILLIMETER-WAVE PLANAR TRANSMISSION LINE STRUCTURES • *Cam Nguyen*

LASER DIODES AND THEIR APPLICATIONS TO COMMUNICATIONS AND INFORMATION PROCESSING • *Takahiro Numai*

FREQUENCY CONTROL OF SEMICONDUCTOR LASERS • *Motoichi Ohtsu (ed.)*

WAVELETS IN ELECTROMAGNETICS AND DEVICE MODELING • *George W. Pan*

OPTICAL SWITCHING • *Georgios Papadimitriou, Chrisoula Papazoglou, and Andreas S. Pomportsis*

MICROWAVE IMAGING • *Matteo Pastorino*

ANALYSIS OF MULTICONDUCTOR TRANSMISSION LINES • *Clayton R. Paul*

INTRODUCTION TO ELECTROMAGNETIC COMPATIBILITY, Second Edition • *Clayton R. Paul*

ADAPTIVE OPTICS FOR VISION SCIENCE: PRINCIPLES, PRACTICES, DESIGN AND APPLICATIONS • *Jason Porter, Hope Queener, Julianna Lin, Karen Thorn, and Abdul Awwal (eds.)*

ELECTROMAGNETIC OPTIMIZATION BY GENETIC ALGORITHMS • *Yahya Rahmat-Samii and Eric Michielssen (eds.)*

INTRODUCTION TO HIGH-SPEED ELECTRONICS AND OPTOELECTRONICS • *Leonard M. Riaziat*

NEW FRONTIERS IN MEDICAL DEVICE TECHNOLOGY • *Arye Rosen and Harel Rosen (eds.)*

ELECTROMAGNETIC PROPAGATION IN MULTI-MODE RANDOM MEDIA • *Harrison E. Rowe*

ELECTROMAGNETIC PROPAGATION IN ONE-DIMENSIONAL RANDOM MEDIA • *Harrison E. Rowe*

HISTORY OF WIRELESS • *Tapan K. Sarkar, Robert J. Mailloux, Arthur A. Oliner, Magdalena Salazar-Palma, and Dipak L. Sengupta*

PHYSICS OF MULTIANTENNA SYSTEMS AND BROADBAND PROCESSING • *Tapan K. Sarkar, Magdalena Salazar-Palma, and Eric L. Mokole*

SMART ANTENNAS • *Tapan K. Sarkar, Michael C. Wicks, Magdalena Salazar-Palma, and Robert J. Bonneau*

NONLINEAR OPTICS • *E. G. Sauter*

APPLIED ELECTROMAGNETICS AND ELECTROMAGNETIC COMPATIBILITY • *Dipak L. Sengupta and Valdis V. Liepa*

COPLANAR WAVEGUIDE CIRCUITS, COMPONENTS, AND SYSTEMS • *Rainee N. Simons*

ELECTROMAGNETIC FIELDS IN UNCONVENTIONAL MATERIALS AND STRUCTURES • *Onkar N. Singh and Akhlesh Lakhtakia (eds.)*

ANALYSIS AND DESIGN OF AUTONOMOUS MICROWAVE CIRCUITS • *Almudena Suárez*

ELECTRON BEAMS AND MICROWAVE VACUUM ELECTRONICS • *Shulim E. Tsimring*

FUNDAMENTALS OF GLOBAL POSITIONING SYSTEM RECEIVERS: A SOFTWARE APPROACH, Second Edition • *James Bao-yen Tsui*

RF/MICROWAVE INTERACTION WITH BIOLOGICAL TISSUES • *André Vander Vorst, Arye Rosen, and Youji Kotsuka*

InP-BASED MATERIALS AND DEVICES: PHYSICS AND TECHNOLOGY • *Osamu Wada and Hideki Hasegawa (eds.)*

COMPACT AND BROADBAND MICROSTRIP ANTENNAS • *Kin-Lu Wong*

DESIGN OF NONPLANAR MICROSTRIP ANTENNAS AND TRANSMISSION LINES • *Kin-Lu Wong*

PLANAR ANTENNAS FOR WIRELESS COMMUNICATIONS • *Kin-Lu Wong*

FREQUENCY SELECTIVE SURFACE AND GRID ARRAY • *T. K. Wu (ed.)*

ACTIVE AND QUASI-OPTICAL ARRAYS FOR SOLID-STATE POWER COMBINING • *Robert A. York and Zoya B. Popović (eds.)*

OPTICAL SIGNAL PROCESSING, COMPUTING AND NEURAL NETWORKS • *Francis T. S. Yu and Suganda Jutamulia*

ELECTROMAGNETIC SIMULATION TECHNIQUES BASED ON THE FDTD METHOD • *Wenhua Yu, Xiaoling Yang, Yongjun Liu, and Raj Mittra*

SiGe, GaAs, AND InP HETEROJUNCTION BIPOLAR TRANSISTORS • *Jiann Yuan*

PARALLEL SOLUTION OF INTEGRAL EQUATION-BASED EM PROBLEMS • *Yu Zhang and Tapan K. Sarkar*

ELECTRODYNAMICS OF SOLIDS AND MICROWAVE SUPERCONDUCTIVITY • *Shu-Ang Zhou*